OXFORD REFERENCE

The
Oxford Dictionary of
NEW WORDS

Sara Tulloch is a Managing Editor in the Oxford Dictionary Department. She took her degree in Russian, French, and Law, and did postgraduate research on the development of new words in Latvian and Russian. She joined Oxford University Press in 1985, contributing to the Second Edition of *The Oxford English Dictionary* (1989). From 1985 to 1991 she directed the international survey of words and meanings in English conducted for the Oxford English Dictionaries Reading Programme, and at the time of writing this Dictionary she was Senior Editor of the New Words Group.

The Oxford Dictionary of
NEW WORDS

A popular guide to words in the news

Compiled by Sara Tulloch

Oxford   New York

OXFORD UNIVERSITY PRESS

Oxford University Press, Walton Street, Oxford OX2 6DP

Oxford New York Toronto
Delhi Bombay Calcutta Madras Karachi
Kuala Lumpur Singapore Hong Kong Tokyo
Nairobi Dar es Salaam Cape Town
Melbourne Auckland Madrid

and associated companies in
Berlin Ibadan

Oxford is a trade mark of Oxford University Press

British Library Cataloguing in Publication Data
Data available

Library of Congress Cataloging in Publication Data
Data available
ISBN 0-19-283077-5

5 7 9 10 8 6 4

Printed in Great Britain by
Clays Ltd.
Bungay, Suffolk

Preface

This is the first dictionary entirely devoted to new words and meanings to have been published by the Oxford University Press. It follows in the tradition of the *Supplement to the Oxford English Dictionary* in attempting to record the history of some recent additions to the language, but, unlike the *Supplement*, it is necessarily very selective in the words, phrases, and meanings whose stories it sets out to tell and it stands as an independent work, unrelated (except in the resources it draws upon) to the *Oxford English Dictionary*.

The aim of the *Oxford Dictionary of New Words* is to provide an informative and readable guide to about two thousand high-profile words and phrases which have been in the news during the past decade; rather than simply defining these words (as dictionaries of new words have tended to do in the past), it also explains their derivation and the events which brought them to prominence, illustrated by examples of their use in journalism and fiction. In order to do this, it draws on the published and unpublished resources of the *Oxford English Dictionary*, the research that is routinely carried out in preparing new entries for that work, and the word-files and databases of the Oxford Dictionary Department.

What is a new word? This, of course, is a question which can never be answered satisfactorily, any more than one can answer the question 'How long is a piece of string?' It is a commonplace to point out that the language is a constantly changing resource, growing in some areas and shrinking in others from day to day. The best one can hope to do in a book of this kind is to take a snapshot of the words and senses which seem to characterize our age and which a reader in fifty or a hundred years' time might be unable to understand fully (even if these words were entered in standard dictionaries) without a more expansive explanation of their social, political, or cultural context. For the purposes of this dictionary, a new word is any word, phrase, or meaning that came into popular use in English or enjoyed a vogue during the eighties and early nineties. It is a book which therefore necessarily deals with passing fashions: most, although probably not all, of the words and senses defined here will eventually find their way into the complete history of the language provided by the *Oxford English Dictionary*, but many will not be entered in smaller dictionaries for some time to come, if at all.

It tends to be the case that 'new' words turn out to be older than people expect them to be. This book is not limited to words and senses which *entered* the language for the first time during the eighties, nor even the seventies and eighties, because such a policy would mean excluding most of the words which ordinary speakers of English think of as new; instead, the deciding factor has been whether or not the general public was made aware of the word or sense during the eighties and early nineties. A few words included here actually entered the language as technical terms as long ago as the nineteenth century (for example, *acid rain* was first written about in the 1850s and the *greenhouse effect* was investigated in the late nineteenth century, although it may not have acquired this name until the 1920s); many computing terms date from the late 1950s or early 1960s in technical usage. It was only (in the first case) the surge of interest in environmental issues and the sudden fashion for 'green' concerns and (in the second) the boom in home and personal computing touching the lives of large numbers of people that brought these words into everyday vocabulary during the eighties.

There is, of course, a main core of words defined here which *did* only appear for the first time in the eighties. There are even a few which arose in the nineties, for which there is as yet insufficient evidence to say whether they are likely to survive. Some new-words dictionaries in the past have limited themselves to words and senses which have not yet been entered in general dictionaries. The words treated in the *Oxford Dictionary of New Words* do not all fall into this category, for the reasons outlined above. Approximately one-quarter of the main headwords here were included in the new words and senses added to the *Oxford English Dictionary* for its second edition in 1989; a small number of others were entered for the first time in the *Concise Oxford Dictionary*'s eighth edition in 1990.

The articles in this book relate to a wide range of different subject fields and spheres of interest, from environmentalism to rock music, politics to youth culture, technology to children's toys. Just as the subject coverage is inclusive, treating weighty and superficial topics as even-handedly as possible, so the coverage of different registers, or levels of use, of the language is intended to give equal weight to the formal, the informal, and examples of slang and colloquialism. This results in a higher proportion of informal and slang usage than would be found in a general dictionary, reflecting amongst other things the way in which awareness of register seems to be disappearing as writers increasingly use slang expressions in print without inverted commas or any other indication of their register. The only registers deliberately excluded are the highly literary or technical in cases where the vocabulary concerned had not gained any real popular exposure. Finally, a deliberate attempt was made to represent English as a world language, with new words and senses from US English accounting for a significant proportion of the entries, along with more occasional contributions from Australia, Canada, and other English-speaking countries. It is hoped that the resulting book will prove entertaining reading for English speakers of all ages and from all countries.

Acknowledgements

I am grateful to John Simpson and Edmund Weiner, Co-Editors of the *Oxford English Dictionary*, for their help and advice throughout the writing of this book, and in particular for their constructive comments on the first draft of the text; to *OED* New Words editors Edith Bonner, Peter Gilliver, Danuta Padley, Bernadette Paton, Judith Pearsall, Michael Proffitt, and Anthony Waddell, on whose draft entries for the *OED* I based much of what I have written here; to Peter Gilliver, Simon Hunt, Veronica Hurst, and Judith Pearsall for help with corrections and additions to the text; to Melinda Babcock, Nancy Balz, Julie Bowdler, George Chowdharay-Best, Melissa Conway, Margaret Davies, Margery Fee, Ken Feinstein, Daphne Gilbert-Carter, Dorothy Hanks, Sally Hinkle, Sarah Hutchinson, Rita Keckeissen, Adriana Orr, and Jeffery Triggs for quotation and library research; and, last but not least, to Trish Stableford for giving up evenings and weekends to do the proofreading.

The subject icons were designed by Information Design Unit.

How to Use this Dictionary

The entries in this dictionary are of two types: **full entries** and **cross-reference entries**.

Full entries

Full entries normally contain five sections:

1 Headword section

The first paragraph of the entry, or *headword* section, gives

- the main headword in **large bold type**

  Where there are two different headwords which are spelt in the same way, or two distinct new meanings of the same word, these are distinguished by superior numbers after the headword.

- the pronunciation in the International Phonetic Alphabet and enclosed in oblique strokes /prəˌnʌnsɪˈeɪʃ(ə)n/ (see the key to the symbols which follows)

- the part of speech, or grammatical category, of the word in *italic type*

  In this book, all the names of the parts of speech are written out in full. The ones used in the book are *adjective, adverb, interjection, noun, pronoun,* and *verb*. There are also entries in this book for the word-forming elements (*combining form, prefix,* and *suffix*) and for abbreviations, which have *abbreviation* in the part-of-speech slot if they are pronounced letter by letter in speech (as is the case, for example, with *BSE* or *PWA*), but *acronym* if they are normally pronounced as words in their own right (*Aids, NIMBY, PIN,* etc.).

  When a new word or sense is used in more than one part of speech, the parts of speech are listed in the headword section of the entry and a separate definition section is given for each part of speech.

- other spellings of the headword (if any) follow the part of speech in **bold type**

- the subject area(s) to which the word relates are shown at the end of the headword section in the form of one or more graphic icons, for example

  ⊗ ▓ (see the key to the icons which follows).

  The subject icons are only intended to give a general guide to the field of use of a particular word or sense. In addition to the icon, the defining section of the entry often begins with further explanation of the headword's application.

2 Definition section

The *definition* section explains the meaning of the word and sometimes contains information about its register (the level or type of language in which it is used) or its more specific application in a particular field; it may also include phrases and derived forms of the headword (in **bold type**) or references to other entries (in SMALL CAPITALS).

3 Etymology

The third section of the entry, starting a new paragraph and set in smaller type, is the *etymology* section. This explains the origin and formation of the headword. Some words or phrases in this section may be in *italic type*, showing that they are the forms under discussion. Cross-references to other headwords in this book may also occur; these are printed in SMALL CAPITALS.

4 History and Usage

The fourth section, in the same small type but starting another new paragraph, gives an account of the *history and usage* of the word or phrase. Here you will find a description of the circumstances under which the headword entered the language and came into popular use. In many cases this section also contains information about compounds and derived forms of the headword (as well as some other related terms), all listed in **bold type**, together with their definitions and histories. As elsewhere in the entry, words printed in SMALL CAPITALS are cross-references to other headwords.

5 Illustrative quotations

The final section of the entry, in condensed type, contains the *illustrative quotations*. These are arranged in a single chronological sequence, even when they contain examples of a number of different forms. The illustrative quotations in this book do not include the earliest printed example in the Oxford Dictionaries word-file (as would be the case, for example, in the *Oxford English Dictionary*); instead, information about the date of the earliest quotations is given in the *history and usage* section of the entry and the illustrative quotations aim to give a representative sample of recent quotations from a range of sources. The sources quoted in this book represent English as a world language, including quotations from the UK, the US, Australia, Canada, India, South Africa, and other English-speaking countries. They are taken for the most part from works of fiction, newspapers, and popular magazines (avoiding wherever possible the more technical or academic sources in favour of the more popular and accessible). There are nearly two thousand quotations altogether, taken from five hundred different sources.

Cross-reference entries

Because this book is designed to provide more information than the standard dictionary and to give an expansive account of the recent history of certain words and concepts, there is some grouping together of related pieces of information in a single article. This means that, in addition to the full entry, there is a need for cross-reference entries leading the reader from the normal alphabetical place of a word or phrase to the full entry in which it is discussed. Cross-reference entries are single-line entries containing only the headword (with a superior number if identical to some other headword), a subject icon or icons to give some topical orientation, the word 'see', and the headword (in SMALL CAPITALS) under which the information can be found. For example:

ESA 🌱 see ENVIRONMENTALLY

A cross-reference entry is given only if there is a significant distance between the alphabetical places of the cross-referenced headword and the full entry in which it is mentioned. Thus the compounds and derived forms of a full headword are not given their own cross-reference entries because these would immediately follow the full entry; the same is true of the words which start with one of the common initial elements (such as *eco-* or *Euro-*) which have their own full entries listing many different formations in which they are used. On the other hand, the forms grouped together by their final element (for example, words ending in *-friendly* or *-gate*) are all entered as cross-reference entries in their normal alphabetical places.

Alphabetical order

The full and cross-reference entries in this book are arranged in a single alphabetical sequence in *letter-by-letter* alphabetical order (that is, ignoring spaces, hyphens, and other punctuation which occurs within them). The following headwords, taken from the letter E, illustrate the point:

E^1
E^2
e^3
earcon
eco
eco-
ecobabble
ecological
ecu
E-free
EFTPOS
enterprise culture
enterprise zone
E number

●●

Note on Proprietary Status

This dictionary includes some words which are, or which are asserted to be, proprietary names or trade marks. Their inclusion does not imply that they have acquired for legal purposes a non-proprietary or general significance, nor is any other judgement implied concerning their legal status. In cases where the Dictionary Department has evidence that a word is used as a proprietary name or trade mark this is indicated in the text of the entry by the words 'trade mark', but no judgement concerning the legal status of such words is implied thereby.

Pronunciation Symbols

In the International Phonetic Alphabet, the characters b, d, f, h, k, l, m, n, p, r, t, v, w, and z have their normal English sound values.

The other consonant sounds are represented by

g	as in *g*o	s	as in hi*ss*	x	as in Scottish lo*ch*
ŋ	as in si*ng*	ʃ	as in *sh*ip	tʃ	as in *ch*in
θ	as in *th*in	ʒ	as in vi*s*ion	dʒ	as in *j*am
ð	as in *th*en	j	as in *y*et	ʔ	as in Cockney bu*tt*er

The vowel sounds of English are represented as follows:

æ	as in f*a*t	ʊ	as in b*oo*k	aʊə	as in s*our*
ɑ:	as in c*ar*t	u:	as in b*oo*t	eɪ	as in f*a*te
ɛ	as in m*e*t	ɜ:	as in f*ur*	ɛə	as in f*air*
ɪ	as in b*i*t	ə	as in *a*go	ɪə	as in p*ier*
i:	as in m*ee*t	aɪ	as in b*i*te	ɔɪ	as in b*oi*l
ɒ	as in g*o*t	aʊ	as in br*ow*	ʊə	as in t*our*
ɔ:	as in p*or*t	aɪə	as in f*ire*	əʊ	as in g*oa*t
ʌ	as in d*u*g				

In addition to these symbols, the normal vowel symbols a, e, i, o, and u are used to represent the 'cardinal' vowels used in some words of foreign origin. The French vowel in d*u* and t*u* is represented by y. Additional symbols, not representing actual sounds, are:

ˈ   the stress mark, which appears before the syllable carrying the main emphasis

ˌ   the secondary stress mark, which appears before a syllable carrying a lesser degree of emphasis

:   the length mark, indicating that the preceding vowel is long (shown above in the explanations of the long vowels)

( )   parentheses, used round any sound that is optional, including word-final *r* which is only pronounced when followed by a word beginning with a vowel

ʹ   prime, indicating (in words of foreign origin, including in this book some words from Russian) that the preceding consonant is palatalized or 'softened'

Subject Icons

The graphic icons at the end of the headword section of each entry indicate the broad subject field to which the headword relates. The icons used are

Drugs: words to do with drug use and abuse

Environment: words to do with conservation, the environment, and green politics

Business World: words to do with work, commerce, finance, and marketing

Health & Fitness: words to do with conventional and complementary medicine, personal fitness, exercise, and diet

Lifestyle & Leisure: words to do with homes and interiors, fashion, the media, entertainment, food and drink, and leisure activities in general

Music: words to do with music of all kinds (combined with [icon] in entries concerned with pop and rock music)

Politics: words to do with political events and issues at home and abroad

People & Society: words to do with social groupings and words for people with particular characteristics; social issues, education, and welfare

Science & Technology: words to do with any branch of science in the public eye; technical jargon that has entered the popular vocabulary

War & Weaponry: words to do with the arms race or armed conflicts that have been in the news

Youth Culture: words which have entered the general vocabulary through their use among young people

A

AAA 🎆 see TRIPLE A

abled /'eɪb(ə)ld/ *adjective* ◀

Able-bodied, not disabled. Also (especially with a preceding adverb): having a particular range of physical abilities; **differently abled, otherly abled, uniquely abled**: euphemistic ways of saying 'disabled'.

Formed by removing the prefix *dis-* from *disabled*.

The word *abled* arose in the US; it has been used by the disabled to refer to the able-bodied since about the beginning of the eighties, and is also now so used in the UK. The euphemistic phrases *differently abled*, *otherly abled*, and *uniquely abled* were coined in the mid eighties, again in the US, as part of an attempt to find a more positive official term than *handicapped* (the official term in the US) or *disabled* (the preferred term in the UK during the eighties). Another similarly euphemistic coinage intended to serve the same purpose was *challenged*. *Differently abled* has enjoyed some success in the US, but all of the forms with a preceding adverb have come in for considerable criticism.

> Disabled, handicapped, differently-abled, physically or mentally challenged, women with disabilities—this is more than a mere discourse in semantics and a matter of personal preference.
>
> Debra Connors in *With the Power of Each Breath* (1985), p. 92

> In a valiant effort to find a kinder term than handicapped, the Democratic National Committee has coined differently abled. The committee itself shows signs of being differently abled in the use of English. *Los Angeles Times* 9 Apr. 1985, section 5, p. 1

> I was aware of how truly frustrating it must be to be disabled, having to deal not only with your disability, but with abled people's utter disregard for your needs.
>
> *San Francisco Chronicle* 4 July 1990, Briefing section, p. 7

ableism /'eɪb(ə)lɪz(ə)m/ *noun* Also written **ablism** ◀

Discrimination in favour of the able-bodied; the attitude or assumption that it is only necessary to cater for able-bodied people.

Formed by adding the suffix *-ism* (as in AGEISM, *racism*, and *sexism*) to the adjective *able* in the sense in which it is used in *able-bodied*.

This is one of a long line of *-isms* which became popular in the eighties to describe various forms of perceived discrimination: see also FATTISM and HETEROSEXISM. *Ableism* was a term first used by feminists in the US at the beginning of the eighties; in the UK, the concept was first referred to as **able-bodism** in a GLC report in 1984 and was later also called **able-bodiedism**. However, *ableism* was the form chosen by the Council of the London borough of Haringey for a press release in 1986, and it is this form which has continued to be used, despite the fact that it is thought by some to be badly formed (the suffix *-ism* would normally be added to a noun stem rather than an adjective). The spelling *ableism* is preferred to *ablism*, which some people might be tempted to pronounce /'æblɪz(ə)m/. In practice, none of the forms has been widely used, although society's awareness of disability was raised during the International Year of Disabled Persons in 1981. The adjective corresponding to this noun is **ableist**, but its use is almost entirely limited to US feminist writing. For an adjective which describes the same characteristics from the opposite viewpoint, see DISABLIST.

> A GLC report . . . referred throughout to a new phenomenon called mysteriously 'able-bodism'—a reference apparently to that malevolent majority, the fully-fit. *Daily Telegraph* 1 Nov. 1984, p. 18

> Able-ist movements of the late-nineteenth and early twentieth centuries regarded disability as problematic for society. Debra Connors in *With the Power of Each Breath* (1985), p. 99

I was at the national convention of the National Organization for Women. I consider myself a feminist ... but I'm ... embarrassed by the hysteria, the gaping maws in their reasoning and the tortuous twists of femspeak. Who else can crowd the terms 'ableism, homophobia and sexism' into one clause without heeding the shrillness of tone?   *San Francisco Chronicle* 4 July 1990, section A, p. 19

ABS 🖳 see ANTI-LOCK

abuse /əˈbjuːs/ *noun* 🔏 🚬

Illegal or excessive use of a drug; the misuse of any substance, especially for its stimulant effects.

In the context of human relationships, physical (especially sexual) maltreatment of another person.

These are not so much new senses of the word as specializations of context; *abuse* has meant 'wrong or improper use, misapplication, perversion' since the sixteenth century, but in the second half of the twentieth century has been used so often in the two contexts mentioned above that this is becoming the dominant use.

Abuse was first used in relation to drugs in the early sixties; by the seventies it was usual for it to be the second element in compounds such as **alcohol abuse**, **drug abuse**, and **solvent abuse**, and soon afterwards with a human object as the first word: see CHILD ABUSE. Interestingly it is not idiomatic to form similar compounds for other types of *abuse* in its traditional sense: *the abuse of power* rather than 'power abuse', for example. This is one way in which the language continues to differentiate the traditional use from the more specialized one, although there have been some recent exceptions (a tennis player who throws his racquet about in anger or frustration can now be cautioned for *racquet abuse*, for example).

This is a setback for the campaign against increasing heroin abuse among the young in all parts of the country.   *Sunday Times* 9 Dec. 1984, p. 3

Just over 30 per cent of the girls questioned said they had tried solvent abuse.
     Daily Express 20 Aug. 1986, p. 2

Asked why she continued diagnosing abuse after three appeals from other agencies to stop because they could not cope, she replied: 'With hindsight, at the time we were trying to do our best for them. In the event, with some children, we were sadly unable to do that.'   *Guardian* 14 July 1989, p. 2

ace /eɪs/ *adjective* 🗨

In young people's slang: great, fantastic, terrific.

The adjectival use has arisen from the noun *ace*, which essentially means 'number one'.

As any reader of war comics will know, during the First World War outstanding pilots who had succeeded in bringing down ten or more enemy planes were known as *aces*; shortly after this, *ace* started to be used in American English to mean any outstanding person or thing, and by the middle of the century was often used with another noun following (as in 'an ace sportsman'). It was a short step from this attributive use to full adjectival status. In the eighties, *ace* was re-adopted by young people as a general term of approval, and this time round it was always used as an adjective ('that's really ace!') or adverbially ('ace!') as a kind of exclamation.

With staff, everything becomes possible. And—ace and brill—they confer instant status on the employer at the same time. A double benefit: dead good and the apotheosis of yuppiedom.
     Daily Telegraph 12 July 1987, p. 21

The holiday was absolutely ace—loads of sailing and mountain walking, and even a night's camping in the hills.   *Balance* (British Diabetic Association) Aug.–Sept. 1989, p. 45

acid house /ˈæsɪd ˌhaʊs/ *noun* 🎵 🗨

A style of popular music with a fast beat, a spare, mesmeric, synthesized sound, few (if any) vocals, and a distinctive gurgling bass; in the UK, a youth cult surrounding this music and associated in the public mind with SMILEY badges, drug-taking, and extremely large parties known as **acid house parties**. Sometimes abbreviated to **acid** (also written **acieeed** or **aciiied**, especially when used as a kind of interjection).

The word *acid* here is probably taken from the record *Acid Trax* by Phuture (in Chicago slang, *acid burning* is a term for stealing and this type of music relies heavily on SAMPLING, or stealing from other tracks); a popular theory that it is a reference to the drug LSD is denied by its followers (but compare *acid rock*, a sixties psychedelic rock craze, which certainly was). *House* is an abbreviated form of *Warehouse*: see HOUSE.

Acid house music originated in Chicago as an offshoot of house music in 1986; at first it was called 'washing machine', which aptly described the original sound. Imported to the UK in 1988, *acid house* started a youth cult during the summer of that year, and soon spawned its own set of behaviour and its own language. The craze for *acid house parties*, at venues kept secret until the very last moment, exercised police forces throughout the south of England, since they often involved trespass on private land and caused a public nuisance, although organizers claimed that they had been maligned in the popular press.

> I suppose that a lot of acid house music is guilty of ... being completely cold and devoid of any human touch.
>
> *Spin* Oct. 1989, p. 18

> Aciiied was a figment of the British imagination. Like British R&B in the Sixties, it was a creative misrecognition of a Black American pop.
>
> *Melody Maker* 23–30 Dec. 1989, p. 34

> Acid House, whose emblem is a vapid, anonymous smile, is the simplest and gentlest of the Eighties' youth manifestations. Its dance music is rhythmic but non-aggressive (except in terms of decibels).
>
> *Independent* 3 Mar. 1990, p. 12

See also WAREHOUSE

acid rain /ˌæsɪd 'reɪn/ *noun* 🌳

Rain containing harmful acids which have formed in the atmosphere, usually when waste gases from industrial emissions combine with water.

Formed by compounding: *rain* with an *acid* content.

The term *acid rain* was first used as long ago as 1859, when R. A. Smith observed in a chemical journal that the stonework of buildings crumbled away more quickly in towns where a great deal of coal was burnt for industrial purposes; this he attributed to the combination of waste gases with water in the air, making the rain acidic. In the early 1970s the term was revived as it became clear that *acid rain* was having a terrible effect on the forests and lakes of North America, Europe, and especially Scandinavia (killing trees and freshwater life). *Acid rain* started to be discussed frequently in official reports and documents on the environment; but it was not until environmental concerns became a public issue in the eighties that the term passed from technical writing of one kind and another into everyday use. With this familiarity came a better understanding of the causes of *acid rain*, including the contribution of exhaust fumes from private vehicles. By the end of the eighties, *acid rain* was a term which even schoolchildren could be expected to know and understand, and had been joined by variations on the same theme: **acid cloud**, a term designed to emphasize the fact that acidic gases could damage the environment even without any precipitation; **acid fallout**, the overall atmospheric effect of pollution; **acid precipitation**, the name sometimes used for snow or hail of high acidity.

> She has a list of favorite subjects, favorite serious subjects—nuclear proliferation, acid rain, unemployment, as well as racial bigotry and the situation of women.
>
> Alice Munro *Progress of Love* (1987), p. 190

> Burning oil will contribute to the carbon dioxide umbrella and the acid rain deposited on Europe.
>
> *Private Eye* 1 Sept. 1989, p. 25

acquired immune deficiency syndrome ⊗ see AIDS

active /'æktɪv/ *adjective* 🖳

Programmed so as to be able to monitor and adjust to different situations or to carry out several different functions; SMART, INTELLIGENT¹.

A simple development of sense: the software enables the device to *act* on the results of monitoring or on commands from its user.

This sense of *active* became popular in the naming of products which make use of developments in artificial intelligence and microelectronics during the late eighties and early nineties: for example, the **Active Book**, the trade mark of a product designed to enable an executive to use facilities like fax, telephone, dictaphone, etc. through a single portable device; the **active card**, a smart card with its own keyboard and display, enabling its user to discover the remaining balance, request transactions, etc.; **active optics**, which makes use of computer technology to correct light for the distortion placed upon it as it passes through the atmosphere; **active suspension**, a suspension system for cars in which the hydraulic activators are controlled by a computer which monitors road conditions and adjusts suspension accordingly; and **active system**, any computerized system that adjusts itself to changes in the immediate environment, especially a hi-fi system.

> The only development that I would class as the 'biggy' for 1980 was the introduction of reasonably priced active systems. *Popular Hi-Fi* Mar. 1981, p. 15

> The company is also pioneering the development of active or supersmart cards, which rivals ... believe to be impractical on several counts. *New Scientist* 11 Feb. 1989, p. 64

> One of our mottos is 'Buy an Active Book and get 20 per cent of your life back'. *Daily Telegraph* 30 Apr. 1990, p. 31

active birth /ˌæktɪv 'bɜːθ/ *noun* 🗙

Childbirth during which the mother is encouraged to be as active as possible, mainly by moving around freely and assuming any position which feels comfortable.

Formed by compounding: *birth* which is *active* rather than *passive*.

The *active birth* movement was founded by childbirth counsellor Janet Balaskas in 1982 as a direct rejection of the increasingly technological approach to childbirth which prevailed in British and American hospitals at the time. Ironically, this technological approach was known as the *active management of labour*; to many of the women involved it felt like a denial of their right to participate in their own labour. The idea of *active birth* was to move away from the view that a woman in labour is a patient to be treated (and therefore passive), freeing her from the encumbrance of monitors and other medical technology whenever possible and handing over to her the opportunity to manage her own labour. The concept has been further popularized in the UK by Sheila Kitzinger.

> The concept of Active Birth is based on the idea that the woman in labour is an active birthgiver, not a passive patient. Sheila Kitzinger *Freedom & Choice in Childbirth* (1987), p. 63

> New Active Birth by Janet Balaskas ... After Active Birth, published in 1983, updated New Active Birth prepares a woman for complete participation in the birth of her child. *Guardian* 1 Aug. 1989, p. 17

active citizen /ˌæktɪv 'sɪtɪz(ə)n/ *noun* 🏛

A member of the public who takes an active role in the community, usually by getting involved in crime prevention, good neighbour schemes, etc.

Formed by compounding: a *citizen* who is *active* in society rather than passively soaking up the benefits of community life.

The term *active citizen* was first used in the name of the **Active Citizen Force**, a White militia in South Africa, set up in 1912 and consisting of male citizens undergoing national service. In a completely separate development, *active citizen* started to be used in the US from the late seventies as a more polite way of saying 'political activist' or even 'future politician'; some *active citizens* even organized themselves into pressure groups which were able to affect local government policies. In the UK, the term *active citizen* and the associated policy of **active citizenship** were popularized by the Conservative government of the eighties, which placed great emphasis upon them, especially after the Conservative Party conference of 1988. The focus of *active citizenship* as encouraged by this government was on crime prevention (including NEIGHBOURHOOD WATCH) and public order, rather than political activism. This put it on the borderline with vigilante activity, a cause of some difficulty in turning the policy into concrete action.

> Pervading the researches will be an effort to plumb individuals' moral convictions, their motives for joining or not joining in active citizenship.
>
> *Christian Science Monitor* (New England edition) 2 June 1980, p. 32

> Intermediate institutions . . . help to produce the 'active citizen' which Ministers such as Douglas Hurd have sought to call into existence to supplement gaps in welfare provision.
>
> *Daily Telegraph* 3 May 1989, p. 18

> 'Active citizens' . . . brought unsafe or unethical practices by their employers to official notice. As their stories reveal, active citizenship carries considerable personal risk. Blacklisting by other employers is a frequent consequence.
> *Guardian* 27 June 1990, p. 23

acupressure /ˈækjʊˌprɛʃə(r)/ *noun* Ⓧ

A COMPLEMENTARY therapy also known as *shiatsu*, in which symptoms are relieved by applying pressure with the thumbs or fingers to specific pressure points on the body.

Formed by combining the first two syllables of *acupuncture* (*acupressure* is a Japanese application of the same principles as are used in Chinese acupuncture) with *pressure*. The word *acupressure* actually already existed in English for a nineteenth-century method of arresting bleeding during operations by applying pressure with a needle (Latin *acu* means 'with a needle'); since no needle is used in *shiatsu* it is clear that the present use is a separate formation of the word, deliberately referring back to acupuncture but without taking into account the original meaning of *acu-*.

Acupressure has been practised in Japan as *shiatsu* and in China as *G-Jo* ('first aid') for many centuries; it was exported to the Western world during the 1960s, but at first was usually called *shiatsu*. During the late seventies and early eighties *acupressure* became the preferred term and the word became popularized, first in the US and then in the UK, as COMPLEMENTARY medicine became more acceptable and even sought after. In the late eighties the principle was incorporated into a popular proprietary means of avoiding motion sickness in which elastic bracelets hold a hard 'button' in place, pressing on an **acupressure point** on each wrist. A practitioner of *acupressure* is called an **acupressurist**.

> Among the kinds of conditions that benefit from acupressure are migraine, stress, and tension-related problems.
> *Natural Choice* Issue 1 (1988), p. 19

> After one two-hour massage that included . . . acupressure, I was addicted.
> Alice Walker *Temple of My Familiar* (1989), p. 292

acyclovir /eɪˈsaɪkləʊvaɪə(r), -vɪə(r)/ *noun* Ⓧ

An antiviral drug that is effective against certain types of herpes, including cytomegalovirus.

Formed by combining all but the ending of the adjective *acyclic* (in its chemical sense, 'containing no cycle, or ring of atoms') with the stem of *viral*.

The drug was developed at the end of the seventies and became the only effective treatment for genital herpes that was available during the eighties. It was widely publicized as a breakthrough in antiviral medicine at a time when genital herpes was seen as the most intractable sexually transmitted disease affecting Western societies (before the advent of AIDS). During the late eighties it was used in combination with AZT (or ZIDOVUDINE) in the management of cytomegalovirus, a herpes virus which affects some people already infected with HIV.

The beauty of acyclovir is that it remains inactive in the body until it comes in contact with a herpes-induced enzyme. The enzyme then activates the drug.
Maclean's 2 Nov. 1981, p. 24

Professor Griffiths said studies in the US have shown the drug Acyclovir to be effective in preventing the side effects of CMV infection.
Guardian 7 July 1989, p. 3

Adam /ˈædəm/ *noun* 🎯

In the slang of drug users, the hallucinogenic DESIGNER DRUG methylenedioxymethamphetamine or MDMA, also known as ECSTASY.

The name is probably a type of backslang, reversing the abbreviated chemical name *MDMA*, dropping the first *m*, and pronouncing the resulting 'word'; it may be influenced by the associations of the first Adam with paradise. A similar designer drug is known in drugs slang as *Eve*.

For history, see ECSTASY.

On the street, its name is 'ecstasy' or 'Adam', which should tell how people on the street feel about it.
Los Angeles Times 29 Mar. 1985, section 5, p. 8

One close relative of MDMA, known as Eve—MDMA is sometimes called Adam—has already been shown to be less toxic to rats than MDMA. Because of a 'designer-drug' law passed in 1986, Eve is banned too.
Economist 19 Mar. 1988, p. 94

additive /ˈædɪtɪv/ *noun* 🌶 ▩

A substance which is added to something during manufacture, especially a chemical added to food or drink to improve its colour, flavour, preservability, etc. (known more fully as a **food additive**).

Additive has meant 'something that is added' since the middle of this century; recently it has acquired this more specialized use, which partly arose from the desire to abbreviate *food additive* once the term was being used frequently.

Public interest in what was being put into foods by manufacturers grew rapidly during the eighties because of the green movement, with its associated diet-consciousness and demand for 'natural' products, and also because of growing evidence of the harmful effects of certain additives (including their implication in hyperactivity and other behavioural problems in children). This interest was crystallized in the mid eighties by new EC regulations on naming and listing additives and the publication of a number of reference books giving details of all the permitted food additives as well as some of the possible effects on health of ingesting them. Possibly the most famous of these was Maurice Hanssen's *E for Additives* (1984); certainly after the publication of this book, *additive* could be used on its own (not preceded by *food*) without fear of misunderstanding. In response to the public backlash against the use of chemical *additives*, manufacturers began to make a publicity point out of foods which contained none; the phrase **free from artificial additives** (bearing witness to the fact that food additives from natural sources continued to be used) and the adjective **additive-free** began to appear frequently on food labels from the second half of the eighties.

Last week Peter turned up at Broadcasting House with the first ever commercially produced non-sweetened, additive-free yoghurt.
Listener 10 May 1984, p. 15

Every human and inhuman emotion magnified itself in New York; thoughts . . . more quickly became action within and beyond the law; some said the cause lay in the food, the additives, some said in the polluted air.
Janet Frame *Carpathians* (1988), p. 103

See also ALAR, E NUMBER, -FREE

advertorial /ˌædvəˈtɔːrɪəl/ *noun* 〰

An advertisement which is written in the form of an editorial and purports to contain objective information about a product, although actually being limited to the advertiser's own publicity material.

Formed by replacing the first two syllables of *editorial* with the word *advert* to make a blend.

The *advertorial* (both the phenomenon and the word) first appeared in the US as long ago as the sixties, but did not become a common advertising ploy in the UK until the mid eighties. *Advertorials* came in for some criticism when they started to appear in British newspapers since there was a feeling of dishonesty about them (as deliberately inducing the reader to read them as though they were editorials or features), but they apparently did not contravene fair advertising standards as set out in the British Code of Advertising Practice:

An advertisement should always be so designed and presented that anyone who looks at it can see, without having to study it closely, that it is an advertisement.

In many cases the page on which an *advertorial* appears is headed **advertising** or **advertisement feature** (a more official name for the *advertorial*), and this is meant to alert the reader to the nature of the article, although the layout of the page often does not. The word *advertorial* is sometimes used (as in the second example below) without an article to mean this style of advertisement-writing in general rather than an individual example of it.

Yes, advertorials are a pain, just like the advertising supplement pages in *Barron's*, but I question whether 'anyone who bought FNN would have to junk the programming'.

Barron's 24 Apr. 1989, p. 34

This will probably lead to a growth in what the industry calls 'advertorial'—a mixture of public relations and journalism, or editorial with bias.   *Sunday Correspondent* 22 Apr. 1990, p. 27

aerobics /ɛəˈrəʊbɪks/ *noun* Ⓧ Ⓧ

A form of physical exercise designed to increase fitness by any maintainable activity that increases oxygen intake and heart rate.

A plural noun on the same model as *mathematics* or *stylistics*, formed on the adjective *aerobic* ('requiring or using free oxygen in the air'), which has itself been in use since the late nineteenth century.

The word was coined by Major Kenneth Cooper of the US Air Force as the name for a fitness programme developed in the sixties for US astronauts. In the early eighties, when fitness became a subject of widespread public interest, *aerobics* became the first of a string of fitness crazes enthusiastically taken up by the media. The fashion for the **aerobics class**, at which aerobic exercises were done rhythmically to music as part of a dance movement called an **aerobics routine**, started in California, soon spread to the UK, Europe, and Australia, and even reached the Soviet Union before giving way to other exercise programmes such as CALLANETICS. Although a plural noun in form, *aerobics* may take either singular or plural agreement.

Aerobics have become the latest fitness craze.

Observer 18 July 1982, p. 25

The air-waves of the small, stuffy gym reverberated with the insistent drum notes as thirty pairs of track shoes beat out the rhythm of the aerobics routine.   Pat Booth *Palm Beach* (1986), p. 31

See also AQUAROBICS

affinity card /əˈfɪnɪtɪ ˌkɑːd/ *noun* Sometimes in the form **affinity credit card** 🗠

A credit card issued to members of a particular affinity group; in the UK, one which is linked to a particular charity such that the credit-card company makes a donation to the charity for each new card issued and also passes on a small proportion of the money spent by the card user.

Formed by combining *affinity* in the sense in which it is used in *affinity group* (an American term meaning 'a group of people sharing a common purpose or interest') with CARD¹. In the case of the charity cards, the idea is that the holders of the cards share a common interest in helping the charity.

Affinity cards were first issued in the US in the late seventies in a wide variety of different forms to cater for different interest groups. These cards were actually issued through the affinity group (which could be any non-profit organization such as a college, a union, or a club), and entitled its members to various discounts and other benefits. When the idea was taken up by large banks and building societies in the UK in 1987, it was chiefly in relation to charities, and the idea was skilfully used to attract new customers while at the same time appealing to their social conscience.

> One alternative [to credit-card charges] is an affinity credit card linked to a charity, although the Leeds Permanent Building Society is considering charging for its affinity cards.
>
> *Observer* 29 Apr. 1990, p. 37

> Affinity cards cannot be used to access any account other than one maintained by a Visa card-issuing financial institution. *Los Angeles Times* 10 Oct. 1990, section D, p. 5

affluential /ˌæflʊˈɛnʃ(ə)l/ *adjective* and *noun*

adjective: Influential largely because of great wealth; rich and powerful.

noun: A person whose influential position in society derives from wealth.

Formed by telescoping *affluent* or *affluence* and *influential* to make a blend.

A US coinage of the second half of the seventies, *affluential* became quite well established (especially as a noun) in American English during the eighties, but so far shows little sign of catching on in the UK.

> *Spa* is the name of the mineral-water resort in Belgium, and has become a word for 'watering place' associated with the weight-conscious affluentials around the world.
>
> *New York Times Magazine* 18 Dec. 1983, p. 13

affluenza /ˌæflʊˈɛnzə/ *noun*

A psychiatric disorder affecting wealthy people and involving feelings of malaise, lack of motivation, guilt, etc.

Formed by telescoping *affluence* and *influenza* to make a blend.

The term was popularized in the mid eighties by Californian psychiatrist John Levy, after he had conducted a study of children who grow up expecting never to need to earn a living for themselves because of inheriting large sums of money. The name *affluenza* had apparently been suggested by one of the patients. By the end of the eighties, the term had started to catch on and was being applied more generally to the guilt feelings of people who suspected that they earned or possessed more than they were worth.

> The San Francisco group also runs seminars that teach heiresses how to cope with guilt, lack of motivation, and other symptoms of affluenza, an ailment she says is rampant among children of the wealthy. *Fortune* 13 Apr. 1987, p. 27

> Also pathogenic is 'affluenza', the virus of inherited wealth, striking young people with guilt, boredom, lack of motivation, and delayed emotional development.
>
> *British Medical Journal* 1 Aug. 1987, p. 324

ageism /ˈeɪdʒɪz(ə)m/ *noun* Also written **agism**

Discrimination or prejudice against someone on the grounds of age; especially, prejudice against middle-aged and elderly people.

Formed by adding the suffix *-ism* (as in *racism* and *sexism*) to *age*.

The word was coined by Dr Robert Butler of Washington DC, a specialist in geriatric medicine, in 1969; by the mid seventies it was fairly common in the US but did not really enter popular usage in the UK until the late seventies or early eighties. Until then, it was often written **age-ism**, displaying a slight discomfort about its place in the language. Along with a number of other *-isms*, *ageism* enjoyed a vogue in the media during the eighties, perhaps partly because of a growing awareness of the rising proportion of older people in society and the need to ensure their welfare. The adjective and noun **ageist** both date from the seventies and have a similar history to *ageism*.

The government campaign against 'ageism' was stepped up this weekend with a call for employers to avoid discrimination against the elderly in job advertisements.

Sunday Times 5 Feb. 1989, section A, p. 4

John Palmer, who had been at that desk for many years, was completely screwed . . . I think that's ageist. *New York* 23 July 1990, p. 29

See also ABLEISM, FATTISM, and HETEROSEXISM

AI /ˌeɪˈaɪ/ *abbreviation* 🖥

Short for **artificial intelligence**, the use of computers and associated technology to model and simulate intelligent human behaviour.

The initial letters of *Artificial Intelligence*.

Attempts to 'teach' computers how to carry out tasks (such as translation between languages) which would normally require a human intelligence date back almost as far as computer technology itself, and have been referred to under the general-purpose heading of *artificial intelligence* since the fifties. This was being abbreviated to *AI* in technical literature by the seventies, and by the eighties the abbreviation had entered the general vocabulary, as computing technology became central to nearly all areas of human activity. The abbreviation is often used attributively, with a following noun, as in **AI technology** etc.

Sales for AI technology will top £719 million this year. *Business Week* 1 July 1985, p. 78

Military research . . . has been both the driving force and . . . paymaster of AI development.
CU Amiga Apr. 1990, p. 89

-Aid /eɪd/ *combining form* Also written **-aid** and without hyphen 🎵

The second element in names of efforts to raise money for charity.

Based on *Band Aid*, the punning name of a rock group formed by Irish rock musician Bob Geldof in 1984 to raise money for famine relief in Ethiopia; *Band-Aid* is also the trade mark of a well-known brand of sticking-plasters. Until Bob Geldof became involved in this area, *aid* had tended to be associated with economic assistance given by one government to another, often with political conditions attached.

The enormous success of Bob Geldof's appeal for Ethiopia, which began with the release of Band Aid's record *Do they know it's Christmas?* in 1984 and continued with a large-scale rock concert called *Live Aid* in 1985, laid the foundations for this new combining element in the language. Whereas in the sixties, fund-raising organizations and events had favoured the word *fund* in their titles, it now became fashionable to use *-Aid* following the name of your group or activity (**School-Aid** for schoolchildren's efforts, **Fashion-Aid** for a charity fashion show, etc.), or after the name of the group being helped (as in **Kurd Aid**, an unofficial name for a Red Cross concert in aid of Kurdish refugees in May 1991).

Sport Aid organizers were yesterday endeavouring to maximize the money raised by Sunday's worldwide Race Against Time in aid of African famine relief. *The Times* 28 May 1986, p. 2

Inspired by the Live Aid rockathon, Willie Nelson staged Farm Aid I in Champaign to help the needy closer to home. *Life* Fall 1989, p. 142

aid fatigue 🎵 see COMPASSION FATIGUE

Aids /eɪdz/ *acronym* Also written **AIDS** ☣

Short for **acquired immune deficiency syndrome**, a complex condition which is thought to be caused by a virus called HIV and which destroys a person's ability to fight infections.

An acronym, formed on the initial letters of *Acquired Immune Deficiency Syndrome*.

The condition was first noticed by doctors at the very end of the seventies and was described under the name *acquired immune deficiency state* in 1980, although later research has shown that a person died from *Aids* as long ago as 1959 and that the virus which causes it may have existed in Africa for a hundred years or more. Colloquially the condition was also sometimes referred to as *GRID* (gay-related immune disease) in the US before the name *Aids* became established. The US Center for Disease Control first used the name *acquired immune deficiency syndrome* and the acronym *Aids* in September 1982, and by 1984 the disease was already reaching epidemic proportions in the US and coming to be known as the scourge of the eighties. At first *Aids* was identified as principally affecting two groups: first, drug users who shared needles, and second, male homosexuals, giving rise to the unkind name *gay plague*, which was widely bandied about in newspapers during the mid eighties. Once the virus which causes the immune breakdown which can lead to *Aids* was identified and it became clear that this was transmitted in body fluids, sexual promiscuity in general was blamed for its rapid spread. These discoveries prompted a concerted and ill-received government advertising campaign in the UK which aimed to make the general public aware of the risks and how to avoid them; this resulted, amongst other things, in the revival of the word CONDOM in everyday English.

The acronym soon came to be written by some in the form *Aids* (rather than *AIDS*) and thought of as a proper noun; it was also very quickly used attributively, especially in **Aids virus** (a colloquial name for HIV) and the adjective **Aids-related**. By 1984 doctors had established that infection with the virus could precede the onset of any symptoms by some months or years, and identified three distinct phases of the syndrome: *lymphadenopathy syndrome* developed first, followed by **Aids-related complex** (**ARC**), a phase in which preliminary symptoms of fever, weight loss, and malaise become apparent; the later phase, always ultimately fatal, in which the body's natural defences against infection are broken down and tumours may develop, came to be known as **full-blown Aids**. Colloquially, the phases before the onset of *full-blown Aids* are sometimes called **pre-Aids**.

The language of *Aids* (**Aidspeak**) became both complex and emotive as the eighties progressed, with the word *Aids* itself being used imprecisely in many popular sources to mean no more than infection with HIV—a usage which, in the eyes of those most closely concerned with *Aids*, could only be expected to add to the stigmatization and even victimization of already isolated social groups. The Center for Disease Control published a carefully defined spectrum of stages, in an attempt to make the position clear: *HIV antibody seronegativity* (i.e. the absence of antibodies against HIV in the blood), *HIV antibody seropositivity* (see ANTIBODY-POSITIVE), *HIV asymptomaticity*, *lymphadenopathy syndrome*, *Aids-related complex*, and *full-blown Aids*. In order to lessen the emotive connotations of some tabloid language about *Aids*, pressure groups tried to discourage the use of **Aids victim** and replace it with **person with Aids** (see PWA). The terminology had become so complex and tricky that those who could find their way about it and understood the issues came to be known as **Aids-literate**. At the time of writing no cure has been found for *Aids*.

> In just one year the list of people at risk from AIDS has lengthened from male homosexuals, drug-abusers and Haitians, to include the entire population [of the USA]. *New Scientist* 3 Feb. 1983, p. 289

> St. Jude Children's Research Hospital in Memphis . . . will look at potential drug treatments in animals for an AIDS-related form of pneumonia, pneumocystis carinii.
> *New York Times* 1 May 1983, section 1, p. 26

> *Buddies'* project is not to examine the construction of gay identity but to take apart the mythology of AIDS as a 'gay plague'. *Film Review Annual* 1986, p. 160

> Of 34 mothers who gave birth to children with Aids at his hospital, only four had any symptoms of the disease or Aids-related complex, a milder form. *Daily Telegraph* 3 Feb. 1986, p. 5

> Like many well-educated professionals who are sexually active, the man had become an AIDS encyclopedia without changing his habits. *Atlantic* Feb. 1987, p. 45

See also SLIM

Aidsline 〖 see -LINE

Aids-related virus ⊗ see HIV

airhead /'ɛəhɛd/ noun ◧

In North American slang, a stupid person; someone who speaks or acts unintelligently.

Formed by compounding: someone whose *head* is full of *air*; perhaps influenced by the earlier form *bubblehead* (which goes back to the fifties).

Airhead has been a favourite American and Canadian term of abuse since the beginning of the eighties, used especially for the unintelligent but attractive type of woman that the British call a BIMBO. At first *airhead* was associated with teenage VALSPEAK, but it soon spread into more general use among all age-groups. Although very common in US English by the mid eighties, *airhead* did not start to catch on in the UK or Australia until the end of the decade.

His comedies of manners are very funny, and the vain airheads who populate his novels are wonderfully drawn.   *Christian Science Monitor* 2 Mar. 1984, section B, p. 12

Mature women . . . left the airheads to be abused by the stuffy, bossy older men and wore shorter skirts than their teenage daughters.   *Indy* 21 Dec. 1989, p. 7

airside /'ɛəsaɪd/ noun ▨

The part of an airport which is beyond passport controls and so is only meant to be open to the travelling public and to bona fide airport and airline staff.

Formed by compounding: the *side* of the airport giving access to the *air* (as opposed to the *landside*, the public area of the airport).

The word *airside* has been in use in the technical vocabulary of civil aviation since at least the fifties, but only really came to public notice during the late eighties, especially after the bombing, over Lockerbie in Scotland, of a Pan-Am passenger jet after it left London's Heathrow airport in December 1988. As a result of this and other terrorist attacks on air travel, a great deal of concern was expressed about the ease with which a person could gain access to *airside* and plant a device, and several attempts were made by investigative reporters to breach security in this way. Tighter security arrangements were put in place. The word *airside* is used with or without an article, and can also be used attributively in **airside pass** etc. or adverbially (**to go airside** etc.).

Far too many unvetted people have access to aircraft . . . No one should get an 'airside' pass without . . . clearance.   *The Times* 27 June 1985, p. 12

For several hours the terminal-building was plunged into chaos. 'Airside' was sealed off by armed police.   *Daily Telegraph* 18 Apr. 1986, p. 36

Alar /'eɪlɑ:(r)/ noun ♣

A trade mark for daminozide, a growth-regulating chemical used as a spray on fruit trees to enable the whole crop to be harvested at once.

Alar has been manufactured under this brand name since the mid sixties and is used by commercial growers to regulate the growth of fruit (especially apples), so that larger, unblemished fruit which remains on the tree longer can be produced. The chemical does not remain on the surface of the fruit, but penetrates the flesh, so it cannot be washed off or removed by peeling. The results of research published in the second half of the eighties showed that, when the apples were subsequently processed (in order to make apple juice, for instance), *Alar* could be converted into unsymmetrical dimethylhydrazine (or UDMH), a potent carcinogen. This discovery brought *Alar* unwelcome publicity during the late eighties: mothers anxious to protect their children from harmful chemicals in foods (among them some famous mothers such as film star Meryl Streep in the US and comedian Pamela Stephenson in the UK) led a campaign to have its use discontinued. *Alar* was voluntarily withdrawn by its manufacturers, Uniroyal, from use on food crops in the US and Australia in 1989; in the UK the Advisory Committee on Pesticides declared it safe.

> Some products which have been publicised as Alar-free by retailers and manufacturers were still found to contain Alar.
> *She* Oct. 1989, p. 1

> Most people are far more frightened of the threat of cancer than of the flulike symptoms that they associate with food poisoning. Fanning their anxieties are frequent alerts: about dioxin in milk, aldicar in potatoes, Alar in apples.
> *New York Times* 7 May 1990, section D, p. 1

alcohol abuse 🔪 📇 see ABUSE

alcohol-free 🔆 see -FREE

Alexander technique /ælɪgˈzɑːndə(r) tɛkˌniːk/ *noun* ⊗

A COMPLEMENTARY therapy which aims to correct bad posture and teach people balanced use of their bodies as an aid to better health.

The name of F. Matthias *Alexander*, who invented the *technique*.

The *Alexander technique* was developed by Alexander, an Australian actor who subsequently devoted his life to physiotherapy, at the end of the nineteenth century, and was promoted by the writer Aldous Huxley in the forties. It was not widely taken up by the general public until the seventies in the US and the early eighties in the UK, when complementary medicine and alternative approaches to health became more socially acceptable than previously. It continued to enjoy a vogue in the late eighties, since it fitted in well with the NEW AGE approach to self awareness. Although not claiming to cure any organic health problems, teachers of the *Alexander technique* maintain that it can relieve or even remove symptoms, notably back pain, as well as helping people to prevent pain and discomfort in later life.

> The Alexander Technique is a very careful, gentle way of increasing awareness; it was a joy to learn how to listen to myself.
> *Out from the Core* Feb. 1986, p. 1

> I saw an ad ... for a cheap introductory course in Alexander technique and as I had poor posture and ... an aching back, I went along.
> *Good Housekeeping* May 1990, p. 1

aliterate /eɪˈlɪtərət/ *adjective* and *noun* 📇

adjective: Disinclined to acquire information from written sources; able to read, but preferring not to.

noun: A person who can read but chooses to derive information, entertainment, etc from non-literary sources.

A hybrid word, formed by adding the Greek prefix *a-* in the sense 'without' to *literate*, a word of Latin origin. The hybrid form was intended to make a distinction between the *aliterate* and the *illiterate* (formed with the equivalent Latin prefix *in-*), who are *unable* to read and write.

The word *aliterate* was coined in the late sixties, but it was not until the eighties that there began to be real evidence that the increasing popularity of television and other 'screen-based' media (including information on computer screens) was having a noticeable effect on people's use of reading and writing skills. This observation came soon after it had been revealed that there were considerable numbers of people leaving school unable to read and write. In the early eighties, the noun **aliteracy** developed as a counterbalance to *illiteracy*; the two terms described these twin problems. As the eighties progressed, graphics and video became even more heavily used to put across information, to teach, and to entertain; *aliteracy* is therefore likely to become increasingly prevalent in the nineties.

> The nation's decision-making process ... is threatened by those who can read but won't, Townsend Hooper, president of the Association of American Publishers, told some 50 persons attending an 'a-literacy' conference.
> *Publishers Weekly* 1 Oct. 1982, p. 34

> According to a recent estimate, 60 million Americans—almost one-third of our entire population—

is illiterate. And a recent report from the Librarian of Congress suggests that we may have at least the same number who are aliterate.   *The Times* 27 Dec. 1985, p. 12

all-terrain bike ⚒ see MOUNTAIN BIKE

alpha test /ˈælfə ˌtɛst/ *noun* and *verb* 💻

noun: A preliminary test of an experimental product (such as computer software), usually carried out within the organization developing it before it is sent out for beta testing.

transitive verb: To submit (a product) to an alpha test.

Formed by compounding. *Alpha*, the first letter of the Greek alphabet, has long been used to denote the first in a series; the *alpha test* is the first *test* in a routine series.

The concept of the *alpha test* comes from the world of computer software development, where it has been used since the early eighties. Its purpose is to iron out as many bugs as possible before allowing the software to be used by outsiders during the second phase of testing (see BETA TEST). A person whose job is to test software in this way for the developer is an **alpha-tester**; the process is known as **alpha testing** and the product at this stage of development is the **alpha-test version**.

> As the operations manager for a large computer equipment manufacturer, Ray Majkut helped oversee the 90-day test of a 200-line private branch exchange, an experience he regarded as more of an alpha test than a beta test.   *Network World* 14 Apr. 1986, p. 35

> Apple set Hypercard 2.0 into alpha test right before the quake, making a spring intro likely.
> *InfoWorld* 23 Oct. 1989, p. 110

Altergate 🏛 see –GATE

alternative /ɔːlˈtɜːnətɪv/ *adjective* and *noun* ⚒

adjective: Offering a different approach from the conventional or established one; belonging to the COUNTER-CULTURE.

noun: An approach that is alternative in this way; also, a follower of alternative culture.

A simple development of sense: *alternative* first meant 'offering a choice between two things', but by the end of the last century could be used to refer to choices involving more than two options. The meaning dealt with here probably arose from the phrase *alternative society* (see below).

The word *alternative* was first used in this sense when the hippie culture of the late sixties, with its rejection of materialism and traditional Western values, was described as an **alternative society**. Almost immediately, anything that served the counter-culture also came to be described as *alternative* (for example the **alternative press**, consisting of those newspapers and magazines that were aimed at radical youth); uses arose from within the counter-culture, too (for example the **alternative prospectus**, which gave the students' view of an educational establishment rather than the official view). Although the term *alternative society* itself had fallen from fashion by the end of the seventies, the adjective enjoyed a new vogue in the eighties as the green movement urged society to seek new approaches to natural resources, fuel sources, etc. and the health and fitness movement became increasingly influential in advocating unconventional medical therapies. The most important *alternatives* of the past decade have been:

alternative birth, birthing 👶 , any method of childbirth that tries to get away from the intrusive, HIGH-TECH approach of modern medicine towards a more natural and homely setting in which the mother has control;

alternative comedy ⚒ , comedy that is not based on stereotypes (especially sexual or racial ones) or on conventional views of humour, but often includes an element of black humour or surrealism and an aggressive style of performance; also **alternative comedian**, **alternative comedienne**, practitioners of this;

alternative energy 🌱 , energy (such as solar power, wind generation, etc.) derived from any source that does not use up the earth's natural resources of fossil fuels or harm the environment;

alternative medicine, therapy ⊗ , any medical technique that aims to promote health and fitness without the use of drugs, often involving the patient in self-awareness and self-help; COMPLEMENTARY medicine;

alternative technology 🌱 💻 , technology deliberately designed to conserve natural resources and avoid harm to the environment, especially by harnessing renewable energy sources.

> Babies are born with as little medical intervention as possible in the hospital's Alternative Birth Center, located on a separate floor from the maternity wing.
> *Money* Dec. 1983, p. 205

> A recent survey of more than 1,000 practitioners, conducted by the Institute for Complementary Medicine, found the number of patients turning to alternative therapies growing at an annual rate of 15 per cent, with a 39 per cent increase in patients visiting homeopaths.
> *Chicago Tribune* 8 Apr. 1985, p. 1

> Jennifer is a 20-year-old *Alternative*, with short platinum hair jelled and sprayed into a cone, bright face, smart casual clothes and heavy worker's boots. *Courier-Mail* (Brisbane) 27 Sept. 1988, p. 17

> The so-called alternative comedy boom was initially compared to the punk phenomenon and ultimately has proved to be equally as impotent. *Arena* Autumn/Winter 1988, p. 163

> Waterfall Vegetarian Food ... is launching its new range of alternative salami slices with its Vegelami slice. *Grocer* 21 Jan. 1989, p. 168

> The ... Trust will invest in companies working to ensure a better cleaner environment (waste processing, alternative energy, recycling, etc). *Green Magazine* Apr. 1990, p. 82

angel dust /ˈeɪndʒ(ə)l ˌdʌst/ *noun* Sometimes written **angels' dust** 💉

In the slang of drug users, the hallucinogenic drug phencyclidine hydrochloride or PCP (see PCP¹).

Formed by compounding. The drug was originally taken in the form of a powder or *dust*; it may be called the dust of *angels* because of the supposedly heavenly visions that it produces, although it has been claimed that the reason is that the drug was first distributed illegally by Hell's Angels.

Angel dust was popular in the drugs subculture of the sixties (when the term was sometimes used to refer to drug mixtures such as cocaine, heroin, and morphine, or dried marijuana with PCP). In the eighties *angel dust* enjoyed a short-lived revival as one of the preferred drugs of the new psychedelia associated with ACID HOUSE; the term became the usual street name this time round for PCP, which also had a large number of other slang names such as *cornflakes*, *goon*, *hog*, *loopy dust*, and *rocket fuel*.

> She could've been on something ... Acid, angel dust. Elmore Leonard *Glitz* (1985), p. 69

> PCP or 'angel dust', a strong anaesthetic which came after LSD in 1960s drug fashions ... has recently emerged anew. Now they call it 'rocket fuel' in Chicago and mix it with peanut butter.
> *Sunday Times* 24 Mar. 1985, p. 12

> 'Angel dust', one of the most dangerous street drugs ever created, may soon have a new role—in treating heart attack and stroke victims. *Observer* 12 Mar. 1989, p. 32

angioplasty /ˈændʒɪəʊˌplæstɪ/, /-ˌplɑːstɪ/ *noun* ⊗

An operation to repair a damaged blood vessel or to unblock a coronary artery.

A compound formed on classical roots: *angio-* is the Latinized form of a Greek word, *aggeion*, meaning 'a vessel'; *-plasty* comes from Greek *plastia*, 'moulding, formation'.

Angioplasty has been known as a medical term since the twenties, but came into the news during the eighties particularly as a result of the development of two new techniques for carrying it out. **Balloon angioplasty,** available since the mid eighties, involves passing a tiny balloon up the patient's arteries and inflating it to remove blood clots or other blockages. **Laser angioplasty,** still in its experimental stages in the late eighties, makes use of lasers to burn away blockages, and is designed to be minimally invasive. The development of these techniques has meant that expensive heart surgery under general anaesthetic can now often be avoided, with *angioplasty* taking place instead under local anaesthetic. *Angioplasty* by these new means has therefore been vaunted in the popular science press as a very significant medical advance.

> Arterial lesions would remain at the center of medical interest in coronary heart disease for decades to come. Cholesterol-lowering diets would aim to slow their growth; bypass surgery would attempt to route blood around them; in angioplasty, a tiny balloon would squeeze the lesions open.
> *Atlantic* Sept. 1989, p. 39

Anglo-Irish agreement /ˌæŋgləʊˌaɪrɪʃ əˈgriːmənt/ *noun* 📷

A formal agreement between the United Kingdom and the Republic of Ireland, signed on 15 November 1985, establishing an intergovernmental conference and providing for greater cooperation between the two countries, especially where the sovereignty and security of Northern Ireland were concerned.

Anglo- is the combining form of *English*, but doubles as the combining form for *British* and 'of the United Kingdom', since neither has a combining form of its own; to describe the *agreement* as *Anglo-Irish* therefore means not just that it was between England and Eire, but between the whole United Kingdom and Eire (and so by implication included Northern Ireland, even though it met with opposition there).

The *Anglo-Irish agreement* was the subject of some considerable speculation in the press long before it was actually signed by British Prime Minister Margaret Thatcher and Irish Taoiseach Garret Fitzgerald at Hillsborough, Co. Down, in 1985: the earliest uses of the term date from the very beginning of the eighties. It became very frequently used in newspapers during the mid eighties, partly as a result of the intense opposition to it raised by Ulster Unionists. They particularly objected to the fact that their political representatives had not been involved in the negotiations and to the implications they saw in it for the sovereignty of Northern Ireland. Attempted Ulster talks in May 1991 sought to involve them first in a new agreement.

> The disagreement goes to the heart of the problem of how to introduce Dublin as a partner in the talks and what role it would have in renegotiating the replacement of the Anglo-Irish Agreement.
> *Guardian* 28 June 1990, p. 2

animal-free ⊗ ✖ see -FREE

animalist¹ /ˈænɪməlɪst/ *noun* 📷

An animal rights campaigner or supporter.

A contraction of *animal liberationist*; formerly, an *animalist* was a follower of the philosophy of animalism or an artist who treated animal subjects.

This snappier term arose in US English during the mid eighties and is as yet barely established in the language. The movement to which it refers, variously known as **animal liberation, animal lib,** and **animal rights,** has a much longer history—the term *animal liberation* goes back to the early seventies—and there is a good case for a term which would be less of a mouthful than **animal liberationist** or **animal rights campaigner,** although this one suffers from possible confusion with the opposite meaning of the adjective *animalist* in the entry below.

> The uproar resulted from a column two weeks ago in which I reported that animalist Barbara Toth was enraged over the possibility that some Asian immigrants in Canoga Park might be turning strays into dog foo young.
> *Los Angeles Times* (Valley edition) 22 July 1985, section 2, p. 7

The dismal sight on Tuesday night of bedraggled 'animalists' distributing protest literature to queues of happy families agog with the expectancy of pure pleasure. *Financial Times* 28 July 1988, p. 21

animalist² /ˈænɪmǝlɪst/ *adjective*

Discriminating against animals; demeaning animals or denying them rights by the way one speaks, thinks, or behaves.

Formed by adding the suffix *-ist* as used in *racist* or *sexist* to *animal*: compare *ageist* (see AGEISM).

Also very new and still rare, this sense of *animalist* is a British usage which promises to give rise to some considerable confusion by creating a situation in which the noun *animalist* and its corresponding adjective carry almost opposite meanings. Ultimately one or other sense must surely survive at the expense of the other—if indeed either catches on.

Animal rights campaigners on Merseyside are urging parents and teachers to stop children using 'animalist' expressions, which they claim demean certain creatures.

Daily Telegraph 27 Oct. 1989, p. 5

animatronics /ˌænɪmǝˈtrɒnɪks/ *noun*

The technique of constructing robots which look like animals, people, etc. and which are programmed to perform lifelike movements to the accompaniment of a pre-recorded soundtrack.

Formed by combining the first three syllables of *animated* with the last two of *electronics* to make a blend.

The idea of *animatronics* (which originally had the even more complicated name *audio-animatronics*, now a trade mark) was developed by Walt Disney during the sixties for use at the World's Fair and later for Disneyland and other THEME PARKS. The movements and gestures of the robots (each of which may be called an **animatron** or an **animatronic**) are extremely life-like, but because they are pre-programmed they cannot be responsive or interactive: for this reason, *animatronics* has been described as being 'like television with the screen removed'. During the eighties, *animatronics* became more widely known as the theme park idea and the robotics technology were exported from the US to other parts of the world. Although it looks plural in form, *animatronics* always takes a singular agreement when it refers to the technique; plural agreement indicates that it is being used for a group of the robots themselves. The adjective used to describe the technology or the robots is **animatronic**.

'How-about-some-you'd-pay-twice-as-much-for-anywhere-else,' yells Stein, his mouth seeming to move independently of the words, like one of those eerie Animatronic Disney robots.

Forbes 12 Nov. 1979, p. 177

Sally Animatronics Pty Ltd has set up shop in Sydney to capitalise on what it perceives to be a boom market in Australia . . . —the production of lifelike robots for theme parks, exhibitions and museums. The robots, known as animatronics, were made famous by Disneyland . . . Designing an animatronic figure is a difficult process. *The Australian* 24 Nov. 1987, p. 58

The animals and acrobats of the popular entertainment will give way to a Disney-style 'animatronic' show, part of a £17.5-million plan to revamp the Tower. *The Times* 28 Sept. 1990, p. 17

antibody-positive /ˌæntɪbɒdɪˈpɒzɪtɪv/ *adjective*

Having had a positive result in a blood test for the AIDS virus HIV; at risk of developing Aids.

Formed by compounding; having a *positive* test for *antibodies* to HIV. Long before Aids, *antibody-positive* was in technical use for the result of any blood test for antibodies to a virus; it is only in popular usage that it has become specialized almost exclusively to the Aids sense.

This sense of *antibody-positive* arose during the mid eighties, when fear of Aids was at its height and much publicity was given to it. Since infection with HIV could precede the onset of any Aids symptoms by a period of years, and only some of those who were tested positive would in fact develop symptoms at any time, health officials emphasized the need to avoid over-reacting to a positive test and tried (with varying degrees of success) to prevent discrimination against those who were known to be *antibody-positive*. The adjective for a person found not to have been infected or a test with a negative result is **antibody-negative**, but this is less commonly found in popular sources.

> Without testing facilities at, say, clinics for sexually transmitted diseases, 'high-risk' donors might give blood simply to find out their antibody status (and possibly transmit the virus while being antibody-negative). *New Statesman* 27 Sept. 1985, p. 14

> This longstanding concentration on the clinical manifestations of AIDS rather than on all stages of HIV infection (i.e., from initial infection to seroconversion, to an antibody-positive asymptomatic stage, to full-blown AIDS) has had the ... effect of misleading the public.
> Susan Sontag *Aids & its Metaphors* (1989), p. 31

anti-choice /ˌæntɪˈtʃɔɪs/ *adjective* Sometimes written **antichoice** ⊗ ⧫

Especially in US English, opposed to the principle of allowing a woman to choose for herself whether or not to have an abortion; a derogatory synonym for *pro-life* (see under PRO-).

Formed by adding the prefix *anti-* in the sense 'against' to *choice*.

The whole issue of abortion has been an extremely contentious one in US politics during the past fifteen years. The term *anti-choice* arose in the second half of the seventies as a label applied to pro-life campaigners by those who had fought for women's rights in the US and resented the erosion of their work by the anti-abortion lobby. As such it is deliberately negative in form (supporters of the rights of the unborn child would describe themselves in more positive terms such as pro-life or RIGHT-TO-LIFE). Although abortion has also been an important issue in the UK in the eighties, the term *anti-choice* has hardly been used in British sources until quite recently.

> She said there are at least three races in the state where a clear anti-choice incumbent is being opposed by a strong pro-choice challenger. *San Francisco Chronicle* 26 June 1990, section B, p. 4

anti-lock /ˌæntɪˈlɒk/ *adjective* 🖳

Of the brakes of a car or other vehicle: set up so as to prevent locking and skidding when applied suddenly; especially in **anti-lock brake** (or **braking**) **system** (**ABS**), a patent system which allows sudden braking without any locking of the wheels.

Formed by adding the prefix *anti-* in the sense 'preventing' to the verb stem *lock*.

Anti-lock braking was developed in the sixties from a similar system which had been applied to aeroplanes (under the name *wheel-slide protection system*). The first application to motor vehicles was Lockheed's *Antilok* (a trade mark); at first it was used mainly for heavy trucks and the like. The term began to appear frequently in car advertising in the early eighties, when the system became generally available on private cars (either as an optional extra or a standard feature), and was used as a strong marketing point. The system works by momentarily releasing the brakes and freeing the locked wheel as often as necessary to avoid skid. *Anti-lock* is occasionally used on its own as a noun as a shortened form of *anti-lock brake system*.

> Unlike car systems, the motorcycle ABS does not allow full application of the brakes while cornering. *Daily Mirror* (Sydney) 21 Oct. 1988, p. 111

> An anti-lock brake system is available. This amazing sports sedan also has a Bumper-to-Bumper warranty that's good for 3 years. *Life* Fall 1989, p. 85

antivirus 🖳 see VACCINE

Aqua Libra /ˌækwə ˈliːbrə/ *noun* 🎱

The trade mark of a health drink containing spring water, fruit juices, and a number of other ingredients, which is promoted as an aid to proper alkaline balance and good digestion.

Latin *aqua* 'water' and *libra* 'balance': literally 'water balance' (compare BALANCE).

Aqua Libra was launched under this name in 1987, at a time when there was a fashion for non-alcoholic drinks, and many smart executives favoured mineral water (see DESIGNER).

> Aqua Libra ... is completely free of alcohol and I like it because it is not as sweet as, say Perrier and orange juice.
> *Financial Times* 31 Dec. 1988, Weekend FT, p. IX

> The smart set in England this season is drinking Aqua Libra. The pale-gold beverage is a blend of sparkling water, passion fruit juice and apple juice, seasoned with sesame, sunflower, melon, tarragon and Siberian ginseng.
> *Forbes* 25 Dec. 1989, p. 48

Aquarobics /ˌækwəˈrəʊbɪks/ *noun* Sometimes written **aquarobics** or **aquaerobics** ⊗ ✖

The trade mark of a fitness programme, including a form of AEROBICS, in which the exercises are done in a shallow swimming pool.

Formed by substituting the Latin word *aqua* 'water' for the first syllable of *aerobics*.

Aquarobics was developed by Georgia Kerns and Judy Mills in the US in 1980 and registered there as a trade mark. By the late eighties it had spread to the UK and was becoming a popular alternative to aerobics, being promoted especially as a form of exercise suitable for people with physical disabilities or those recovering from operations.

> The movable floor can be lowered from 1.5 feet to 10 feet and is used for such water exercise classes as aquarobics and aquafitness.
> *Business First of Buffalo* 9 Mar. 1987, p. 30

> Many ... handicapped people said how beneficial the Aquarobics Exercises had been.
> *Keep Fit* Autumn 1989, p. 7

arb /ɑːb/ *noun* 〰

In financial jargon, a dealer in stocks who takes advantage of differing values in different markets to make money; especially on the US stock exchange, a dealer in the stocks of companies facing take-over bids.

A colloquial shortened form of *arbitrageur*, a French word borrowed into English in the late nineteenth century for any stock dealer who makes his money from buying stock in one market and selling in another.

Although the practice of *arbitrage* (the simultaneous buying and selling of large quantities of stock in different markets so as to take advantage of the price difference) is well established—it dates from the late nineteenth century—the word *arbitrageur* was not shortened to *arb* in print until Wall Street risk arbitrageurs started buying up large quantities of stock in companies facing take-over bids in the late seventies. These take-overs attracted considerable media interest, and the word *arb* started to appear frequently in the financial sections of newspapers from about the beginning of the eighties.

> For a start you often have to make use of the 'arbs', very useful gentlemen indeed in a bid battle.
> *Sunday Telegraph* 25 Mar. 1984, p. 19

> It should have been the risk arbitrageurs' finest year ... Instead, in the wake of archrival Ivan F. Boesky's admission of insider trading, the arbs are being battered. *Business Week* 8 Dec. 1986, p. 36

ARC ⊗ see AIDS

aromatherapy /əˌrəʊməˈθɛrəpɪ/ *noun* Sometimes in the form **aromatotherapy** ⊗

A COMPLEMENTARY therapy which makes use of essential oils and other plant extracts to promote a person's health, general well-being, or beauty.

Actually borrowed from French *aromathérapie*, although the formation of the English word is self-explanatory: *therapy* based on *aromatic* oils.

Aromatherapy was promoted by the French chemist René-Maurice Gattefossé in the thirties, but was not widely taken up in English-speaking countries until the seventies, when the search began for natural remedies to replace the increasingly intrusive techniques of traditional medicine. There was nothing new, of course, in the use of plant extracts for medicinal purposes; it was the therapeutic effect of inhaling the aromatic oils or massaging them into the skin that Gattefossé claimed to have discovered anew. During the eighties, when alternative therapies proliferated and there was a premium on the use of natural ingredients, *aromatherapy* graduated from fringe status to a reasonably respected technique, especially for the relief of stress-related symptoms. A practitioner of *aromatherapy* is called an **aromatherapist**; the adjective used to describe an oil which has some use in *aromatherapy* is **aromatherapeutic**.

> Today in Britain most therapists and their clients use aromatherapy as a form of relaxation with some benefits to minor medical conditions. *Here's Health* June 1988, p. 89

> For details of a qualified aromatherapist in your area contact the International Federation of Aromatherapists. *Prima* Aug. 1988, p. 74

artificial intelligence 🖳 see AI

ARV ⊗ see HIV

asset /ˈæsɛt/ noun ∿

The first word of a number of compounds fashionable in the business and financial world, including:

asset card, a US name for the debit card (see CARD¹);

asset management, the active management of the assets of a company so as to optimize the return on investments; the job of an **asset manager**;

asset-stripping, the practice of selling off the assets of a company (especially one which has recently been taken over) so as to make maximum profit, but without regard for the company's future; the activity of an **asset-stripper**.

The word *assets*, which originally came from Anglo-French *assets* (modern French *assez* enough) was reinterpreted as a plural noun with a singular *asset* by the nineteenth century; however, it was only in the late twentieth century that it acquired compounds based on this singular form.

All three compounds entered the language through US business usage in the mid seventies; *asset-stripping* had been practised since the fifties, but did not become widely known by this name until the seventies. *Asset management* and *asset-stripping* have been widely used in the UK during the eighties, even moving into non-technical usage. By the end of the decade, though, *asset-stripping* had become an unfashionable name for an activity which financiers now preferred to call *unbundling*: see UNBUNDLE.

> Guinness Peat's chief executive . . . reckons that institutions in the post Big Bang City will take one of three forms—bankers, traders or asset managers. *Investors Chronicle* 1 Nov. 1985, p. 54

> The solution . . . —moving the $2 billion asset card business to . . . South Dakota—ushered in a new era in interstate banking. *US Banker* Mar. 1986, p. 42

> One of the large mutual fund families . . . offers not only a variety of funds but an asset management account that would give you a monthly record of all transactions, including reinvestment of dividends. *Christian Science Monitor* 20 Feb. 1987, section B, p. 2

> A more relevant description of Hanson's strategy would be asset-mining rather than asset-stripping; that is, the development of undervalued assets for hidden value. *National Westminster Bank Quarterly Review* May 1987, p. 27

They were returning . . . from visiting a foundry in Derby that had been taken over by asset-strippers.
David Lodge Nice Work *(1988), p. 154*

ATB ▓ see MOUNTAIN BIKE

ATM /ˌeɪtiːˈɛm/ *abbreviation* 〰

Short for **automated teller machine**, a machine which carries out banking transactions automatically. (Usually known colloquially in the UK as a cashpoint or cash dispenser, although it may be capable of carrying out transactions other than cash dispensing.)

The initial letters of *automated* (or *automatic*) *teller machine*.

The full term *automated teller machine* was first used in the mid seventies, when the machines were put into mass operation in US banks; by 1976 this had been abbreviated to *ATM*, which has remained the standard term for the increasingly versatile machines in the US as well as Australia and other English-speaking countries. In the UK, they were available from the middle of the seventies but not used by the mass of the British public until the mid eighties. Consequently, the name *ATM* has tended to be used mostly in official circles, while *cash dispenser*, *cash machine*, and *cashpoint* have been the more popular names. Even though the machines are now capable of registering deposits, providing statements, etc., it seems unlikely that *ATM* will become the regular term in the UK as well.

Bill payments and loan repayments can be made through ATMs . . . 80 per cent of all ATM transactions were withdrawals, 10 per cent were inquiries and 10 per cent were deposits.
Sunday Mail Magazine (Brisbane) 12 Oct. 1986, p. 16

Need cash at midnight? Hit the ATM.
Life Fall 1989, p. 49

See also CASH DISPENSER

audio-animatronics 🖵 see ANIMATRONICS

autogenic training /ˌɔːtəʊˌdʒɛnɪk ˈtreɪnɪŋ/ *noun* ⊗ ▓

A relaxation technique in which the patient is taught a form of self-hypnosis and biofeedback as a way of managing stress.

A translation of the German name, *das autogene Training*. *Autogenic*, an adjective which has been used in English since the late nineteenth century, literally means 'self-produced'. It is not the *training* that is self-produced, though; *autogenic training* is designed to teach people how to produce a feeling of calm and well-being in themselves in stressful circumstances. A more accurate (though long-winded) name would be *training in autogenic relaxation*.

Autogenic training was invented in Germany and first popularized by psychiatrist and neurologist Johannes Schultz from the thirties until the fifties. It is the first of three stages in a method which is known in its entirety as **autogenic therapy**. Although it has reputedly been used by East German athletes for decades, it only became widely practised outside Germany in the seventies and eighties. The technique is particularly useful for athletes because it offers the possibility of bringing about positive changes in one's own physical state (such as lowering blood pressure or reducing heart-rate). **Autogenics** is an alternative name for *autogenic therapy* or *autogenic training*; although plural in form, this noun (like AEROBICS) can take singular or plural agreement.

A new study indicates that autogenics—a form of mental press-ups—are as good for reducing stress . . . as physical exertions.
She July 1985, p. 115

Liz Ferris uses autogenic training with athletes. This discipline is designed to help switch off the body's stress mechanisms.
Observer 6 May 1990, p. 21

automated teller machine 〰 see ATM

aware /ə'wɛə(r)/ *adjective* 🍏 🎚

Of a person, social group, etc.: fully informed about current issues of concern in a particular field. Of a product: designed, manufactured, or marketed in such a way as to take account of current concerns and attitudes. (Often with a preceding adverb indicating the field of concern, as **ecologically** or **environmentally aware**, **socially aware**, etc.)

Formed by increasingly elliptical use of the adjective: first, people were described as being *aware of* certain issues, then they were simply described as *socially* (etc.) *aware*, and finally their quality of *awareness* was ascribed to the products which resulted from their concerns.

People have been described as *socially* or *politically aware* since the early seventies; as the green movement gained momentum in the late seventies and early eighties it became increasingly important to be *ecologically* or *environmentally aware* as well. The adjective started to be applied to things as well as people in the early eighties; this usage remains limited in practice to *environmentally aware* products and activities and sometimes appears to mean only that some part of the profit on the sales is to be donated to a green cause.

Most of the machines described as being 'environmentally aware' will also cost you over £400.

Which? Jan. 1990, p. 49

The main dessert component was one of the few ecologically aware trademarked foods, the 'Rainforest Crunch' ice cream made by Ben & Jerry's, which donates some of the profits from this flavor to a rain forest preservation fund.   *Los Angeles Times* 21 June 1990, section E, p. 8

awesome /'ɔːs(ə)m/, in the US /'ɑs(ə)m/ *adjective* 🔲

In North American slang (especially among young people): marvellous, great, stunningly good.

Awesome originally meant 'full of awe', but by the end of the seventeenth century could also be used in the sense 'inspiring awe, dreadful'. The apparent reversal of meaning that has now taken place started through a weakening of the word's meaning during the middle decades of the twentieth century to 'staggering, remarkable'; this was then further weakened and turned into an enthusiastic term of approval in the eighties.

Within the youth culture, terms of approval come into fashion and go out again quite rapidly. After becoming frequent in its weakened sense of 'mind-boggling' during the sixties and seventies, *awesome* was taken up in the eighties as one of the most fashionable words of general approval among young Americans. In particular it was associated with the speech of *preppies* and the New York smart set, and often seemed to be part of a fixed phrase, preceded by *totally*. Surprisingly, it has remained popular among young people into the nineties, and has spread outside the US to Canada and Australia. It has been used in British English in this sense too, but really only in caricatures of US speech.

Stuck in a rut . . . the kid was at the end of his rope when out of the blue . . . *kaboom* . . . 'Awesome!! The Acclaim remote for Nintendo!'   *Captain America* Nov. 1989, p. 7

Roxanne Shante is quite simply the baddest sister around, and teamed with Marley Marl at the mixing desk she is awesome.   *Number One* 8 Nov. 1989, p. 43

That night I *freebased* a *fractal* of *crack* and *blissed out* on *E*. It was *awesome*. It was *ace*. It was *wicked*, *bad* and *def*. It was twenty quid. OUCH!   *Blitz* Dec. 1989, p. 130

Azeri /ə'zɛəri/ *noun* and *adjective* Sometimes written **Azari** 🎚

noun: A member of a Turkic people living mainly in Azerbaijan, Armenia, and northern Iran; an Azerbaijani. Also, their language.

adjective: Of or belonging to this people or their language.

The Turkish form (*azeri*) of what was originally a Persian word for fire; the place-name *Azerbaijan* is a compound meaning 'fire-temple'. *Azeri* is apparently the preferred form among those of Azeri ethnic origin, since it preserves a distinction between the Turkic people and anyone who lives in Azerbaijan (*Azerbaijani* can mean either).

Although used in ethnographical and linguistic works since at least the last century, *Azeri* was not a word that the average reader of English newspapers would have recognized until the late eighties. Then ethnic unrest on the border between the Armenian and Azerbaijan Republics was widely reported in the newspapers. Since the trouble was partly caused by the fact that large numbers of ethnic Armenians lived within the borders of Azerbaijan and Azeris in Armenia, it was necessary for journalists to make the distinction between the inhabitant of Azerbaijan (an *Azerbaijani*) and the *Azeri*.

> At least two civilians, one Armenian and one Azeri, attacked Armenian homes . . . Azeri mobs had burned 60 houses . . . Three Azeris were shot and killed by troops.　　*Observer* 27 Nov. 1988, p. 23

AZT /eɪzɛdˈtiː/, in the US /eɪziːˈtiː/ *abbreviation*

Short for **azidothymidine**, a drug used in the treatment of AIDS to stop the virus HIV from replicating itself within the patient's body; now officially known as ZIDOVUDINE.

The first two letters of *azido-* combined with the initial letter of *thymidine*.

Azidothymidine was developed in the US during the mid seventies, before Aids became a problem, but was always intended as a RETROVIRUS inhibitor. When HIV was identified as the probable cause of Aids in the mid eighties, its applicability to this virus was tested and it was found that it could prolong the life of Aids patients by preventing the virus from copying itself and so reducing the patients' susceptibility to infections. This discovery led to its being promoted in the press as a 'wonder drug' and even as a cure for Aids, although its testers continued to emphasize the fact that it was only capable of slowing down the development of the disease. Once the drug was in use for treating Aids, the name *azidothymidine* was usually abbreviated to *AZT*. This is still the name by which the drug is known colloquially, despite the fact that its official name has been changed to *Zidovudine*.

> The company has been sharply criticized for the cost of AZT, and recently cut the price by 20 per cent. An adult with AIDS now pays about $6,500 a year for the drug.
> *New York Times* 26 Oct. 1989, section A, p. 22

● ●

B

-babble /ˈbæb(ə)l/ *combining form*

The jargon or gobbledegook that is characteristic of the subject, group, etc. named in the first part of the word:

ecobabble 🌱 , environmental jargon; especially, meaningless GREEN jargon designed to make its user sound environmentally AWARE;

Eurobabble 🏛 , the jargon of European Community documents and regulations;

psychobabble 🧠 , language that is heavily influenced by concepts and terms from psychology;

technobabble 💻 , technical jargon, especially from computing and other high-technology areas.

The noun *babble* means 'inarticulate or imperfect speech, especially that of a child': the implication here is that these jargon-ridden forms of the language sound like so much nonsense to those who are not 'in the know'. In these words *babble* has been added on to the combining form of *ecological* etc. like a suffix: compare the earlier use of *-speak* in this way, after George Orwell's *Newspeak* and *Oldspeak* in the novel *1984*.

Psychobabble was coined in the US in the mid seventies, when various forms of psychoanalysis and psychotherapy were fashionable and the terms of these subjects were often bandied about by laypeople who only partly understood them. In 1977, Richard Rosen devoted a whole book to the subject of Americans who used this language of analysis. It was not long before other forms using *-babble* started to appear in the language: *Eurobabble* arrived soon after Britain's entry into the EC and *ecobabble* followed in the mid eighties as the green movement gained momentum.

> Is the environmental hoopla resonating through the halls of American business 'mere corporate ecobabble intended to placate the latest group of special-interest loonies'?
> *Los Angeles Times* 1 Feb. 1990, section E, p. 1

> No matter that the Kohl–Mitterrand accords might amount to no more than Eurobabble. They, and many British voters, see a Continental future in which ever more business is ordained without British involvement. *The Times* 27 Apr. 1990, p. 13

baby boomer ⁅⁆ see BOOMER

baby buster ⁅⁆ see BUSTER

Bach /bɑːx/ *proper noun* ⊗

In **Bach** (or **Bach's**) **flower remedies** (sometimes simply **Bach remedies**): a COMPLEMENTARY therapy related to homoeopathy, in which a number of preparations of plant origin are used to relieve emotional states which (according to the inventor of the remedies, Edward Bach) underlie many physical illnesses.

The name of Edward Bach combined with *flower remedies*.

Dr Edward Bach (1886–1936) was a Harley Street specialist who became interested in homoeopathy and developed the remedies as his own contribution to the discipline. According to his theory, the mind and body can be in a positive state (ease) or degenerate into a negative one (disease). He developed 38 different remedies, each designed to produce the positive state of ease for a particular personality type. *Bach flower remedies* were not widely known or used until the middle of the eighties, when they suddenly became fashionable, perhaps as a result of the general upsurge of interest in homoeopathy and alternative therapies at this time.

> The key to the Bach Remedies is that they are chosen not for the symptoms of the illness, but for the underlying emotional state of the client. *Out from the Core* Feb. 1986, p. 14

backward masking /ˌbækwəd ˈmɑːskɪŋ/ *noun* ♪

A technique in music recording in which a disguised message is included in such a way as to be audible only when the disc is spun backwards, although it may allegedly be perceived subliminally during normal playing. Also, the message itself.

Formed by compounding: *masking* a message that has to be played *backwards* to be heard. In psychology, *backward masking* is a technical term used since the sixties to mean 'disruption of a stimulus by a second, similar stimulus which closely follows it'.

The idea of hiding a backward message on a rock record was first tried by the Beatles as long ago as the sixties, but the term *backward masking* only became widely known during the early eighties as a result of attempts by Christian fundamentalist groups to have the practice banned. They claimed that a number of rock groups were including satanic messages on their records using this technique, and that these messages had a subliminal effect on the listener. In parts of the US, legislation was passed in the mid eighties making warning notices compulsory on all records carrying *backward masking*, and by the early nineties one rock band had even been sued (unsuccessfully) for compensation after two teenagers committed suicide while listening to a record said to contain hidden messages.

> In the last two years, Styx has been targeted by fundamentalist religious groups for the 'backward masking' of satanic messages on its albums.   *New York Times* 27 Mar. 1983, section 2, p. 27

bad /bæd/ *adjective* 🔲

In young people's slang, especially among Blacks in the US: excellent, spectacular, full of good qualities.

A reversal of meaning: compare WICKED and the earlier use of *evil* in this sense.

This sense of *bad* originated among Black jazz musicians in the US in the twenties and by the seventies had spread into more general use among US Blacks. It was taken up by the young in general during the eighties as a favourite term of approval, especially preceded by the adverb *well*: anything that was described as **well bad** had really gained the highest accolade. Its use among White British youngsters is an example of the spread of Black street slang as a cult language in the late eighties, with the popularity of HIP HOP culture etc. When used in this sense, *bad* has the degrees of comparison **badder** and **baddest** rather than *worse* and *worst*.

> We ran into some of the baddest chicks, man, we partied, we had a nice time.
> Gene Lees *Meet Me at Jim & Andy's* (1988), p. 203

> Roxanne Shante is quite simply the baddest sister around, and teamed with Marley Marl at the mixing desk she is awesome.   *Number One* 8 Nov. 1989, p. 43

bad-mouth /'bædmaʊθ/ *transitive verb* Also written **badmouth** 🔲

In US slang (especially among Blacks): to abuse (someone) verbally; to put down or 'rubbish' (a person or thing), especially by malicious gossip.

The verb comes from the Black slang expression *bad mouth* (a literal translation of similar expressions in a number of African and West Indian languages), which originally meant 'a curse or spell'.

The earliest use of *bad-mouth* as a verb in print is an isolated wartime use by James Thurber in 1941, although it was almost certainly in spoken use before this. By the sixties it had become fairly common in US Black English, but it was not until the late seventies that it acquired any currency in British slang. In the eighties it started to appear in respectable journalistic sources without quotation marks or any other sign of slang status. The corresponding verbal noun **bad-mouthing** is also common.

> The dealing fraternity and the auctioneers, despite the fact that they never cease bad-mouthing each other, are mutually dependent.   *The Times* 16 Nov. 1981, p. 10

> Jo-Anne was a bitter enemy who could be relied on to bad-mouth her at every opportunity.
> Pat Booth *Palm Beach* (1986), p. 180

bag people /'bæg ˌpiːp(ə)l/ *plural noun* 🔲

Homeless people who live on the streets and carry their possessions in carrier bags.

Formed by compounding (*people* whose main characteristic is the *bags* they carry) after the model of *bag lady* (see below). A tramp who carries his personal effects in a bag has been called a *bagman* in Australian English since the end of the nineteenth century.

The earliest references to *bag people* come from New York City in the seventies, and are in the form **bag lady** (sometimes written **baglady**) or **shopping-bag lady**; at that time it was mostly elderly homeless women who piled their belongings into plastic carrier bags and lived on the streets. By the mid eighties both the phenomenon and the term had spread to other US cities and to the UK, and sensitivity to sexist language had produced **bag person** along with its plural form *bag people*.

> They even had a couple of black-clad bagladies sitting silently on straight chairs by the door.
> Martin Amis *Money* (1984), p. 105

Peterson saw The Avenue's funky charm and its cast of misfits as inspirations for his painting. 'I like the bag people and the alcoholics and the street people.'

Los Angeles Times (Ventura County edition) 12 May 1988, section 9, p. 2

bagstuffer /ˈbægstʌfə(r)/ *noun* 〰️ ▩

A piece of promotional literature handed out to shoppers in the streets or put into shopping bags at the checkout.

Formed by compounding: these leaflets are usually treated as so much waste paper with which to *stuff* one's *bag*.

The *bagstuffer* (originally called a **shopping-bag stuffer**) was invented in the seventies in the US as a variation on the *flyer*. It became a widespread advertising ploy in the eighties, despite environmentalists' concern about wasteful use of paper and the destruction of rainforests.

As the vote approaches, soda bottlers have begun airing television commercials against it. Supermarkets have opposed it through 'bagstuffer' leaflets in their stores.

New York Times 23 Apr. 1982, section B, p. 1

You have to market your pharmacy to supermarket customers through coupons and bagstuffers; to the community through ads in flyers, and by offering free services.

Supermarket News 15 May 1989, p. 43

bailout /ˈbeɪlaʊt/ *noun* Sometimes written **bail-out** 〰️

Financial assistance given to a failing business or economy by a government, bank, etc. so as to save it from collapse.

The noun *bailout* is derived from the verbal phrase *bail out*, which has a number of distinct meanings. In this case, it is questionable whether it is a figurative use of the nautical sense 'to throw water out of (a boat) so as to prevent it from sinking' or the legal sense 'to get (a person) released from custody by providing the money needed as security (bail)'.

The financial sense of *bailout* comes originally from the US, where the practice was first written about in the seventies. *Bailouts* occurred with increasing frequency in other parts of the English-speaking world as the eighties progressed and the economic climate became more difficult even for large businesses; in the UK, though, the Conservative government of the eighties opposed government *bailouts*. The word *bailout* is often used attributively, with another noun following, especially in **bailout loan** and **bailout plan**.

Governments have to avoid protectionism, bailouts that cannot work and subsidies just to keep industries alive.

Toronto Star 28 May 1986, section A, p. 16

The executive branch is collaborating with Congress in putting part of the savings and loan bailout 'off-budget', thereby raising . . . the real cost of it.

Washington Post 1 Oct. 1989, section D, p. 7

Baker day /ˈbeɪkə ˌdeɪ/ *noun* ◖

Colloquially in the UK, any one of several days in the normal school year statutorily set aside for in-service training of teachers and mainly intended as a preparation for teaching the NATIONAL CURRICULUM.

Named after Kenneth *Baker*, who was the Education Secretary responsible for introducing them.

Compulsory in-service training for teachers was introduced in 1987 as part of a drive towards greater accountability in the teaching profession (see INSET); the five days set aside during the school year 1987–8 to prepare for the national curriculum had already been nicknamed *Baker days* by children and teachers alike by early 1988. *Baker days* were popular with children (for whom they meant an extra day off school), but did not meet with universal approval from teachers and parents.

A Leeds delegate told the conference . . . the Baker Days were 'universally hated and resented' within staffrooms.

Daily Telegraph 18 Apr. 1990, p. 2

balance /'bæləns/ noun ⊗

In the language of ALTERNATIVE or COMPLEMENTARY medicine: a harmonious relationship of body, mind, and spirit, which it is claimed can only be achieved by treating the whole person.

Balance has been used in the general figurative sense of 'equilibrium' for several centuries (its original and literal sense is 'scale(s)'); the recent movement towards therapies that take a holistic approach has meant that it is now commonly applied in this context, often without further explanation (not *balance of* anything, but simply *balance*).

The rise of alternative therapies in general from 'fringe' to respectable COMPLEMENTARY status during the eighties brought this use of *balance* to public notice; in particular, techniques such as biofeedback which aim to put the patient more in touch with the natural rhythms of life and increase self-awareness, as well as the growing NEW AGE culture, have stressed this concept of *balance* as a central precept for health. This view has been further reinforced by the GREEN movement, with its emphasis on maintaining ecological *balance* so as not to upset the natural rhythms there: human life and health are seen as inextricably linked with the balance of nature as a whole. Marketers and copywriters had noticed this development by the middle of the eighties, and had begun using the word *balance* liberally in descriptions of a wide variety of products, including food and drink, beauty preparations, etc.

> This 'holistic' perspective on the essence of healing presents us with a practical challenge: How can we best utilize the knowledge and services encompassed by Western medicine while maintaining a 'healthstyle' attuned to principles of order, balance, and self-reliance?
>
> Michael Blate *Natural Healer's Acupressure Handbook* (1978), p. viii

> The body is used as a source of ideas about 'wholeness', 'balance' and 'harmony', involving both the body and the mind ... Nature is deduced from the hypothesis of the instinct of the body for health. But health is only found by discovering an inner balance and harmony.
>
> Rosalind Coward *The Whole Truth* (1989; paperback ed. 1990), p. 32

balloon angioplasty ⊗ see ANGIOPLASTY

band /bænd/ verb ⨌ ▌▌

To arrange (pay scales, taxes, interest rates, etc.) in graduated bands. Also as an adjective **banded**; noun **banding**.

A figurative application of the sense of the verb 'to mark with bands or stripes'; the noun has long had a corresponding figurative sense 'a range of values'.

Although practised in areas such as income tax for a long time, the principle of *banding* became topical during the discussion of the COMMUNITY CHARGE ('POLL TAX') in the UK in 1990, when pressure was put on the government to introduce a *banded* rate based on people's ability to pay; the new council tax proposed in 1991 included this feature. It was also applied to a practice among some local authorities in the UK of grouping children by ability, so as to ensure that all schools got at least some of the brighter children.

> This limited banding, which would need legislation, would be intended to respond to complaints about the unfairness of the lump-sum tax.
>
> *Economist* 31 Mar. 1990, p. 27

> With Downing Street denying reports that Mrs Thatcher had herself now accepted that the poll tax was unfair, the Prime Minister has already rejected any plan for 'banding' the tax.
>
> *Financial Times* 28 Apr. 1990, section 1, p. 22

Band Aid ♪ ▌▌ see -AID

bandog /'bændɒg/ noun ▌▌

A fighting-dog specially bred for its strength and ferocity by crossing aggressive breeds such as the American pit bull-terrier, rottweiler, and various breeds of mastiff.

The word *bandog* has existed in the English language since the fifteenth century: originally, it was any dog that had to be tied up to guard a house or because of its ferocity (*band* in its historical sense 'fastening' combined with *dog*). Its use was soon generalized to cover any ferocious dog (such as a mastiff or bloodhound); the practice of breeding these cross-breeds for secret dog-fights has led to its being revived and specialized in meaning.

The news that ferocious cross-breeds were being produced and used in the UK both for illegal dog-fighting and as a way of keeping police at bay while other crimes were committed was reported by the RSPCA in early 1990. This followed public concern about a number of attacks on children by rottweilers and other ferocious dogs which had become increasingly popular as pets. Legislation in May 1991 ensured that the most dangerous *bandogs* became *banned dogs*.

> The Kennel Club said yesterday it would discipline any member who rears bandogs—American pit bull terriers crossed with rottweilers, mastiffs or Rhodesian ridgebacks.
> *Daily Telegraph* 8 Mar. 1990, p. 3

bang 〰 see BIG BANG

bankable /'bæŋkəb(ə)l/ *adjective* 〰 ▓

Certain to bring in a profit; good for the box office (said of a production which is sure to succeed or of a star whose name alone will ensure the success of the venture).

Formed by adding the adjectival suffix *-able* to *bank*. The adjective *bankable* already existed in the sense 'receivable at a bank'; this show-business use rests on a pun, in that the producer can *bank on* a profit which in turn can be *banked*.

Bankable has been used in this sense in Hollywood jargon since the fifties. During the seventies it increasingly featured in popular magazine articles about film-making and became popularized still further in the eighties by wider reporting of the processes which precede the actual making of a film. As the Hollywood-style hype was applied to other areas of the arts (writing, music, etc.), it became commonplace to read about *bankable* names in these fields as well.

> Sales of the chosen book may rocket. I say 'may' deliberately because I am not so sure how bankable all the shortlist are. *Bookseller* 20 Oct. 1984, p. 1705

> Becoming highly bankable, Allen discovered, meant becoming instantly popular with incipient entrepreneurs. *New Yorker* 29 Apr. 1985, p. 61

Barbour /'bɑːbə(r)/ *noun* ▓

Short for **Barbour jacket**, the trade mark of a well-known brand of WAXED JACKET.

> This autumn [the shop] is developing a rather Sloane country image due to the run on its Barbours and Cricket jackets. *Financial Times* 10 Sept. 1983, section 1, p. 13

> The Seventies brought introspection, and the fashion of 'me' emerged in the Thatcher Eighties. In 1989, clad in designer clothes and Barbour jacket, the student programmed a Filofax to ensure that no problems would frustrate the quest for that coveted job in the City. *The Times* 20 Jan. 1990, p. 36

bar-code /'bɑːkəʊd/ *noun* and *verb* Also written **barcode** or **bar code** 〰 ▣

noun: A machine-readable code consisting of a series of lines (bars) and spaces of varying width, used for stock control on goods for sale, library books, etc.

transitive verb: To label (goods, etc.) with a bar-code.

Formed by compounding: a *code* based on the width of *bars*.

The *bar-code* was invented as long ago as the early sixties and was quite widely used by public libraries for their book-issuing systems by the mid seventies. The code has to be 'read', and in the early days this was usually done using a light pen. With the introduction of computerized tills and EPOS during the eighties, *bar-codes* became seemingly ubiquitous on goods of all kinds, and a variety of types of **bar-code reader** could be seen (and heard bleeping) at the tills. By the early nineties the *bar-code* had been put to more inventive uses still: television-programme magazines published them on their pages so that videos could be programmed direct from the code, and scientists used them to label the subjects of their experiments (in one case, *bar-codes* were stuck to the hairs on the backs of hundreds of bees). The adjective used to describe goods which carry a *bar-code* is **bar-coded**; the practice of providing goods with them is **bar-coding**.

> Bar-code reader... comes with a sheet of bar codes... You set the timer by running the reader over the appropriate bar codes for day, time and channel required. *Which?* Sept. 1989, p. 450

> The electronic supermarket check-out, which bleeped and flashed up the cost of items taken from the bar codes on the packets, also warranted some attention. *Good Food* Jan./Feb. 1990, p. 26

basically /ˈbeɪsɪk(ə)lɪ/ *adverb*

In short, putting it bluntly, actually. (Usually in speech and often used at the beginning of a sentence or clause.)

A weakened sense of the adverb, which originally meant 'essentially, fundamentally, at root'. The weakening arises as much from the way in which the word is used (a 'sentence' adverb) as from the context; the result is a word which in most cases is redundant, adding nothing to the sense and simply giving the speaker time to think. Purists object to it in much the same way as they do to *hopefully* used at the beginning of a clause.

Although it had been in use in speech for some decades, it only became really fashionable to use *basically* in this almost meaningless way during the late seventies, when it took over from *actually* as a favourite 'filler'. The fashion may have been reinforced by the increased influence of the recorded television interview: the interviewee, anxious to reply succinctly enough to be sure of having the whole answer broadcast but also wanting to make it clear that this was not all that *could* be said on the subject, would prefix the reply with *basically*. Whether or not it once had a legitimate purpose, *basically* used in this way fast became a cliché and passed from spoken English into the written language as well.

> I'm not political, you know, *basically* I don't know the first thing about *politics* or *economics* or all that LSE-type crap, despite what you think. Stephen Gray *Time of Our Darkness* (1988), p. 142

> 'Basically I got served off the court,' she admitted. 'She served unbelievably well. I couldn't get the ball back in that last set.' *Guardian* 10 July 1989, p. 15

> In a few cases, Western women who were told to report with their husbands to pick up their exit visas had to watch the men taken away by security officials, presumably adding to Saddam's human shield. 'They basically traded the husband for the visa,' said a Western diplomat. *Washington Post* 2 Sept. 1990, section A, p. 1

basuco /bəˈzuːkəʊ/, /bəˈsuːkəʊ/ Also written **basuko**, **bazuco**, or **bazuko**
noun 🖉

A cheap, impure form of cocaine, made by mixing coca paste with a variety of other substances, which is extremely addictive when smoked for its stimulant effects.

A Colombian Spanish word; perhaps connected with Spanish *basura* 'sweepings, waste' (since the drug is made from the waste products of refined cocaine) or with *bazucar* 'to shake violently'. Another suggestion is that there have actually been two stages of borrowing here: first the English weapon-name *bazooka* was borrowed into Spanish, then it was applied figuratively to the drug (with its explosive effect), and finally the word was re-borrowed into English in a slightly altered form.

Basuco is the South American equivalent of CRACK, and has been smoked in Latin American countries for some time. The drug first appeared in the English-speaking world in the mid eighties and at first was also known as *little devil* or *Suzuki*, but *basuco* now seems to be its established name.

> There's a big internal market; a lot of coke and basuko used by the street boys. Charles Nicholl *The Fruit Palace* (1985), p. 67

Police and drug enforcement agencies [in Florida] believed basuco had the potential to create a bigger problem than crack ... The cost of using basuco was as little as $1 a dose.

Courier-Mail (Brisbane) 15 Dec. 1986, p. 6

While it takes two years of regular cocaine use to become addicted, it takes only a few weeks to become hooked on *bazuko*, a mind-blowing mix of coca base, marijuana and tobacco containing such impurities as petrol, ether and even sawdust. *The Times* 14 Sept. 1987, p. 10

battlebus /ˈbæt(ə)lbʌs/ *noun* 🏛

A bus used as a mobile centre of operations by a politician during an election campaign.

Formed by compounding: a *bus* in which one goes into *battle*, figuratively speaking.

The *battlebus* was a feature of the British general election campaign fought by the Liberal–SDP Alliance in 1983; the buses even bore the name *battlebus* on their sides. By the time of the next general election in 1987, the *battlebus* had become an established feature of election campaigning and was used by other parties as well.

She said the message to Mrs Thatcher from the by-election was loud and clear: 'It's time to go.' Then, taking her own advice, she zoomed off in the Sylvia Heal Battlebus for a lightning victory lap around the constituency. *Financial Times* 24 Mar. 1990, p. 1

bazuco, bazuko 🔫 see BASUCO

beat box /ˈbiːt ˌbɒks/ *noun* Also written **beat-box** or **beatbox** 🎵 👾

In colloquial use among musicians, a drum machine (an electronic device for producing a variety of drum-beats and percussion sounds as backing for music or rapping: see RAP); hence a style of music with a throbbing electronic drum-beat which often also accompanies interludes of rapping. Also, another name for a GHETTO BLASTER.

Formed by compounding: a *box* which produces the *beat*.

The *beat box*, which is essentially a percussion synthesizer, became a popular alternative to the conventional drum kit during the early eighties, when synthesized sounds in general opened up new possibilities for many bands. It was really the increased popularity of rap and its spread outside the Black music scene that led to the development of a distinct style of music called *beat box* by the mid eighties. A *beat box* is an expensive piece of equipment, so it is perhaps not surprising that some youngsters tried to imitate the sound without actually using a *beat box*; this led to the development of a new action noun **beatboxing**, the activity of making percussion noises like those of a *beat box* using only one's mouth and body.

How do you compare an album like that to ... the sparse beat-box music and intensely engaging call-and-response served up by today's leading rap group, Run-D.M.C.?
New York Times 9 Jan. 1985, section C, p. 14

Booming out of beat boxes on the street and bounced to in aerobics classes, the 'Big' beat sounds like the next equal-play anthem for American women. *Washington Post* 19 Mar. 1985, section C, p. 1

They usurp rap and beatbox, scratching their own frequently wild guitar marks on top.
Q Mar. 1989, p. 72

Beaujolais Nouveau /ˌbəʊʒəleɪ nuːˈvəʊ/ *noun* 🗺

Beaujolais wine that is sold while still in the first year of a vintage.

French for 'new Beaujolais'.

Beaujolais Nouveau was made commercially available in the early seventies, and, although it had been allowed no time to mature and in consequence struck some wine-lovers as very acidic, it proved an instant success. Its popularity led to the development of a new sport in the hotel and catering world: the race to be the first to have the new year's vintage in stock. Some wine bars and restaurants even went to the lengths of having stocks flown in by helicopter so as to pip others at the post. As the eighties progressed, signboards saying 'The Beaujolais Nouveau has arrived' became a common sight on pavements outside these places in mid November. **Beaujolais Primeur** (literally 'early-season Beaujolais') is the correct term for Beaujolais sold during the first few months of the vintage (from mid November until the end of January), and is sometimes used interchangeably with *Beaujolais Nouveau*, but *Beaujolais Nouveau* is much better known in English.

A wine shipper telephoned that he'd reserved me fifty cases of Beaujolais Nouveau for November 15th . . . I never waited for the Nouveau to be delivered but fetched it myself.

Dick Francis *Proof* (1984), p. 76

becu 〰 see ECU

bell /bɛl/ *noun*

In the British colloquial phrase **give** (someone) **a bell**: to ring (someone) up, to contact by telephone.

A variation on the theme of *give* (someone) *a ring* and *give* (someone) *a tinkle*, phrases which go back to the thirties.

Although probably in use in spoken British English for some time, this phrase did not start to appear in print until the early eighties. When it did start to spread it was perhaps under the influence of such television series as *Only Fools and Horses* and *Minder* (both of which popularized the working-class speech of London's East End). Certainly at about that time it became a popular phrase in the youth press as a less formal way of saying 'ring up'. It is curious that it should have caught on in this way at a time when fewer and fewer telephones actually had bells; during the eighties telephone bells were largely replaced by electronic tones, warbles, chirps, etc.

DJ Sammon gave me a bell *and* wrote me a letter (thorough chap) about his shows.

Rave! 6 Mar. 1990, p. 18

bells and whistles /ˌbɛlz ənd 'wɪs(ə)lz/ *noun phrase* 🖥

In colloquial use in computing, additional facilities in a system, program, etc. which help to make it commercially attractive but are often not really essential; gimmicks.

An allusion to the old fairground organs, with their multiplicity of *bells* and *whistles*; the *bells* of a computer are actually a range of electronic bleeps.

There are more than 600 microsystems on the market so it is hardly surprising that the manufacturers have taken to hanging a few bells and whistles on to their machines to get them noticed.

Sunday Times 26 Aug. 1984, p. 49

belly-bag, belt-bag ✖ see BUM-BAG and FANNY PACK

best before date /ˌbɛst bɪ'fɔː ˌdeɪt/ *noun phrase* ✖

A date marked on a food package (usually preceded by the words 'best before') to show the latest time by which the contents can be used without risk of deterioration.

Formed by combining the statutory words *best before* with *date*: the *date before* which the food is in *best* condition.

The use of *best before dates* was codified in the UK in 1980, when new food labelling regulations stipulated that perishable foods should carry some indication of their durability including the words *best before* and a date; very perishable foods must carry a SELL-BY DATE or some other indication of the shelf-life of the product within the store. After outbreaks of salmonella poisoning and listeriosis at the end of the eighties, it was felt that for high-risk perishables *best before* was a rather ambiguous label, suggesting that the goods would be *best* consumed before the date given but could safely be eaten for some time afterwards (whereas in some cases this would actually have been quite dangerous). This led to the wider use of an unambiguous USE-BY DATE on foods most likely to cause illness if stored too long. The *best before date* has now become so commonplace that it has acquired a figurative use among City personnel: one's *best before date* is the age beyond which one will be considered past one's best by prospective employers.

> Date marking is now required on most pre-packed foods (with a few exceptions, such as frozen foods, wine and vinegar) unless they have a shelf-life of at least 18 months . . . This is expressed as *either* a best before date (day, month, year) [etc.]
>
> Maurice Hanssen *The New E for Additives* (1987), p. 17

> Their colleagues in Eurobond dealing and corporate finance have 'sell by' and 'best before' dates (in most jobs, at age 35) as career markers. *Observer* 29 Mar. 1987, p. 51

Betamax /'bi:təmæks/, in the US /'beɪtə-/ *noun* ✖ 📷

The trade mark of one of the two standard formats for video and videotapes; also abbreviated to **Beta**.

The name is not (as popularly supposed) derived from the Greek letter name *beta*, but from the Japanese word *beta-beta* 'all over' and English *max* (short for *maximum*: see MAX); however, the inventors were making conscious and deliberate use of the pun with Greek *beta* to create an English-sounding product name.

The first home-video systems were developed by Sony in the sixties; the immediate predecessor of the *Betamax* was the *U-Matic*, developed in the late sixties. In order to create a smaller machine using smaller tapes, a new method of recording was invented for the *Betamax*, known as *beta* or 'all over' recording because it did away with the tape structure of guard bands and empty spaces which had previously been employed, and instead used the whole area of the tape. The Sony *Betamax* video system was first available in the mid seventies, but at first it was not possible to buy pre-recorded cassettes in this format. However, the policy soon changed and by the mid eighties video rental had become an important market in which two formats competed: *Betamax* and *VHS*. VHS eventually became the standard format for home video, although *Betacam*, a derivative of *Betamax*, is used for television news-gathering worldwide.

> If you plan to watch a lot of pre-recorded films . . . there may be difficulties getting a wide choice on Beta; VHS versions are much more common. *What Video* Dec. 1986, p. 95

> When Betamax was introduced, our first task was to help people understand why video systems were important in the home . . . We beat our brains, and finally came up with the phrase 'Time Shift'. We were explaining the concept . . . all over the world with such catch phrases as; 'For the first time, the world of TV is in your hands with Betamax', or 'Look at your TV just like a magazine'.
>
> Sony Corporation *Betamax 15th Anniversary* (1990), p. 8

beta test /'bi:tə ˌtɛst/, in the US /'beɪtə-/ *noun* and *verb* 📷

noun: A test of an experimental product (such as computer software), carried out by an outside organization after alpha testing by the developer (see ALPHA TEST) is complete.

transitive verb: To submit (a product) to a beta test.

Formed by compounding. *Beta*, the second letter of the Greek alphabet, has long been used to denote the second in a series; the *beta test* is the second *test*, carried out only after successful alpha testing.

For history see ALPHA TEST. A person whose job is to test software in this way for a separate developer is a **beta-tester**; the process is known as **beta testing** and the product at this stage of development is the **beta-test version**.

> Problem solving together with alpha and beta testing of new products require a minimum of 2 years experience. *The Times* 21 Mar. 1985, p. 39

bhangra /'bæŋgrə/ or /'bɑːŋgrə/ *noun* and *adjective* Also written **Bhangra** ♪ 💀

noun: A style of popular music mainly intended for dancing to, which fuses elements of Punjabi folk music with features of Western rock and disco music.

adjective: Belonging to this style of music or the subculture surrounding it.

A direct borrowing from Punjabi *bhāngṛā*, a traditional Punjabi folk dance associated with harvest.

Bhangra music originated in the Asian community in the UK in the early eighties, when pop musicians with a Punjabi ethnic background started to experiment with Westernized versions of their parents' musical traditions. At first it was only performed for Asian audiences, but by the end of the eighties had attracted a more general following. It is sometimes called **bhangra beat**.

> This was not the middle of a feverish Saturday night, but a Wednesday mid-afternoon excursion for devotees of the Bhangra beat, the rhythm of the Punjabi pop . . . An up and coming group . . . turned in a performance which set the seemingly incompatible rhythmic stridency of funk and Bhangra dance to a compulsive harmony. *Independent* 30 June 1987, p. 12

> This is a bhangra 'all-dayer', part of a booming sub-culture that has sprung up around an English-born hybrid of Punjabi folk and Western rock music. *Sunday Telegraph Magazine* 22 May 1988, p. 36

bicycle moto-cross 🔀 💀 see BMX

big bang /ˌbɪg 'bæŋ/ *noun* Frequently written **Big Bang** 〰

In financial jargon, the deregulation of the Stock Exchange in London on 27 October 1986. Hence, any far-reaching reform.

Big bang literally means 'a great explosion' and has been used since the forties to refer especially to the theory that the universe was formed as a result of a single huge explosion. Since the deregulation was to involve several significant changes in trading practices which would all be introduced at once, the whole process was likened to this explosive supposed moment of creation.

The deregulation of the Stock Exchange resulted from a restrictive practices suit brought by the Office of Fair Trading against the Stock Exchange in 1978; this case was dropped after the Stock Exchange agreed, in 1983, to do away with minimum commissions. However, the abolition of these made it difficult for the Stock Exchange to maintain the distinction between stock-brokers and stock-jobbers, and it became clear that further changes would be needed. The term *big bang* was in use from about that time, as financiers discussed the respective merits of a phased introduction of the changes and a *big bang* approach. The main areas of change were the creation of a single category of broker-dealer to replace stockbrokers and stock-jobbers, the admission of institutions as members, and the introduction of a new electronic dealing system known as SEAQ (Stock Exchange Automatic Quotation System). *Big bang* is sometimes used without a preceding article ('after Big Bang', etc.); it is also sometimes abbreviated to **bang**, especially in **post-bang**, an adjective meaning 'belonging to the period after *big bang*'. Since the London *big bang*, the term has also been used in a transferred sense, for example in discussions of EMU', with reference to economic reforms in Eastern Europe, and even to describe the new financial basis of the Health Service in the UK.

> In the wake of the City's Big Bang, American and Japanese banks are chasing each other to occupy the few high-tech buildings. *City Limits* 19 Feb. 1987, p. 10

> Less than three months after Big Bang, the start of the Solidarity-led government's package of strict austerity and radical market reforms, Poland is in ruins. *Economist* 24 Mar. 1990, p. 65

> The scale of the 'big bang' reflects the Government's determination to push through far-reaching health reforms. *Sunday Express* 16 Sept. 1990, p. 5

See also MARKET MAKER

bike /baɪk/ *noun*

In the British slang phrase **on your bike** (frequently written **on yer bike**): go away, push off, get away with you. Also, get on with it, 'pull your finger out'.

Originally a Cockney expression and typically graphic: the hearer should 'push off', and, in order to get away faster, should pedal, too.

Although almost certainly in spoken use since the early sixties, the phrase *on your bike* did not start appearing in print at all frequently until the eighties, when it suddenly became a fashionable insult. It was probably made the more popular by a speech which Norman Tebbit (then UK Employment Secretary) made at the Conservative Party Conference in October 1981, pointing out that his father had not rioted in the 1930s when unemployed, but had 'got on his bike and looked for work'. This speech was also the cause of some confusion in the meaning of the phrase: whereas before it had always been a ruder (but not obscene) way of telling someone to push off or indicating that you did not believe a word of what they were saying (the senses in which it continued to be used by those in the know), it was now taken up by the press as a favourite cliché to be used in stories about anyone who was unemployed, and acquired the secondary meaning 'get on with it, make an effort'. In this secondary sense it is sometimes used as an adjectival phrase rather than an exclamation, to describe the attitude which Tebbit's remark betrayed.

> The first ever Tory prime minister who truly believes in pull-yourselves-up-by-your-bootlaces, she wants upwardly mobile, self-helping, on-yer-bike meritocrats. *Financial Times* 12 Sept. 1984, p. 24

> On your bike Jake, I said, this joke has gone far enough. *Punch* 16 Oct. 1985, p. 44

> 'Wally son, it's Pim.' 'On your bike. Pim's doing five in Durham.' Tom Barling *The Smoke* (1986), p. 115

Billygate 🗂 see –GATE

bimbo /ˈbɪmbəʊ/ *noun* 🚹

In media slang, an attractive but unintelligent young woman (especially one who has an affair with a public figure); a sexy female AIRHEAD.

This was originally a direct borrowing from Italian *bimbo* 'little child, baby'. The word was in use in English in other senses before this one developed (see below); in all of them the original Italian meaning has been lost, but in this case there may be some connection with the use of *baby* for a girlfriend, and possibly some influence from *dumbo* as well.

Bimbo first came into English in the early twenties, when it was used on both sides of the Atlantic (although mainly in the US) as a contemptuous term for a person of either sex; ironically, P. G. Wodehouse wrote in the forties about 'bimbos who went about the place making passes at innocent girls after discarding their wives'. By the end of the twenties it had developed the more specific sense of a stupid or 'loose' woman, especially a prostitute. During 1987, *bimbo* started to enjoy a new vogue in the media, this time without the implication of prostitution: journalists claimed that the *bimbo* was epitomized by young women who were prepared to 'kiss and tell', ending their affairs with the rich and famous by selling their stories to the popular press. In the US *bimbos* cost politicians their careers; Britain also had its own 'battle of the bimbos' in 1988, when the affairs of certain rich men were exposed and the lifestyle of the *bimbo* was discussed in court. The word started to acquire derivatives: a teenage *bimbo* came to be known as a **bimbette** and a male *bimbo* as a **bimboy** (but see also HIMBO), while having an affair with a *bimbo* was even described as **bimbology** in one paper.

> In the strict sense the bimbo exists on the fringes of pornography, and some cynics might say she has the mental capacity of a minor kitchen appliance. *Independent* 23 July 1988, p. 5

> A gathering of playboys just wasn't a party unless there was at least one ... scantily clad bimbette swimming around in a bathtub of shampoo. *Arena* Autumn/Winter 1988, p. 157

> Actor Rob Lowe was at the Cannes Film Festival, expressing frustration with his reputation as the Brat Pack's leading bimboy. *People* 5 June 1989, p. 79

Still, Smith, and Gans are not bimbos and understandably bristle at accusations that they are chatty-cathies for their white male superiors. *New York Woman* Nov. 1989, p. 66

bio- /ˈbaɪəʊ/ *combining form* 🌱 ⊗ 💻

Part of the words *biology* and *biological*, widely used as the first element of compounds relating to biology or biotechnology; frequently used as a shortened form of *biological(ly)*.

Formed by abbreviating *biology* and *biological*; in both words this part is ultimately derived from Greek *bios* 'life'.

Compounds relating to 'life' have been formed on *bio-* in English for over three centuries, and even the ancient Greeks used it as a combining form. During the second half of the twentieth century, however, advances in BIOTECHNOLOGY and the increasing interest in GREEN issues caused a proliferation in popular language of compounds in these areas, alongside the continuing use of *bio-* in scientific terminology. Like ECO-, *bio-* was particularly productive in the late sixties and early seventies, and many of the compounds which had been well known then came back into fashion during the eighties, often undergoing further development. The development of plastics and other synthetic products which were **biodegradable**, that is, those that would decompose spontaneously and hence not become an environmental hazard, led during the eighties to the verb **biodegrade**. **Biomass**, originally a biologists' term for the total amount of organic material in a given region, was later also used of fuel derived from such matter (also called **biofuel**, or, in the case of the mixture of methane and other gases produced by fermenting biological waste, **biogas**; this was burnt to produce what became known as **bioenergy**). By contrast, **biofeedback**, the conscious control of one's body by 'willing' readings on instruments (such as heart-rate monitors) to change, reappeared in the eighties as one of the techniques used in AUTOGENIC TRAINING. Computer scientists continued to speculate that micro-organisms could be developed that would function like the simple logic circuits of conventional microelectronics, thus paving the way for **biocomputing** with **biochips**. Biological warfare, a more disturbing application of biotechnology, became sufficiently familiar to be abbreviated as **biowar**. Concern about the effect of even peaceful technology on the **biosphere** (the component of the environment consisting of living things) was expressed in the philosophy of **biocentrism**, in which all life, rather than just humanity, is viewed as important (much as in GAIA theory). Direct and sometimes violent opposition to such aspects of biological research as animal experimentation and genetic engineering was organized by **biofundamentalists** (see also ANIMALIST[1] and FUNDIE). As a result of the Green Revolution, the public was made more aware of the threat posed by intensive cultivation of particular species to **biodiversity**, the richness of variety of the biosphere.

Towards the end of the decade *bio-* began to be used indiscriminately wherever it had the slightest relevance, either frivolously or because of its advertising potential (just as *biological* had once been a glamorous epithet for washing powder). The prefix is sometimes even used as a free-standing adjective in this sense, meaning little more than 'biologically acceptable'. Examples include **biobeer**, **biobottom** (an 'eco-friendly nappy cover'), **bio home**, **bio house**, **bioloo**, **bioprotein**, and **bio yoghurt**.

The term bio-chip, coined only about four years ago, already means different things to different people. In the United States, where the word arose, researchers generally use it to refer to chips in which the silicon transistors would be replaced by single protein-like molecules. Such a molecule could be stable in one of at least two different forms of ... charge distribution, depending on its external environment. But some scientists, particularly in Europe, now seem to use bio-chip more widely to refer to any 'smart' system small enough to interact with a cell.

The Age (Melbourne) 28 Nov. 1983, p. 5

Even medical insurance companies are now beginning to recognize the value of a veritable A-to-Z of 'holistic' therapies ..., including acupuncture, biofeedback and chiropractic.

John Elkington & Julia Hailes *The Green Consumer Guide* (1988; paperback ed. 1989), p. 260

The bio-diversity campaign is an attempt to bring the seriousness of the global situation to the attention of people in all walks of life. *The Times* 31 Mar. 1989, p. 5

German architect Joachim Ebler has designed a range of 'bio homes' ... The buildings are made with timbers from sustainable sources and are not treated with chemical preservatives.

Green Magazine Oct. 1989, p. 14

Therapeutic properties ... are ascribed to the presence of the live lactic acid bacteria, particularly in the bio-yoghurts, said to promote the friendly bacteria in the gut which can be affected by the overuse of antibiotics. *Healthy Eating* Feb./Mar. 1990, p. 37

The 43-year-old Californian has chosen to have a second child because her teenage daughter has leukaemia and will die without a transplant of bone marrow ... Biofundamentalists claim emotively that she wants to use the baby as 'a spare part' ... Bone marrow will be extracted for implanting into her 17-year-old sister. *Daily Telegraph* 9 Apr. 1990, p. 16

biotechnology /ˌbaɪəʊtɛkˈnɒlədʒɪ/ *noun* 🖳

The branch of technology concerned with the use of living organisms (usually microorganisms) in industrial, medical, and other scientific processes.

Formed from the combining form BIO- and *technology*.

Micro-organisms are capable of carrying out many chemical and physical processes which it is not possible or economic to duplicate: varieties of cheese and wine, for example, are given their distinctive flavours and appearances by the action of bacteria and fungi, and antibiotics such as penicillin could originally only be produced from cultures of particular micro-organisms. During the seventies and eighties the increasing sophistication of genetic engineering, in particular recombinant DNA technology, made it possible for a **biotechnologist** to 'customize' micro-organisms capable of producing important or useful substances on a large scale. Insulin, interferon, and various hormones and antibodies have been produced by this method, as well as foodstuffs such as mycoprotein. Strains of bacteria which digest oil spills and toxic wastes have also been developed. The commercial importance of *biotechnology* was recognized in 1980 when the US Supreme Court ruled that such genetically engineered micro-organisms could be patented: during the eighties a number of firms appeared which specialized in the manufacture of substances by these means. Such a business is known as a **biotech company** or **biotech**. The potential of these companies as investments was recognized in 1982 by the editors of the science journal *Nature*, who began publishing performance statistics for the stocks of some representative US companies operating in the field.

Conventional brewing and wine making are not usually regarded as biotechnology but many other fermentation processes are. *The Times* 9 June 1983, p. 22

To an extent, the biotech companies have taken over from the high-techs as the main vehicle for investors' 'risk' dollars. *Courier-Mail* (Brisbane) 30 June 1986, p. 28

A biotechnologist in London has found a way to make the natural stimulant which triggers the 'immune system' of plants. *New Scientist* 23 June 1988, p. 48

black economy /ˌblæk ɪˈkɒnəmɪ/ *noun* 〰️

The underground economy of earnings which are not declared for tax purposes, etc.

Formed by applying the *black* of *black market* to the *economy*.

The *black economy* was first so named at the end of the seventies, when it was revealed that undeclared earnings accounted for an increasing proportion of the national income in several Western countries. The trend continued throughout the eighties.

Part-time jobs have tended to be filled either by new entrants to the workforce, or in the 'black economy'—by people on the dole who do not declare their earnings. *The Times* 24 June 1985, p. 17

Black Monday /ˌblæk ˈmʌndeɪ/ 〰️

In the colloquial language of the stock-market, the day of the world stock-market crash which began in New York on Monday 19 October 1987 and resulted in great falls in the values of stocks and shares on all the world markets.

Any day of the week on which something awful happens can be given the epithet *black*; the name *Black Monday* had, in fact, already been used over the centuries for a number of Mondays, notably (since the fourteenth century) for Easter Monday. *Black Tuesday* was a term already in use on Wall Street to refer to Tuesday 29 October 1929, the worst day of the original Wall Street crash.

Within days of the dramatic drop in share prices which started in New York and sent panic all over the world, the financial press was describing the event as *Black Monday*. The crash had important economic consequences in several countries, so *Black Monday* is likely to remain a meaningful financial nickname for some time.

> The Dow Jones, once up 712 points for the year, drops 508 points on Black Monday. Paper losses total $500 billion.
> *Life* Fall 1989, p. 28

> Many institutions and individual investors have shied away from stock-index futures, blaming them for speeding the stock market crash on Black Monday two years ago.
> *Wall Street Journal* 17 Oct. 1989, section C, p. 29

See also MELTDOWN

black tar /ˌblæk ˈtɑː(r)/ *noun*

In the slang of drug users, an exceptionally pure and potent form of heroin from Mexico. Also known more fully as **black-tar heroin** or abbreviated to **tar**.

Formed by compounding: this form of heroin is dark (*black*) in colour and has the consistency of *tar*; *tar* had also been a slang word for opium since the thirties.

Black tar first became known under this name to drug enforcement officials in Los Angeles in 1983 (though it may in fact be the same thing as *black stuff*, slang for brown Mexican heroin since the late sixties); its abuse had become a serious and widespread problem in various parts of the US by 1986. It is made and distributed only from opium-poppy crops in Mexico using a process which makes it at the same time very pure and relatively cheap. *Black tar* has a large number of other slang names, including those listed in the *Economist* quotation given below.

> DEA officials blame the low price of 'black tar' for forcing down other heroin prices, causing the nation's first general increase in overall heroin use in more than five years.
> *Capital Spotlight* 17 Apr. 1986, p. 22

> Black tar, also known as bugger, candy, dogfood, gumball, Mexican mud, peanut butter and tootsie roll ... started in Los Angeles and has since spread to 27 states ... What makes black tar heroin unique is that it has a single, foreign source—Mexico—and finds its way into Mexican-American distribution networks, often via illegal immigrants.
> *Economist* 7 June 1986, p. 37

blanked /blæŋkt/ *adjective*

In young people's slang: ignored, cold-shouldered, out on a limb.

This is presumably a figurative use: a person who is *blanked* apparently no longer exists—he or she might as well be a *blank* space.

This usage seems to have originated as a verb **blank** (someone or something) in the world of crime several decades ago (compare *blank out*, meaning literally 'to rub out'). As a verb it was apparently used by both criminals and policemen; in his book *The Guvnor* (1977), Gordon F. Newman uses it several times, for example 'He also blanked Scotch Pat's next suggestion, about calling a couple of girls.' It has only recently emerged as an adjective among young people.

> Are you blanked? Safe? Or lame?
> *New Statesman* 16 Feb. 1990, p. 12

blip /blɪp/ *noun and verb* 〰️

noun: A temporary movement in statistics (usually in an unexpected or unwelcome direction); hence any kind of temporary problem or hold-up; a 'hiccup'.

intransitive verb: (Of figures, as on a graph etc.) to rise suddenly; (of a business, an economic indicator, etc.) to suffer a temporary 'hiccup'.

A figurative use of an existing sense of *blip* in radar: the small bump on a financial graph which represents the temporary change looks rather like the apparent rise and fall of the *blip* as it appears on the even trace on a radar screen.

Blip started to be used figuratively in this way, particularly in economics and finance, during the seventies. In the UK it was largely limited to economic or business jargon until September 1988, when Nigel Lawson, then Chancellor of the Exchequer, was widely quoted as having announced that a significant increase in the Retail Price Index was to be regarded only as a 'temporary blip' and not as a sign that the government's anti-inflation policies were failing. After this, the word became fashionable in the British press and it was common to find it applied more widely, outside the field of finance, to any temporary problem. As was the case with Mr Lawson, it is not unusual to find that the person who describes a sudden change as a *blip* is not yet in a position to know whether it will, in the end, prove to be only temporary. This adds a certain euphemistic tinge to the usage.

> Nigel Lawson's dilemma is the Conservative Party's also. Is the first tremor on its happy political landscape merely 'a blip', as the Chancellor has called the storm that has gradually engulfed him?
> *Listener* 2 Mar. 1989, p. 10

> Prices moved higher during overnight trading, and blipped a shade higher still following the release of the G.N.P. figures.
> *New York Times* 27 Apr. 1989, section D, p. 19

BMX /biːɛmˈɛks/ *abbreviation* ✖️ 🏍️

Short for **bicycle moto-cross**, a sport involving organized cycle-racing and stunt-riding on a dirt track. Also applied to the particular style of sturdy, manoeuvrable cycle used for this.

The initial letters of *Bicycle* and *Moto-*, with *X* representing the word *cross*.

BMX developed in the US in the late seventies, when youngsters pressed for special tracks where they could race each other on their bikes without interfering with normal road traffic or pedestrians. It quickly became popular in several countries, and, by the mid eighties, ownership of a distinctive *BMX* bike had become a status symbol among young people, whether or not they actually intended to take part in the sport. The main characteristics of the cycles are their manoeuvrability (making possible some very daring stunts in **freestyle BMX**), small colourful wheels, and brightly-coloured protective pads fixed on the tubular frame. A wide variety of other *BMX* merchandise (such as racing suits, helmets, and gloves) became available during the eighties as manufacturers cashed in on the popularity—and the dangers—of the sport. By the end of the eighties, organized *BMX* on tracks had waned, although the bikes and stunts remained popular.

> Danny and the Mongoose Team promote the 'fastest growing youth sport in the country'—BMX bike racing—with a single called 'BMX Boys'.
> *Sounds* 3 Dec. 1983, p. 6

> Up on the far top corner of camp lies the BMX track. A very fast downhill track with four turns and jumps ... adds up to a fun and competitive track.
> *BMX Plus!* Sept. 1990, p. 36

boardsailing /ˈbɔːdseɪlɪŋ/ *noun* Also written **board sailing** or **board-sailing** ✖️

Another (more official) name for WINDSURFING.

Formed by compounding: *sailing* on a *board*.

The name *boardsailing* was first used in the US at the very beginning of the eighties for a water sport which had developed out of surfing, involving a board (a *sailboard*) similar to a surfboard but using wind in a small sail rather than waves for its power. The sport developed during the seventies and at first was also known as *sailboarding*. Particularly since it became an Olympic demonstration sport in 1983, it has been known officially as *boardsailing*, although most people probably know it colloquially as WINDSURFING. A person who practises this sport is known as a **boardsailor** or **boardsailer** (officially, that is: *sailboarder* and *windsurfer* also exist!).

A more contentious point is whether HRH and his fellow enthusiasts are wind surfers, sailboarders, boardsailers or simply bored sailors.   *Daily Mail* 9 Apr. 1981, p. 39

After scoring seven firsts in as many pre-Olympic boardsailing regattas this year, ... Penny Way is fast becoming Britain's hottest Olympic hopeful.   *The Times* 8 June 1990, p. 42

body mousse 💥 see MOUSSE¹

body-popping /ˈbɒdɪpɒpɪŋ/ *noun* Also written **body popping** or **bodypopping** 💥 🔲

A style of urban street dancing featuring jerky robotic movements, made to music with a disco beat; abbreviated in street slang to **popping**.

Formed by compounding: the *popping* part is probably a reference to the jerkiness of the dance's movements in response to the popping beat of the music, which is reminiscent of the electronic bleeps of a computer monitor. There may also be some influence from West Indian English *poppy-show* 'an ostentatious display' (itself ultimately related to *puppet show*). Certainly the idea is to perform mechanical movements like those of a robot or doll, punctuated by a machine-gun rhythm.

Body-popping developed on the streets of Los Angeles in the late seventies and became popular in other US cities, especially among teenagers in the Bronx area of New York, by the early eighties. Along with BREAK-DANCING, with which it gradually merged to become one of the styles of street dancing contributing to HIP HOP culture, *body-popping* proved to be one of the most important dance crazes of the decade. By the middle of the eighties it had spread throughout the English-speaking world, and CREWS of dancers (both Black and White) had been formed in the UK and elsewhere. The verb (**body-**)**pop** and agent noun (**body-**)**popper** date from about the same time as *body-popping*.

The Pop is very characteristic of the Electric Boogie. Because of the popping nature of Breakdance music, your Boogie will be fresh if you can Pop with all your moves. It is as if the music were Popping you.   Mr Fresh with the Supreme Rockers *Breakdancing* (1984), p. 68

Kids on the rough, tough streets of the Bronx used to beat each other up until they began to have battles in 'break dancing' and 'body popping'.   *The Times* 2 Feb. 1985, p. 9

'What's the difference between breaking and popping?' 'When they popping, they be waving, you know, doing their hands and stuff like that. When they breaks, they spins on the floor, be going around.'   *American Speech* Spring 1989, p. 32

body-scanner /ˈbɒdɪˌskænə(r)/ *noun* ⊗ 🔲

A scanning X-ray machine which uses computer technology to produce cross-sectional pictures (tomographs) of the body's internal state from a series of X-ray pictures.

Formed by compounding: a *scanner* which produces pictures of the whole *body*.

The *body-scanner* (at first called a *whole-body scanner* or *total body scanner*) was developed by EMI in 1975, using the same technology as had been used to produce the brain scanner a few years earlier. It was immediately welcomed as a powerful diagnostic tool, especially since it was capable of showing up tumours in all parts of the body while they were still at an early stage of development. During the eighties the *body-scanner* became commonplace in the US, but its high price made it a rarer acquisition in the National Health Service in the UK. As the technology of ultrasound and magnetic resonance imaging (see MRI) have developed, the term *body-scanner* has been extended in colloquial use to cover all kinds of machines which scan the body and compute cross-sectional pictures of its inside.

> The studies could also give a better understanding of crystals, which are widely used in electronics, and of magnetism, which is exploited in many body scanners.
>
> *Sunday Times* 6 May 1990, section D, p. 15

body-snatching 〰 see HEADHUNT

bodysuit /'bɒdɪs(j)uːt/ *noun* 🔲

A close-fitting stretch all-in-one garment for women, used mainly for exercising and sports.

Formed by compounding: a *suit* (something like a swimsuit in fabric structure) to cover the whole *body*.

The *bodysuit* first appeared as a fashion garment in the late sixties (when it was usually an all-in-one body garment fastened with snap fasteners at the crotch); in the late seventies and eighties it enjoyed a new lease of life as a skin-tight all-in-one sports garment, benefiting from the craze for exercise regimes and the fashion for sportswear outside the gymnasium and sports stadium.

> Before he changes into his tight red Spandex bodysuit with the plunging neckline, there is the quick hint of a tattoo lurking beneath the rolled-up sleeve on his right arm.
>
> *Washington Post* 13 May 1982, section C, p. 17

> Four schoolgirls stunned spectators and officials by wearing 'Flo Jo' bodysuits at Victoria's most prestigious schools' athletics meeting at the weekend.   *Courier-Mail* (Brisbane) 31 Oct. 1989, p. 3

> The eye-boggling bodysuit . . . is a style trend that has been taken up by designers.
>
> *New York Times* 5 Aug. 1990, section 6, p. 38

boff 🍺 see BONK

boggling /'bɒglɪŋ/ *adjective*

In colloquial use: staggering, stunning, overwhelming.

Formed by dropping the word *mind* from *mind-boggling*, itself a fashionable expression since the mid sixties.

Boggling started to be used following nouns other than *mind*, and also on its own, in the mid seventies. By the end of the eighties, *mind-boggling* seemed quite dated, while *boggling* was commonly used, especially to describe a very large statistic or sum of money—in fact anything that would make you boggle-eyed with amazement or surprise. Although essentially a colloquial usage, *boggling* is found in print, especially in journalism.

> Per-mile costs fell fractionally as a result of the additional travel, whose total was a boggling 1.526 trillion miles.   *New York Times* 18 Aug. 1985, section 5, p. 9

> Serious damage can mean even more boggling bills, but at least your insurance should cover it.
>
> *Which?* Mar. 1990, p. 144

bomb factory /'bɒm ˌfæktərɪ/ *noun* 🔲

In the colloquial usage of police press releases: a place where terrorist bombs are made illegally or materials for their manufacture are secretly stored.

Formed by compounding: an unofficial *factory* for *bombs*.

The term *bomb factory* seems to have been invented by the police, who have used it in press releases announcing the detection of terrorist bomb manufacture since the mid seventies. The term was taken up enthusiastically by the press—especially the tabloids, for whom it satisfied all the requirements of headline material (short words, the use of nouns in apposition, and emotiveness).

He had no idea the four people in the room were turning it into a bomb factory.

The Times 21 June 1986, p. 3

A senior police officer described the hoard—one of the biggest ever found—as 'practically the entire contents of a bomb factory'.

Daily Mirror 12 Nov. 1990, p. 2

bonk /bɒŋk/ *verb* and *noun* ▮

transitive or *intransitive verb*: In young people's slang, to have sex with (someone); to copulate.

noun: An act of sex.

Bonk originally meant 'to hit resoundingly' and the corresponding noun was an onomatopoeic word for the abrupt thud that is heard when something hard hits a solid object (such as the head); it was used fairly typically in the school-playground joke 'What goes ninety-nine *bonk*?'—'A centipede with a wooden leg', which has been told for at least half a century. The transition from 'to hit resoundingly' to the present use was made by way of an intransitive sense 'to make a bonking noise, to thud'. The slang use has parallels in the *bang* of *gang-bang* and in the American slang equivalent *boff* (noun and verb). A less likely theory is that it is backslang for *knob*, also a vulgar slang way of saying 'have sex'.

This sense of *bonk*, which is really a humorous euphemism, has apparently been in spoken use among young people (especially, it seems, at a number of public schools) since the fifties and first appeared in print in the seventies. Although middle-class slang, it is coarse enough not to have been used in print at all frequently until the middle of the eighties. Then it was brought into vogue by journalists unable to resist the pun with *bonk* as the onomatopoeic word for the sound a tennis ball makes in contact with the racquet: in the 1987 season, the defending Wimbledon champion Boris Becker was giving disappointing performances, something which the tabloids put down to too much *bonking*. This episode was followed by much journalistic speculation about the origin of the word (including a street interview on the consumer programme *That's Life*) and considerably increased use of it in print, often with heavy innuendo. As is often the case with words taken up by the media in this way, interest in it died down within a short time, but by then it had acquired a respectability that allowed it to be used even in the quality newspapers. The corresponding action noun is **bonking**; agent noun **bonker**.

The Fleet Street rags had their angle after the Doohan victory: BONKED OUT; TOO MUCH SEX BEATS BIG BORIS.

Sports Illustrated 6 July 1987, p. 21

Flaubert bonked his way round the Levant, his sense of sexual adventure unquenched by the prospect, soon realised, of catching unpleasant diseases.

Independent 28 May 1988, p. 17

Police took away ... a 'little black' book containing the names of thousands of women with whom the legendary Belgian bonker is said to have had steamy love romps.

Private Eye 15 Sept. 1989, p. 23

boom box /'buːm ˌbɒks/ *noun* ▨ ♪

In US slang, the same thing as a GHETTO BLASTER.

Formed by compounding: a *box* which *booms*.

For history, see GHETTO BLASTER.

How about a law against playing 'boom boxes' in public places?

Washington Post 26 June 1985, section C, p. 10

boomer /'buːmə(r)/ *noun* ▮

In US slang, short for **baby boomer**: a person born as a result of the *baby boom*, a sharp increase in the birth rate which occurred in the US at the end of the Second World War and lasted until the mid sixties.

Formed by dropping the word *baby* from *baby boomer*. Before this, *boomer* had meant 'a person who pushes or boosts an enterprise' in US English.

The term *baby boom* has been in use in US English since the forties, but it was only when the children born as a result of the postwar boom reached maturity in the seventies and eighties that *baby boomers* started to be referred to frequently in the American press. This generation was by then so numerically significant in US society that advertisers, businesses, and politicians considered them an essential group to cater for. So frequent did the name *baby boomer* become that by the end of the eighties it could be abbreviated to *boomer* without fear of misunderstanding, and *boomer* itself became the basis for compounds such as **boomer-age** and **post-boomer**.

> The post-boomers have also had to deal with the more recent sellout of the baby boom generation.
> *Globe & Mail* (Toronto) 27 May 1989, section D, p. 5

> The script is ambitiously constructed, tracing the relationships of several boomer-age parents with their kids, their siblings, and their own parents.
> *New Yorker* 18 Sept. 1989, p. 28

> The boomer group is so huge that it tends to define every era it passes through, forcing society to accommodate its moods and dimensions.
> *Time* 16 July 1990, p. 57

See also BUSTER

boot /buːt/ *verb* 🖳

transitive: To start up (a computer) by loading its operating system into the working memory; to cause (the system or a program) to be loaded in this way. *intransitive*: (Of a computer) to be started up by the loading of the operating system; (of a program) to be loaded.

An abbreviated form of *bootstrap* 'to initiate a fixed sequence of instructions which initiates the loading of further instructions and, ultimately, of the whole system'; this in turn is named after the process of *pulling oneself up by one's bootstraps*, a phrase which is widely supposed to be based on one of the eighteenth-century *Adventures of Baron Munchausen*. Despite the traditional practice of getting sluggish machines to work by giving them a surreptitious kick, there is no connection whatever between this verb and *boot* meaning 'to kick'.

Bootstraps have been used in computing since the fifties, but it was not until personal computers became widespread in the seventies and eighties that the noun *bootstrap* and the corresponding verb were abbreviated to *boot*. The verb is often used with *up*; the action noun for this process is **booting (up)**.

> If a computer does not have a hard drive and must be booted from a floppy, one should boot from a 'write-protected' disc that cannot be altered.
> *New Scientist* 4 Mar. 1989, p. 42

> At last the Amiga can boast a game you'll be proud to boot up when your crystal analyst comes round to listen to your collect of Brian Eno LPs.
> *CU Amiga* Apr. 1990, p. 57

born-again /ˈbɔːnəˌgɛn/, /ˌbɔːnəˈgɛn/ *adjective* 🎚

Full of the enthusiastic zeal of one recently converted or reconverted to a cause; vigorously campaigning. Also, getting a second chance to do something.

A figurative application of the adjective, which originally developed from the verbal phrase *to be born again* (after the story of Jesus and Nicodemus in St John's Gospel, chapter 3) and was properly used to apply to an evangelical Christian who had had a conversion experience of new life in Christ and made this experience the basis for all later actions.

The adjective *born-again* has been used to refer to fundamentalist or evangelical Christians (especially in the Southern States of the US) since at least the sixties. Probably the most influential factor leading to the development of a figurative sense was the election of Jimmy Carter to the Presidency of the United States in 1977; the connection between his *born-again* Baptist background and the policies that he put forward was made much of in the press at the time, as were the hopes of fundamentalist 'Bible Belt' Christians for his Presidency. Another (quite separate) influence was the rise of fundamentalism within the Islamic world during the early eighties and the zeal with which it was presented to the West. By the end of the eighties, the figurative use was well established and could be applied to virtually any convert to a cause, however trivial; it had also started to be used to describe anyone who had been given a second chance to do something (another 'life' in the language of games).

> Duncan and Jeremy are born-again northerners. They saw the northern light last year, when they turned their backs on London. *Sunday Express Magazine* 9 Aug. 1987, p. 23

> In March 1988 I was a born-again student, having got my PPL in 1954 . . . then having to let the licence go at the end of 1956 when marriage came along. *Pilot* Nov. 1988, p. 26

bottle /'bɒt(ə)l/, but often pronounced Cockney-fashion as /'bɒʔəw/ *noun*

In British slang: courage, spirit, guts. Usually in phrases such as **have (got) a lot of bottle**, to be spirited or courageous; to have guts; **lose one's bottle**, to lose one's nerve (and so as a phrasal verb **bottle out**, to lose one's nerve; to pull out, especially at the last minute).

The phrase *no bottle* has been used in underworld slang to mean 'no use, worthless' since the middle of the nineteenth century; it is likely that this was reinterpreted this century to mean 'lacking substance or spirit', and that from there *bottle* started to be used on its own and eventually to be incorporated into new phrases. The rhyming slang expression *bottle and glass* for 'arse' is often assumed to have something to do with these expressions (in which case *bottle* would be more strictly 'guts'), but this may be no more than popular speculation.

These phrases, which are essentially part of the spoken language, started to appear in written sources in the sixties as representations of Cockney or underworld speech. Their use was reinforced by a milk marketing campaign in the early eighties, the caption for which read 'It's gotta lotta bottle', and by television series such as *Minder*, in which Cockney expressions were brought to a wide audience. *Bottle out* did not appear in the written language at all until the very end of the seventies (at about the same time as this series was first shown).

> Goodness, was I going to give her a bad time! Of course, when it got down to it, I bottled out completely. Robert McLiam Wilson *Ripley Bogle* (1989), p. 162

> You appear not to have the bottle, courtesy or wherewithal to actually approach her in person. *Just Seventeen* Dec. 1989, p. 22

> Some of the warders lost their bottle and just fled. *News of the World* 8 Apr. 1990, p. 6

bottle bank /'bɒt(ə)l ˌbæŋk/ *noun* 🌳

A collection point to which empty bottles and other glass containers can be taken for recycling.

Formed by compounding; whereas in *blood bank*, *sperm bank*, etc. the metaphor extends to deposits and withdrawals, the recycling *bank* accepts deposits only.

An early manifestation of public interest in conservation, the *bottle bank* scheme started in the UK in 1977. The covered skips or plastic bells normally used for this purpose had become a familiar sight in supermarket car parks by the end of the eighties—often overflowing, since there proved to be more enthusiasm among the public than capacity to recycle the glass.

> Why not take your old, non-returnable glass bottles to your local bottle bank instead of throwing them away? *Which?* Aug. 1984, p. 355

bought deal /ˌbɔːt 'diːl/ *noun* 📈

In financial jargon, an arrangement for marketing an issue of bonds or shares, in which a securities house buys up all the stock (often after tendering against other houses) and then resells it at an agreed price.

Formed by compounding; the issuer of the shares can be sure that the whole *deal* will be *bought* in advance.

A practice which originated in the US in the early eighties, the *bought deal* soon proved attractive to companies in the UK as well as an alternative to the standard rights issue; however, the legal right of shareholders to first refusal on new issues of shares in the UK gave it limited applicability.

The American 'bought deal' might become the norm for equity issues as well as for fixed interest loans.   *The Times* 11 Sept. 1986, p. 23

bovine spongiform encephalopathy ⊗ ✖ see BSE

boy toy ▮ see TOYBOY

brat pack /'bræt ˌpæk/ *noun* ✖

In media slang, a group of young Hollywood film stars of the mid eighties who were popularly seen as having a rowdy, fun-loving, and pampered lifestyle and a spoilt attitude to society; more generally, any precocious and aggressive clique.

Formed by compounding; deliberately made punningly like *rat pack*, a slang name for a group of rowdy young stars led by Frank Sinatra in the fifties.

The term was coined by David Blum in *New York* magazine in 1985 in an article about the film *St Elmo's Fire*, and quickly caught on in the media. At a time when rich young stars of sport as well as films were gaining a reputation for bad behaviour in public places, it became a kind of shorthand for the young who had been spoilt by early success and thought the whole world should be organized to suit them. Blum's article also coined the term **brat packer** for a member of the original Hollywood *brat pack*; this, too, is used more widely to refer to members of other *brat packs*, from professional tennis players to young, successful authors.

The Brat Packers act together whenever possible.   *New York* 10 June 1985, p. 42

Border hit back at an Indian newspaper report, which dubbed the Australian cricket team a 'brat pack', notorious for uncouth behavior.   *Brisbane Telegraph* 21 Oct. 1986, p. 2

Young guns. A new generation rediscovers an old genre: brat-packers Estevez, Sutherland, Sheen and Lou Diamond 'La Bamba' Phillips in a rollicking re-run of the Billy The Kid legend.
          Q Mar. 1989, p. 119

break-dancing /'breɪkdɑːnsɪŋ/ *noun* Also written **breakdancing** or **break dancing** ✖ ◉

A very individualistic and competitive style of dancing, popularized by Black teenagers in the US, and characterized by energetic and acrobatic movements performed to a loud insistent beat; abbreviated in the slang of those who dance it to **breaking**.

Formed by compounding: the *dancing* that was developed specifically to fill the *break* in a piece of rap music (i.e. an instrumental interlude during which the DJ would be busy mixing, sampling, etc.). In Jamaican English, *to broke up* has meant 'to wriggle the body in a dance' since at least the fifties; in the Deep South of the US a *breakdown* has been the name for a riotous dance or hoedown (with an associated verbal phrase *to break down*) since the middle of the nineteenth century, but the connection between rap music and the development of *break-dancing* in New York was so close that these older dialectal uses are unlikely to have had much influence.

This style of dancing was pioneered during the late seventies by teams of Black teenage dancers (notably the 'Rock Steady Crew') on the streets of the south Bronx area of New York; each team (or CREW) worked in parallel with graffiti artists, and the combination of music, art, and street entertainment that they developed formed the core of the new Black street culture called HIP HOP. By 1982 the phenomenon had been taken up by the press and widely publicized (to such an extent that by the mid eighties there was talk of over-exposure in the media and **break-sploitation**, an alteration of the more familiar word *blaxploitation* 'exploitation of Blacks'). To connoisseurs, *breaking* is only one of a number of styles of movement making up the highly competitive dance culture; others include BODY-POPPING, the *lock*, and the *moonwalk*. In *breaking* itself, dancers spin on the ground, using the body like a human top, and pivoting on a shoulder or elbow, the head, or the back. The craze quickly spread to other parts of the world and began to lose its association with Black culture. The noun *break-dancing* was quickly followed by the verb **break-dance** (simply **break** in Black slang use) and both these forms also exist as nouns; a person who break-dances is a **break-dancer** (or **breaker**).

> While Freddy lays down chanting, talking, rhythmic rap, the Break Dancers break, trying to out-macho one another. They jump in the air and land on their backs, do splits and flip over.
> *Washington Post* 4 June 1982, Weekend section, p. 5

> They are young street dudes, nearly all of them black, anywhere from 10 to 23 years old, and what they are doing is a new style of dancing known as 'breaking' or 'break dancing'.
> *Daily News* 23 Sept. 1983, p. 18

> In Leningrad the Juventus Health and Sports Club has activities from Aikido wrestling, skateboarding and break-dancing to tennis.
> *The Times* 5 Apr. 1989, p. 46

> It seems any moment they will break from this 4,000-year-old tradition and spin off into a lively break-dance.
> *Burst of Excitement* (California Institute of Technology) Mar. 1990, p. 3

briefcase 🎵 see GHETTO BLASTER

brilliant /ˈbrɪljənt/ *adjective* 🔫

In young people's slang: great, fantastic, really good. Often abbreviated to **brill**.

A weakening of the original meaning (in much the same way as *great*, *fantastic*, etc. had been weakened by earlier generations of young people), followed in the case of *brill* by clipping of the ending (like the earlier *fab* etc.)

Although the literal meaning of *brilliant* is 'shining brightly', the adjective had been used figuratively for two centuries and more before being taken up as a cult word by young people; these earlier figurative uses often described some kind of spectacle, or a person with abnormal talents. From about the end of the 1970s, though, *brilliant* began to be used to express approval of just about anything. When used in this way, it is sometimes pronounced as a three-syllable word with the primary stress shifted to the final syllable: /ˌbrɪlɪˈənt/. *Brill* appeared in the early eighties. Both are considered a little dated by the very young, but they still seem to be going strong in comics and children's television programmes.

> I allowed Pandora to visit me in my darkened bedroom. We had a brilliant kissing session.
> Sue Townsend *The Growing Pains of Adrian Mole* (1984), p. 15

> I think your magazine is brill.
> *Music Making* July 1987, p. 11

brilliant pebbles /ˌbrɪljənt ˈpɛb(ə)lz/ *plural* noun Also written **Brilliant Pebbles** 🎖

A code-name for small computerized heat-seeking missiles designed to intercept and destroy enemy weapons; part of the US Strategic Defense Initiative (or STAR WARS). Also, the technology used to produce these.

One of a series of names making a word-play out of the idea of *smart* weaponry. The largest, heaviest, and least *intelligent* weapons (see INTELLIGENT¹) were spoken about by scientists as *moronic mountains*, smaller and more intelligent ones as *smart rocks* (a term coined by SDI chief scientist Gerald Yonas: see SMART), and yet smaller and smarter ones as *brilliant pebbles*; a fourth category in the series was *savant sand*.

Brilliant pebbles were the idea of US scientist Lowell Wood, who proposed in 1988 that existing smart-rocks technology could simply be 'shrunk' to smaller weapons. Work then started on developing *brilliant pebbles* in place of the space-based interceptor originally planned for Star Wars. Their *brilliance* is explained by the fact that each would carry a microchip frozen to superconducting temperatures and as powerful as a supercomputer.

> The SDI organization has funded assembly of brilliant pebbles hardware at the laboratory, and tests to demonstrate the concept are planned in the near future.
> *Aviation Week* 11 July 1988, p. 37

The Pentagon has been pushing the smart rocks, while Congress has been championing the ground-based missiles. Mr Edward Teller advocates 'brilliant pebbles'. *Economist* 4 Feb. 1989, p. 44

Brixton briefcase /ˌbrɪkstən ˈbriːfkeɪs/ *noun* 🗙 ♪

In British slang, the same thing as a GHETTO BLASTER. (Considered by some to be racially offensive.)

For etymology and history, see GHETTO BLASTER.

The other five had on their laps large stereo portable radios which, I believe, are colloquially spoken of as Brixton briefcases. *The Times* 22 July 1986, p. 13

Frank asked someone to fetch his briefcase from his car ... but ... all they could see was a ghetto blaster. So they went back and told Frank. 'That WAS my briefcase man—my Brixton briefcase,' said Frank. *Fast Forward* 28 Mar. 1990, p. 6

broker-dealer 〰 see BIG BANG

BSE /ˌbiːɛsˈiː/ *abbreviation* ⊗ 🗙

Short for **bovine spongiform encephalopathy**, an incurable viral brain condition in cattle which causes nervousness, staggering, and other neurological disorders, and eventually results in death. Known colloquially as MAD COW DISEASE.

The initial letters of *Bovine Spongiform Encephalopathy*. *Bovine* because it affects cattle; *spongiform* in that it produces a spongy appearance in parts of the brain tissue; *encephalopathy* is a word made up of Greek roots meaning 'disease of the brain'.

Bovine spongiform encephalopathy was first identified in the UK in 1986, and quickly started to affect a considerable number of cattle in different parts of the country. The discovery in May 1990 that it was possible for it to be transmitted to cats, possibly through pet foods containing brain tissue or offal from cattle, led to international public concern over the safety of British beef for human consumption. The disease has a long incubation period—a number of years—so it was difficult for experts to be sure that no cases in humans would occur in the future; but a government inquiry found that it was extremely unlikely. Steps were taken to ensure that meat from affected cattle did not enter the food chain, and the public panic over beef began to die down.

Bovine spongiform encephalopathy (BSE) twists the tongues of vets and wrecks the brains of cows. It is also new and baffling. Since the first case of the disease was diagnosed in December 1986, it has struck down 120 animals from 71 herds. *Economist* 14 Nov. 1987, p. 92

The disease in cows is similar to Scrapie which occurs in sheep, and it's possible that BSE may have been transferred to cattle from sheep. *Which?* Sept. 1989, p. 428

BSE-free ⊗ 🗙 see -FREE

B two (B2) bomber ✈ see STEALTH

bubblehead 🙂 see AIRHEAD

buddy /ˈbʌdɪ/ *noun and verb* ⊗ 🙂

noun: Someone who befriends and supports a person with Aids (see PWA) by volunteering to give companionship, practical help, and moral support during the course of the illness.

intransitive verb: To do this kind of voluntary work. Also as an action noun **buddying**.

A specialized use of the well known American sense of *buddy*, 'friend'. The American film *Buddies*, released quite early in the Aids era (1985), was surely influential in popularizing this specialized use.

For several generations children in the US have been encouraged to follow the **buddy system**—never to go anywhere or take part in any potentially dangerous activity alone, but to take a *buddy* who can bring help if necessary; a similar practice is followed by adults in dangerous situations. The scheme to provide *buddies* for people with Aids, started in late 1982 in New York, is an extension of that system, recognizing that these people need friendship that is often denied them once they are diagnosed as having the condition.

> Our greatest priority is to ensure that no person who has contracted an AID related disease is without some kind of personal support ... It is therefore our aim to create a buddy system.
> *New York Native* 11 Oct. 1982, p.

> I suppose the book wouldn't have been written if I hadn't buddied, because I wouldn't have had sense of knowing the reality of Aids. *The Times* 29 June 1987, p.

> When one of the members crossed the Rubicon from HIV to Aids, Helpline always appointed two three buddies to 'see the person through'. *Independent* 21 Mar. 1989, p.

bum-bag /'bʌmbæg/ *noun* Also written **bumbag** or **bum bag** 🔲

A small pouch for money and other valuables, attached to a belt and designed to be worn round the waist or hips; a British name for the FANNY PACK.

Formed by compounding; skiers wear them with the pouch to the back, above the bottom (the '*bum*'), although as fashion accessories they are normally worn with the pouch in front, where the contents can best be protected from pickpockets.

The *bum-bag* has been well known to skiers, motorcyclists, and ramblers for some decades as a useful receptacle for sandwiches, waterproofs, and other bits and pieces; being worn round the waist, it leaves the hands free. In the late eighties the *bum-bag* made the transition from a piece of sports equipment to a fashion item: perhaps because of the risk of bag-snatching in busy city streets, it became fashionable to wear a *bum-bag* for shopping and everyday use, and in 1990 was considered one of the main fashion 'accents' in the UK. As such, it is probably only a temporary item in the more general language.

> The most brilliant accessory is the bum-bag. Slung around the waist, it doubles as a belt and secure place for valuables. *Indy* 21 Dec. 1989, p.

buppie /'bʌpɪ/ *noun* Also written **Buppie** or **buppy** 🔲

A Black urban (or upwardly-mobile) professional; a YUPPIE who is Black.

Formed by substituting the initial letter of *black* for the *y-* of *yuppie* (see YUPPIE).

The word *buppie* was invented by the US media in 1984 as one of several variations on the theme of *yuppie*. Unlike some of the others—such as GUPPIE, *juppie* (a Japanese yuppie), and *puppie* (a pregnant yuppie)— this one caught on: perhaps this was because it identified a distinct group which was obviously rejecting its 'roots' culture in favour of the values and aspirations of a yuppie peer group.

> Bryant Gumbel and Vanessa Williams are both Buppies. Of course, it wouldn't be Yuppie to be Miss America unless you are the first black one. *People* 9 Jan. 1984, p.

> Old Harrovian and self-confessed buppie, with a fifth-in-a-row hit, Danny D's entrepreneurship is about to go global. *Evening Standard* 1 May 1990, p.

burn-bag /'bɜːnbæg/ *noun* 🔲

In the jargon of US intelligence, a container into which classified (or incriminating) material is put before being destroyed by burning. Also sometimes known as a **burn basket**.

Formed by compounding: a *bag* or *basket* for what is to be *burned*.

The word has been used in US intelligence circles since at least the sixties, but did not come to public notice until the political scandals of later decades: first Watergate (1972) and then the Iran–contra affair (1986: see CONTRA). In relation to these two incidents it was used especially to refer to the means which allowed prominent politicians to dispose of incriminating documents allegedly linking them with the scandal; the chairmen of relevant inquiries could not then require them to be produced.

'I frankly didn't see any need for it at the time,' he [John Poindexter] said of the document, known as an intelligence finding. 'I thought it was politically embarrassing. And so I decided to tear it up, and I tore it up, put it in the burn basket behind my desk.   *New York Times* 16 July 1987, section A, p. 10

burn-out /'bɜːnaʊt/ *noun* Frequently written **burnout**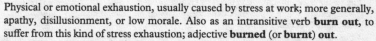

Physical or emotional exhaustion, usually caused by stress at work; more generally, apathy, disillusionment, or low morale. Also as an intransitive verb **burn out**, to suffer from this kind of stress exhaustion; adjective **burned** (or **burnt**) **out**.

A noun formed on the verbal phrase *burn oneself out*, meaning 'to use up all one's physical or emotional resources'; the noun *burn-out* already existed in the more literal sense of the complete destruction of something by fire, as well as in two technical senses.

The *burn-out* syndrome, which is thought to be a direct result of the high-stress lifestyles of the past two decades, was first identified and named in the mid seventies by American psychotherapist Herbert J. Freudenberger. Once the preserve of those in jobs requiring a high level of emotional commitment (such as charity work, medicine, and teaching), *burn-out* soon started affecting professional sportspeople, executives, and entertainers, too. In the late eighties, the word remained very fashionable, taking over from the more old-fashioned terms *depression* (imprecise except as a clinical term) and *nervous breakdown* (for cases of complete *burn-out*).

The most moderate form of burnout occurs when the sufferer endures a heavy stressload.
Management Today July 1989, p. 122

She may find herself trapped into trying to please everybody and do everything, failing to set boundaries to her role, which leads to chronic overwork and burn-out.
Nursing Times 29 Nov.–5 Dec. 1989, p. 51

Addled with divorce headaches and post-*Born* burnout, Cruise isn't doing press; but would you like to talk to Don and Jerry, perhaps?   *Premiere* June 1990, p. 92

burster /'bɜːstə(r)/ *noun*

A machine for separating or **bursting** continuous stationery (such as computer listing paper) into individual sheets.

Formed by adding the agent suffix *-er* to *burst*; originally, a *burster* was a charge of gunpowder for bursting a shell.

The word has existed in the technical jargon of office machinery since the fifties, but has only become widely known since the advent of computers and listing paper to nearly all offices, with the attendant nuisance of separating printout into pages.

Users who work through a heavy load of fan-fold may find that a 'burster' . . . is a useful accessory.
Susan Curran *Word Processing for Beginners* (1984), p. 45

buster /'bʌstə(r)/ *noun*

In US slang, short for **baby buster**: a person born in the generation after the baby boom (see BOOMER), at a time when the birth rate fell dramatically in most Western countries.

Formed by dropping the word *baby* from *baby buster*, following the model of BOOMER. In economic terms (especially in US English), a *bust* is a slump, that is the opposite of a *boom*.

The *busters*—children born from the late sixties onwards—are becoming an important force in Western economies now that they are adults. These economies, once able to grow continuously, must now shrink if the smaller population is not to *bust* them.

Busters may replace boomers as the darlings of advertisers.
headline in *Wall Street Journal* 12 Nov. 1987, p. 41

bustier /'bʌstɪeɪ/, in French /bystje/ *noun* 🔀

A short, close-fitting bodice (usually without straps), worn by women as a fashion top.

A direct borrowing from French *bustier* 'bodice'. The garment helps to define the *bust*, and so makes its wearer appear *bustier*, but this is a popular misunderstanding of the origin of the word.

The *bustier* came into fashion in the early eighties; one of its most famous devotees is the rock star Madonna, who has probably done much to keep the fashion going by regularly making public appearances in a *bustier*.

> Delicately edged suede jackets and bustiers in scarlet and black sat atop wafts of brightly coloured chiffon skirts for evening.
> *London Evening News* 17 Mar. 1987, p. 18

buyout /'baɪaʊt/ *noun* Sometimes written **buy-out** 〰️

The purchase of a controlling share in a company, either by its own employees or by another company.

The noun is formed on the verbal phrase *to buy* (someone) *out*.

The word originated in the US in the mid seventies, when there was a marked rise in company take-overs and tender offers. In some buyout schemes it was the company's own employees who were encouraged to buy up sufficient stock in the firm to retain control; other variants are the **management buyout** or **MBO**, in which the senior directors of a company buy up the whole stock, and the **leveraged buyout** (pronounced /'lɛvərɪdʒd/: see LEVERAGE) or **LBO**, practised mainly in the US, in which outside capital is used to enable the management to buy up the company. Although originally American, the *buyout* soon reached UK markets as well; by the mid eighties there were firms of financial advisers on both sides of the Atlantic specializing in this subject alone. Variations on the same theme are the **buy-back**, in which a company repurchases its own stock on the open market (often as a defensive ploy against take-overs), and the **buy-in**, in which a group of managers from outside the company together buys up a controlling interest.

> Leveraged buyouts are commonly used in the United States to defeat hostile takeover bids, but have yet to be successfully tested in Britain.
> *The Times* 2 May 1985, p. 21

> Latest statistics show buyouts and buy-ins by outside managers running at a record level this year.
> *Daily Telegraph* 30 Oct. 1989, Management Buyouts Supplement, p. i

> Lifting the veil of secrecy was ordinarily enough to kill a developing buyout in its cradle: Once disclosed, corporate raiders or other unwanted suitors were free to make a run at the company before management had a chance to prepare its own bid.
> Bryan Burrough & John Helyar *Barbarians at the Gate* (1990), p. 8

buzzword see FUZZWORD

bypass /'baɪpɑːs/ *noun* Also written **by-pass** ⊗

A permanent alternative pathway for a blood vessel, artery, etc. (especially near the heart or brain), created by transplanting a vessel from elsewhere in the body or inserting an artificial one. Also, the operation by which this is achieved or the artificial device that is inserted.

A figurative use of the word *bypass*, which was regularly used in the sixties and seventies for an alternative road built to route traffic round a bottleneck such as a large town; the medical *bypass*, too, is often created to avoid an obstruction or constriction in the existing network.

The art of *bypass* surgery was developed during the sixties and seventies and was becoming routine by the eighties. By an interesting reversal of linguistic roles, new roads were often called *arterials* rather than *bypasses* in the eighties, and the medical sense of *bypass* showed signs of becoming the dominant meaning of the word. It is often used attributively, in **bypass operation, bypass surgery**, etc.

> Sir Robin Day was yesterday 'progressing very nicely' after his heart by-pass operation in a London hospital. *News of the World* 3 Mar. 1985, p. 2

> The findings may have far-reaching implications . . . offering patients a low-risk alternative to cholesterol-lowering drugs, bypass operations and angioplasty, a technique in which clogged arteries are opened with a tiny balloon that presses plaque against the artery walls. *New York Times* 14 Nov. 1989, section C, p. 1

• •

C

cable television /ˌkeɪb(ə)l ˈtɛləvɪʒ(ə)n/ *noun* ❌

A system for relaying television programmes by cable (rather than broadcasting them over the air), usually into individual subscribers' homes; also, collectively, the stations and programmes that make use of this system. Often abbreviated to **cable tv** or simply **cable**.

Formed by compounding; a straightforward combination of the existing nouns *cable* and *television*.

The first experiments with *cable television* were carried out in the US in the early sixties, but at first the system was officially known as *community antenna television*, since the signal is picked up by a shared antenna before being cabled to individual receivers. The snappier name *cable tv* or *cable television* was first used in the mid sixties in the US, competing for a time with **Cablevision** (a trade mark which belonged to one of the larger companies operating the system there). After unsuccessful experiments here too in the fifties, *cable television* was finally adopted in the UK at the beginning of the eighties, giving rise to much speculation about its probable effect on the quality and choice of programmes in conventional broadcasting; in the event it enjoyed a smaller take-up than SATELLITE television. Once established in any individual country, *cable tv* has tended to be abbreviated further to *cable* alone (without a preceding article); the word is often used to refer to the stations or programmes available rather than the system. There is also a verb **cable**, 'to provide (a home, area, etc.) with cable television'.

> Reports that the government will soon approve plans to bring cable television to Britain have appeared in almost every newspaper. *New Scientist* 9 Sept. 1982, p. 674

> Even *Coronation Street* . . . failed to catch on when it was shown on a New York channel in 1976 and on nationwide cable in 1982. *Listener* 4 Dec. 1986, p. 29

> Cabling a typical 100,000-home franchise takes four to five years, costs £35 million—£350 for each home passed by the fibre-optic link which carries the signals. *Business* Apr. 1990, p. 100

cache /kæʃ/ *noun and verb* 🖳

noun: Short for **cache memory**, a small high-speed memory in some computers which can be used for data and instructions that need to be accessed frequently, instead of the slower main memory.

transitive verb: To place (data, etc.) in a separate high-speed memory. Adjective **cached**, action noun **caching**.

A figurative use of *cache*, which originally meant 'a hiding place' (borrowed into English at the end of the eighteenth century from French *cache*, related to *cacher* 'to hide'); from here it went on to mean 'a temporary store' (Arctic explorers, for example, put spare provisions in a *cache*, and the verb *to cache* also already existed for this activity). A computer *cache* is, in effect, only another kind of temporary store.

The *cache* memory was invented by IBM in the late sixties, but the verb and its derivatives appear not to have developed until the early eighties.

> Window images are normally cached in a form to allow fast screen redraw.
> *Personal Computer World* Nov. 1986, p. 17

> If the information is held in the cache, which can be thought of as a very fast on-chip local memory, then only two clock cycles are required. *Electronics & Wireless World* Jan. 1987, p. 10

Callanetics /ˌkælə'nɛtɪks/ *plural noun* (but usually treated as *singular*) ⊗ ✖

The trade mark of a physical exercise programme originally developed in the US by Callan Pinckney and based on the idea of building muscle tone through repeated tiny movements using deep muscles.

Formed by combining the woman's name *Callan* with *-etics*, after the model of *athletics*; probably also influenced by *callisthenics*, a nineteenth-century word for gymnastics for girls, designed to produce the 'body beautiful' (itself formed on Greek *kallos* 'beauty').

One of a long line of exercise programmes and workout routines popular in the eighties, *Callanetics* was made the subject of a book of the same name in the US in 1984. Despite claims that Callan Pinckney had 'stolen' exercises from the workout routines of her own teachers, the programme was hailed as a new approach to exercise and by 1988 was proving extremely successful commercially. When the book *Callanetics* was first published in the UK in 1989 it started a new exercise craze, helped on by reports that the Duchess of York had used the programme to get herself back into shape after the birth of her daughter Beatrice. Pinckney herself claims that the unique feature of *Callanetics* is the way in which it works out deep muscles through movements of only half an inch in each direction from a starting position.

> Callanetics requires only two hour-long work-out sessions a week.
> *Sunday Times Magazine* 5 Mar. 1989, p. 2

camcorder /'kæmkɔːdə(r)/ *noun* Occasionally written **cam-corder** ✖

A portable video camera with a built-in sound recorder, which can produce recorded video cassettes (and in some cases also play them back).

A clipped compound, formed by combining the first syllable of *camera* with the last two of *recorder*.

Prototype *camcorders* were produced almost simultaneously by several Japanese companies at the beginning of the eighties; the word was first used in English-language sources in 1982. By the end of the eighties it had become almost a household word, as video took over from cine and home movies for recording family occasions, travel, etc.

> If you want to use a video camera simply to record events in the school year then the camcorder might be for you. *Times Educational Supplement* 30 Nov. 1984, p. 29

> The eight-millimetre camcorders (eight-millimetre refers to the width of the tape) ... produce tapes that cannot be used with the VHS format. *New Yorker* 24 Nov. 1986, p. 98

camp-on /'kæmpɒn/ *noun* 〰 ▣

A facility of electronic telephone systems which allows an unsuccessful caller to 'latch on' to a number so that the call is automatically connected once the receiving number is available.

The noun is formed on the verbal phrase *to camp on to*, which in turn is a figurative use of the verb *to camp*: the caller stakes claim to a place in the queue, and this 'pitch' is automatically registered by the system.

First used in the mid seventies, the *camp-on* became increasingly widespread with the rise in popularity of push-button electronic telephones during the eighties.

A Thorn Ericsson PABX can provide over twenty aids to efficient communications. Here is one of them: Camp-on busy. An incoming call for an extension that is already engaged (busy) . . . can be 'camped' on to the engaged extension.   *Daily Telegraph* 10 Mar. 1977, p. 2

campylobacter /ˌkæmpaɪləʊˈbæktə(r)/ *noun* ⊗

A bacterium occurring in unpasteurized dairy produce and other everyday foods and capable of causing food poisoning in humans.

The bacterium takes its name from the genus name *Campylobacter*, which in turn is formed from a Greek word *kampulos* 'bent, twisted' (the bacteria in this family being twisted or spiral in shape) and the first two syllables of *bacterium*.

Campylobacter is an important cause of non-fatal cases of food poisoning. The word, first used in the early seventies, would probably have remained known only to bacteriologists had it not been for public interest in—and concern about—food safety in the UK in 1989–90.

60 per cent of all poultry carcasses were infected with either salmonella or campylobacter.
                    The Times 2 Mar. 1990, p. 2

can bank /ˈkæn ˌbæŋk/ *noun* 🌱

A collection point to which empty cans may be taken for RECYCLING.

For etymology, see BOTTLE BANK.

With increasing consumption of fizzy drinks from ring-pull cans in the eighties, the *can bank* was a natural development of the recycling idea started by the *bottle bank*.

So far there are less than 200 'can banks' operated by 60 local authorities in Britain. One big problem is that it isn't easy enough to distinguish steel from aluminium.
                    John Button *How to be Green* (1989), p. 112

Candida /ˈkændɪdə/ *noun* ⊗

Short for *Candida albicans*, a yeastlike fungus which causes inflammation and itching in the mouth or vagina (commonly known as *thrush*), and is also thought to cause digestive problems when it multiplies in the digestive tract. Also, loosely, the set of digestive problems caused by excessive quantities of *Candida* in the gut; candidiasis.

A shortened form of the Latin name *Candida albicans*; popularly, the genus name *Candida* (which is formed on the Latin word *candidus* 'white') is used to refer to the particular species *Candida albicans* (whose name is a sort of tautology, meaning 'white-tinged white').

The effects of *Candida* in the mouth and vagina (thrush) have been well known since the thirties. The theory that the fungus can get out of control in the gut (especially on a Western diet high in refined sugars) and cause digestive illness is one that has only been given any credence in the past decade, and is still not fully supported in traditional medicine.

Bill Wyman . . . tours the world . . . while she stays in Britain suffering from an agonising allergy . . . He spoke of his wife's painful illness, Candida . . . Candida's a yeast allergy that usually affects the stomach . . . Certain food's OK for the Candida, but bad for the liver.   *News of the World* 8 Apr. 1990, p. 9

cap /kæp/ *verb* and *noun* 📰

transitive verb: To impose a limit on (something); specifically, of central government: to regulate the spending of (a local authority) by imposing an upper limit on local taxation.

noun: An upper limit or 'ceiling', especially one imposed by central government on a local authority's spending.

This sense arises from the image of placing a cap or capping on the top of something (a general sense of the verb which has existed since the seventeenth century), and may be related more specifically to the *capping* of oil wells as a way of controlling pressure. As such, it is almost opposite in meaning to the colloquial sense of the verb, 'to exceed or excel, to outdo'.

This type of **capping** became topical in the mid eighties with the UK government's *capping* of local authority spending (first in the form of **rate-capping**, and in 1990 as **charge-capping** or **poll-capping**). Councils on which this was imposed, or the taxes they could levy, were described as **capped** (**rate-capped**, **charge-capped**, etc.).

> The major cost would come in lost interest on cash flow because most people would delay paying until the lower, charge-capped, demand arrived. *Independent* 20 Mar. 1990, p. 8

> The Court of Appeal yesterday dismissed the second stage of the legal campaign by 19 Labour local authorities against the Government's decision to cap their poll tax levels and order cuts in their budgets. *Guardian* 28 June 1990, p. 2

> A council once famous for getting disadvantaged people into further education has abolished all discretionary maintenance grants because it has been charge-capped. *Times Educational Supplement* 7 Sept. 1990, p. 6

capture /ˈkæptʃə(r)/ *noun* and *verb* 🖳

noun: The process of transferring information from a written, paper format to machine-readable form (on a computer). Known more fully as **data capture**.

transitive verb: To convert (data) in this way, using any of several means (such as punched tape, keyboarding, optical character readers, etc.).

The noun and verb arose at about the same time, probably through specialization of a figurative sense of the verb *to capture* meaning 'to catch or record something elusive, to portray in permanent form' (as, for example, a likeness might be captured in a painting or photograph).

A technical term in computing from the early seventies onwards, *capture* entered the more general language in the eighties and became one of the vogue words in journalistic articles about any computerization project and in advertising copy for even minimally computerized products.

> About 70% of all data captured is reentered at some future point. *ABA Banking Journal* Dec. 1989, p. 74

> Unmatched range of edit/capture facilities simply not offered by other scanners at this unbeatable price. *CU Amiga* Apr. 1990, p. 68

carbon tax 🌱 see GREENHOUSE

card¹ /kɑːd/ *noun* 〰

A thin rectangular piece of semi-rigid plastic carrying the membership details of the owner and used to obtain credit, guarantee cheques, activate cash dispensers, etc.

Although made of plastic, this kind of *card* closely resembles in size, shape, and purpose a business or membership *card* (itself named after the material from which it was traditionally made); in the electronic age, size, shape, and recorded data (usually on a magnetic strip) are the important characteristics, for they determine whether or not the *card* may be used in the appropriate machinery.

In the UK, the stiff plastic *card* was first widely used by banks as a method of guaranteeing payment on cheques from the late sixties onwards; this kind of *card* was generally known as a **cheque card**. The huge increase in consumer credit facilities which took place in the US during the sixties and in the UK during the seventies meant that the embossed **credit card** or **charge card** became very common. By the eighties it was not unusual for an individual **cardmember** to carry a whole range of *cards* for different purposes, including the types mentioned above and the **store option card** (or simply **option card**) giving interest-free credit for a limited period on goods from a specified store. Some people even considered that PLASTIC had taken over from money in the US and the UK. This view was reinforced by the introduction in 1982 of a plastic *card* to replace coins in public telephone boxes (see PHONECARD), the increasing popularity of the CASH DISPENSER (which allows people to use a **cash card** as a means of obtaining cash, discovering their bank balance, etc.), and the introduction of the **debit card** (which uses electronic point-of-sale equipment to debit the cost of goods direct from the customer's bank account, without the intervention of cheques or credit facilities). *Card* technology became a growth area during the eighties with the need to increase card-users' protection against theft and misuse; the **chip card**, a *card* which incorporates a microchip to store information about the transactions for which it is used, was one of the proposed solutions to this problem. With the proliferation of different kinds of *cards*, machinery was needed which could 'read' the information stored on the magnetic strip quickly and efficiently; by the end of the eighties, the **card-swipe**, a reader similar to an electronic eye, across or through which the *card* is 'wiped' rapidly, was widely used for this purpose. The term (*credit-*)*card* (short for (*credit-*) *card-sized*) began to occur in attributive position in the mid eighties to describe the thing named by the following noun as being the same size as, or in some other way similar to, a *card* (see the last quotation below).

> I reported the missing credit cards . . . but I did not call my bank that evening, trusting that nobody could use that card without the PIN code. *New York Times* 21 Nov. 1989, section A, p. 24

> Forstmann Little would receive senior debt rather than junior debt—roughly the difference between an American Express card and an IOU.
> Bryan Burrough & John Helyar *Barbarians at the Gate* (1990), p. 292

> UK Banks and building societies . . . are vigorously promoting the advantages of the new style three-in-one card covering cheque guarantee, cashpoint and debit card facilities.
> *Observer* 22 Apr. 1990, p. 35

> The British Heart Foundation has leaflets on angina and other heart conditions as well as credit card guides to pacemaker centres. *Daily Telegraph* 26 June 1990, p. 13

See also AFFINITY CARD, GOLD CARD, and SWITCH

card² /kɑːd/ *noun* 🖳

A printed circuit board (see PCB²) similar in appearance to a credit card and having all the circuitry required to provide a particular function in a computer system.

So named because of its resemblance to a credit card; just as a small piece of *board* (or *card-board*) is a *card*, so too a small circuit *board* is punningly called a *card*.

Slot-in *cards* providing extra facilities for a computer system (at first known almost exclusively as **expansion cards**) became a popular feature of the PCs of the eighties. The word *card* is often preceded by another word explaining the function (as in **graphics card** or **EGA card**, a *card* upgrading a computer to display enhanced graphics); this sometimes results in rather cryptic names such as **hard card**, a *card* upgrading the memory of a computer to the equivalent of hard-disc storage capacity. Because it provides the user with any of a number of new options without the need to buy a new computer, this kind of *card* is sometimes known as an **option card**.

> VideoFax comes as a pair of circuit boards, or 'cards', which plug into the back of a personal computer. *New Scientist* 21 Jan. 1989, p. 39

> No matter how reliable, how well engineered or how many options your intelligent multiport card claims to offer, . . . it will severely limit the numbers of users your system will support.
> *UnixWorld* Sept. 1989, p. 36

cardboard city /ˌkɑːdbɔːd ˈsɪtɪ/ *noun* 🏙

An area of a large town where homeless people congregate at night under makeshift shelters made from discarded cardboard boxes and other packing materials.

Formed by compounding: a *city* made from *cardboard*.

A phenomenon of the eighties, and an increasing problem in large cities both in the UK and in the US. Sometimes written with capital initials, as though it were a place-name in its own right.

This is not a country where families can live under bridges or in 'cardboard cities' while the rest of us have our turkey dinner. *Washington Post* 23 Dec. 1982, section A, p. 16

In *The Trackers of Oxyrhyncus* . . . the people of Cardboard City erupt on to the stage. These are the men and women, some old and some very young, who live beneath the arches on the South Bank. *Independent Magazine* 19 May 1990, p. 14

Cardiofunk /'kɑːdɪəʊfʌŋk/ *noun* ⊗ ✖

The trade mark of a cardiovascular exercise programme which combines aerobic exercises with dance movements.

Formed from the combining form *cardio-* 'heart' (Greek *kardia*) and *funk*, a type of popular music (see FUNK).

A development of AEROBICS, *Cardiofunk* was invented in the US in 1989 and imported to the UK in 1990.

Cardiosalsa and Cardiofunk classes are jammed at the five Voight Fitness and Dance Centers. *USA Today* 4 Jan. 1990, section D, p. 1

Tessa Sanderson . . . is a fan of cardiofunk and has got together with Derrick Evans to present the video Cardiofunk: the Aerobic programme. *Company* June 1990, p. 25

cardphone ✖ 🖳 see PHONECARD

carer /'kɛərə(r)/ *noun* 🔢

Someone whose job involves caring; especially, a person who looks after an elderly, sick, or disabled relative at home and is therefore unable to take paid employment.

Formed by adding the agent suffix *-er* to *care*; the word had existed in the more general sense of 'one who cares' since the seventeenth century.

This sense arose out of the concept of *caring professions* (see below) and the realization that much unpaid caring was being done by relatives who could not or would not entrust their elderly or sick loved-ones to professional care. The word was first used in this way towards the end of the seventies and became very fashionable in the mid eighties as increasing efforts were made to provide *carers* with the support they need. When used on its own, without further qualification, *carer* now usually means a person who cares for someone unpaid at home (also called a **care-giver** in the US); **professional carer** is often used for a member of the caring professions.

When a son is the primary care-giver, it is usually by default: either he is an only son or belongs to a family of sons. *New York Times* 13 Nov. 1986, section C, p. 1

Ms Caroline Glendinning, who made the study while a research fellow at York University, called yesterday for increased benefit rates for carers and for a non-means tested carer's costs allowance. Carers also needed opportunities for part-time work, flexi-time employment, and job sharing. There are an estimated six million carers. *Guardian* 12 July 1989, p. 8

caring /'kɛərɪŋ/ *adjective* 🔢

Committed, compassionate; of a job: involving the everyday care of elderly, sick, or disabled people.

Formed by turning the present participle of the verb *care* into an adjective.

Caring was first used as an adjective (in the sense 'committed, compassionate') in the mid sixties. By the end of the seventies there had been much talk in the UK of the need for a **caring society** supported by a strong welfare state, and certain professions (such as medicine, social work, etc.) had been recognized as **caring professions**. With the change of emphasis towards individual responsibility and away from the NANNY STATE in the eighties, the *caring society* based on the welfare state received less attention, but the government put forward the idea of **caring capitalism** instead. After the conspicuous consumption of the eighties, journalists identified a change of ethos in Western societies which prompted them to christen the new decade the **caring nineties**.

> A lot of people seemed to have come from the so-called caring professions—social work, psychotherapy, and so on. *New Yorker* 22 Sept. 1986, p. 58

> The Government had long urged local authority social service departments to act in an enabling and not just a providing capacity. They would be responsible, after consulting agencies such as doctors and other caring professions, for assessing individual needs, designing care arrangements, and ensuring that they were properly administered. *Guardian* 13 July 1989, p. 6

> His major driving force is 'caring capitalism', showing that making money does not always mean exploiting others. *Today* 13 Mar. 1990, p. 6

carphone /ˈkɑːfəʊn/ *noun* Also written **car phone** or **car-phone** ❌ ▣

A radio telephone which can be fitted in and operated from a car.

Formed by compounding: a *phone* used in a *car*.

The *carphone* has been available since the sixties, but only really became popular in the late eighties as less expensive and more reliable models came on to the market. Their popularity, especially among the YUPPIE set, with whom they were considered a status symbol, has led to concern about the safety of one-handed driving. This was possibly influential in the British government's decision to tax their use more heavily in the April 1991 budget.

> 'Darling can you keep next Friday free for our appointment at the amniocentesis clinic,' Nicola chirps down the Cellnet (Yuppiespeak for car phone). *Today* 21 Oct. 1987, p. 36

> The carphone, that symbol of success that says you are so much in demand that you cannot afford to be incommunicado for a moment. *The Road Ahead* (Brisbane) Aug. 1989, p. 19

See also CELLULAR and VODAFONE.

Cartergate ⌂ see –GATE

cascade /kæsˈkeɪd/ *noun* 〰️

In business jargon, the process of disseminating information within an organization from the top of the hierarchy downwards in stages, with each level in the hierarchy being briefed and in turn briefing the next level down; a meeting designed to achieve this.

A figurative use of the word *cascade*, in which the information is seen as falling and spreading like a waterfall. It has parallels in a technical sense of the word in transport: the process of relegating rolling stock etc. to successively less demanding uses before decommissioning it altogether.

Cascade was a fashionable marketing and business term which found its way into other professions, such as education, during the eighties. The opposite effect, in which those at the bottom of the hierarchy feed back their views to the higher echelons, has jokingly been called 'splashback'.

> An elaborate training programme has been arranged, spread over four phases in what is called a 'cascade'. Heads of department are trained so that they can go back into schools and train the teachers. *The Times* 25 Apr. 1986, p. 10

cash card 〰️ see CARD[1]

cash dispenser /ˈkæʃ dɪsˌpɛnsə(r)/ *noun* 〰️ ▣

A machine from which cash can be obtained by account-holders at any time of day or night by inserting a cash card and keying in a PIN.

Formed by compounding: a *dispenser* of *cash*.

Cash dispensers were introduced in the sixties, but made much more versatile (and therefore more popular) during the seventies and eighties, when the name **cashpoint** started to take over from *cash dispenser*. Also sometimes called a **cash machine**. For further history see ATM.

> Ian first noticed the mystery debits one weekend when he tried to withdraw money from a cashpoint, and couldn't. *Which?* Sept. 1989, p. 411

> With an Abbeylink card you can also have round-the-clock access to a national network of cash machines ... Problems with cash dispensers are the biggest cause for complaint [to the Building Societies Ombudsman], followed by building societies that charge home owners an administration fee if they refuse to take out buildings insurance through them.
> *Good Housekeeping* May 1990, pp. 18 and 191

Cassingle /kə'sɪŋg(ə)l/ *noun*

The trade mark of an audio cassette carrying a single piece of (usually popular) music, especially one which needs no rewinding; the cassette version of a single disc.

Formed by combining the first syllable of *cassette* with *single* to make a blend.

The *Cassingle* was introduced in the UK in the late seventies and in the US at the beginning of the eighties, when the popularity of the single disc in the popular music world was waning and much popular music was listened to on tape. In the UK it started purely as a promotional device, given away to radio stations and disc jockeys to encourage them to give airtime to singles; by the end of the eighties, though, *Cassingles* were commercially available.

> Singles ... recently introduced by CBS (which introduced the two-sided disc back in 1908); the cassingle, which lists for $2.98 and goes totally against the idea of convenience.
> *Washington Post* 31 Oct. 1982, section L, p.1

> All the figures tell the same story. Single and LP records are on the way out. Within 10 years, we will all be buying 'cassingles', cassettes and compact discs. *Independent* 20 Feb. 1987, p. 14

casual /'kæzjʊəl/, /'kæʒʊəl/ *noun* Frequently written **Casual**

In the UK, a young person who belongs to a peer group favouring a casual, sporty style of dress and soul music, and often characterized by right-wing political views, aggressively or violently upheld.

Named after their characteristic style of dress, which is studiedly *casual* (but certainly not untidy—for example, sports slacks rather than jeans).

Successors to the Mods of earlier decades, the first groups of *casuals* seem to have been formed in the early eighties. By 1986 they were firmly associated with football violence, having been described in the Popplewell report on crowd safety and control at sports grounds as groups which attached themselves to particular teams, 'bent on fighting the opposition fans in order to enhance their own prestige'. The subculture also exists outside the football ground, though, especially in wealthier areas.

> Politics just aren't that important for 90 per cent of skinheads. And you're more likely to get violence from the Casuals at football matches than any of us. *Independent* 23 Jan. 1989, p. 14

casual sex /ˌkæzjʊəl 'sɛks/ *noun*

Sexual activity between people who are not regular or established sexual partners.

Formed by compounding: *sex* which is *casual*.

A change in public attitudes towards sexual activity was the essential prerequisite for sexual activity to be described as *casual sex*, since the description implies that sex with a diversity of partners is conceivable—a view which, however much it may have been held by individuals, was not much aired in public before the 'swinging' sixties. During the seventies significant numbers of people began to question the conventional wisdom that only husband and wife, or those in a 'steady relationship', should have sexual intercourse. However, the idea that sex could become a transaction between any two (or more) otherwise unacquainted people remained controversial, despite the existence of such long-established forms of *casual sex* as prostitution. Use of the expression steadily increased, possibly indicating more widespread acceptability for the concept, and by the late seventies *casual* could also be applied to sexual partners. What brought the phrase to unprecedented prominence during the eighties was the AIDS crisis, which made non-judgemental plain speaking about the reality of people's sexual behaviour essential.

> The length of the list might suggest that Auden was in the habit of 'cruising'—picking up boys for casual sex. *Humphrey Carpenter W. H. Auden (1981), p. 97*

> The advice is to either avoid casual sex or to use a condom. *New Musical Express 14 Feb. 1987, p. 4*

See also SAFE SEX

CAT¹ /kæt/ *acronym* ⊗ 🖳

Short for **computerized axial tomography**, a medical technology which provides a series of cross-sectional pictures of internal organs and builds these up into a detailed picture using an X-ray machine controlled by a computer.

An acronym, formed on the initial letters of *Computerized Axial Tomography*; sometimes expanded as *Computer-Aided* or *Computer-Assisted Tomography*.

The technique was developed by EMI in the US in the mid seventies and was at first known as **CT scanning** (an alternative name which is still widely used, especially in the US). By producing detailed pictures of the inside of the body (and in particular of brain tissue) it revolutionized diagnostic procedures, often doing away with the need for exploratory surgery. *CAT* is normally used attributively, like an adjective: the image produced is a **CAT scan**; the equipment which produces it is a **CAT scanner**; the process is **CAT scanning** rather than *CAT* alone.

> Voluntary groups have raised the money . . . to buy CAT scanners for their local hospitals.
> *Listener 28 Apr. 1983, p. 2*

> Very soon after meeting Gabriel, I sent him to get a CT scan of his head and discovered a medium-sized tumor in his brain. *Perri Klass Other Women's Children (1990), p. 222*

cat² /kæt/ *noun and adjective* 🌳

noun: Short for **catalytic converter**, **catalyst**, or **catalyser**, a device which filters pollutants from vehicle exhaust emissions, thereby cutting down air pollution.

adjective: Catalysed; fitted with a catalytic converter (used especially in **cat car**).

Formed by shortening *catalytic converter*, *catalyst*, or *catalyser* to its first syllable.

Catalytic converters were first developed in the fifties, but the abbreviation *cat* did not start to appear frequently in print until about 1988, when the first models of car fitted with a *cat* as a standard option became available in the UK. Although quite separate from the issue of UNLEADED fuel, the desirability of *cat cars* has tended to be discussed in connection with the widespread switch to lead-free petrol, since a *cat* can only do its job—to 'scrub' carbon monoxide, nitrogen oxide, and hydrocarbons from the exhaust—in cars which run on unleaded fuel. At first, new models were produced in both *cat* and **non-cat** versions, but **cat-only** models look increasingly likely in the nineties.

> Unusually, Ford have been completely wrong-footed on this one by arch-rival Vauxhall, who are to start supplying cat cars in the UK this autumn. *Performance Car June 1989, p. 20*

> The new Turbo's exhaust system . . . features a metallic-element catalytic converter, while even the wastegate tailpipe is equipped with a cat and a muffler. *Autocar & Motor 7 Mar. 1990, p. 13*

> 'Cats' are like honeycombs with many internal surfaces . . . covered with precious metals which react with harmful exhaust gases. *Independent 3 Aug. 1990, p. 2*

CD /siːˈdiː/ *noun* 🖳

Short for **compact disc**, a small disc on which audio recordings or other data are recorded digitally and which can be 'read' optically by the reflection of a laser beam from the surface.

The initial letters of *Compact Disc*.

CD technology was invented by Philips for audio recording towards the end of the seventies as the most promising medium for the accurate new DIGITAL recordings. By 1980 Philips had pooled their resources with Sony and it was clear that the *CD* was to become the successor to the grooved audio disc. During the early eighties the *optical disc* (another name for the *CD*) was also vaunted as the medium of the future for other kinds of data, since the storage capacity was vastly greater than on floppy—or even hard—discs; a number of large reference works and commercial databases became available on **CD ROM** (compact disc with read-only memory), the form of *CD* used for data of this kind. The sound and data are recorded as a spiral pattern of pits and bumps underneath a smooth protective layer; inside the special **CD player** or **CD reader** needed to 'read' each of these kinds of disc, a laser beam is focused on this spiral. By 1990 the *CD* had become the established medium for high-quality audio recordings and new forms of *CD* were being tried: the **photo-CD**, for example, was suggested as a permanent storage medium for family photographs, the digitized images being 'read' by a *CD player* and viewed on a television screen. **CD video** (or **CDV**) applies the same technology to video. *Multimedia CDs*, including **CDI** (**Compact Disc Interactive**) and *DVI* (*Digital Video Interactive*) offer the possibility of combining text, sound, and images on a single disc. **CDTV** allows the viewer to interact with recorded television.

Whatever you want—get it on CD Video from your record or Hi Fi dealer.
Sky Magazine Apr. 1990, p. 14

The CDTV system involves a unit the same size as a video recorder which plugs into a standard television set. *Daily Telegraph* 13 Aug. 1990, p. 4

CDI . . . emphasises the fact that it is a world standard. This is a claim that can only be equalled by records, tapes and audio CDs . . . To achieve this Philips and Sony developed a new system and a new CD format for text, graphics, stills, and animation. *Information World Review* Sept. 1990, p. 20

The Kodak Photo CD system, jointly developed by Kodak and Philips of the Netherlands, digitally stores images from negatives or slides on compact discs. The pictures can then be shown on ordinary television or computer screens with a Photo CD player that also plays audio CDs.
Chicago Tribune 19 Sept. 1990, section C, p. 4

Ceefax /'si:fæks/ *noun*

In the UK, the trade mark of a teletext system (see TELE-) operated by the BBC.

A respelling of *see* (as in *seeing*) combined with *fax* (see FAX[1,2]): *seeing facsimile*, on which you may *see facts*.

Ceefax was introduced in the early seventies and is now a standard option on most new television sets in the UK.

Telesoftware is carried by teletext—in other words, it is part of the BBC's Ceefax service.
Listener 16 June 1983, p. 38

See also ORACLE

cellular /'sɛljʊlə(r)/ *adjective*

Being part of a mobile radio-telephone system in which the area served is divided into small sections, each with its own short-range transmitter/receiver; **cellular telephone**, a hand-held mobile radio telephone for use in this kind of system.

This kind of radio-telephone system is termed *cellular* from the small sections, called *cells*, into which the operating area is divided. The same frequencies can be used simultaneously in the different *cells*, giving greater capacity to the system as a whole.

This kind of mobile telephone became available in the late seventies and was considerably more successful than the more limited non-cellular radio telephone. By the mid eighties *cellular* was often abbreviated to **cell-**, as in **cellphone** for *cellular telephone* and **Cellnet**, the trade mark of the cellular network operated by British Telecom in the UK (and also of a similar service in the US), sometimes also used to mean a *cellphone*.

It will soon be possible to use either of the two cellular networks started this year off almost the entire south coast. *The Times* 15 Feb. 1985, p. 37

The mobile phone is the perfect symbol, if not of having arrived, then at least of having the car pointed in the right direction. It would no doubt come as a surprise to most cellphone users that their conversations are in the public domain, as it were, available to anyone with a scanning receiver, a little time to kill, and a healthy disregard for personal privacy. Fortunately for cellphone users, it's very difficult for us eavesdroppers to 'lock in' on one conversation for more than a few minutes. *Guardian* 14 July 1989, p. 7

CFC /ˌsiːɛfˈsiː/ *abbreviation*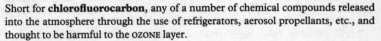

Short for **chlorofluorocarbon,** any of a number of chemical compounds released into the atmosphere through the use of refrigerators, aerosol propellants, etc., and thought to be harmful to the OZONE layer.

The initial letters of the elements which make up the chemical name *chlorofluorocarbon*: compounds of *chlorine, fluorine,* and *carbon.*

CFCs have been in use as refrigerants, in aerosols, and in the plastics industry for some decades, but came into the public eye through the discovery that they were being very widely dispersed in the atmosphere and that chlorine atoms derived from them were contributing to ozone depletion. The experimental work showing this to be the case was carried out during the seventies; by the early eighties, environmental groups were trying to publicize the dangers and some governments had taken action to control the use of *CFCs*, but it was not until the end of the decade that *CFC* became an almost universally known abbreviation in industrialized countries and manufacturers started to produce large numbers of products labelled **CFC-free.** If not followed by a number or in a combination such as **CFC gases,** the term is nearly always used in the plural, since there is a whole class of compounds of similar structure and having similar effects on the ozone layer, although some are more harmful than others.

Shoppers are told that meat and eggs are packaged in CFC-free containers.
Daily Telegraph 2 May 1989, p. 17

India alone estimates its bill for replacing CFCs over the next 20 years will be £350 million. Mrs Thatcher said it was essential that all nations joined the process of ridding the world of CFCs otherwise the health of the people of the world and their way of life would suffer.
Guardian 28 June 1990, p. 3

Du Pont has . . . promised to suspend production of ozone-destroying CFCs by 2000.
News-Journal (Wilmington) 9 July 1990, section D, p. 1

chair /tʃɛə(r)/ *noun*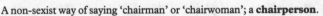

A non-sexist way of saying 'chairman' or 'chairwoman'; a **chairperson**.

Formed by dropping the sex-specific part of *chairman* etc. An impersonal use of *Chair* (especially in the appeal of *Chair! Chair!* and in the phrase *to address the chair*) had existed for centuries and provided the precedent for this use.

A usage which arose from the feminist movement in the mid seventies. Although disliked by some, it has become well established. It is interesting, though, that it has not produced derivatives: one finds **chairpersonship** of a committee, but only very rarely **chairship**.

On the more general aspects of the *arriviste's* upward trajectory, however, such as the craft of . . . chairpersonship, he has much less to say. *Nature* 9 Dec. 1982, p. 550

She has annoyed the Black Sections by refusing to resign as chair of the party black advisory committee. *Tribune* 12 Sept. 1986, p. 7

challenged ⊗ ▮▮ see ABLED

Challenger 💻 see SHUTTLE

chaos /'keɪɒs/ noun 💻

A state of apparent randomness and unpredictability which can be observed in the physical world or in any dynamic system that is highly sensitive to small changes in external conditions; the area of mathematics and physics in which this is studied (also called **chaos theory** or **chaology**).

A specialized use of the figurative sense of *chaos*, 'utter confusion and disorder' (a sense which itself goes back to the seventeenth century). Although actually determined by tiny changes in conditions which have large consequences, the processes which scientists call *chaos* appear at first sight to be random, utterly confused, and disordered.

The serious study of *chaos* began in the late sixties, but it was only in the mid seventies that mathematicians started to call this state *chaos* and not until the mid eighties that the study of these phenomena came to be called *chaos theory*. It is relevant to any system in which a very small change in initial conditions can make a significant difference to the outcome; a humorous example often quoted is the *butterfly effect* in weather systems—these systems being so sensitive to initial conditions that it is said that whether or not a butterfly flaps its wings on one side of the world could determine whether or not a tornado occurs on the other side. By the beginning of the nineties the study of **chaotic** systems had already proved to offer important insights to all areas of science—and indeed to our understanding of social processes—partly because it views systems as dynamic and developing rather than looking only at a static problem. A person who studies *chaos* is a **chaologist, chaos theorist,** or **chaoticist**.

> When the explorers of chaos began to think back on the genealogy of their new science, they found many intellectual trails from the past . . . A starting point was the Butterfly Effect.
> *James Gleick Chaos: Making a New Science* (1988), p. 8

> Chaos theory presents a Universe that is deterministic, obeying fundamental physical laws, but with a predisposition for disorder, complexity and unpredictability. *New Scientist* 21 Oct. 1989, p. 24

> One of the tasks facing students of complex chaotic systems . . . is to investigate fully the range of predictability in each case. *The Times* 9 Aug. 1990, p. 13

charge-capping 📇 see CAP

charge card ✍ see CARD¹

chase the dragon /ˌtʃeɪs ðə ˈdrægən/ *verbal phrase* 🗡

In the slang of drug users, to take heroin (or heroin mixed with another smokable drug) by heating it on a piece of folded tin foil and inhaling the fumes.

The phrase is reputed to be translated from Chinese and apparently arises from the fact that the fumes move up and down the piece of tin foil with the movements of the molten heroin powder, and these undulating movements resemble the tail of the dragon in Chinese myths.

This method of taking heroin comes from the Far East, as does the imagery of the phrase. It has been practised in the West since at least the sixties; in the eighties, with the threat of contracting Aids from used needles, it became more popular than injecting and the phrase became more widely known.

> Probably the stuff was now only twenty per cent pure. Still, good enough for 'chasing the dragon' Hong Kong style with match, silver foil, and paper tube. *Timothy Mo Sour Sweet* (1982), p. 50

> A hundred men or more lay sprawled 'chasing the dragon'—inhaling heroin through a tube held over heated tinfoil. *The Times* 24 May 1989, p. 13

> A smokable dollop of heroin costs about $10, about the same as a 'rock' of crack, which means that one can 'chase the dragon' for $20. *Sunday Telegraph* 18 Feb. 1990, p. 17

chatline ⁅⁆ see -LINE

chattering classes /ˌtʃætərɪŋ ˈklɑːsɪz/ *noun* ⁅⁆

In the colloquial language of the media in the UK, educated members of the middle and upper classes who read the 'quality' newspapers, hold freely expressed liberal political opinions, and see themselves as highly articulate and socially aware.

A catch-phrase (apparently coined by the journalist Frank Johnson in the early eighties and popularized by Alan Watkins of the *Observer*), after the model of *working classes*—the main characteristic of the group being readiness to express social and political opinions which are nevertheless seen by those in power as mere *chatter*.

According to an article by Alan Watkins in the *Guardian* (25 November 1989), the term was coined by Frank Johnson in conversation with Watkins in the late seventies or early eighties, when the two journalists lived in neighbouring flats. Certainly it was Watkins who subsequently popularized this apt description and turned it into a useful piece of shorthand for a well-known British 'type'. According to Watkins, the most important characteristics of the *chattering classes* at the time were their political views (usually including criticism of the then Prime Minister, Margaret Thatcher), their occupations (social workers, teachers, journalists, 'media people'), and their preferred reading matter (newspapers such as the *Guardian, Independent,* and *Observer*).

> Does anybody really care who is elected Chancellor of the University of Oxford? Only the chattering classes are exercised.
> *Daily Telegraph* 7 Mar. 1987, p. 14

cheque card ⟿ see CARD¹

child abuse /ˈtʃaɪld əˌbjuːs/ *noun* ⁅⁆

Maltreatment of a child, especially by physical violence or sexual interference.

Formed by compounding. The specialized sense of *abuse* here had already been in use for some time before the (sexual) abuse of children came to public attention during the eighties, and is common in other combinations: see ABUSE.

Child abuse was first used as a term in the early seventies, but mostly to refer to crimes of physical violence ('baby battering') or neglect. During the eighties (and particularly as a result of the public enquiry into the large numbers of children diagnosed as sexually abused in Cleveland, NE England, in 1987) it became clear that the sexual abuse of children, often by a parent or other family member, was much more widespread than had previously been thought, even though many of the original diagnoses were later discredited. Since then, the term *child abuse* has been used especially to refer to sexual interference with a child, and seems to have taken over from the older term *child molesting*. In 1990 the subject gained widespread publicity once again in the UK as police investigated the suspected abuse of children by adults allegedly involved in satanic rituals (known as **ritual abuse** or **satanic abuse** as well as *child abuse*).

> Child abuse occurs in all walks of life ... Doctors and lawyers, too, batter their kids.
> *New York Times* 6 Jan. 1974, p. 54

> Grave disquiet was expressed ... about the conclusions drawn from diagnostic sessions held at the Great Ormond Street Hospital child abuse clinic in those cases where there was doubt whether a child had been sexually abused.
> *The Times* 16 July 1986, p. 36

Childline ⁅⁆ see -LINE

China syndrome /ˈtʃaɪnə ˌsɪndrəʊm/ *noun* ⬚

A hypothetical sequence of events following the MELTDOWN of a nuclear reactor, in which so much heat is generated that the core melts through its containment structure and deep down into the earth.

Formed by compounding: the idea is that the *syndrome* ultimately results in the meltdown's reaching *China* (from the US) by melting through the core of the earth.

The *China syndrome* was always a fictional concept. It began as a piece of the folklore of nuclear physics but was widely popularized by the film *The China Syndrome* (produced in the US in 1979), which dealt with a fictional case of the official cover-up of an operational flaw in a nuclear reactor. Partly as a result of this film and partly because of the near meltdown which occurred at Chernobyl in the Ukraine in 1986, the idea of the *China syndrome* came to symbolize people's fears about the increasing use of nuclear power, even though the actual sequence of events in the fictional *China syndrome* was obviously far-fetched. The phrase had become sufficiently well known by the late eighties to be applied punningly by journalists in a number of other contexts, notably in relation to mass pro-democracy demonstrations in Beijing in 1989 and their subsequent violent suppression by the Chinese government.

Mr. Velikhov's announcement gave no clear indication just how close the Chernobyl disaster came to creating the so-called 'China Syndrome'. *The Times* 12 May 1986, p. 1

For at least a decade, government and business leaders around the world have based their Asian thinking on the belief that China was an economically developing, politically stable giant. Now all that has been stood on its head. There is a new China syndrome. *Business Week* 26 June 1989, p. 76

China white 🖉 see DESIGNER DRUG

chip card ∿ see CARD¹

chlorofluorocarbon 🌡 see CFC

chocolate mousse 🌡 see MOUSSE²

cholesterol-free ⊗ ✖ see -FREE

citizen-friendly 🖿 see -FRIENDLY

claimant /ˈkleɪmənt/ *noun* ▮▮

A person claiming a state benefit (especially unemployment benefit).

A specialized use of the word *claimant*, which has been used in the more general sense of 'one who makes a claim' since the eighteenth century.

The term has been used in official documents since the twenties, but was taken up by the claimants themselves in the seventies as a word offering solidarity; claimants' unions were formed and soon the word started to appear in new contexts such as notices announcing discounts.

The administration argues that its tough program—reviewing records of claimants and actually cutting off benefits from persons deemed able to work—stems from a 1980 law. *Christian Science Monitor* 27 Mar. 1984, p. 17

A new and unneccessary hurdle for the thousands of claimants who have been unfairly thrown off the disability rolls. *New York Times* 26 Mar. 1986, section A, p. 22

See also UNWAGED

clamp /klæmp/ *transitive verb* ✖

To immobilize (an illegally parked car) by attaching a WHEEL CLAMP to it. Also, to subject (a person) to the experience of having his or her car clamped.

A specialized use of the verb, which has existed in the general sense 'to make fast with a clamp' since the seventeenth century.

For history and usage, see WHEEL CLAMP.

In the first eight weeks 4,358 vehicles were clamped with the Denver shoe. *Daily Telegraph* 14 July 1983, p. 19

We've been clamped!! One just can't avoid *every* potential hazard!! *Holiday Which?* Mar. 1990, p. 73

classist /'klɑ:sɪst/ *adjective* and *noun* 🔃

adjective: Discriminating against a person or group of people because of their social class; class-prejudiced.

noun: A person who holds class prejudices or advocates class discrimination.

Formed by adding the suffix *-ist* (as in *racist* and *sexist*) to *class*; the corresponding *-ism* (*classism*) is a much older word, going back to the middle of the nineteenth century.

This word belongs to the debate about social attitudes and motivations which resulted from the feminist movement of the second half of the seventies.

> The user called another participant in the conversation 'a classist' for arguing that (particular) middle class values and behaviors were superior. *American Speech* Summer 1988, p. 183

Clause 28 /ˌklɔːz ˌtwentɪ'eɪt/ *noun* 📁 🔃

In the UK, a clause of the Local Government Bill (and later Act) banning local authorities from 'promoting homosexuality', and thereby imposing restrictions on certain books and educational material, works of art, etc.; hence also used allusively for the loss of artistic freedom and mood of HOMOPHOBIA seen by many as the sub-text of this legislation. Sometimes referred to simply as **the Clause**.

Formed by compounding: the *clause* numbered *28* in the original Local Authority Bill. Although the Bill became an Act in mid 1988, and the *clause* therefore became a *section*, the term *Section 28* did not gain much currency outside government or legal circles.

Clause 28 was discussed in Parliament for the first time at the end of 1987 and was welcomed by a large number of Conservative MPs as an expression of their party's commitment to 'traditional family values' and its pledge to tackle the problem of the 'permissive society' which had resulted from increased sexual freedom in the seventies and early eighties. From the opposite side of the political spectrum, though, the emergence of measures like *Clause 28* in the late eighties was interpreted as being symptomatic of a growing institutionalized homophobia in the post-Aids era. It was largely the opponents of *Clause 28* who continued to use the term— after the Bill became an Act in mid 1988—to allude to this perceived mood of artistic censorship and repressiveness.

> The homeless, the loss of artistic freedom (Clause 28), the unemployment figures and the cuts in arts funding were the subjects discussed. *Independent on Sunday* 18 Nov. 1990, p. 23

> In the years immediately following 1967 there was a tripling of the prosecutions for homosexual offences. What is happening today follows the same logic, reshaped by a decade of new right dominance, the impact of aids, and the climate that brought us Clause 28. *Gay Times* Apr. 1991, p. 3

click /klɪk/ *intransitive* or *transitive verb* 🖱

In computing, to press one of the buttons on a MOUSE; to select (an item represented on-screen, a particular function, etc.) by so doing.

Click, like ZAP, began as an onomatopoeic word for any of various small 'mechanical' sounds, such as finger-snaps or the cocking of a gun. The same word was also used as a verb, meaning either 'to make, or cause to make, this sound' or (a later development) 'to operate (a device which clicks)'. The mouse is simply the latest in a succession of possible objects for this later transitive sense.

> Prodigy uses the mouse extensively ... In place of a GEM double click, you have to click both buttons. *Music Technology* Apr. 1990, p. 36

> It allows you to browse until you find the file you're looking for, and, assuming you're in 'recover' mode, click on its name to request the server to deliver it back to your client at the desktop.
> *UnixWorld* Jan. 1991, p. 54

clock /klɒk/ *transitive verb*

In slang, to take notice of (a person or thing), to spot; also, to watch, to stare at.

Probably derived from the practice of clock-watching, which involves repeated glancing at the clock.

This word has been in use in underworld or criminal slang since about the forties, but has recently been taken up by journalists and moved into a rather more respectable register.

> This is the one rhythm machine that puts you back in the driving seat. Clock the SBX-80 at Roland dealers now.
> *International Musician* June 1985, p. 86

> Our waiter . . . was so busy clocking him that he spilt a bottle of precious appleade over the tablecloth.
> *Sunday Express Magazine* 3 Aug. 1986, p. 33

clone /kləʊn/ *noun* 🖳

A computer which deliberately simulates the features and facilities of a more expensive competitor; especially, a copy of the IBM PC.

A specialization of the figurative sense of *clone* which originated in science fiction: from the early seventies, a *clone* was a person or animal that had developed from a single somatic cell of its parent and was therefore genetically an identical copy. The computer *clones* were designed to be identical in capability to the models that inspired them (and, in particular, to run the same software).

A usage which arose during the eighties, as a number of microcomputer manufacturers attempted to undercut the very successful IBM personal computer (and later its successor, the PS2). Also widely used for other cut-price copies (for example, of cars and cameras as well as other computers).

> Amstrad [is] leading the cut price clones attacking IBM personal computers on price.
> *Marketing* 11 Sept. 1986, p. 5

> The company is a major porter to Far Eastern clone makers, who are developing copies of Sun Microsystems' SPARC-based workstations.
> *UnixWorld* Jan. 1991, p. 68

cocooning /kəˈkuːnɪŋ/ *noun* 🏠

In the US, the practice of nurturing one's family life by spending leisure time in the home with one's family; the valuing of family life and privacy above social contact and advancement. Also as a verb **cocoon** and an agent noun **cocooner**.

This specialized sense derives from the idea of a *cocoon* as a protective layer or shell: Americans are seen as deliberately retreating from the stressful conditions of life outside the home into the cosy private world of the family. Towards the end of the seventies in his book *Manwatching*, the anthropologist Desmond Morris had observed a similar protective device among people who live or work in crowded places where privacy is difficult to achieve:

> Flatmates, students sharing a study, sailors in the cramped quarters of a ship, and office staff in crowded workplaces, all have to face this problem. They solve it by 'cocooning'. They use a variety of devices to shut themselves off from the others present.

Cocooning can be seen as one step on from the *nesting* which is characteristic of new parents.

The word was apparently coined by Faith Popcorn—a New York trend analyst—in 1986, after analysis of socio-economic trends had shown that people in the US were going out and travelling less, ordering more takeout food to eat at home, doing more of their shopping from catalogues rather than in person, and showing more interest in traditional pastimes (such as craft work) which could be done at home. Within a few years this had had a significant commercial effect in the US—but it remains to be seen whether the trend will be limited to affluent Americans. *Cocooning* is seen by some as an up-market way of saying 'being a COUCH POTATO'.

> We are benefitting from 'cocooning'. Everyone wants to spend more time at home with family. Crafts like cross-stitching and fabrics for children and home decorating have experienced tremendous growth.
> *Fortune* 30 July 1990, p. 132

You could be … what Americans call a 'cocooner'—a rich yuppie who escapes the violence of society by shutting himself up with his designer wife and baby behind a screen of security alarms.

Sunday Express 16 Sept. 1990, p. 25

cohabitation /ˌkəʊhæbɪˈteɪʃ(ə)n/, sometimes /kɔabitasjɔ̃/ *noun* 📷

Coexistence or co-operation in government between members of opposing parties, especially when one is the President and the other the Prime Minister. Hence, by extension, the coexistence of different currencies in a single monetary system. Also as an intransitive verb, **cohabit**.

Borrowed into English from French *cohabitation*. In both languages, this is a figurative use of *cohabitation* in the sense 'living together as though man and wife, although not actually married'. Political *cohabitation* is seen as a marriage of inconvenience brought about by the fickleness of the voting public.

The word was first used in this sense in English in a report of a speech made by French President Valéry Giscard d'Estaing in 1978, during a period of coalition government in France. As the eighties progressed, the French voting public tended to favour a Socialist President (François Mitterrand) in combination with a conservative Prime Minister, making *cohabitation* a fact of life in French politics. During the discussion of EMS and EMU' in the late eighties, the word was used by journalists in a transferred sense to refer to the coexistence of different standards for European currencies.

Like France, Portugal is adjusting to the 'cohabitation' of a Socialist president and a conservative Prime Minister.

Economist 5 Apr. 1986, p. 57

Via EMS, the D-mark became Europe's leading currency, while the yen and the dollar cohabited.

Business Apr. 1990, p. 43

cold call /ˌkəʊld ˈkɔːl/ *verb and noun* 📈

In marketing jargon,

transitive verb: To make an unsolicited telephone call or visit to (a prospective customer) as a way of selling a product.

noun: A marketing call on a person who has not previously expressed any interest in the product. Also as an action noun **cold calling**.

Formed by compounding: the *call*, whether by telephone or in person, is made *cold*, without any previous warm-up, or preparation of the ground.

The term was first used in the early seventies as a more jargony equivalent for 'door-to-door selling' (and at that time *cold calling* was mostly done door-to-door); in the eighties the rise of telemarketing (see TELE-) and the emphasis on 'hard sell' has meant a huge increase in *cold calling* by telephone.

On the first cold call I ever made I started saying what I had been trained to say when to my astonishment the person I had rung said 'yes'.

Marketing 11 Sept. 1986, p. 20

We've never been happy with 'cold calling' and are very disappointed that the FSA extended it further. People don't make calm, rational decisions if they're smooth-talked into signing by strangers in their homes.

Which? Jan. 1990, p. 35

Financial salesmen will be able to 'cold call' customers and sell investment trust savings schemes.

The Times 30 Mar. 1990, p. 23

collectable /kəˈlɛktəb(ə)l/ *noun* Also written **collectible** (especially in the US) 🔧

Any article which might form part of a collection or is sought after by collectors, especially a small and relatively inexpensive item or one expressly produced for collectors.

Formed by turning the adjective *collectable* into a noun. In its more general sense the adjectiv simply means 'that may be collected', but it has been used by collectors to mean 'worth col lecting, sought after' since the end of the last century.

Not a particularly new word—even as a noun—among collectors themselves, but one which has enjoyed increased exposure in the past decade, partly through the boom in collecting as hobby. The noun is nearly always used in the plural.

> What distinguishes all these catalog 'collectibles' is that they are at once ugly, of doubtful value, an expensive.   Paul Fussell *Class* (1983), p. 11

> The wonderful thing about 'collectables' is that anyone with just a few extra pounds can become collector.   *Miller's Collectables Price Guide 1989–90*, volume 1, p.

colourize /ˈkʌləraɪz/ *transitive verb* Written **colorize** in the US 🔲 🖥

To add colour to (a black-and-white film) by a computerized process called **Col-orizer** (a trade mark). Also as an adjective **colourized**; noun **colourization**.

The verb has existed in the sense 'to colour' since the seventeenth century, but was rarely used until the invention of the *Colorizer*. This use of the verb is likely to be a back-formation from *Colorizer* rather than a straightforward sense development.

The *Colorizer* program has been used in Canada since the early eighties; the name was regis tered as a trade mark in the mid eighties. Also during the mid eighties, the practice of *colouriz ing* classic black-and-white films (especially for release as home videos) caused considerable controversy, with one side claiming that a company which had bought the rights to a particu lar film should be allowed to do as it wished with it, and the other maintaining that classic films were works of art not to be tampered with in any way.

> 'Colorizing' great movies such as Casablanca ... is like spray-painting the Venus de Milo.
> *Time* 5 Nov. 1984, p.

> Rather than legislate directly against the business interests that stood to profit from colorization Congress approved provisions under which films could be given landmark status and protected ..
> When broadcast recently on TBS, colorized pictures have been labeled as such.
> *Philadelphia Inquirer* 20 Sept. 1989, section A, p.

commodification /kəˌmɒdɪfɪˈkeɪʃ(ə)n/ *noun* 〰

The process of turning something into a commodity or viewing it in commercial terms when it is not by nature commercial; commercialization.

Formed by adding the process suffix *-ification* to the first two syllables of *commodity*.

Coined in the seventies, *commodification* has become a fashionable word to describe the eight ies' increasingly commercial approach to the Arts and to services (such as health care) which would not previously have been regarded as marketable. In financial sources, the word has also been used to refer to the tendency in the late eighties for money to be traded as though it were a commodity.

> [Artists] have made conscious attempts over the last decade to combat the relentless commodifi-cation of their products.   Lucy Lippard *Overlay* (1983), p. 6

community antenna television 🔲 see CABLE TELEVISION

community charge /kəˈmjuːnɪtɪ ˌtʃɑːdʒ/ *noun* 〰 🔲

In Great Britain, a charge for local services at a level fixed annually by the local au thority and in principle payable by every adult resident; the official name for the tax popularly known as the POLL TAX.

Formed by compounding: a *charge* for *community* services, and payable by every adult resident of the *community* who is not specially exempted.

The government announced its intention to replace the system of household rates with a *community charge* in 1985; the original plan was for a flat-rate charge of £50 per person. The plan was first put into effect in Scotland in 1989 and in the rest of Great Britain (but not Northern Ireland) in 1990. In both places it met with considerable opposition and a campaign of non-payment, not least because of the high level of tax fixed by many local authorities, the large discrepancies from one area to another, and the absence of any kind of means testing from the system (although those on low incomes could apply for rebates). The government's decision to CAP the tax in high-spending areas only compounded the problem, since bills had already been issued by many of the local authorities affected. *Community charge* is the official term used by the government and some local authorities; popularly, though, and in some literature issued by non-Conservative local authorities, it is known as *poll tax*. In April 1991, the government announced the result of its review of the *community charge*, which, it said, would be replaced after consultation by a property-based *council tax* by 1993.

> You don't pay the *personal* charge if you're ... a prisoner, unless you're inside for not paying the community charge or a fine. *Which?* Oct. 1989, p. 476

> This week's violent community charge agitation has sparked a dramatic resurgence in the fortunes of Militant Tendency and other Trotskyite groups. *The Times* 8 Mar. 1990, p. 5

compact disc 📀 see CD

compassion fatigue /kəmˈpæʃ(ə)n fəˌtiːg/ 🚩

A temporarily indifferent or unsympathetic attitude towards others' suffering as a result of overexposure to charitable appeals.

Formed by compounding: *fatigue* affecting one's capacity for *compassion*.

Compassion fatigue was first written about in the US in the early eighties, and at first was used mainly in the context of refugee appeals and the resulting pressure on immigration policy there. In the UK *compassion fatigue* was first mentioned when famines in Ethiopia in 1984-5 became the subject of graphic television appeals, followed by large-scale fund-raising events such as Band Aid (see -AID). It was feared that the British public could only stand the sight of so many starving children before 'switching off' emotionally to their suffering, but in the event the response to these appeals was good and it seemed that the issues most vulnerable to *compassion fatigue* were the ones generally perceived as 'old news'. The same effect on governmental agencies has been described as *aid fatigue*.

> Geldof, the Irish rock musician who conceived the event and spearheaded its hasty implementation, said that he 'wanted to get this done before compassion fatigue set in', following such projects as the African fund-raising records 'Do They Know It's Christmas?' and 'We Are the World'.
> *New York Times* 22 Sept. 1985, section 2, p. 28

> It is a chilling vision, a cataclysm. Compassion fatigue be damned. There is no doubt that we in Britain, without ceasing to wage our domestic battle against Aids, should be careful not to forget Africa, fighting its far more savage war.
> *Independent on Sunday* 1 Apr. 1990, Sunday Review section, p. 10

complementary /ˌkɒmplɪˈmɛntərɪ/ *adjective* 🏵

Of a therapy or health treatment: intended to complement orthodox medical practices; ALTERNATIVE, naturopathic. Also of a practitioner: not belonging to the traditional medical establishment.

A specialized application of *complementary* in its normal sense, 'forming a complement', the idea being that the alternative therapies do not compete with traditional medicine, but form a natural complement to it. This is the successor to the earlier and more dismissive 'fringe medicine', which saw these techniques as being on—or even beyond—the fringe of conventional medicine.

The term **complementary medicine** was coined by Stephen Fulder and Robin Munro in a report on the use of these techniques in the UK, published in 1982:

After extensive consideration of titles such as 'alternative medicine', 'fringe medicine' or 'natural therapeutics' we have decided to use the term '*complementary medicine*' to describe systems ... which stand apart from but are in some ways complementary to conventional scientific medicine.

Since then it has become very common, reflecting the change in public attitudes to these techniques during the decade (from 'fringe' or even 'quack' medicine to an accepted approach). Apart from *complementary medicine*, the adjective is used in **complementary therapist, complementary practitioner**, etc.

The Research Council for Complementary Medicine (RCCM) was set up to find research methods acceptable to both complementary and conventional practitioners.

Practical Health Spring 1990, pull-out section, p. 5

The plight of Mrs S wishing to fight cancer with complementary medicine before surgery ... but rejected for this reason by five doctors is sad indeed. She could no doubt be helped by more than one complementary therapy. *Kindred Spirit* Summer 1990, p. 38

computer-aided tomography, computer-assisted tomography ⊗ ▣ see CAT[1]

computerate /kəm'pju:tərət/ *adjective* ▣

Proficient in the theory and practice of computing; computer-literate.

Formed by combining *computer* and *literate* into a blend, taking advantage of the shared syllable *-ter-*. There was a precedent for this concept in the words *numeracy* and *numerate* (mathematically literate), which in the late fifties introduced the idea of a range of skills modelled on *literacy/literate*.

When computing skills became sought after in the job markets in the seventies, there was much discussion of *computer literacy* and the need to provide a general education which would produce *computer-literate* individuals. It was a short step from this metaphor to the blend *computerate*, which started to appear in the early eighties. The corresponding noun *computeracy* has been used colloquially since the late sixties, but also attained a more general currency during the eighties. A similar, but less successful, coinage is the punning adjective **computent**, competent in the use of computers (coined by Richard Sarson in the mid eighties), along with its corresponding noun **computence**.

Chapman and Hall are looking for a numerate and computerate person with publishing experience.
New Scientist 30 Aug. 1984, p. 59

Computeracy will not solve all your problems. headline in *Guardian* 28 Feb. 1985, p. 25

Andy's computence did not make him a philosopher or a captain of industry ... But he passed on some of his computence to me, for which I will always be grateful ... Computent Andy, illiterate and innumerate in the eyes of the educational system though he may be, has made me computent, and thereby more literate and numerate than I was. *The Times* 19 Apr. 1988, p. 33

computer-friendly ▣ see -FRIENDLY

computerized axial tomography ⊗ ▣ see CAT[1]

computer virus ▣ see VIRUS

condom /'kɒndɒm/, /'kɒndəm/ *noun* ⊗

A sheath made of thin rubber and worn over the penis during sexual intercourse, either to prevent conception or as a prophylactic measure.

Of unknown origin; often said to be the name of its inventor, although this theory has never been proved.

The word has been used in this sense in English since the early eighteenth century. It is included here only because it acquired a renewed currency—and a new respectability—in the language as a direct result of the spread of Aids in the 1980s. Whereas *sheath* or trade marks such as *Durex* were the only terms (apart from slang expressions) in widespread popular use in the UK immediately before the advent of Aids, it was *condom* that was chosen for repeated use in government advertising campaigns designed to explain the concept of SAFE SEX to the general public in the mid eighties. Soon the word had become so widespread that there were even reports of schoolchildren who had invented a new version of the playground game *tag* in which the safe area was not the 'den' but the *condom*. The pronunciation with full quality given to both vowels /'kɒndɒm/ belongs only to this twentieth-century use (in the past it had been pronounced /'kɒndəm/ or /'kʌndəm/, to rhyme at the end with *conundrum*) and possibly reflects the unfamiliarity of the word to the speakers of the government advertisements. In 1988 there was an attempt to introduce a *condom* for women to wear; meanwhile, the buying of the male version was presented very much as a joint duty for any Aids-conscious couple. This emphasis in advertising, as well as the generally permissive attitude to sexual relationships of any orientation in the eighties, led to the development of the nickname **condom culture**, used especially by those who favoured stricter sexual morals.

> More women should buy, carry and use condoms to help stop the spread of Aids, according to the organisers of National Condom Week, which starts today. The intention is to encourage people to get used to buying and carrying the contraceptives without embarrassment or inhibition.
>
> *Guardian* 7 Aug. 1989, p. 5

> The government has promoted a 'condom culture' of sex without commitment as part of a dismal record on support of family life, the National Family Trust claims today.
>
> *Daily Telegraph* 11 Aug. 1989, p. 2

> Everyone on the docks has ... condoms ... Pull a kid aside ... and he'll tell you he doesn't need them ... Does it *sound* to you like I need to put on a bag? *Village Voice* (New York) 30 Jan. 1990, p. 34

connectivity 🖾 see NEURAL

consumer terrorism 🚹 see TAMPER

contra /'kɒntrə/ *noun* Sometimes written **Contra** 🖺

A member of any of the guerrilla forces which opposed the Sandinista government in Nicaragua between 1979 and 1990; often written in the plural **contras**, these forces considered collectively.

An abbreviated form of the Spanish word *contrarrevolucionario* 'counter-revolutionary', probably influenced by Latin *contra* 'against'.

The word appeared on the US political scene at the very beginning of the eighties and became an increasingly hot issue in view of the US presidential administration's desire to aid the overthrow of the Sandinista government in Nicaragua. This reached its peak in the **Iran–contra** affair of 1986, when it was alleged that profits from US arms sales to Iran had been diverted to aid the *contras*, even though legislation had by then been passed to prevent any material aid from being sent; the ensuing Congressional hearings made the word *contra* known throughout the English-speaking world even if reporting of the long civil war in Nicaragua itself had not. Despite a plan agreed by Central American leaders in August 1989 to 'disband' the rebels, even the end of the Sandinista government after the elections in 1990 did not immediately bring an end to guerrilla activity from the *contras*.

> Oliver North, the ex-Marine colonel at the heart of the Iran–contra affair, whom Ronald Reagan dubbed 'a true American hero', was yesterday spared a prison term. *Guardian* 6 July 1989, p. 20

> The scenario clearly involved some kind of trade-off of contra aid and drugs and money.
>
> *Interview* Mar. 1990, p. 42

contraflow /'kɒntrəfləʊ/ *noun* 🏁

In the UK, a temporary traffic flow system (for example during carriageway repairs on a motorway) in which traffic is diverted on to the outer lane or lanes of the opposite carriageway, so that the carriageway which remains fully operational is in effect a temporary two-way road.

Contraflow has existed as a word meaning 'flow in the opposite direction' since the thirties; the traffic use is a specialized application of this sense.

The first *contraflow* systems on British roads—at least, the first to be called *contraflow*—appeared in the seventies. As the country's system of motorways began to age in the eighties, the *contraflow* became a seemingly ubiquitous sight and one was reported on radio traffic news almost every day. Sometimes *contraflow* is used on its own to signify the whole traffic-flow system; often, though, it is used attributively in **contraflow system**, etc.

> Resurfacing . . . has meant closing the northbound section and funnelling traffic into a contraflow system of two lanes each way on the southbound side. *The Times* 9 Apr. 1985, p. 3

> A spokesman said the contraflow was working smoothly at the time of the crash and visibility was good. *Daily Telegraph* 7 Sept. 1987, p. 4

Contragate 🔒 see -GATE

cook-chill /ˌkʊkˈtʃɪl/ *adjective and noun* 🔆

adjective: Of foods: sold in a pre-cooked and refrigerated form, for consumption within a specified time (usually after thorough reheating). Also in the form **cook-chilled**.

noun: The process of pre-cooking and refrigerating foods for reheating later.

Formed by compounding: the principle is first to *cook* and then to *chill* the food.

The system was invented as an offshoot of partially cooked frozen meals, and had become popular in institutional catering by the early eighties. The term was widely popularized in the UK in 1989, when there was an increase in cases of listeriosis thought to be caused at least in part by failure to store *cook-chill* foods correctly or reheat them thoroughly.

> The Department of Health has already advised people in at-risk groups not to eat cook-chill foods cold, and—if you buy one to eat hot—to make sure that it's reheated until it's 'piping hot'. *Which?* Apr. 1990, p. 206

core wars /ˈkɔː ˌwɔːz/ *plural noun* 🖳

In computing jargon, a type of computer game played by programming experts, in which the object is to design and run a program which will destroy the one designed and run by the opponent.

Formed by compounding; *core* is a reference to the old ferromagnetic *cores* which made up the memory elements of computers used in the fifties and sixties, before the advent of semiconductor chips. Active memory is still sometimes referred to as *core* memory, even in modern computers.

The 'sport' of *core wars* originated among computer scientists at Bell Laboratories in the US in the late fifties and sixties and was originally the proper name of a program developed by the computer-games group there. It was popularized in the US in the mid eighties, probably as a more respectable offshoot of the interest in mischievous programs such as the computer VIRUS and WORM and in defensive programming techniques which could be used to protect software from attack. By 1986 it had been raised to the level of international competition, but remains a minority interest.

> Robert Morris Sr. . . . . played a game based on a computer virus over 40 years ago . . . Called Core Wars, the game centered around the design of a program that multiplied and tried to destroy other players' programs. *Personal Computing* May 1989, p. 92

corn circle 🌱 see CROP CIRCLE

cornflakes 🗡 see ANGEL DUST

corn-free ✖ see -FREE

corpocracy /kɔːˈpɒkrəsɪ/ *noun* 〰

Corporate bureaucracy: bureaucratic organization in large companies (or in a particular company), especially when excessively hierarchical structures lead to overstaffing and inefficiency. Such companies are described as **corpocratic**; a director of one is a **corpocrat**.

Formed by combining the first two syllables of *corporate* with the last two of *bureaucracy* to make a blend.

The word was coined by American economist Robert Heller in his book *The Common Millionaire* (1974), but was still sufficiently unfamiliar in the mid eighties for John S. Berry and Mark Green to present it as a new coinage in *The Challenge of Hidden Profits: Reducing Corporate Bureaucracy and Waste* (1985). In the UK the word—although not the phenomenon—was popularized by financier Sir James Goldsmith. *Corpocracy* was presented as an important reason for the uncompetitiveness of British and American businesses during the eighties.

> It doesn't believe much in hierarchy, rule books, dress codes, company cars, executive dining rooms, lofty titles, country club memberships or most other trappings of corpocracy.
>
> *Forbes* 23 Mar. 1987, p. 154

> Such a complete change of direction is not likely to be welcomed by directors who I would describe as complacent or entrenched in their current 'corpocratic' culture.
>
> Sir James Goldsmith in *First*, 3.3 (1989), p. 18

corporate makeover 〰 see MAKEOVER

couch potato /ˈkaʊtʃ pəˌteɪtəʊ/ *noun* ✖ 📺

In slang, a person who spends leisure time passively (for example by sitting watching television or videos), eats junk food, and takes little or no physical exercise.

Formed by compounding; a person with the physical shape of a *potato* who spends as much time as possible slouching on the *couch*. The original humorous coinage by Californian Tom Iacino relied on a pun: because of their love for continuous viewing of the television (known in US slang as the *boob tube*, unlike British slang, which uses the term for a skimpy stretch bodice), these people had formerly been called *boob tubers*; for their emblem, cartoonist Robert Armstrong therefore drew the best known *tuber*—a *potato*—reclining on a *couch* watching TV, formed a club called *The Couch Potatoes*, and later went on to register the term as a trade mark.

The US trade mark registration for the term *couch potato* claims that it was first used on 15 July 1976. Robert Armstrong (who is really responsible for popularizing the term and maintaining the cult) has claimed that this coinage was not his, attributing it instead to Tom Iacino, another 'Elder' of the cult, who used it when asking to speak to a fellow Elder (known only as 'The Hallidonian') on the telephone. The Couch Potatoes club which Armstrong formed aimed to raise the self-esteem of *tubers*, and provided a counterbalance to the cult of physical fitness which was by then a dominant influence in American society. With the growth of the domestic video market, the *couch potato* cult became very popular during the eighties and resulted in much merchandising—*couch potato* teeshirts, dolls, stationery, books, etc. designed to promote pride in the *tuber* culture. Many variations on the term developed too: the obvious **couch potatoing** and **couch potatodom** and a whole range of words based on *spud*, such as *vid spud*, *telespud*, *spud suit*, and *spudismo*. With the coining of the trend analyst's term COCOONING in 1986, *couch potatoes* felt that their way of life was being officially recognized; however, a National Children and Youth Fitness Study carried out in the US in 1987 made it clear that it was not to be officially condoned, criticizing parents for not getting children to take outdoor exercise and for raising a nation of *couch potatoes*. The *couch potato* concept and merchandising reached the UK in the late eighties, although the lifestyle had existed without a name for some time before that.

> Though Mr. Armstrong's brainchild has yet to make him rich, he is still undaunted, spreading the Couch Potato gospel: 'We feel that watching TV is an indigenous American form of meditation. We call it "transcendental vegetation".'   *Parade* 3 Jan. 1988, p. 6

> The economy could be thrown into recession because of the couch potato's penchant for staying home with the family, watching TV and munching on microwave popcorn.   *Atlanta* Oct. 1989, p. 61

council tax 〰 ⬛ see COMMUNITY CHARGE

counter-culture /'kaʊntəkʌltʃə(r)/ *noun* Also written **counter culture** or **counterculture** ⬛ ⬛

A radical, ALTERNATIVE culture, especially among young people, that seeks out new values to replace the established and conventional.

Formed by adding the prefix *counter-* (an anglicized form of the Latin *contra* 'against') to *culture*: something that rebels *against* established *culture*.

The *counter-culture* has, in a sense, always been with us, since the younger generation in each succeeding age rebels against the values of its parents and tries to establish a new lifestyle; but the word *counter-culture* was first used in the US to describe the hippie culture of the sixties by those who looked back on it from the end of the decade. The concept was popularized by Theodore Roszak in his book *The Making of a Counter-Culture* (1969). *Counter-culture* has come to be used especially to refer to any lifestyle which attempts to get away from the materialism and consumption of the post-war Western world; in the eighties, it has tended to give way to the word ALTERNATIVE, especially in British English. A follower of the *counter-culture* is a **counter-culturalist**.

> The counter-culture ponytail is gone, sacrificed to the heat of arena lights and the sizzling sweat of the fast-break pace.   *Time* 30 May 1977, p. 40

> It was the counter-culture, the alternative society, a middle-class movement, an explosion of creative energy, a bunch of unwashed, stoned-out air heads.   *Observer* 23 Oct. 1988, p. 43

> The fact that so many counter-culturalists have now cut their hair ... and ... become green 'rainbow warriors', is a point which seems to have been overlooked.   *Films & Filming* Mar. 1990, p. 50

courseware ⬛ see -WARE

Cowabunga /ˌkɑːwəˈbʌŋgə/, sometimes pronounced /ˌkaʊə-/ Originally written **kowa-bunga** or **Kawabonga**; now also **cowabunga** *interjection* ⬛

In young people's slang (originally in the US), an exclamation of exhilaration or satisfaction, or sometimes a rallying cry to action: yippee!, yahoo!, yabbadabba doo!

The word was originally used in the fifties (in the form *kowa-bunga* or *Kawabonga*) as an exclamation of anger by the cartoon character Chief Thunderthud in *The Howdy Doody Show*, written by Eddie Kean. By the sixties, it had entered surfing slang as a cry of exhilaration when riding the crest of a wave. Since the surfers of the sixties had been the children for whom *The Howdy Doody Show* was written, it is easy to see how the word made this transition; it is less clear how Eddie Kean came upon it. Chief Thunderthud used the expression when annoyed, or if something went wrong; when things went well, he said *Kawagoopa*. Although Thunderthud was meant to be an American Indian, there had been early speculation that *cowabunga* might come from the Australian or South Seas surfing world; interestingly, *kauwul* is recorded as an aboriginal word in New South Wales for 'big', *bong* for 'death', and *gubba* for 'good', but this is surely no more than a curious coincidence.

As mentioned above, *Cowabunga* was in use as an exclamation among Californian surfers by the sixties. It reached a wider audience through a series of films about a surfer called *Gidget* in the sixties, through its use by the cookie monster in the children's television series *Sesame Street* in the seventies, and more particularly from 1990, when it was taken up as the rallying cry of the Teenage Mutant TURTLES. In the book of *Teenage Mutant Ninja Turtles: the Movie*, the turtles are searching for a suitable cry:

> They turned to Donatello, who struggled to come up with the perfect word to describe their exploits. But Donatello was at a loss. His brothers continued to top each other: 'Tubular!' 'Radical!' 'Dyna-mite!' At last Splinter raised a finger and brought an end to the debate. 'I have always liked', he said quietly, 'cowabunga.' The turtles stared at him, grinning, then laid down high-threes all around. 'Cow-a-*bung*-a!' they cried in unison. And the battle-cry was born.

The word soon crossed the Atlantic as part of turtlemania, with the result that one could hear the cry of 'Cowabunga, dudes!' from British children apparently unaware that, as far as their parents were concerned, they were speaking a foreign language.

> 'Hey, Mike, I didn't know that you could drive!' 'Me neither . . . cowabunga!'
> *Teenage Mutant Hero Turtles* 10–23 Feb. 1990, p. 20

> Marketers are betting that youngsters will have the same reaction as American kids: Cowabunga!
> *Newsweek* 16 Apr. 1990, p. 61

crack /kræk/ *noun*

In the slang of drug users, a highly addictive, crystalline form of cocaine made by heating a mixture of it with baking powder and water until it is hard, and breaking it into small pieces which are burnt and smoked for their stimulating effect.

The name arises from the fact that the hard-baked substance has to be *cracked* into small pieces for use, as well as the *cracking* sound the pieces make when smoked.

The substance itself first came to the attention of US drug enforcement agencies in 1983, but at that time was generally known on the streets as ROCK or FREEBASE. The name *crack* appeared during 1985 and by 1986 had become established as the usual term, both among drug users and by the authorities; since 1988, the fuller term **crack cocaine** has tended to replace *crack* alone in official use. *Crack*'s appearance on the US drug market coincided with a marked rise in violent crime, testifying to its potency and addictiveness, with users prepared to go to almost any lengths to get more. The word *crack* quickly became the basis for compounds, notably **crack-head** (in drugs slang, a user of *crack*) and **crack house** (a house where *crack* is prepared or from which it is sold). The phrasal verb **crack (it) up** has also acquired the specialized meaning in drugs slang of smoking *crack*.

> In New York and Los Angeles drug dealers have opened up drug galleries, called 'crack houses'.
> *San Francisco Chronicle* 6 Dec. 1985, p. 3

> 'Crack it up, crack it up,' the drug dealers murmur from the leafy parks of the suburbs to New York City's meanest streets. *Time* 4 Aug. 1986, p. 27

> Charlie and two fellow 'crackheads' took me to a vast concrete housing estate in South London where crack is on sale for between £20 and £25 a deal. *Observer* 24 July 1988, p. 15

> Some crack users [in Washington DC], unable to work for a living, will go out with a lead pipe or a bat and hit defenceless women. *Japan Times* 19 May 1989, p. 20

See also WACK

cracking   see HACK

crank /kræŋk/ *verb*

In the slang of drug users in the UK: to inject (a drug). Often as a phrasal verb **crank up**.

A figurative use of the verb which normally means 'to start a motor by turning the crank'; a synonym in drugs slang for *jack (up)*, which follows a similar type of metaphor.

A word which has been used by drug users in the UK since about the beginning of the seventies, *crank* seems to be a rare example of a piece of drugs slang which is exclusively British. US drug slang has *crank* as a noun for methamphetamine and *cranking* for repeated use of methamphetamine, but the verb is apparently not used at all. In Britain, it is normally used in the context of heroin injection.

> 'Where do you inject?' 'Me feet, me arms, me hands.' 'Would you give up cranking?' 'No, it's the needle I'm into.'
> *Sunday Telegraph* 29 Oct. 1989, p. 1

creative /kriːˈeɪtɪv/ *adjective* 〰

Used euphemistically in the language of finance: exploiting loopholes in financial legislation so as to gain maximum advantage or present figures in a misleadingly favourable light; ingenious or inventive.

A figurative extension of meaning: *creative* had been used of writing that was inventive or imaginative since the early nineteenth century, and in context frequently meant no more than 'fictional'. The *creative* accountant's task is to interpret the figures imaginatively, with the result that a largely fictional picture of events is often presented.

Used in the business world (especially in **creative accountancy** or **creative accounting** since the early seventies, the euphemism was popularized in the mid eighties, when it was rumoured that the technique had been used in presenting both central and local government figures. At this time *creative accounting* also became the subject of a number of books published for people running small businesses or working on their own.

> Mr Nicholas Ridley, the Secretary of State for the Environment, is today expected to warn high spending councils that he is ready to take tough new action to stamp out 'creative accounting'.
> *The Times* 21 Nov. 1986, p.

cred¹ /krɛd/ *noun* 🗯

In young people's slang: credibility, reputation, peer status.

Formed by abbreviating *credibility* to its first syllable.

The emphasis on *cred* in the early nineties arises from the concept of *street credibility* which developed at the very end of the seventies. *Street credibility* (which by the early eighties was being abbreviated to *street cred*) originally involved popularity with, and accessibility to, members of the urban street culture, who were seen as representing ordinary people. Before long, though the term had come to mean familiarity with contemporary fashions—or the extent to which a person was 'hip'. Once the concept was established, the word *street* was often dropped, leaving *cred* alone.

> 'Cred' was achieved by your rhetorical stance and no one had more credibility than the Clash.
> Bob Geldof *Is That It?* (1986), p. 12

> 'They've got to have total cred,' Boxall insisted, when listing the special qualities he is looking for [in a magazine editor].
> *Sydney Morning Herald* 1 Feb. 1990, p. 2

cred² /krɛd/ *noun* 〰

In colloquial use (originally in the US): financial credit.

Formed by abbreviating *credit* to its first syllable.

A natural development in view of the boom in the use of credit facilities during the late seventies and eighties. Also used in combinations, especially **cred card**.

> Neat trick, eh? Cash and cred all in one bundle.
> *The Face* Jan. 1989, p. 6

credit card 〰 see CARD¹

crew /kruː/ *noun* 🗯

In HIP HOP culture, a group of rappers, break-dancers, graffiti artists, etc. working together as a team. Also, loosely, one's gang or POSSE.

A specialized use of *crew* in the sense of 'a body or squad of people working together', which goes back to the seventeenth century. In this case, there is probably a conscious allusion to the Rock Steady Crew: see BREAK-DANCING.

Originally used mainly of groups of rappers (from about 1982 in the US), the term was soon applied to street groups using other hip-hop forms of expression such as break-dancing and graffiti (see TAG²) and by the end of the decade had been adopted more generally by groups of youngsters.

> To kids out of the South Bronx and Harlem, what the top crews make is big bucks. For a one-night gig ... a dancer takes home $150 to $300.   *Village Voice* (New York) 10 Apr. 1984, p. 38

> He and four friends, members of a crew of graffiti artists who call themselves the L.A. Beastie Boys, gathered at the park.   *Los Angeles Times* 22 Oct. 1987, section 10 (Glendale), p. 1

crop circle /'krɒp ˌsɜːk(ə)l/ *noun* 🎵

A (usually circular) area of standing crops which has been inexplicably flattened, apparently by a swirling, vortex-like movement.

Formed by compounding: a *circle* of flattened *crop*.

The puzzling phenomenon of *crop circles* (sometimes also called **corn circles**) has been perplexing scientists for about a decade. Since the early eighties increasing numbers of circles and other patterns have been reported in areas as far apart as the South of England, the farming belt of the US, and Australia, often appearing overnight. A number of theories—ranging from meteorological changes or fungi to alien spaceships or the activity of hoaxers—have been put forward to explain them, but none has been conclusive.

> They are the result not of the supernatural but of an everyday, common garden variety of fungi, according to biologists Mr Michael Hall and Mr Andrew Macara, who have been conducting a study into the crop circle conundrum.   *Sunday Telegraph* 11 Mar. 1990, p. 5

> Could the enormous increase in the perplexing crop circles be anything to do with the Earth's vital energies?   *Kindred Spirit* Summer 1990, p. 26

crossover /'krɒsəʊvə(r)/ *noun* and *adjective* Sometimes written cross-over 🎵 🎻

noun: The process of moving from one culture (or especially from one musical genre) to another; something or someone that has done this (specifically, a musical act or artist that has moved from a specialized appeal in one limited area of music into the general popular-music charts).

adjective: (Of a person) that has made this transition from one culture or genre to another; (of music, an act, etc.) appealing to a wide audience outside its genre, sometimes by mixing musical styles.

The noun is formed on the verbal phrase *cross over* and has been used in a number of specialized senses in English since the eighteenth century. The cultural sense here is perhaps in part a figurative application of the genetic *crossover* (one of the word's specialized senses, in use since the early years of this century), in which the characteristics of both parents are displayed as a result of the *crossing over* of pairs of chromosomes.

Since the sixties, *crossover* has been used in politics (especially in the US) in relation to the practice or tactic of switching votes from the party with which one is registered to another party—for instance in a State primary. Within the music industry *crossover* was being used by the mid seventies in relation to records in the country charts which were tending to *cross over* into popular music generally, and it was not long before this process became more generalized, for example as various Black sounds acquired a more general appeal to White audiences. In the eighties, *crossover* was one of the favourite words of the music industry and there was plenty of scope for its use, as soundtracks from films and television series increasingly figured in the charts and the big names of classical music ventured into middle-of-the-road and easy

CRUCIAL

listening recordings. In the broader cultural context sociologists use *crossover* to refer to the way in which people from one ethnic background consciously leave their ROOTS culture for another, more prestigious one; this has led to an extended use of *crossover* in relation to fashion, as ETHNIC cultures acquired high prestige and became fashionable in Western society. Other extended uses of the word included actresses crossing over from theatre to films and even a supermarket which had gone over to wholefoods to cash in on the new green culture of the late eighties.

> 'I think the crossover has already started happening', says Salman Ahmed. 'This year I've noticed a lot of white and coloured kids at the shows . . .' Within the world of bhangra there are mixed reactions to the idea of crossover.
> *Sunday Telegraph Magazine* 22 May 1988, p. 38

> It showed the group making the crossover from deft-but-faceless R&B outfit to 'far out' funkers.
> *Q* Dec. 1989, p. 169

> Blame prefigured what fashion mood critics would soon call 'crossover culture'—the white mainstream's fresh infatuation with black style.
> *Vogue* Sept. 1990, p. 87

crucial /ˈkruːʃ(ə)l/ *adjective*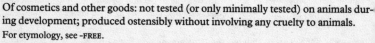

In young people's slang: very good or important, great, fantastic.

An example of the way in which meaning is weakened and trivialized in the idiom of young people: compare ACE, AWESOME, and RAD.

Crucial belongs to the slang usage of the very young (largely the pre-teenage group) in the late eighties. It was popularized especially by children's television presenters and other media personalities, notably the comedian Lenny Henry, who devoted a whole book to the subject. As often happens with such slang words, the respectability which *crucial* gained by being used in print caused it to go out of fashion rather among the youngsters who were using it.

> Martha (aged seven): 'Lenny Henry, he wrote the "guide to cruciality", so we don't say crucial no more.'
> *New Statesman* 16 Feb. 1990, p. 12

> The very latest buzz-word, after last year's favourite sayings like 'mental, mental', 'crucial' and 'wicked', is 'raw'.
> *Daily Star* 20 Mar. 1990, p. 13

> I have worn out three sets of trainers running around telling my friends how crucial Young Eye is.
> *Private Eye* 26 Oct. 1990, p. 21

cruelty-free /ˌkruəltɪˈfriː/ *adjective*

Of cosmetics and other goods: not tested (or only minimally tested) on animals during development; produced ostensibly without involving any cruelty to animals.

For etymology, see -FREE.

This is a term which started to appear in the late eighties as a natural consequence of the increasingly well-publicized animal liberation movement—a movement whose arguments seemed to get a more sympathetic hearing once green views in general became acceptable. *Cruelty-free* often appears on the labels of cosmetics, pharmaceuticals, and other everyday products which have hitherto been routinely tested on laboratory animals but are now produced without actual cruelty (although the interpretation of 'actual cruelty' evidently still varies); vegetarians also sometimes use it to refer to animal-free food products.

> Mary Bonner showed over 50 people how enjoyable a cruelty-free Christmas can be with her celebration roast, mushroom stuffing and red wine sauce, vegan Christmas Cake and mince pies.
> *Vegetarian* Mar./Apr. 1988, p. 42

> Pamphlets that bring news of . . . where they can purchase 'cruelty-free' soaps and shampoos.
> *Forbes* 20 Mar. 1989, p. 44

crumblie /ˈkrʌmblɪ/ *noun*

In young people's slang: an old or senile person (older than a WRINKLIE).

Formed by treating a figurative sense of the adjective as a noun; the metaphor relies on the assumption among the young that all elderly people will eventually 'crack up' and become senile. This process of *crumbling*, they suppose, is the natural next step after going *wrinkly*.

Used mainly by children and teenagers from about the late seventies, and apparently limited to British English.

> The growing fashion among teenagers is to describe their parents as 'wrinklies' and their grandparents as 'crumblies'. A reader, however, tells me how she countered this when . . . she described her own children, in their earshot, as 'pimplies'. *Daily Telegraph* 26 Jan. 1987, p. 17

cryo- /ˈkraɪəʊ/ *combining form*

Widely used in compounds relating to extreme cold, especially when this is an artificial means of preserving tissue.

From the Greek *kruos* 'frost, icy cold'.

Early words formed with this combining form concerned temperatures not much below the freezing point of water. However, as it became possible to create lower and lower temperatures artificially, *cryo-* came to be associated with the sort of intense cold that could only be achieved with the aid of 'cold-creating' or **cryogenic** equipment, such as apparatus for liquefying nitrogen or other gases. During the sixties and seventies the creation of such temperatures began to find applications in electronics and surgery: below a certain point some materials become *superconductors*, that is to say they lose all electrical resistance, which makes them very useful in a wide range of applications (in BRILLIANT PEBBLES, for example), while **cryosurgery** uses intense cold to remove or destroy tissue just as effectively as heat. Until the late seventies **cryonics** (or **cryopreservation**), the use of extreme cold to preserve living tissue, had remained at an experimental stage because of the tendency of water to expand when frozen—making the formation of ice crystals within living cells lethally damaging. However, study of the few animals which can survive freezing led to the development of substances which circumvent some of the problems (**cryoprotectants**). During the eighties it became possible to **cryopreserve** an increasingly wide range of tissues for future use: sperm may be stored in a **cryobank**, and frozen embryos may now be thawed out for **cryobirth**. The lack of any reliable means of freezing and thawing the entire human body without severe damage has not prevented **cryonicists**, mostly on the West coast of the US, from setting up businesses offering **cryonic suspension** to those willing to pay for it, especially the incurably ill (who may wish to be 'thawed out' when a treatment for their condition arrives).

> Once a month, she goes to the Southern California Cryobank, a commercial sperm bank in Los Angeles, pays $38 for a syringe of sperm packed in dry ice, which she either takes back to the health center for insemination, or takes home. *New York Times* 20 July 1980, section 6, p. 23

> Still others call for these pre-embryos to be cryopreserved—frozen for months, years and perhaps indefinitely. Once the pre-embryos are thawed out, they can be used as if they were fresh.
> *Washington Post* 12 Apr. 1988, section Z, p. 14

> Cryonicists . . . talk . . . of storing the brains of the frozen hopeful in the bodies of anencephalic babies. *Independent* 1 Aug. 1988, p. 13

> Mr Thomas Donaldson, 46, wants his head cryonically suspended in the anticipation that a way will be found to attach it to a healthy body and cure his brain disorder. *Daily Telegraph* 3 May 1990, p. 12

> A mathematician from Sunnyvale, California, has filed a lawsuit in America for the right to 'cryonic suspension' before death. *The Times* 27 Oct. 1990, p. 3

crystal healing /ˌkrɪst(ə)l ˈhiːlɪŋ/ *noun*

An ALTERNATIVE therapy popular in NEW AGE culture and based on the supposed healing power of pulsar crystals. Sometimes also called **crystal therapy** or **crystal treatment**.

Formed by compounding: *healing* by *crystals*.

The idea of harnessing the healing power which—according to the **crystal healer**—emanates from some crystals is not new: its supporters claim that it goes back to the practices of the ancient Greeks. However, it only gained any real popularity with the rise of the New Age movement in California. By the end of the eighties this idea had spread outside the US to other English-speaking countries but was still regarded by many as being on the fringe of serious healing.

> For the esoteric set, crystal healing, extraterrestrials and transchanneling will be summer pursuits
> *Los Angeles Times* 29 May 1987, section 5, p.

> Ben says something called crystal healing is one of the new fads brought in by what he calls 'weirdos' from the United States. *Sunday Mail Magazine* (Brisbane) 10 Apr. 1988, p. 1

crystal meth ✗ see ICE

CT ✗ ▨ see CAT[1]

cursor /'kɜːsə(r)/ *noun* ▨

A distinctive symbol on a computer screen (such as a flashing underline or rectangle which shows where the next character will appear or the next action will take effect and which can usually be moved about by using a **cursor key** on the keyboard or a MOUSE.

From Latin *cursor* 'runner' (the agent-noun formed on the verb *currere* 'to run'). When first used in English (until the middle of the seventeenth century) the word meant a runner or messenger; it then came to be used for a part of a mathematical instrument etc. that moved backwards and forwards (for example, the transparent slide with a hair-line which forms part of a slide-rule). It was a logical step to its present use in the computer age, since it is the *cursor* which 'runs' round the screen.

The first uses of the word *cursor* in computer technology are associated with the development of a MOUSE in the mid sixties, although the idea had been invented (and described using other names such as *marker*) by John Lentz of IBM in the fifties. Even though the *cursor* had first been thought of in connection with mouse technology, the principle of having a *cursor* which was controlled using keys on the keyboard was well-established in home computing in the late seventies, before windows and mice (see WIMP[2]) became widespread. With the increased popularity of home computing and word-processing in the eighties, *cursor* has passed from the technical vocabulary into everyday currency.

> Cursor movement is particularly important in word processing, and well laid-out cursor keys are a real boon. Susan Curran *Word Processing for Beginners* (1984), p. 31

> For home use you may not mind if the cursor is a bit slow to move on occasions.
> *Which?* Nov. 1988, p. 524

cuss ▨ see DISS

cutting edge ▨ see LEADING EDGE

cyberpunk /'saɪbəpʌŋk/ *noun* Sometimes written **Cyberpunk** ▨

A style of science fiction writing combining high-tech plots (in which the world is controlled by artificial intelligence) with unconventional or nihilistic social values. Also, a writer of (or sometimes a character in or follower of) cyberpunk.

Formed by combining the first two syllables of *cybernetics* (the science of control systems) with *punk* (probably as an allusion to the hard, aggressive character of punk music, with which *cyberpunk* has much in common, particularly in its harshness and deliberate attempt to shock).

Although only a few years old, *cyberpunk* has grown into a leading genre of science fiction. The word may have been coined by Gardner Dozois to describe the work of a number of writers in the mid eighties, notably William Gibson and Bruce Sterling. William Gibson's book *Neuromancer* (1984) is seen as a foundational influence; so much so, in fact, that another name for the writers of this type of fiction is *Neuromantics*. They have also been called *outlaw technologists* or the *mirror-shades group*, while the genre has been called *technopunk* or *radical hard SF* as well as *cyberpunk*. Outside the world of science fiction only *cyberpunk* has been widely popularized, especially as a result of the television adaptation of *Neuromancer*, *Max Headroom*. In 1991 *Cyberpunk* was the title of Peter von Brandenburg's documentary film on the genre, which itself used some of the techniques characteristic of *cyberpunk* writing.

> The purveyors of bizarre, hard-edged, high-tech stuff, who have on occasion been referred to as 'cyberpunks'... They are the '80s generation. *Washington Post* 30 Dec. 1984, p. 9

> It's the Rhetoric of the New. Pitched somewhere between the SF genre of cyberpunk and the mainstream brat novel. *Listener* 4 May 1989, p. 29

· ·

D

dairy-free see -FREE

daisy chain[1] /'deɪzɪ ˌtʃeɪn/ *noun and verb* ∿

noun: In financial jargon, a string of buyers who concentrate their dealings on a particular stock in order to raise its price artificially.

transitive verb: To raise (prices) artificially in this way.

A specialized use of the figurative sense of *daisy chain*, which has been used as a noun since the middle of the last century to refer to any linking together of people or things in the fashion of a real daisy chain.

A practice which began with strings of traders in crude oil who bought and sold to each other on paper in the seventies, the *daisy chain* became a shady and only semi-legal activity on the wider market in the mid eighties. The conspirators make a show of activity in their chosen market, thereby pushing up the price and attracting unsuspecting investors. They then pull out, leaving the new investors with overpriced stock. Most countries have tried to curb the practice legally.

> They have been buying crude from resellers who illegally inflated the prices and supplying products to brokers whose only function was to 'daisy chain' the prices.
> *Washington Post* 31 May 1979, section A, p. 11

> Can order be brought to the daisy chain market? *The Times* 19 Feb. 1986, p. 17

> Lincoln traded junk bonds with other daisy chain members at 'artificial and escalating prices so that both parties could recognize artificial and improper profits', the suit said.
> *Los Angeles Times* (Orange County edition) 10 Feb. 1990, section D, p. 11

daisy chain[2] /'deɪzɪ ˌtʃeɪn/ *transitive verb* 🖥

To link (computers and other electronic devices used with them) to each other in series, forming a chain which is connected to a single controlling device.

Daisy chain had come to be used as a verb meaning 'to join things together in the manner of a daisy chain' during the middle years of the century; the computing sense is a specialization of that use.

> Occupying a full-size slot, each SCSI device lets you daisy-chain other devices to it.
> *PC World* Oct. 1989, p. 80

> Twenty or more players can be daisy-chained to one card. *Guardian* 18 Jan. 1990, p. 29

daisy wheel /ˈdeɪzɪ ˌwiːl/ *noun* Also written **daisy-wheel** or **daisywheel** 🖳

A removable printing unit in some computer printers and electronic typewriters, consisting of a disc of spokes extending radially from a central hub, each spoke having a single printing character at its outer end.

Formed by compounding: a *wheel* which in some ways resembles a *daisy* with its radiating 'petals'.

The *daisy wheel* type of printer was introduced in the late seventies and proved a popular alternative to dot-matrix printing in cases where clear, typewriter-like quality was needed. The wheel revolves to position the next character in front of a single hammer (a process which in the early machines was both slow and noisy, although this was improved in later models). The wheels are removable, allowing a number of different scripts or founts to be used on a single printer, but only text can be printed (a limitation which does not apply to the cheaper, poorer-quality dot-matrix or the more expensive, top-quality laser printers—both can also print graphics such as charts and graphs).

> As I write, an IBM word processor with daisywheel sits malevolently waiting for me in a customs shed. Anthony Burgess *Homage to QWERTYUIOP* (1986), p. xii

damage limitation /ˈdæmɪdʒ lɪmɪˌteɪʃ(ə)n/ *noun* 📄

The action or process of minimizing the damage to one's cause (usually a political one) after an accident, mistake, etc. has occurred. Also sometimes called **damage control**.

Formed by compounding.

The term *damage limitation* was first used in the mid sixties to refer to a policy in US politics of planning for the disaster of nuclear war, so as to have mechanisms in place for minimizing the damage to the US of a first strike by the enemy; *damage control* originated in international shipping law and later came to be used figuratively in politics. Both terms were applied in new contexts in the eighties as a series of political scandals and mistakes involving individual politicians or whole parties threatened to affect the polls unless **damage-limiting** measures were taken.

> The meeting decided to put Lord Whitelaw in charge of a 'damage limitation' exercise. Part of this would be a speech by Mrs Thatcher distancing the government from the [Channel] tunnel.
> *Economist* 14 Feb. 1987, p. 19

daminozide 🌱 see ALAR

DAT /dæt/ or /diːeɪˈtiː/ *acronym* Also written **dat** ✖ 🖳

Short for **digital audio tape**, a kind of audio tape on which sound is recorded digitally, equivalent in quality to a DIGITAL recording on CD. Also, a piece or cassette of digital audio tape.

An acronym, formed on the initial letters of *Digital Audio Tape*.

Digital audio tape was developed experimentally at the beginning of the eighties and had started to be called *DAT* outside technical trade sources by 1985. It was widely used in recording studios as a convenient form of high-quality master tape. However, when commercial production was first talked about in the mid eighties there was near panic among some record producers (called **DATphobia** by one music paper), since *DAT* was expected to pose a considerable threat to the growing compact disc market, and to be much more difficult to protect from copying and piracy. After a lull in the late eighties, the word came back into the news in 1990 as companies talked of making *DAT* commercially available in 1991.

> Compact Discs have been marketed as the ultimate in sound. If DAT allows you to copy CDs ... with absolutely no loss in that quality, where does this put the major record houses currently investing sharp-intake-of-breath sized sums on CD pressing plants? *Q* Oct. 1986, p. 18

The introduction of DAT has been bitterly fought here by record companies fearing unstoppable competition to compact discs.   *Music & Musicians International* Feb. 1988, p. 14

During a visit to Japan a year or so ago, I was convinced the year for consumer DAT is '91. I still believe that to be the case.   *Music Week* 23 June 1990, p. 4

data capture 🖳 see CAPTURE

Data Discman 🖳 see WALKMAN

data massage 〰️ 🖳 see MASSAGE

data tablet 🖳 see TABLET

dawn raid /ˌdɔːn ˈreɪd/ *noun* 〰️

In financial jargon, a swift buying operation carried out at the beginning of the day's trading, in which a substantially increased shareholding is obtained for a client, often as a preliminary to a take-over.

A figurative use of a compound which comes originally from military contexts but had become something of a journalistic cliché in reports of police operations during the twentieth century: the media often reported that a *dawn raid* had been carried out on a house occupied by suspected drug dealers or other criminals.

A phenomenon which began at the very beginning of the eighties, the *dawn raid* offers a 'predator' company the chance to take an intended victim by surprise, and is therefore a popular preliminary to a take-over. The proportion of shares which may be bought up in this way by a **dawn raider** has been successively limited during the eighties so as to give a fairer chance to the target company.

Market lethargy has brought out the dawn raiders again, despite the recent stock exchange report on such practices.   *Economist* 26 July 1980, p. 84

Its shares rose 14p to 235p, 5p below the new terms, as Blue Circle picked up a 29.5 per cent stake in a dawn raid on the stock market.   *Guardian* 3 Aug. 1989, p. 11

ddI /diːdiːˈaɪ/ *abbreviation* Also written **DDI** ⊗

Short for **dideoxyinosine**, a drug which has been tested for use in the treatment of AIDS.

The initial letters of *Di-*, *Deoxy-*, and *Inosine*.

The compound *dideoxyinosine* was first synthesized in the mid seventies in connection with cancer research; in the late eighties it was suggested that it should be tried as an alternative to AZT (ZIDOVUDINE) in treating people with Aids. It was successfully tested in clinical trials in the US in 1989 and trials in the UK followed in 1990. Like AZT, *ddI* prevents the Aids virus HIV from replicating itself within the body.

Almost 20 times as many people have flocked to free distributions of the new drug DDI than have signed up for the clinical trial.   *New York Times* 21 Nov. 1989, section A, p. 1

The UK trial of ddI will be accompanied by a similar trial in France.   *Lancet* 10 Mar. 1990, p. 596

DDI may offer an alternative treatment to the many people with AIDS who cannot tolerate zidovudine.   *New Scientist* 26 May 1990, p. 32

deafened /ˈdɛfənd/ *adjective* ⊗ 🍴

Of a person: having lost the faculty of hearing (although not deaf from birth) to such an extent as to have to rely on visual aids such as lip-reading in order to understand speech. The corresponding noun for the state of being deafened is **deafenedness**.

A specialized use of the adjective, which has existed since the seventeenth century in the more

general sense 'deprived of hearing', but has usually referred to temporary deafening (as, for ex
ample, by a loud noise).

The distinction between the *deaf* (who have never been able to hear) and the *deafened* (who los
their hearing after having acquired normal language skills) has been made in medical literature
for some time, often with an adverb making the situation absolutely clear, as *pre-lingually dea*
and *post-lingually deafened*. In popular usage, though, *deaf* has tended to serve both functions
as well as being used frequently to mean 'hard of hearing' (for which the official term is now
hearing-impaired). The term *deafened* was brought into wider usage—partly as an attempt to
alert the public to this important distinction and make them aware of the special problems of
the *deafened*—by the formation of the National Association for Deafened People in 1984.

> Deafened people share many problems with those born deaf, but there is a gulf between us in terms
> of lifestyle.
> *Good Housekeeping* Sept. 1986, p. 4!

> Lip-reading ... confounds crucial distinctions between the hard of hearing, the profoundly deafene
> and the pre-lingually profoundly deaf. The hard of hearing and the deafened have ... been ... sup
> porters of oralism; and the born deaf have retaliated by speaking as if they alone were the true deaf
> *Independent* 16 May 1989, p. 1!

death metal ♪ 👾 see THRASH

debit card 〰 see CARD¹

debrezhnevization 📷 see DECOMMUNIZE

debt counselling /'dɛt ˌkaʊnsəlɪŋ/ *noun* Written **debt counseling** in the US 〰 🍴

Professional advice and support provided for those who have fallen into debt and are
unable to meet their financial commitments. The work of a **debt counsellor**.

Formed by compounding: *counselling* about *debt*.

The term was first used in technical sources as long ago as the late sixties, but did not become
at all common in general usage until the late seventies in the US and the eighties in the UK. The
successive problems of the credit boom (leading to credit-card debt) and high interest rates
(causing people to default on mortgage payments) have made it increasingly common since
then.

> As debt counselors all over the state can attest: The woods around here are full of people who can't
> handle a single credit card without getting into deep, deep trouble.
> *Los Angeles Times* 30 Jan. 1986, section 5, p. 14

> For homeowners forced into debt by rising interest rates, the Portsmouth Building Society has set up
> a free debt counselling phoneline ... manned by staff trained in debt counselling.
> *Daily Telegraph* 10 Feb. 1990, p. 34

decommunize /diːˈkɒmjʊnaɪz/ *transitive verb* 📷

To remove the communist basis from (a country, its institutions or economy), especially in Eastern Europe; loosely, to democratize. Also as a noun **decommuniza-
tion**, the process of dismantling communism; adjective **decommunized**.

Formed by adding the prefix *de-* (in its commonest sense of removal or reversal) and the verbal
suffix *-ize* to the root *commun-*.

The word has been in use since the early eighties, when the first signs emerged of a willingness
in communist countries to allow a small amount of private enterprise in some areas of their
economies. Its use became more frequent in the late eighties—first in relation to Poland and
Hungary and later to all former Warsaw Pact countries, as the whole edifice of Marxism in
Eastern Europe began to be replaced by varying degrees of democracy and capitalism. The
verb is sometimes used intransitively, in the sense 'to become decommunized'. The noun *de-*

communization covers all the processes, both economic and political, which contribute to the dismantling of communism, whereas *democratization* and its Russian equivalent *demokratizatsiya* really refer only to the political process. *Debrezhnevization* was used for a short time to describe the personal discrediting of Leonid Brezhnev and his style of government, a process which took place during the mid eighties, shortly after Mikhail Gorbachev came to power in the Soviet Union.

> The momentum of decommunization is likely to carry most of the successor states of the Soviet Union quite far to the right. *The Times* 24 Feb. 1990, p. 10

> 'We cannot decommunize a whole society overnight,' says Friedrich Magirius, superintendent of Leipzig's Protestant churches, who notes that East Germany was 'a typical dictatorship'. *Time* 9 July 1990, p. 75

deepening 🏛 see WIDENING

deep green 🌱 see GREEN

deep house ♪ 🎭 see GARAGE and HOUSE

def /dɛf/ *adjective* 🎭

In young people's slang (originally in the US): excellent, great, 'cool'. Often used in the phrase **def jam**, brilliant music.

Usually explained as a clipped form of *definite* or *definitive* (in its slang sense 'the last word in . . .'); compare RAD and *brill* (see BRILLIANT). However, it seems more likely to be connected with the use of *def* (derived from *death*) as a general intensifying adjective in West Indian English. This is borne out by a number of early uses of *def* in RAP lyrics, where *death* can be substituted more readily than *definite* or *definitive* (words which would not anyway be appropriate in this context).

Def belongs originally to HIP HOP, where it started to be used by rappers in about the mid eighties; the US record label *Def Jam* dates from about that time. The word soon became extremely fashionable among both Black and White youngsters in the US and the UK. A series of programmes for a teenage audience on BBC2 from 9 May 1988 onwards was given the general heading 'DEF II'. For further emphasis, the suffix *-o* may be added, giving **deffo**.

> Further def vinyl to look out for includes deejay Scott La Rock's album. *Blues & Soul* 3–16 Feb. 1987, p. 30

> Shot in super-slick black and white, with a half-hour colour 'behind the scenes' documentary, this is actually quite a funky lil' package. And a deffo must for all Jan fans. *P.S.* Dec. 1989, p. 27

deforestation 🌱 see DESERTIFICATION

dehire ⬛ see DESELECT

deleverage 〰 see LEVERAGE

democratization, demokratizatsiya 🏛 see DECOMMUNIZE

deniability /dɪnaɪə'bɪlɪtɪ/ *noun* 🏛

Ability to deny something; especially, in the context of US politics, the extent to which a person in high office is able to deny knowledge of something which is relevant to a political scandal.

Formed by adding the noun suffix *-ability* to *deny*, giving a noun counterpart for the adjective *deniable*.

Deniability is one of those potential words which the building blocks of affixation would make it possible to form at any time, and in fact it was first used in its more general sense at the beginning of the nineteenth century. The special political sense, though, dates from the political scandals of the late twentieth century in the US—first the Watergate scandal of 1972–4, and later the Iran–contra affair of 1986 (see CONTRA). This special sense seems to have originated in CIA jargon, where it was sometimes used in the phrase **plausible deniability**. It was popularized at the time of the Watergate scandal by an article by Shana Alexander in *Newsweek* in 1973, entitled 'The Need (Not) To Know', and indeed the whole point of this concept is the perceived need to protect the President (or another high official) from knowledge of some shady activity, so that he will be able to tell any ensuing inquiry that he knew nothing about it.

> The concept of 'plausible deniability' was devised by the late CIA director, Mr William Casey, by having Israeli arms brokers as middlemen.
> *Daily Telegraph* 11 July 1987, p. 6

> I made a very definite decision not to ask the President so that I could insulate him from the decision and provide some future deniability . . . The buck stops here with me.
> John Poindexter quoted in *Time* 27 July 1987, p. 24

> The government is rendering itself less competent, preparing a more thoroughgoing deniability.
> Marilynne Robinson *Mother Country* (1989), p. 182

Denver boot, shoe ✖ see WHEEL CLAMP

desaparecido /ˌdɛzæpærəˈsiːdəʊ/ *noun* 🔒 ◀

Any of the many people who disappeared in Argentina during the period of military rule there between 1976 and 1983; by extension, anyone who has disappeared in South or Central America under a totalitarian regime.

A direct borrowing from Spanish *desaparecido* 'disappeared', the past participle of the verb *desaparecer* 'to disappear'.

The plight of the *desaparecidos*, also called in English **the disappeared** or **disappeared ones**, was much discussed in the newspapers in the US and the UK from about the late seventies. Many were never seen again after being arrested by the army or police, and can only be presumed killed in detention; many others were children who were taken away from their arrested parents and placed with other families without any consent. Since the end of the military regime, the *desaparecidos* have remained in the news from time to time, and some of those formerly in detention have reappeared. The effort continues to trace as many of the displaced children as possible and return them to their real families. Recently the word has been extended in use to anyone who has suffered a similar fate in Spanish America.

> People whose children or husbands or wives were *desaparecidos*—'disappeared ones'—would go to Cardinal Arns, and the Cardinal would stop whatever he was doing and drive to the prisons, the police, the Second Army headquarters.
> *New Yorker* 2 Mar. 1987, p. 62

> The Grandmothers of the Plaza de Mayo are assembling a genetic databank on grandparents whose grandchildren are still missing, and on children who suspect that they are desaparecidos but whose grandparents have yet to be identified.
> *Nature* 18 June 1987, p. 553

deselect /ˌdiːsɪˈlɛkt/ *verb* 🔒

Of a local constituency party in the UK: to reject (an established candidate, especially a sitting Member of Parliament) as its constituency candidate for an election.

Formed by adding the prefix *de-* (indicating reversal) to the verb *select*. This kind of formation with *de-* is characteristic of euphemistic verbs like *deselect*—compare *dehire* for 'sack' in the US (where *deselect* has also been used as a euphemism for 'dismiss').

The verb has been used in this sense in British politics since the very end of the seventies, when the Labour Party's reselection procedure made **deselection** a real danger for a number of Labour MPs. The practice was particularly common during the middle years of the eighties, and the word came to be used in other contexts (such as local government) at that time.

> Mr Woodall, MP for 12 years . . ., launched a bitter attack on his opponents in the NUM and local party who, he said, had 'connived' to deselect him.
> *Daily Telegraph* 24 Feb. 1986, p. 24

Echoes of a more turbulent past also emerged from the NEC's monthly meeting in the long-running dispute over Frank Field's deselection as Birkenhead's sitting MP. *Guardian* 28 June 1990, p. 20

desertification /dɪˌzɜːtɪfɪˈkeɪʃ(ə)n/ *noun*

The changing of fertile land into desert or arid waste, especially as a long-term result of human activity. Also sometimes known as **desertization**.

Formed by adding the process suffix *-ification* to *desert*.

The process of *desertification* was recognized as a world environmental problem as long ago as the mid seventies, but it was not until the late eighties that the word became widely known as a result of the GREEN movement and increased awareness of environmental issues generally. The problem is exacerbated by destruction of forests (*deforestation*), erosion of the topsoil, and GLOBAL warming (which involves formerly fertile areas in drought). As the process takes place, the affected land is first termed *arid*, then **desertified**.

Some 6.9 million sq. km. of Africa ... were under direct threat of desertification in 1985, according to UN estimates. *The Annual Register 1985* (1986), p. 395

The very processes of extracting Third World resources result in environmental disasters—deforestation, massive soil-erosion and desertification. *New Internationalist* May 1987, p. 13

designer /dɪˈzaɪnə(r)/ *adjective*

Originally, of clothes and other fashion items: bearing the name or label of a famous designer, and therefore (by implication) expensive or prestigious. Later extended to describe anything fashionable among YUPPIEs and the smart set generally; also applied to anything that can be designed individually for or by a particular user.

An attributive use of the noun *designer* which has become so common in recent years that it is now regarded by many as an adjective.

This use of *designer* began with the **designer scarf** (also known as a *signature scarf*) back in the mid sixties, but did not really take off in the language until the late seventies. Then denim jeans were elevated from simple workaday clothing to high fashion by the addition of the **designer label** on the pocket, which made them **designer jeans** and therefore comparatively expensive. The trend spread to other areas of fashion (notably **designer knitwear**) in the early eighties; by the middle of the decade the word had become one of the advertising industry's favourites, and anything associated with the smart and wealthy class targeted by these advertisers could have the *designer* tag applied to it ironically (for example, overpriced sparkling mineral water served by trendy wine bars came to be called **designer water**). A distinct branch of meaning started to develop in the second half of the eighties, perhaps under the influence of the same advertisers and fashion writers. Whereas before this, *designer* items had to be created by a designer (or at least bear the name of a designer: the name was often licensed out on goods which the designer had never seen), the emphasis was now on designing for the individual customer, and in some cases the consumers were even encouraged to do the designing themselves. This was the era of such things as **designer stubble** (a carefully nurtured unshaven look) and **designer food** (inspired by the chef-artists of *nouvelle cuisine*). The concept has been used outside the world of 'lifestyle' and fashion as well, for example in popular descriptions of genetic engineering.

Small wonder Perrier is called Designer Water. My local wine bar has the cheek to charge 70p a glass. *The Times* 4 Sept. 1984, p. 12

I mean Ah'd ... got into ma designer tracksuit just to be casual like.
Liz Lochhead *True Confessions* (1985), p. 72

Designer stubble of the George Michael ilk has also run its bristly course. Hockney thinks that the only people who can get away with it are dark, continental men whose whiskers push through evenly. *Guardian* 7 Aug. 1989, p. 17

Altering the shape of plants is another possibility—what Professor Stewart calls designer plants ... In some cases they could be made to grow a canopy across the bare earth to keep in gases like carbon dioxide.   *Guardian* 5 Mar. 1990, p. 6

'Designer' pianos in coloured finishes, veneers and marquetries now form about 5 per cent of the market.   *Ideal Home* Apr. 1990, p. 84

See also DESIGNER DRUG

designer drug /dɪˈzaɪnə ˌdrʌg/ *noun* 🖋

A drug deliberately synthesized to get round anti-drug regulations, using a structure which is not yet illegal but which mimics the chemistry and effects of an existing, banned drug; hence any recreational drug with an altered structure.

For etymology, see DESIGNER. The ultimate in made-to-measure kicks, the *designer drug* was also *designed* to keep one step ahead of anti-drugs laws.

Designer drugs were being made privately as early as 1976; the first designer 'look-alikes' of heroin appeared on the streets in the late seventies under the names *China White* and *new heroin*. The term itself was coined several years later when Professor Henderson of the University of California at Davis investigated the large number of deaths and Parkinsonian symptoms among users of China White in California. Despite attempts to limit them by legislation, *designer drugs* mimicking prohibited amphetamines enjoyed an explosion in the late eighties, as drug users looked for ways of avoiding heroin use with its associated Aids risk. With the new legislation came a development in the sense of the term: any recreational drug which deliberately altered the structure of an existing drug could be called a *designer drug*, as could a drug used by a sports competitor hoping to avoid falling foul of random tests.

The legality of the designer drugs is only one of the many powerful economic incentives working to make them the future drugs of abuse.   *Science* Mar. 1985, p. 62

Some of these people obviously also use cocaine, marijuana and some exotic designer drugs.
                                                                                                  New York Times 23 Sept. 1989, p. 23

desk organizer 🏏 see ORGANIZER

desk-top /ˈdɛsktɒp/ *noun* and *adjective* Also written **desktop** 🖥

noun: A personal computer which fits on the top surface of a desk (short for **desk-top computer**). Also, a representation of a desk-top on a VDU screen.

adjective: Using a desk-top computer system to produce printed documents to a publishable standard of typesetting, layout, etc.; especially in the phrase **desk-top publishing** (abbreviation **DTP**).

A specialized use of the transparent compound *desk-top*.

The *desk-top computer* goes back to the seventies, but only started to be called a *desk-top* for short in the mid eighties. At about the same time, computer manufacturers whose systems made use of ICONs and other features of WIMPS (see WIMP²) started to use *desk-top* widely as a way of referring to the representation of the top of a working desk that appeared on the screen. *Desk-top publishing* depends on software packages that were only first marketed in the mid eighties. Essentially it makes available to the computer user a page make-up and design facility which makes it possible to create any arrangement on the 'page' of text and graphics output from other packages such as word processing and spreadsheets, using a wide variety of different type-styles and sizes. The design can then be printed using a laser printer. These systems proved very popular for the production of documents on a small scale, bypassing the cost of commercial typesetting and design. By 1990 the dividing line between *desk-top* and conventional typesetting systems had blurred: this book, for example, was typeset using *DTP* software, but output on a high-quality imagesetter.

Given today's low cost desktop publishing systems, almost anyone could set up as a newsletter publisher, working from home.   *Guardian* 10 Aug. 1989, p. 29

There's nothing remotely hostile about a desktop with icons for both Unix and DOS applications.
                                                                                    PC User 11 Oct. 1989, p. 203

It was in fact set on a personal computer DTP system (feel the quality, never mind the width!).
                                                                                    Creative Review Mar. 1990, p. 47

des res /ˌdɛz ˈrɛz/ *noun* Also written **des. res.** ❌

Colloquially in the UK (originally among estate agents), a desirable residence; an expensive house, usually in a 'sought-after' neighbourhood.

Formed by abbreviating *desirable* and *residence* to their first three letters.

Des res belongs originally to the highly abbreviated and euphemistic language of estate agents' newspaper advertisements, where the cliché has been in use for some years. During the mid eighties, though, it moved into a more general colloquial idiom, often used rather ironically. *Des res* is sometimes used as an adjective—again, often ironically.

The days of the 'des res' that clearly isn't are set to end for estate agents.
                                                                                    The Times 20 Apr. 1990, p. 2

WDS make many practical suggestions as to how women's toilets could be improved; if all were adopted, they'd become highly des res.   *Guardian* 11 July 1990, p. 17

For those for whom the genuine article is not beyond reach, the Georgian country house (right) is one typical English version of the des res.   *Independent* 22 Dec. 1990, p. 33

device /dɪˈvaɪs/ *noun* 🌱

Euphemistically, a bomb.

Formed by shortening the earlier euphemism *explosive device*.

The word was used as long ago as the late fifties in **nuclear device**, a euphemism for *atom bomb*, but this term was rarely shortened to *device* alone. In the age of international terrorism, the euphemism was taken up in police jargon, at first often in the longer form **explosive device** or **incendiary device**, and widely used in press releases describing terrorist attacks in which explosives were used. During the course of the eighties *device* seems to have become an established synonym for *bomb* in news reports.

After sprinkling them with an unidentified liquid, an explosive charge was put on top of the human pile. The device detonated as planned.   *Washington Post* 3 Jan. 1981, section A, p. 1

February 24: A device pushed through a letter box wrecked an army careers office in Halifax, West Yorkshire.   *Guardian* 11 June 1990, p. 2

diddy goth 🦉 see GOTH

dideoxyinosine ⊗ see DDI

dietary fibre ⊗ ❌ see FIBRE

differently abled 🎗 see ABLED

digital /ˈdɪdʒɪt(ə)l/ *adjective* ❌ 🖥

(Of a recording) made by digitizing, or turning information about sound into a code of numerical values or digits, and storing this.

A straightforward development of the adjective *digital* in the computing sense 'operating on data in the form of digits'; first the method of recording was described as *digital*, and then the adjective was also applied to a recording or piece of music reproduced in this way.

The technology for *digital* recording was developed as early as the sixties, but it was not until the late seventies that the first **digital discs** became commercially available. The sound information that is stored includes millions of coded pulses per second; until the advent of the CD there was no suitable medium for this mass of information. This method of recording is considerably more faithful to the original sound than analogue recording (the audio method previously used) and the recording does not deteriorate so quickly; as a result, *digital* recording has more or less taken over the classical market (where fidelity of sound is especially important) and is also widely used for popular music. The process of translating a signal into coded pulses is called **digitization** (or **digitalization**); older analogue recordings are often re-recorded using the *digital* technique and are then described as **digitally remastered**.

> The performances could hardly be more authentic, with magnificent playing and an ample resonance in this fine digital recording. *Sunday Times* 14 Oct. 1984, p. 40

> In their day (1957–59) these recordings stood as superior examples of the conducting and engineering art. They sound even more impressive today in RCA's digitally remastered version. *Chicago Tribune* 22 Apr. 1990, section 13, p. 22

digital audio tape ✖️ 🖥️ see DAT

digital video interactive ✖️ 🖥️ see CD

DINK /dɪŋk/ *acronym* Also written **Dink, dink, Dinkie, Dinky**, etc. 🔢

Colloquially, either partner of a career couple with no children, both of whom have an income from work and who are therefore viewed as affluent consumers with few drains on their resources.

Formed on the initial letters of *Double* (or *Dual*) *Income No Kids*; in the variant forms *Dinkie* or *Dinky*, the diminutive suffix -*ie*, -*y* is added in imitation of YUPPIE, although *Dinky* is sometimes explained as *Double Income No Kids Yet*.

DINK is one of a line of humorous terms (often acronyms) for social groupings that followed in the wake of the successful *yuppie* in the mid eighties. It owes its existence to the trend analysts and marketing executives of the US and Canada, who in 1986 identified and targeted this group as an increasingly important section of the American market. Typically, the partners in a *DINK* couple are educated to a high level and each is committed to a high-paid career; the social trend underlying the coinage is that women with high educational qualifications tend to have fewer children, and to have them later in their careers than was previously the case. For two or three years, *DINK* appeared to be almost as successful a coinage as *yuppie* (despite its confusability with the US slang word *dink* 'penis', also used as a personal term of abuse); derivatives included **dinkdom** and the adjective **undink** (not characteristic of a *DINK*). Less successful variants on the theme, such as *OINK* (One Income No Kids), *Nilkie* (No Income Lots of Kids), and *Tinkie* (Two Incomes, Nanny and Kids) came and went during 1987. A later attempt was *SITCOM* (Single Income, Two (K)ids, Outrageous Mortgage), which appeared in 1989, but this also failed to make much impression.

> These speedy high-rollers are upper-crust DINKs . . . They flourish in the pricier suburbs as well as in gentrified urban neighborhoods. *Time* 20 Apr. 1987, p. 45

> The wolf is looming through the smoked-glass door even for many hard-working Dinkie . . . couples. *The Times* 2 May 1990, p. 10

direct broadcasting by satellite ✖️ 🖥️ see SATELLITE

dirty dancing ✖️ 📼 see LAMBADA

dis 📼 see DISS

disablist /dɪsˈeɪblɪst/ *adjective* Also written **disable-ist** or **disableist** 🔢

Showing discrimination or prejudice against disabled people; characterized by ABLEISM.

Formed by adding the adjectival suffix *-ist* to the root form of *disabled*, after the model of *ageist* (see AGEISM), *racist*, and *sexist*.

The word was coined in the mid eighties as the adjectival counterpart for *ableism*. At first it was sometimes written *disableist* or even *disable-ist*, but *disablist* now seems to be becoming established as the usual form. **Disablism**, which represents the opposite side of the coin from ableism (discrimination against the disabled rather than in favour of the able-bodied) very rarely occurs as a term.

> I am not apologising for SM and believe that in itself it is neither racist, classist, disablist nor anti-semitic. *Spare Rib* May 1986, p. 6

> Labour has promised to infuse racist, sexist, 'disablist', and 'ageist' criteria into higher education, like those that are making an academic mockery of some American institutions.
> *Daily Telegraph* 8 Nov. 1989, p. 20

See also ABLED

disappeared (ones) see DESAPARECIDO

Discman see WALKMAN

disco /'dɪskəʊ/ *noun* Also written distco

A power-distribution company; any of the twelve regional companies set up in 1989 to distribute electricity in England and Wales.

Formed by combining the first syllable of *distribution* with *co*, a long-established abbreviation of *company* which had already been used as a suffix in company and brand names (for example, *Woolco* for a Woolworths brand).

Disco was used in company names in the US before becoming topical in the UK because of the government's reorganization of the electricity supply in the late eighties and their plans to sell off the *discos* as part of their privatization strategy. *Distco* seems to be the officially preferred form, although *disco* is commoner in the newspapers (despite confusability with the musical *disco*). The sale of the distribution companies took place in 1990.

> It is argued that smaller distcos, such as Manweb and South Wales, will have lower growth prospects to push down costs. *Observer* 18 Mar. 1990, p. 57

> The discos have much better growth prospects than the water companies, while the gencos generate a unique 'fuel'. *Daily Telegraph* 25 July 1990, p. 23

> Lloyds pitched for the business of arranging the loans . . . for three discos, with two of whom it already enjoyed a relationship as a clearing bank. *Daily Telegraph* 17 Aug. 1990, p. 17

See also GENCO

disco-funk see FUNK

dish see SATELLITE

diss /dɪs/ *verb* Also written dis

In young people's slang (originally in the US): to put (someone) down, usually verbally; to show disrespect for a person by insulting language or dismissive behaviour. Also as an action noun **dissing**.

Formed by abbreviating *disrespect* to its first syllable.

Diss originated in US Black English and has been popularized through the spread of HIP HOP. In Black culture, insults form an important part of the peer-group behaviour known as *sounding* or *playing the dozens*, in which the verbal repartee consists of a rising crescendo of taunts and abuse. The concept of *dissing* moved outside Black culture through its use in RAP, and is now widely known among Whites both in America and in the UK; even children interviewed in an

Inner London school playground in 1990 practised this trading of insults, referring to them as *cusses*.

> The victim, according to detectives, made the mistake of irritating Nuke at a party. 'He dissed him' Sergeant Croissant said.
> *New York Times* 15 Nov. 1987, section VI, p. 52

> The gladiatorial rapping, the sportswear, the symbolic confrontations ('dissing') are all about self-assertion.
> *Weekend Guardian* 11 Nov. 1989, p. 20

> While taking a dispute to someone's home is the ultimate in 'dissing' ... there are other insults that can be just as deadly ... 'You dis, you die,' some youths say.
> *Boston Globe* 2 May 1990, p. 12

distco ⚋ see DISCO

doc, docu- /dɒk/, /ˈdɒkjuː/ *combining forms* ✕

Parts of the word *documentary*, used in **docudrama** (also called **dramadoc** or **drama-doc**) and **docutainment** to show that a film or entertainment contains an element of documentary (or at least that real events have formed the basis for it).

Doc, which also exists as a free-standing colloquial abbreviation of *documentary*, is used as the second part of an abbreviated compound; when the *documentary* element comes first, the *-u-* is kept as a link vowel.

The dramatized documentary (*dramadoc, docudrama*) suddenly became a fashionable form of television entertainment at the end of the seventies in the US, and this was a fashion which lasted through the eighties both in the US and in the UK. The proportions of fact and dramatic licence in these programmes is variable, whereas the *docutainment* (a word which dates from the late seventies and appears to be a Canadian coinage) is more likely to be factual, but designed both to inform and entertain: compare *infotainment* (at INFO-).

> This two-part production about the life and times of Douglas MacArthur is no docudrama. It is instead a documentary or, more precisely, five hours of 'docutainment', a fascinating ... biography based on William Manchester's book about America's most intriguing, epic soldier.
> *Los Angeles Times* 3 Mar. 1985, p. 3

> While the film is not a 'docu-drama', immense pains have been taken to achieve authenticity.
> *Daily Telegraph* 8 Mar. 1990, p. 18

See also FACTION

donutting ✕ 🔒 see DOUGHNUTTING

doom and gloom ⚋ 🔒 see GLOOM AND DOOM

doorstep /ˈdɔːstɛp/ *verb* 🔒

intransitive: Of a politician: to canvass support by going from door to door, talking to voters on their doorsteps; also as an action noun **doorstepping** and agent noun **doorstepper**.

transitive: Of a journalist, campaigner, etc.: to 'stake out' the doorstep of (a person in the news, someone in a position of authority or power in a particular area, etc.) in the hope of getting a statement or story from them.

Formed by treating the noun *doorstep* as though it were a verb. This shift originally took place at about the beginning of this century, when door-to-door salesmen carried out their trade by *doorstepping*.

The intransitive, political sense goes back at least to the sixties, when door-to-door canvassing took over from public debate as the most important means of winning voters to one's cause—but *doorstepping* and *doorstepper* are later developments. The media use of the verb belongs to the eighties, when investigative journalism and straightforward intrusions of privacy on the part of journalists came in for some considerable criticism. The staying power of some journalists and press photographers became so widely publicized that the transitive verb started to develop a transferred sense: a person who was determined to get a decision or change of policy on a particular issue would talk of *doorstepping* the person responsible in order to achieve this (in much the same way as one might speak of *lobbying* one's MP).

> The journalists are often the last ones to see him before he goes to bed or the first to see him when he gets up in the morning, spending late nights at his house after his day is over and doorstepping him next morning.
> *The Times* 13 Jan. 1988, p. 30

> Some say it is time for a new approach, with bands of scientific inspectors doorstepping laboratories around the world.
> *New Scientist* 4 Aug. 1988, p. 31

> *Hard News*... will doorstep editors and reporters, if necessary, to get a reply.
> *Independent* 5 Apr. 1989, p. 17

double zero option ⬚ see ZERO

doughnutting /'dəʊnʌtɪŋ/ *noun* Also written **donutting** ✖ ⬚

In television jargon, the clustering of politicians round a speaker during a televised parliamentary debate so as to fill the shot and make the speaker appear well supported.

Formed by adding the suffix *-ing* to *doughnut*—presumably alluding to the ring shape of some doughnuts as resembling the ring of supporters, or to the jam in the middle as representing the speaker, surrounded by the apparently substantial dough of his support.

The word is often said to have been used in connection with the first televised debates from the federal parliament in Ottawa, but Canadian newspaper reports of the time do not bear this out (describing the practice, but not using the word). When the British parliament began to be televised, and particularly when House of Commons debates first appeared on TV screens in 1989, the word enjoyed a brief vogue in the press amid speculation that members would attempt to fill the seats immediately behind the speaker so as to make the chamber appear full, even when in fact a debate had attracted only a handful of MPs. Its use in popular sources promises to be shortlived.

> Mr Kirkwood did have a little ring of fellow-Liberals around him. But this practice of 'doughnutting', as Canadian parliamentarians call it, exhausts the nutters more than it fools the viewers.
> *Daily Telegraph* 24 Nov. 1989, p. 14

dozens 🔲 see DISS

dramadoc ✖ see DOC, DOCU-

drive-by /'draɪvbaɪ/ *noun* Plural **drive-bys** 🔲

In the US, a criminal act (usually a shooting) carried out from a moving vehicle. Also known more fully as a **drive-by shooting**.

Formed by dropping the word *shooting* from *drive-by shooting* and treating what remains as a noun.

The *drive-by* represents a reappearance in American crime of the gang-led murder carried out from a moving car, something which many would associate with the twenties rather than the eighties. In its new manifestation in the late eighties and early nineties it is particularly associated with rival teenage gangs, but the gun is often shot randomly into a crowd, endangering innocent passers-by as well as the gang targets.

> The task force suggested increased penalties for drive-by shootings and other gang-related homicides, and for the possession and sale of controlled substances, including phencyclidine.
> *New Yorker* 3 Nov. 1986, p. 128

> In Chicago, 'drive-bys' contributed to a 22 per cent leap in the youth murder rate last year.
> *The Times* 7 Feb. 1990, p. 10

drug abuse 🗡 🍴 see ABUSE

DTP 💻 see DESK-TOP

dude /duːd/ 😺

In urban street slang (originally in the US): a person, a guy, one of the 'gang'. Often used as a form of address: friend, buddy.

Dude is a slang word of unknown origin that was first used in the US in the 1880s to mean 'a dandy, a swell' or (as a Western cowboys' word) 'a city-dweller'. By the early 1970s it had been taken up in US Black English to mean 'a man, a cool guy or cat' (and later 'any person'), losing its original negative connotations.

This more general use of *dude* was popularized outside Black street slang through the *blaxploitation* films of the late seventies and, more particularly, through the explosion of HIP HOP during the eighties. Its spread into British English idiom, at least among children, was finally ensured by repeated use among the Teenage Mutant TURTLEs and other US cartoon characters in comic strips, cartoons, and games.

> Dudes like that, they're totally dialled in. They can earn a quarter of a million a year, serious coin.
> Richard Rayner *Los Angeles Without a Map* (1988), p. 68

> It is the teenage Bart who has caught the public's imagination. With his skateboard and, touchingly, his catapult, he is a match for anyone, not least because of his streetwise vocabulary. 'Yo, dude!' he says; 'Aye caramba!' and—most famously—'Eat my shorts!' *Independent* 29 July 1990, p. 17

dumping /'dʌmpɪŋ/ *noun* 🌱

The practice of disposing of radioactive or toxic waste by burying it in the ground, dropping or piping it into the oceans, or depositing it above ground in another country.

A specialized use of the verbal noun *dumping*, which literally means 'throwing down in a heap'.

It was only in the late seventies that environmentalists began to expose the scale of *dumping* by all the industrialized nations over the previous decade and the environmental disasters that this could cause. Hazardous waste had been buried in landfill sites on which houses were later built, sent off to Third World countries desperate for revenue, and pumped into rivers and oceans. *Dumping* became a topical issue in the UK in the eighties first because of public resistance to plans to bury radioactive waste in British landfill sites and later when the UK fell foul of European Community directives on clean beaches because of the large quantities of raw sewage being pumped out to sea from British shores.

> Dumping increases the input of nutrients such as nitrogen and phosphorus into the marine environment.
> Steve Elsworth *A Dictionary of the Environment* (1990), p. 243

> Waste trichloroethene probably gets into the tap water because of careless dumping.
> *Which?* Aug. 1990, p. 433

Dutch house 🎵 😺 see HOUSE

Dutching /'dʌtʃɪŋ/ *noun* Also written dutching 🌱 🏴󠁧󠁢󠁳󠁣󠁴󠁿

In the jargon of the British food industry, the practice of sending substandard food intended for the UK market for irradiation in the Netherlands (or some other European country where irradiation is permitted) so as to mask any bacterial contamination before putting it on sale in British shops.

Formed by making a 'verbal' noun from the adjective *Dutch* (since the irradiation is normally carried out in the Netherlands) and the suffix *-ing*; a similarly euphemistic expression for the same process is 'sending on a holiday to Holland'.

The practice of *Dutching* was exposed in a Thames television documentary in 1985, but it was not at that time given this name. Both the word and the practice became topical in 1989 during discussions of the proposed legalization of food irradiation. At a time when there was widespread public concern over food-related illnesses, many people were shocked to discover that bad food was already being passed off as good in this way.

> A dealer...talked about 'Dutching' to a Sunday Times reporter posing as a potential buyer. Asked if the prawns would pass health tests at a British port...: 'Well, they won't if they come into England directly. But if they went into Holland and Belgium, yes.' *Sunday Times* 6 Aug. 1989, section 1, p. 3

See also IRRADIATION

DVI ⬛ ⬛ see CD

dweeb /dwiːb/ *noun* ⬛

In North American slang: a contemptible or boring person, especially one who is studious, puny, or unfashionable; a 'nerd'.

Of unknown origin; probably an invented word influenced by *dwarf*, *weed*, *creep*, etc.

The term has been in use since the early eighties, and may have originated in US prep school slang. The corresponding adjective is **dweeby**.

> Norman, a research dweeb with a rockabilly hairdo.
> *Kitchener-Waterloo Record* (Ontario) 9 Nov. 1989, section C, p. 22

> Nathan Hendrick, 9, is wonderfully nerdy as Leonard Digbee, a dweeb's dweeb whose only goal in life is to one-up Harriet. *Los Angeles Times* 19 July 1990, p. 6

> 'These Val guys are totally gross. They think they're real, but you can tell they're Barneys.' She says 'dweeby types' often 'snog right up' to her when she's wearing her 'floss', or thong-back bikini.
> *Wall Street Journal* 27 Sept. 1990, section A, p. 1

dynamize /ˈdaɪnəmaɪz/ *transitive verb* ⬛

To increase the value of (a pension) by taking inflation into account in the calculations of final salary on which the pension is based; to calculate (final salary) by adding the value of inflation in successive years to a real salary some years before retirement. Such a pension or salary is **dynamized**; the calculation involved is **dynamization**.

The verb *to dynamize* has been in use in financial contexts with the more general meaning 'make more dynamic or effective' since the seventies. The use in relation to pensions is a specialization of this.

The *dynamized* pension is an approved way of avoiding the Inland Revenue's maximum allowable pension rule (that a pension may not be worth more than two-thirds of final salary) and dates from the late seventies.

> Norwich Union...cannot dynamise the pension without the trustees' approval.
> *Daily Telegraph* 14 Oct. 1989, p. 31

●●

E

E¹ 🌳 ⬛ see E NUMBER

E² see ECSTASY

E

e³ 🔲 see ELECTRONIC

earcon 🔲 see ICON

Earth-friendly 🌳 see -FRIENDLY

eco *adjective* 🌳 see ECO- below

eco- /'iːkəʊ/ *combining form* 🌳

Part of the words *ecology* and *ecological*, widely used as the first element of compounds and blends which relate in some way (sometimes quite tenuously) to ECOLOGY, the environment (see ENVIRONMENT¹), or GREEN issues. Hence as a free-standing adjective: ecological, environment-friendly.

The first two syllables of *ecology* and *ecological*; in both words this part is ultimately derived from Greek *oikos* 'house' (ecology being, properly speaking, the study of the 'household' or community of organisms).

One of the most fashionable combining forms of the late eighties, *eco-* had already enjoyed a vogue in the late sixties and early seventies, especially in US English. As a formative element of scientific terminology (for example in words like **ecoclimate, ecosphere, ecospecies, ecosystem,** and **ecotype**), it goes back to the twenties and thirties; scientists have also used it as a kind of shorthand for 'ecological and . . .' (for example in **ecocultural, ecogenetic, ecogeographical, ecophysiological,** etc.). The explosion of non-technical uses arises from the increasing influence of the green view of politics, and represents a shift in meaning which had also taken place in the use of the full forms *ecology* and *ecological*: *eco-* in these words can signify a range of different connections with 'the environment' or with environmental politics, but not usually (if ever) with the community of organisms studied by ecology proper. At the furthest extreme of this development are the words in which *eco-* is synonymous with *environment-friendly* (see -FRIENDLY) and often operates as a free-standing adjective (see the quotations below).

Among the formations of the earlier vogue period were **eco-activist, eco-catastrophe** (or **ecodisaster**), and **ecofreak** (also called an **eco-nut** or **eco-nutter**). Many of these seventies formations betray a lack of sympathy with environmental action groups and others who were already campaigning against the destruction of the environment; the formations of the eighties and early nineties, on the other hand, tended to have much more positive connotations, as green politics became acceptable and even desirable. Some of the earlier forms were now telescoped into blends: *eco-catastrophe*, for example, became **eco-tastrophe**. Many *ad hoc* formations using *eco-* have appeared in only one or two contexts (especially when it is used as a type of adjective); a few of these are illustrated in the quotations below.

Among the more lasting *eco-* words (some originally formed by the environmental campaigners of the seventies, others new to the eighties or early nineties) are: **eco-aware(ness)**; *ecobabble* (see under -BABBLE); **ecocentric** (and **ecocentrism**); **ecoconsciousness; ecocrat; ecocrisis; ecodoom** (and **-doomster, -doomsterism**); **ecofeminism; eco-friendly; ecolabel(ling)** (see also ENVIRONMENTAL); **ecomania** (sometimes called **ecohysteria**); **ecopolitics** (also **ecopolicy, ecopolitical**); **ecoraider; ecorefugee; ecosocialism** (and **ecosocialist**); **ecotage** (also called **ecoterrorism**) and **ecoteur** (also an **eco-guerrilla** or **ecoterrorist**); **ecotechnology** (and **ecotechnological**); **Ecotopian** (as an adjective or noun, from *Ecotopia*, an ecologically ideal society or environmental Utopia); **eco-tourism** and **eco-tourist**.

> Whew, the day certainly had a funny colour to it—a harp light, but livid, bilious, as if some knot of eco-scuzz still lingered in its lungs.
> *Martin Amis Money* (1984), p. 43

> Among the measures called for are . . . introduction of 'ecomark' labels for products that have little adverse effect on the environment.
> *Nature* 25 May 1989, p. 242

Tom Cruise will wear a shock of bright green hair in his next movie, fighting such evil characters as Sly Sludge ... in an effort to wipe out those 'eco-villains who pollute the earth'.

Sunday Mail Magazine (Brisbane) 11 Feb. 1990, p. 42

Four eco-warriors risk their lives as Greenpeace attempts to prevent a ship dumping waste in the North Sea. *Sky Magazine* Apr. 1990, p. 3

Oiling the wheels of eco progress. *Times Educational Supplement* 11 May 1990, section A, p. 12

What scientists call an 'eco-tastrophe' [on Mount St Helen's] has witnessed a remarkable recovery by nature. *Guardian* 18 May 1990, p. 12

Lex Silvester is no Crocodile Dundee, but dedicated to eco-tourism, blending sightseeing with conservation. *The Times* 2 June 1990, p. 29

The 'Eco house', in its own acre garden, will demonstrate how we can live in a more environmental friendly way with highly efficient insulation, solar heating, energy efficient appliances and organic gardening. *Natural World* Spring/Summer 1990, p. 9

The Department of the Environment produced a useful discussion paper on eco-labelling back in August 1989, and after some lengthy consultation set up an Advisory Panel. *She* Aug. 1990, p. 122

An overwhelming groundswell of support transformed Greenpeace from a daring but ragtag band of eco-guerrillas into the largest environmental organization in the world in barely over a decade.

New York Times Book Review 25 Nov. 1990, p. 14

As products with specious 'eco-friendly' claims multiply on store shelves, the need for substantiated product information has intensified. *Garbage* Nov.–Dec. 1990, p. 17

ecobabble see -BABBLE

ecological / ˌiːkəˈlɒdʒɪk(ə)l/ *adjective*

Concerned with ECOLOGY or GREEN issues; hence, environment-friendly, ENVIRONMENTAL.

For etymology, see ECO- and ECOLOGY.

Ecological has developed in very much the same way as ENVIRONMENTAL during the past ten years, developing the sense 'concerned with environmental issues' in the seventies (see ECOLOGY below) and the more elliptical sense 'environment-friendly' in the early eighties.

It seems it can already be economical (though surely not ecological) to fly cargo to London for onward trucking to Paris and points east, and vice versa. *Guardian* 19 June 1990, p. 15

ecology /iːˈkɒlədʒɪ/ *noun*

Conservation of the environment (see ENVIRONMENT[1]); GREEN politics. Often used attributively, in **Ecology Party** etc., in much the same sense as the adjectives ENVIRONMENTAL and GREEN.

A sense development of the noun *ecology*, which is formed on the Greek word *oikos* 'house', and originally referred only to the branch of biology which has to do with the 'household' or community of organisms and how they relate to their surroundings. Since it was the potential destruction of habitats (including the human one) that first focused political attention on green issues, *ecology* came to be used popularly to refer to the protection of the natural world from the effects of pollution.

The transformation of *ecology* from scientific study to political cause was foreseen by the writer Aldous Huxley in his paper *The Politics of Ecology* (1963), in which he wrote:

Ecology is the science of the mutual relations of organisms with their environment and with one another. Only when we get it into our collective head that the basic problem confronting twentieth-century man is an ecological problem will our politics become realistic ... Do we propose to live on this planet in symbiotic harmony with our environment?

The word *ecology* was popular throughout the seventies as the **ecology movement** gained momentum. In the eighties, though, *ecology* has tended to be replaced in its attributive use by *green*—the Ecology Party in the UK officially changed its name to the Green Party in 1985, for example—and by *the environment* elsewhere.

> The strongest organised hesitation before socialism is perhaps the diverse movement variously identified as 'ecology' or 'the greens'.　　　　　　　　　　　*New Socialist* Sept. 1986, p. 36

> The Polish Ecology Club was the second independent organisation to be established after Solidarity, and has several thousand members.　　　　　　　　　　　*EuroBusiness* June 1990, p. 14

economic and monetary union 📖 see EMU¹

Ecstasy /'ɛkstəsɪ/ *noun* Also written ecstasy or XTC 💉

In the slang of drug users, the hallucinogenic DESIGNER DRUG methylenedioxymethamphetamine or MDMA, also known as ADAM. Sometimes abbreviated to **E** (and used as a verb, in the sense 'to freak out on Ecstasy').

The name refers to the extreme feelings of euphoria and general well-being which the drug induces in its users. The word *ecstasy* has been used in the sense of 'rapturous delight' since the sixteenth century; 'street chemists' in the eighties have simply applied it in a more specialized and concrete sense.

It has been claimed that the drug was first made in the early years of this century as an appetite suppressant and patented in 1914 by the pharmaceutical company Merck; according to the chemical literature it was first synthesized in 1960 and did not become known as MDMA until the seventies. It was not until 1984, though, that it was made as a designer drug; by 1985 it had appeared on the streets in the US and was being called *Ecstasy* or *Adam*. It soon acquired a reputation as a drug of the smart, wealthy set; it was *Ecstasy* that the media most associated with the introduction of ACID HOUSE culture to the UK in 1988, claiming that the drug, in the form of small tablets, could easily be sold at crowded acid house parties, and lent itself to being 'pumped' down with fizzy drinks and the energetic style of dancing practised there. Despite claims by psychotherapists that it had a legitimate therapeutic use in releasing the inhibitions of some psychiatric patients, research showed that prolonged use could do irreversible damage to nerve cells in the brain, and it was banned in both the US and the UK. It remains one of the most popular illicit drugs of the eighties and early nineties; its users are sometimes known as **Ecstatics**.

> If cocaine and angel dust were the drugs of the 70s, Ecstasy may be the escape of the 80s.
> 　　　　　　　　　　　*Courier-Mail* (Brisbane) 31 May 1985, p. 4

> It is 3,4-methylenedioxymethamphetamine, MDMA, ADAM, Decadence, Essence, XTC, Ecstasy. Ecstasy! Paradise induced. And as of July, by emergency order of the Drug Enforcement Administration, illegal.　　　　　　　　　　　*Washington Post* 1 June 1985, section D, p. 1

> Police fear Acid House parties . . . provide an ideal opportunity for professional criminals to sell drugs, particularly the 'designer' drug Ecstasy favoured in the Acid house culture.
> 　　　　　　　　　　　*Independent* 7 Nov. 1988, p. 2

> The really great thing was three years ago, the Ecstasy explosion, when everybody started E'ing all over the place, there was all these different sorts of music getting mixed up.
> 　　　　　　　　　　　*Melody Maker* 23–30 Dec. 1989, p. 38

ecu /'eɪk(j)uː/, /'ɛk(j)uː/ *acronym* Also written Ecu or ECU 〰

Short for **European Currency Unit**, a unit of account used as a notional currency within the EMS and in Eurobond trading, and intended as the future common currency of EC countries under EMU¹. Also, a coin denominated in ecus.

An acronym formed on the initial letters of *European Currency Unit*, but influenced by and deliberately referring back to the French word *écu*, a name for a historical French gold or silver coin worth different amounts in different periods. This influence explains the fact that most English speakers use an anglicized version of the French pronunciation /eky/ rather than spelling out /iːsiːˈjuː/.

Ecu was adopted as the name for the European Community's currency unit in the early seventies (after a short period during which it was known as the *EMU*, or *European Monetary Unit*). In the UK the word was hardly known outside financial markets until the late eighties, when it became a central subject in discussions of EMS and EMU. The value of the *ecu* is based on a weighted average of a 'basket' of European currencies. The Delors report provided for the *ecu* to become the single European currency in the third stage of development of EMU, replacing the existing national currencies of EC member states. The UK government in particular opposed this implied loss of national sovereignty, and the Chancellor John Major put the issue at the centre of his counter-proposals for EMU in June 1990, suggesting an intermediate stage when Europe would use a **hard ecu** alongside national currencies, moving on to the *ecu* as a single currency unit only if individual member states decided they wanted this. *Ecu* coins were minted as collectors' items in some countries, including Belgium, where they have been legal currency since 1987, but are rarely used. *Ecus* were increasingly popular for business transactions, travellers' cheques, and as a stable currency for mortgages before the UK's entry to the ERM in October 1990. A million *ecus* make one **mecu** and a billion *ecus* one **becu**, although neither term is in common use.

> Charcol has launched a mortgage in ECUs . . . because ECUs should be less volatile than a single currency.   *Sunday Times* 19 Feb. 1989, Business section, p. 15

> 'I think that really it will become a reality when that currency exists,' he says, pulling an ECU coin out of his pocket.   *Financial World* 7 Mar. 1989, p. 40

> The 1989 budget was adopted on 15 December 1988 and provides for total Community expenditure of 44.8 becu (£29.9 bn) in payment appropriations.   *Accountancy* June 1989, p. 43

> Another clever aspect of Mr Major's scheme is that the EMF would manage the ecu so that it was never devalued at a currency realignment: it would be a 'hard ecu'.   *Economist* 23 June 1990, p. 64

E-free 🌳 ✄   see E NUMBER

EFTPOS /ˈɛftpɒs/, /ˈɛftpɒz/ *acronym* Also written **Eftpos, eft/pos,** or **EFT-Pos** 〰 ▣

Short for **electronic funds transfer at point of sale**, a method of paying for goods and services by transferring the cost electronically from the card-holder's account to the retailer's using a card such as a credit or debit card and a special terminal at the cash-desk.

The initial letters of *Electronic Funds Transfer at Point Of Sale*; the formation is modelled on the earlier acronyms EPOS and *POS*, point of sale.

EFTPOS was heralded in the late seventies as the facility which would ensure a cashless society within a decade. In practice, it was not officially announced in the UK until 1982, and was only generally introduced in the second half of the eighties. The rather cumbersome abbreviation, which does not lend itself very readily to being pronounced as a word, is used mainly in business circles; popularly, *EFTPOS* facilities in the UK are usually known by the proper names Switch and Connect, while in the US *EFTPOS* is often referred to as simply as **EFT** (an abbreviation which has a longer history than *EFTPOS*).

> While Publix was launching its p.o.s. debit card system last week, Abell and other EFT experts suggested that any debit card system be considered carefully before a supermarket company invests in joining bank-controlled switch networks.   *Supermarket News* 2 July 1984, p. 20

> A trial of some 2,000 EFT-Pos terminals is set to take place, some time in the autumn of 1988, in retailers in Southampton, Leeds and Edinburgh.   *Daily Telegraph* 29 May 1987, p. 19

> EFTPOS . . . will save you the hassle of writing a cheque or carrying cash around. You hand over a debit card like Switch and Connect cards, which deduct money straight from your bank account.   *Which?* Feb. 1990, p. 69

EGA card 🖳 see CARD²

electro /ɪˈlɛktrəʊ/ *combining form, adjective,* and *noun* 🎵 🖳

combining form and *adjective*: (Of popular music) making heavy use of electronic instruments, especially synthesizers and drum machines.

noun: A style of popular dance music with a strong and repetitive electronic beat and a synthesized backing track.

Electro- started life as a combining form of *electric* or *electronic*, as in familiar scientific terms such as *electromagnetism*. In the musical sense it developed from combinations with the names of popular-music styles (*electrobeat, electro-disco,* etc.) to become an adjective in its own right, and eventually to be used as a noun to describe a particular style of dance music.

The first combinations of *electro-* with the names of other popular-music styles date from the early eighties, when synthesized and electronically produced sounds were becoming very important in a number of different areas of pop. One of the earliest and most enduring combinations is **electrofunk**, which expresses just one of the new directions that FUNK has taken in the eighties. More temporary combinations have included **electro-disco** (perhaps the most important, especially in Belgium), **electrobeat, electro-bop, electro-country,** and **electrojazz**. By the mid eighties the music papers had begun to use *electro* on its own, both as an adjective and as a noun. Sometimes this was used as another name for *electric boogie*, the music played on ghetto blasters as an accompaniment to BREAK-DANCING in the street, and a style which ultimately fed into HIP HOP.

> Pianist Herbie Hancock . . . played a sterling set totally unlike his tarted-up electro-funk of recent years.
> *Maclean's* 29 Mar. 1982, p. 66

> No dress restrictions, music policy is well 'ard with P. Funk, House, Go-Go and Electro cutting in.
> *Blues & Soul* 3 Feb. 1987, p. 34

> You get bored with the happening hardcore electro groove business.
> *New Musical Express* 25 Feb. 1989, p. 43

See also TECHNO

electrobash 🔀 🖳 see TECHNOSTRESS

electronic /ɪlɛkˈtrɒnɪk/ *adjective* 🖳

In machine-readable form; existing as data which must be read by a computer. Especially in:

electronic mail (often abbreviated to **email** or **e-mail**), the transfer of messages or files of data in machine-readable form from one user to one or more others by means of a computer network; also, the messages that are sent and received using this facility;

electronic publishing, the publication of text in machine-readable form (on tape, discs, CD-ROM, etc.) rather than on paper; texts published in this way;

electronic text (sometimes abbreviated to **etext**), the machine-readable version of a text, which is created by data CAPTURE.

A development of the adjective *electronic* in the sense 'operated by the methods, principles, etc. of electronics' in which a subtle shift from active to passive has taken place: whereas in the original term *electronic data processing* (a synonym for *computing* in the sixties), *electronic* referred principally to the processing rather than to the data, now it is applied also to the 'soft' copy of the text, the object of the processing. Instead of being operated by electronics, these *electronic* media may only be operated upon by electronic equipment (in practice, specifically by computer). This shift is evident within the development of the term *electronic mail* itself, which at first only referred to the system (operated electronically), but later came to be used also of the messages (existing in a form which meant that they had to be operated upon by the computer). In general during this period *electronic* has tended to become a synonym for *computerized*.

Electronic mail, which relies upon data transfer across telecommunications networks, began in the late seventies and by the mid eighties was frequently abbreviated to *email* or *e-mail*. *Electronic publishing* had begun during the seventies, but did not acquire this name until 1979 and only became a growth industry in the mid eighties; it tends to be popularly confused with conventional publishing using electronic techniques (especially DESK-TOP publishing). The proliferation of *electronic text* was a natural result of the growth of electronic publishing and increasing use of computers for editing and research work during the eighties.

> When our coded file arrives, PPI's Atex computer merges electronic text and digitized artwork into a complete page.
> *Chemical Week* 28 July 1982, p. 7

> The first Electronic Publishing conference was held at Wembley four years ago.
> *Daily Telegraph* 13 June 1988, p. 27

> We read and respond to e-mail as it pleases us, not at our correspondent's convenience.
> *New Scientist* 6 May 1989, p. 66

> Just now the Soviet people are getting into networking. They are not yet used to the idea of electronic mail.
> *Guardian* 3 Aug. 1989, p. 20

electronic funds transfer at point of sale 〰️ 💻 see EFTPOS

electronic keyboard ♪ 💻 see KEYBOARD

electronic point of sale 〰️ 💻 see EPOS

electronic tablet 💻 see TABLET

electronic tagging {} 💻 see TAG¹

email 💻 see ELECTRONIC

EMS /iːɛmˈɛs/ *abbreviation* 〰️

Short for **European Monetary System**, a financial arrangement which consists primarily of an exchange-rate mechanism (**ERM**) linking the currencies of some EC member countries to the ECU so as to limit excessive fluctuations in exchange rates, and common credit facilities.

The initial letters of *European Monetary System*.

The *EMS* was set up in the late seventies, after the failure of the 'snake' to regulate currency fluctuations in Europe. It grew out of dissatisfaction among politicians from some EC countries (notably the former British Chancellor of the Exchequer Roy Jenkins, Helmut Schmidt of West Germany, and Valéry Giscard d'Estaing of France) with the slow progress of plans for economic and monetary union (see EMU¹ below). By the time *EMS* was formally accepted by the European Council in 1978 and put into effect in March 1979, the British government was not prepared to participate fully in it, declining to take part in the exchange rate mechanism which is the core of the system. *EMS* was widely discussed in the British newspapers during the late eighties, as plans for EMU began to move forward, the single European market of 1992 approached, and pressure increased on the UK to join *EMS*. There was a concentration of uses of the term during 1988-9, when it was reported that the then Chancellor of the Exchequer Nigel Lawson favoured British participation as a way of controlling inflation, but could not break Prime Minister Margaret Thatcher's opposition to it. This deadlock eventually contributed to Mr Lawson's resignation in October 1989. His successor, John Major, took the UK into the *ERM* in October 1990, even though the so-called MADRID CONDITIONS had not been met.

Given the existence of the EMS, our continuing non-participation in the ERM cannot fail to cast practical doubt on that resolve [to beat inflation].

Nigel Lawson quoted in *The Times Guide to 1992* (1990), p.107

Sterling quickly lost the big early gains that followed ERM entry. But its ability to hold pre-EMS levels is no mean feat. *Financial Times* 5 Nov. 1990, section 1, p. 19

EMU¹ /iːɛmˈjuː/; occasionally pronounced /ˈiːmjuː/ *abbreviation* Also written **Emu** ⚐

Short for **economic and monetary union**, a programme for full economic unity in the EC, based on the phased introduction of the ECU as a common currency.

Now nearly always explained as the initial letters of *Economic* (*and*) *Monetary Union*, although during earlier discussions (see below) it was intended to stand for *European Monetary Union*, and this expansion is still sometimes given.

EMU is by no means a new abbreviation, the idea having been proposed as early as 1970 as a way of solving currency difficulties in France and Germany. The original plan envisaged that the full union of EC currencies should be achieved by 1980 and be based on a European monetary unit (see ECU). Little progress towards this aim had taken place by 1978, when the European Monetary System (see EMS) was adopted by eight member states as the EC's financial system, incorporating a mechanism for controlling exchange rates. A new impetus for *EMU* was the publication in April 1989 of the Delors report, a three-stage plan for introducing a common currency and aligning the economies of the Twelve. This was discussed at summits in Madrid and Strasburg during 1989, with Britain (or principally Prime Minister Margaret Thatcher) standing out against acceptance of the plan as it stood—despite the enthusiasm of other member states—because of the implied threat to national sovereignty; stage one was, however, adopted. In June 1990, Chancellor of the Exchequer John Major made a counterproposal for the phased introduction of a common currency, designed to minimize the effect on sovereignty (see ECU). One result of all this discussion has been the very widespread use of the abbreviation in newspapers and the media generally during the late eighties and early nineties.

The EC's main debate a few months ago centered on 'EMU', or how to achieve economic and monetary union after 1992. *International Management* Mar. 1990, p. 21

EC monetary officials interpreted Mr Major's emphasis on the elements of agreement between the British government and the other EC countries on crucial aspects of the plan for EMU as a deliberate signal of a new line in London. *Guardian* 2 Apr. 1990, p. 8

EMU² ⚐ see ECU

enterprise culture /ˈɛntəpraɪz ˌkʌltʃə(r)/ *noun* ⚐ ⚏

A capitalist society in which entrepreneurial activity and initiative are explicitly encouraged; a culture founded on an individualistic, go-getting economic ethic.

Formed by compounding: a *culture* founded on (business) *enterprise*. In general, *enterprise* has been a favourite word in the economic vocabulary of the Conservative government in the UK during the eighties and nineties: see also ENTERPRISE ZONE below.

Put forward by Sir Keith Joseph and other prominent Conservatives from the early eighties in the UK, the *enterprise culture* was modelled on the spirit of free enterprise which characterized US society. In the UK it found its expression principally in various schemes to encourage small businesses and financial self-reliance, as well as in the fostering of a more individualistic and materialistic atmosphere in British society.

At the age of 27 she has embraced the enterprise culture and established Upstage Theatre.

Blitz Jan. 1989, p. 11

They are required to ... review their courses and explain how they are going to alter them in the light of the career prospects of their students, the enterprise culture, 1992 ... and, for all I know, the end of the world. *Modern Painters* Autumn 1989, p. 78

enterprise zone /ˈɛntəpraɪz ˌzəʊn/

An area in which a government seeks to stimulate new enterprise by creating financial incentives (such as tax concessions) for businesses.

Formed by compounding: a *zone* in which *enterprise* is actively fostered.

Enterprise zones were first discussed in the late seventies, principally as a way of revitalizing economically depressed areas of inner cities, where there tended to be high levels of unemployment and relatively little investment. The idea has been tried in various parts of the world during the past ten years, including the US, the UK, and Australia.

> The enterprise zone ... development will become the norm in Wales, as more service industries requiring office space move to the area. *Building Today* 22 June 1989, p. 26

E number /ˈiː ˌnʌmbə(r)/ *noun*

A code number in the form of the letter E followed by a group of digits, used as a standard way of referring to approved food additives when listing ingredients on food or drink labels under EC regulations; by extension, an additive (especially the additive to which a particular code refers). Sometimes abbreviated to **E**, an additive.

The initial letter of *Europe(an)* in a compound with *number*.

The European Commission recommended in 1977 that all food additives should be declared by their name or their *E number*; by 1986 this was compulsory except in the case of flavourings. As the eighties progressed, and particularly after the publication in 1984 of Maurice Hanssen's book *E for Additives*, public awareness of *E numbers* grew steadily in the UK. By the early nineties, *E number* was often abbreviated to *E* alone and both terms were popularly used to refer to the additives themselves rather than the codes (a point which was picked up and exploited in a number of food-advertising campaigns). This resulted in labelling and advertising copy which used **E-free** as a synonym for *additive-free*.

> Apparently the effect of Es on Yuppie kids is dramatic. A simple glass of orange squash or a packet of crisps can bring them out in a rash or drive them barmy. *Today* 21 Oct. 1987, p. 36

> It's not so long since we learned the link between eating certain 'E' numbers and the behaviour of highly disruptive children. *She* Oct. 1989, p. 2

environment[1] /ɪnˈvaɪərənmənt/ *noun*

Usually with the definite article, as **the environment**: the sum of the physical surroundings in which people live; especially, the natural world viewed as a unified whole with a pre-ordained interrelationship and balance among the parts which must be conserved. Hence sometimes used in an extended sense: conservation of the natural world; ecology.

A specialized use of *environment*, which literally means 'surroundings', and had been used in the sense of the particular set of physical features surrounding a person or thing since the early nineteenth century.

This sense of *environment*, which in the late eighties and early nineties has been the dominant general sense, grew out of the concern about the natural world—particularly the effects upon it of industrialization and pollution—which was first expressed in any concerted way in the sixties. By the early seventies, some governments were taking enough notice of these concerns to appoint a *Minister* (or *Secretary*) *for the Environment* (colloquially *environment minister, secretary*); but the real vogue for this word only came in the second half of the eighties, after green politics took off in Europe and politicians in general realized that *the environment* promised to be the central political concern of the nineties. From the late eighties onwards, *environment* was frequently used in combinations, too, the most important being *environment-friendly* (see -FRIENDLY). The playfully formed opposite of this is *environment-unfriendly* (see UNFRIENDLY[2]) or **environment-hostile**; other combinations include **environment-conscious(ness)** and **environment-minded(ness)**.

> President Bush said that the environment was now on the 'front burner' and that no other subject, except the anti-drugs campaign, had aroused such fervour among his summit colleagues.
> *Guardian* 17 July 1989, p. 20

> A campaign is being launched to encourage sustainable development within our cities. The status 'Environment City' will be awarded to the four coming nearest to the ideal.
> *Natural World* Spring/Summer 1990, p. 7

> We have to have a government-backed labelling scheme before consumers throw up their hands in horror and revert to their old 'environment-hostile' ways. *She* Aug. 1990, p. 122

environment² /ɪnˈvaɪərənmənt/ *noun* 🖳

In computing jargon, the overall structure (such as an operating system, a collection of software tools, etc.) within which a user, a computer, or a program operates or through which access can be gained to individual programs.

Another specialized use of the sense described above; the *environment* is still the sum total of the surrounding structure, but limited to the restricted world of the computer system. This metaphor of a restricted world is often extended to refer to the ability of a computer user to communicate only in one programming or operating language while in that language's *environment*, as if in a foreign country where only that language is spoken.

Computer scientists have spoken of an integrated structure of tools or an operating system as an *environment* since at least the early sixties. What brought the term into popular use was the rapid development of home and personal computing in the late seventies and eighties.

> In Applications-by-Forms, the 4GL development environment, the interface includes a visual catalog for ease of use. *UnixWorld* Sept. 1989, p. 142

> Designed with the user in mind, the A500 features a friendly WIMP environment and comes supplied with a free mouse. *CU Amiga* Apr. 1990, p. 93

environmental /ɪnˌvaɪərənˈmɛnt(ə)l/ *adjective* 🌳

Concerned with the conservation of the environment (see ENVIRONMENT¹); hence, serving this cause: not harmful to the environment, environment-friendly.

A sense development of the adjective which arises directly from the use of *environment* as a kind of shorthand for 'conservation of the environment'.

The use of *environmental* in this sense seems to have begun in the US towards the end of the seventies, when advertisers first attempted to climb on to the bandwagon of concerns about the environment. In its more general sense 'to do with the conservation of the environment' it is used in a great variety of grammatical constructions; one of the recent ones, **environmental labelling**, is even more elliptical than most, contracting 'to do with the effects of the thing labelled on the conservation of the environment' to a single word. In local government and also in the private sector the term **environmental services** (first used as long ago as the late sixties) seems to have become the fashionable way to refer to the upkeep of the local environment, such as parks and public gardens, waste disposal (including the management of hazardous wastes), and street cleaning. See also *environmental friendliness* (under -FRIENDLY).

> Right Guard spray deodorant . . . now directs itself toward ecological armpits with the epithet 'new environmental Right Guard'. *American Speech* Spring 1983, p. 94

> The Labour Party is planning to issue a 'Green Bill' later this year, setting out its plans for tackling atmospheric pollution, and its proposals for environmental labelling, litter control, handling hazardous waste, and improving water quality. *Guardian Weekly* 30 July 1989, p. 4

> An environmental meeting in Bergen at which ministers from ECE's member countries discussed practical steps to promote 'sustainable growth', the catch-phrase . . . for economic growth that does not destroy the environment. *EuroBusiness* June 1990, p. 64

environmentalism /ɪn͵vaɪərən'mɛntəlɪz(ə)m/ *noun*

Concern with, or support for, the preservation of the environment (see ENVIRON-MENT¹); GREEN politics or consumerism.

A new sense of *environmentalism* which also arises directly from the recent use of *environment*; previously, *environmentalism* was the name of the psychological theory that it is our *environment* ('nurture') rather than our inborn nature that determines individual or national character.

The term *environmentalism* was first used in this sense in the US in the early seventies, at a time when the ECOLOGY movement was starting to gain some public support, but was still widely considered to be the concern of freaks and hippies. In its early uses, the word therefore had a rather derogatory nuance; this was completely turned round in the late eighties, as green ideas became both acceptable and desirable as a replacement for the conspicuous consumption of the first half of the decade. **Environmentalist**, which is used both as an adjective and as a noun, has a longer history than *environmentalism* but has enjoyed the same transformation from negative to positive connotations in the media.

> Even some politicians on the other side of the trenches felt the need to identify themselves with environmentalism. *Sports Illustrated* 15 Nov. 1982, p. 24

> The kind of environmentalism that is finding favour with Bush and his friends in industry has a new slant, substituting the power of market forces for moral outrage and blanket control measures. *Nature* 22 June 1989, p. 570

> Environmentalism is the new religion for the 'us generation' replacing the 'me generation', according to a report released this week. *Courier-Mail* (Brisbane) 4 May 1990, p. 50

environmentally /ɪn͵vaɪərən'mɛntəlɪ/ *adverb*

As regards the conservation of the environment (see ENVIRONMENT¹); used especially to qualify an adjective, as in:

environmentally aware, of a person or group: informed about contemporary concerns for the environment; sensitive to the effect upon the environment of a product, activity, etc.;

environmentally friendly, environment-friendly (see -FRIENDLY);

environmentally sensitive, of a geographical area: officially recognized as containing a habitat for rare species or some other natural feature which should be protected from destruction;

environmentally sound, of a product: having no harmful effects on the environment; environment-friendly.

Most of these formations use *environmentally* in a way which can be predicted from the developments in the use of *environment* (see above); the exception is *environmentally friendly*, which involves a grammatical development as well. The original term *environment-friendly*, modelled on USER-FRIENDLY in computing, implies a dative construction: 'friendly to the environment'. Once the hyphen was dropped and the free-standing adjective *friendly* also acquired the meaning 'harmless', it had to be qualified by an adverb—hence *environmentally friendly*.

Work on *environmentally sensitive areas* (abbreviation **ESA**) began in Canada in the mid seventies and soon spread to other industrialized countries; government regulations ensured that economic development, agricultural practices, etc. were not allowed to destroy the natural beauty of these areas. *Environmentally friendly*, by far the commonest of the other combinations, was first used in the US during the mid eighties; it owes its popularity in part to the enthusiasm with which manufacturers began labelling their products with it, sometimes with little foundation—a practice which in the UK led to calls for government regulation of ECO-labelling. New formations with *environmentally* are cropping up all the time: the ones mentioned here are some of the more important and lasting.

One has to be reasonable. The factory means jobs. There is no factory without emissions. It just has to be as environmentally friendly as possible. *Christian Science Monitor* 6 Apr. 1984, p. 9

Under new proposals from the European Commission, member states are empowered to pay farmers to continue with or revert to traditional farming methods in environmentally sensitive areas. *New Scientist* 15 May 1986, p. 30

Nobody can deny that there are occasions on which the careful guiding of a river along its course requires some bank reinforcement. However, there are plenty of sensible materials to hand for the environmentally aware river engineer. Jeremy Purseglove *Taming the Flood* (1989), p. 191

Environmentally friendly household products are big news on the shopping front. *Health Shopper* Jan./Feb. 1990, p. 7

EPOS /'iːpɒs/ *acronym* Also written **Epos** or **epos** 🐾 🖳

Short for **electronic point of sale**, a computerized system of stock control in shops, in which BAR-CODES on the goods for sale are scanned electronically at the till, which is in turn linked to a central stock-control computer.

The initial letters of *Electronic Point Of Sale*; its inventors probably chose to add *E* (for ELECTRONIC) to the already existing *POS*, point of sale.

EPOS was introduced in the early eighties and by 1990 was widely used in the larger chains of stores. In order for *EPOS* to be used, all goods must carry a bar-code and special electronic tills must be installed, making the changeover an expensive business; one large chain even uses *EPOS* as a verb meaning 'to convert (goods, a shop, etc.) to an *EPOS* system'.

The barcoding of books by their publishers is crucial to the success of the WHS epos system. *Bookseller* 1 Mar. 1986, p. 819

All of the supermarkets (except Waitrose) now have some branches with the EPOS [Electronic Point of Sale] system. *Which?* Feb. 1990, p. 69

I Eposed Oxford—that's where the grey hairs came from. *Bookseller* 26 Apr. 1991, p. 1232

See also EFTPOS

ERM 🐾 see EMS

ESA 🌳 see ENVIRONMENTALLY

etext 🖳 see ELECTRONIC

ethical investment /ˌɛθɪk(ə)l ɪnˈvɛstmənt/ *noun* 🐾

In financial jargon, investment which takes account of the client's scruples by screening the companies to be invested in for their business morality and social outlook.

A transparent combination of *ethical* and *investment*.

The demand for *ethical investment* began in the US in the early eighties and was a natural consequence of the drive to involve ordinary people in capital investment; clearly some customers would not feel happy about handing over their portfolios only to find that they were unwittingly supporting companies whose principles they were unable to agree with. Investments which customers have wanted to avoid have included the politically questionable (notably companies with South African connections), the armaments industry, and companies making 'unhealthy' products (especially tobacco and alcohol). *Ethical investment* became fashionable in the UK and Australia during the second half of the eighties.

The latest craze to be imported from America is for 'ethical investment'. Almost every week, there seems to be a new unit trust launched which promises to invest your money only in 'socially screened' firms. *Daily Telegraph* 25 Sept. 1987, p. 20

Labor backbencher Mr Hayward told Parliament last night that Queensland should legislate to attract 'ethical investment' by superannuation and other funds. *Courier-Mail* (Brisbane) 29 Sept. 1988, p. 26

ethnic /'ɛθnɪk/ *adjective* 🎵 🎭

Of pop and rock music: inspired by, or incorporating elements of, the native music of a particular ethnic group. Especially in **ethnic pop** or **ethnic rock**, pop or rock music which fuses native musical traditions with Western rock styles.

A development of the adjective *ethnic* in the sense 'of or pertaining to (a particular) race'; by the mid sixties the adjective was already being used in the more general sense of 'foreign', and this development is simply an application of that sense in a particular context.

The adjective *ethnic* has been applied to folk and modern music for some decades, but the fashion for *ethnic* elements in pop and rock music dates from the late seventies. The distinction between *ethnic music* and WORLD MUSIC is often not clearly drawn.

> As majors attempt to follow Island's commendable packaging of ethnic music, they rely on yet another promotional push to find Africa's Bob Marley. *Blitz* Jan. 1989, p. 35

> Shanachie, the New Jersey-based record company that has specialized in funky international ethnic pop, recently put out two Mahlathini albums. *Washington Post* 15 June 1990, section 2, p. 17

Euro¹ /'jʊərəʊ/ *noun* 📇

Either a European *or* a Eurocommunist (see EURO-).

Formed by shortening *European*, probably under the influence of the combining form *Euro-* used as a free-standing adjective; compare *Brit* used as a noun.

These two rather different uses have been current since the mid eighties; the sense 'a Eurocommunist' really belongs to the jargon used by Communists among themselves, while the more general sense 'a European' is a colloquial nickname for all Europeans (including the British) in the US, but largely limited to continental Europeans (or those in favour of European integration) when used by the British. In this latter use it was particularly topical during the debate about European integration (see EMU¹).

> I'm the only person I know that tries to persuade both Euros and Tankies to join the Labour Party. *Marxism Today* May 1985, p. 9

> Why didn't we assert British Rule and make the Euros change to furlongs and chains, bushels and pecks? *Listener* 6 Feb. 1986, p. 43

> There are the chic Euros on holiday, the armies of retired people, and the smart 'Miami Vice' clones. *Newsday* 5 Jan. 1989, p. 2

> A dense fog of rhetoric in which the Thatcherites insist on their commitment to co-operation and the Euros insist on their devotion to British sovereignty. *Spectator* 20 May 1989, p. 6

Euro² /'jʊərəʊ/ *noun* 📈

Colloquially in finance (especially in the US): a Eurobond, Eurodollar, Eurodollar future, or other item traded on the Euromoney markets (see EURO-).

Formed by abbreviating *Euromoney* or any of the other financial terms formed on *Euro-*.

Although probably in spoken use for some time, *Euro* in this sense did not start to appear in print until the early eighties, at first as a shorthand for *Eurodollar future*. These futures were traded especially at the Chicago Board of Trade, the New York Futures Exchange (from 1981), and the London International Financial Futures Exchange (from 1982). By the end of the eighties the abbreviated form *Euro* had become very common in financial writing and was no longer limited to Eurodollar futures.

> Euros have a very good correlation with domestic CDs—so good, in fact, that maybe the market will not need both contracts. *American Banker* 9 July 1981, p. 11

> Euros tend to remain liquid for a longer period ... If people would downgrade the definition of liquidity ..., you would find a lot of Eurobonds are liquid. *Institutional Investor* May 1988, p. 105

Euro- /ˈjʊərəʊ/ *combining form* 📷

The first part of the name *Europe* and the adjective *European*, widely used in compounds and blends relating to Europe, the European Community, or the 'European' money market. Hence as a free-standing adjective: European, conforming to EC standards or belonging to a European institution.

The first two syllables of *Europe* or *European*, *Euro-* began as a regular adjectival combining form with the function of linking two adjectives together, as in *Euro-American*, *Euro-African*, etc.

Like ECO-, *Euro-* has enjoyed two fashionable periods in English, the first during the sixties (when British membership was first under discussion) and the second more recently, as EC institutions and standards have begun to impinge more on the British way of life and a greater degree of European integration has been under discussion. When the European Common Market was first set up in the late fifties, it was nicknamed *Euromarket* or *Euromart* by some (perhaps in imitation of *Eurovision*, which had begun in the early fifties), and this began the earlier fashion for formations with *Euro-*. The *Euro-* words of the sixties included **Eurocrat** (a European bureaucrat), **Europarliament**, **Eurofarmer**, and several terms to do with the *Euromarket* in the sense of the 'European' financial markets (such as **Eurobond** and **Euro-issue**). In the seventies came (amongst others) **Eurocentrism** (or **Eurocentricity**), **Euro-MP**, **Eurosummit**, and **Eurocredit**.

The rapid growth of the market in **Eurocurrencies** (some of which are exemplified below) and in *Eurobond* trading has meant that *Euro-* has been one of the most fashionable combining forms for financial terms during the eighties and early nineties (examples include **Euroconvertible**, an adjective or noun applied to Eurobonds which can be converted into another type of security, and **Euroequity**, an international equity issue).

By the late seventies it had also become a fashionable combining form for all consumer products, packaging, etc. produced to EC standards (including **Eurobottle**, **Euro-pack**, **Europass**, and **Eurocode**) as well as for the standards themselves (**Eurostandards**). Europe has also been blamed (although perhaps unfairly) for the design of the large wheeled rubbish bin known as a **Eurobin** or WHEELIE BIN. EC standards and regulations themselves came in for some criticism for their use of gobbledygook, which came to be known as *Eurobabble* (see -BABBLE), **Eurojargon**, **Eurolingo**, or **Eurospeak**. The apparent inability of EC countries to cope with the commercial challenges of new technology gave rise to the term **Eurosclerosis** in the early eighties, but this tended to die out in the late eighties as the single European market of 1992 approached and a more optimistic view was taken of the economies of the Twelve.

Nevertheless there was much discussion of the pros and cons of European integration in the late eighties, and the issue certainly contributed to the downfall of Margaret Thatcher, who was considered Britain's leading **Euro-sceptic**. Quite independently of the EC, an important political development of the second half of the seventies was the rise of **Eurocommunism**, a brand of communism which emphasized acceptance of democratic institutions and sought to influence European politics from within; in the mid eighties the **Eurocommunists** and **Eurosocialists** sought to resolve their differences and re-form under the more general heading of the **Euroleft**. The music scene also had a vogue for *Euro-* words, with **Eurodisco**, **Europop**, and **Eurorock**. In the late seventies and eighties there was opposition to the deployment of **Euromissiles** and heated discussion in the US over **Eurosubsidies** given to European firms setting up business or marketing products there.

From the beginning *Euro-* was popular in proper names (for organizations, projects, etc.)—examples include **Eurocontrol** for air-traffic control from the early sixties, **Eurotransplant** for an international file of potential donors in the early eighties, and more recent formations such as **EuroCypher**, an encryption system for satellite transmissions, and **Eurotunnel**, the Anglo-French consortium which undertook the building of the channel tunnel—and in these cases the capital initial was usually kept. In other *Euro-* words, though, there is a tendency for the capital to be replaced by a lower-case initial once the word becomes established, and for hyphenated forms to be joined up into a solid word. Occasionally *Euro* (or *euro*) is used as a free-standing word operating as an adjective and simply meaning 'European' (see the examples below).

Mrs Thatcher is seen in most of the EEC as a Euro-sceptic at best.　　*The Times* 30 June 1986, p. 9

A maximum fine of £1,000 is proposed for owners of all lawnmowers which fail to 'produce a noise of acceptable EEC standard, or Euronoise'.　　*Independent* 4 Dec. 1986, p. 1

Though far larger than the domestic stockmarket, the eurodollar market does not directly involve the general public.　　Michael Brett *How to Read the Financial Pages* (1987), p. 2

Investors in Industry . . . yesterday made its first foray into the Euroyen market with the issue of a 12 billion yen . . . bond, only the third conventional Euroyen issue by a British company.　　*The Times* 14 Feb. 1987, p. 18

The Euro terrorists announced . . . that they had set up a 'Western European Revolutionary offensive'.　　*Evening Standard* 24 Mar. 1987, p. 7

While outside influences transform Euro-pop, white America sticks to some well-tested styles.　　*Guardian* 7 July 1989, p. 33

The Communists meanwhile have split into two separate groups; a 28-strong 'Euro' tendency led by the Italian PCI, and an 'orthodox' grouping of French, Greek and Portuguese communists and the single Irish Workers' Party member.　　*Guardian* 24 July 1989, p. 3

The name Britannia had been dropped from the deal because its nationalistic connotations could have obvious drawbacks in a pan-Euro venture.　　*European Investor* May 1990, p. 57

It would be very regrettable if anyone sought to divert the party down a Euro-sceptic path.　　*Daily Telegraph* 29 Nov. 1990, p. 2

How Euro are you?　　*Radio Times* 18 May 1991, p. 72

Eurobabble 　see –BABBLE

European Currency Unit 　see ECU

European Monetary System 　see EMS

Eve 　see ADAM

exchange rate mechanism 　see EMS

Exocet /ˈɛksəsɛt/ *noun* and *verb*

noun: The trade mark of a kind of rocket-propelled short-range guided missile, used especially in sea warfare. Used figuratively: something devastating and unexpected, a 'bombshell'.

transitive or *intransitive verb*: To deliver a devastating attack on (something) with, or as if with, an Exocet missile; to move as if hit by a missile, to 'rocket'.

A direct borrowing from French *exocet*, literally 'flying fish'; the missiles are made by a French company and they skim across the surface of water like flying fish, making them virtually impossible to detect and destroy.

The name has been registered as a trade mark in the UK since 1970, but came to prominence during the Falklands war of 1982. In particular, the destruction of Royal Naval ships by Argentinian *Exocet* missiles during that conflict helped to establish the figurative use of the word, both as a noun and as a verb.

Then he produced his Exocet: a copy of your most recent readership survey.　　*New Statesman* 27 Sept. 1985, p. 13

The full range of missiles—notably the Exocet, whose very name . . . has become synonymous with highly efficient death and destruction—will be on display.　　*The Times* 10 June 1987, p. 20

Burton's family are furious at Sally's decision to sell the family home . . . Their Exocet reply is to back a critical biography of the late screen hero.   *Telegraph* (Brisbane) 6 Jan. 1988, p. 5

I presented the bristle end of a broom to the back end of the pony, which exoceted up the ramp into the trailer.   *Daily Telegraph* 16 Dec. 1989, Weekend section, p. vii

expansion card see CARD²

expert system /ˌɛkspɜːt ˈsɪstəm/ *noun*

A computer system using software which stores and applies the knowledge of experts in a particular field, so that a person using the system can draw upon that expertise to make decisions, inferences, etc.

Formed by compounding: although not itself *expert*, the *system* is founded on expert knowledge, proving the truth of the maxim that a computer system can only be as good as the input it receives (a principle in computing that is known by the acronym *GIGO* /ˈɡaɪɡəʊ/, or *garbage in, garbage out*).

The first *expert systems* were developed in the second half of the seventies; they have proved very successful and popular, especially in diagnostic work, because of their ability to consider large numbers of symptoms or variables at one time and reach logical conclusions.

The technology of expert systems is said to have now matured to a point where it can help manufacturers improve productivity and hence their competitive position.   *British Business* 14 Apr. 1989, p. 9

explosive device see DEVICE

• •

F

F   see FIBRE

faction /ˈfækʃ(ə)n/ *noun*

A blend of fact and fiction, especially when used as a literary genre, in film-making, etc.; documentary fiction. Also, a book, film, etc. that uses this technique.

Formed by telescoping the words *fact* and *fiction* to make a blend.

The word was invented in the late sixties, when there was a fashion for novels based on real or historical events. In the eighties, the term was also applied to the dramatized television documentaries sometimes called *docudramas* or *drama-docs* (see DOC, DOCU-). The adjective used to describe a work of this kind is **factional** or **factionalized**; the process of combining fact and fiction into a narrative is **factionalization**.

His Merseyside is vivid enough, every bit as 'real' as those fictionalised documentaries we are learning to call 'faction'.   *Listener* 30 June 1983, p. 16

Factional drama will be discussed in detail at a BBC seminar.   *The Times* 13 July 1988, p. 1

Humphrey's . . . *No Resting Place* . . . offers a factionalised account of Indian history.   *Literary Review* Aug. 1989, p. 14

factoid /ˈfæktɔɪd/ *noun* and *adjective*

noun: A spurious or questionable fact; especially, something that is popularly supposed to be true because it has been reported (and often repeated) in the media, but is actually based on speculation or even fabrication.

adjective: Apparently factual, but actually only partly true; 'factional' (see FACTION above).

Formed by adding the suffix *-oid* (from Latin *-oides* and ultimately derived from Greek *eidos* 'form') to *fact*; the implication is that these spurious pieces of information have the form or appearance of facts, but are actually something quite different.

The word was coined by the American writer Norman Mailer in 1973. In his book *Marilyn* (a biography of Marilyn Monroe), he defined *factoids* as

> facts which have no existence before appearing in a magazine or newspaper, creations which are not so much lies as a product to manipulate emotion in the Silent Majority.

Since it so aptly described the mixture of fact and supposition that often characterized both biography and journalism in the seventies and eighties, *factoid* established a place for itself in the language as a noun and as an adjective.

> Santa Fe is full of writers, which is what he has now become. His speciality is big fat factoids full of real people, especially his old boss.   *The Times* 19 Mar. 1987, p. 17

> The vast bulk of it is devoted to a somewhat breathless and awestruck factoid account of how these difficulties will work themselves out to an inevitable, or at least dauntingly probable, finale.
> *Spectator* 4 July 1987, p. 31

factor VIII /ˌfæktər 'eɪt/ *noun* Also written **factor eight** Ⓧ

A substance in blood which is essential to the coagulation process and is deficient in haemophiliacs.

Substances which contribute to the blood-clotting process have been called *factors* since the early years of this century, and were assigned a series of identifying Roman numerals by medical researchers. This is the eighth in the series.

Although congenital *factor VIII* deficiency had been identified as the cause of haemophilia by the fifties, the term did not become widely known until the AIDS era. In the mid eighties, before the implications of Aids for the blood donor system were fully understood, thousands of haemophiliacs worldwide were infected with the Aids virus HIV as a result of receiving injections to boost their levels of *factor VIII*. This, and the subsequent actions for damages, brought the term *factor VIII* to public attention.

> Doctors, unaware of the cause of his illness, pumped him with huge doses of Factor VIII … But with AIDS becoming a public issue … both he and Elizabeth were aware that the massive transfusions of blood could well have exposed him to the virus.   *New Idea* (Melbourne) 9 May 1987, p. 8

> More than 1,200 haemophiliacs were infected with the Aids virus after treatment with contaminated Factor VIII, a blood-clotting agent that was administered through the NHS.
> *Sunday Times* 30 Sept. 1990, p. 1

fanny pack /'fænɪ ˌpæk/ *noun* Also written **fannypack** ▨

The US slang name for a BUM-BAG.

Formed by compounding; in US slang, *fanny* is the equivalent of British slang *bum* and has none of the sexual connotations of the British English *fanny*.

Fanny pack has a similar history in US English to that of *bum-bag* in British English, arising as long as twenty years ago as a term used by skiers, motorcyclists, etc. (sometimes with variations on the name, such as *fanny bag* or *fanny belt*) and moving into the more general vocabulary when the idea was taken up by the fashion world in the late eighties. As a fashion accessory in the US, the *fanny pack* has also been called a *belly-bag*, reflecting the fact that it is worn at the front rather than the back (see BUM-BAG) or *belt bag*, avoiding all reference to human anatomy.

> I've hurt myself and my cameras numerous times … but I've never had a problem, even doing an egg-beater at full speed, with my gear tucked away inside a fannypack.   *Sierra* Jan.–Feb. 1985, p. 45

> Christin Ranger … says her company put out six versions this year (compared with only two last

year), including larger fanny packs that hold lunches or tennis shoes and front-loaders with just enough room for a wallet.
Newsweek 5 Dec. 1988, p. 81

fast-food /ˌfɑːstˈfuːd/ *adjective*

Of substances other than food, especially drugs: instant; quick and easy to make, obtain, and use. Also occasionally of non-material things: intellectually accessible; easy to present or understand.

A figurative use of *fast food*, a term which has been used since the fifties in the US and the seventies in the UK for food which is kept hot or partially prepared in a restaurant and so can be served quickly when required. The term *fast food* was used attributively (in *fast-food service*, *fast-food outlet*, etc.) before being used as a compound noun in its own right, so it is hardly surprising that it should now be perceived and used as an adjective, replacing *instant* in some contexts.

Fast-food was first used in this figurative way in the late seventies and was applied to drugs from the middle of the eighties, when the rapid spread of CRACK on the streets of US cities could be attributed to the fact that it was easily made, cheap to buy, and instantly smokable—it seemed to drug enforcement agencies that anyone who wanted to obtain the drug could do so as easily as buying a hamburger. The description provides a useful distinction between the *fast-food* drugs offering instant gratification (like crack and ICE) and the more complex DESIGNER DRUGS, and so has stuck. The term can be applied in its figurative sense also to consumable but non-material things (such as broadcasting or the arts); this is the more established figurative use and may yet prove to be the most enduring as well.

> If he does talk, listen. Do not respond with 'fast-food' answers such as 'Heck, it can't be so bad', or 'Why don't you take the afternoon off?'
> *Industry Week* 9 Mar. 1981, p. 45

> Fast-food opera that will face an anniversary judgment.
> headline in *Guardian* 3 July 1989, p. 19

> A few years ago, all the talk was about more complex, more expensive 'designer drugs'. Ironically it has turned out to be the fast-food drugs like crack and ice . . . that are tearing us apart.
> *People* 13 Nov. 1989, p. 13

fast track /ˈfɑːst ˌtræk/ *noun, adjective,* and *verb* Also written **fast-track** when used as an adjective or verb

noun: A hectic lifestyle or job involving rapid promotion and intense competition; also called the **fast lane**.

adjective: High-flying, enjoying or capable of rapid advancement.

transitive verb: To promote (a person) rapidly, to accelerate or rush (something) through.

A figurative use of the horse-racing term *fast track* (which dates from the thirties), a race-track on which the going is dry and hard enough to enable the horses to run fast; *track* has a long history in US terms to do with careers, for example in the concept of a *tenure track* for academics.

The figurative use of *fast track* in business arose in the mid sixties; it may owe its popularity to US President Richard Nixon, who claimed at that time that he preferred New York to California because it was the *fast track*. Certainly it became a vogue word in US business circles during the seventies, in all its grammatical uses, and developed a number of derivatives: the agent-noun **fast-tracker** (and even **fast-tracknik**), a person who lives or works in the *fast track*; also the verbal noun **fast-tracking**, the practice of promoting staff rapidly or accelerating processes. In the eighties this vogue has spread to British English, although in the UK *fast lane* is still probably better known as the name for the hectic, competitive lifestyle of the YUPPIE.

> Some of the fast trackers seem so preoccupied with getting ahead that they don't always notice the implications of what they do.
> *Fortune* June 1977, p. 160

Many a thrusting young manager or fast-track public servant has had his hopes dashed.

The Times 15 Dec. 1984, p. 7

An assurance was given to 'fast track' the required planning procedures.

Stock & Land (Melbourne) 5 Mar. 1987, p. 3

fatigue [[see COMPASSION FATIGUE

fattism /'fætɪz(ə)m/ *noun* Also written fatism [[

Discrimination against, or the tendency to poke fun at, overweight people.

Formed by adding the suffix *-ism* (as in *racism* and *sexism*) to *fat*.

Fattism is one of a large number of formations ending in *-ism* which became popular in the eighties to describe perceived forms of discrimination (see also ABLEISM, AGEISM, and HETERO-SEXISM). This one belongs to the second half of the eighties, a time when general diet-consciousness and an emphasis on physical fitness in Western societies made being overweight almost into a moral issue. It was coined by American psychologist Rita Freedman in the book *Bodylove* (1988), in which she points out the insidious influence of one's personal appearance on others (in particular the notion that obese people are lazy or undisciplined):

Looksism gives birth to *fatism*, another cruel stereotype that affects us all.

It is usually used only half-seriously, though, as is the corresponding adjective **fat(t)ist**. The adjective appears to be becoming more established in the language than the noun at present, but neither promises to be permanent.

Fatist is a refreshing new word to me, as opposed to fattest which is much more familiar.

Spare Rib Oct. 1987, p. 5

Dawn French makes no apologies about her size, and any frisson of incipient fattism is instantly quashed in her commanding presence.   *Sunday Express Magazine* 25 Mar. 1990, p. 18

Now Ms Wood looks smarter and has lost so much weight, some of her fattist pieces lose their credibility.   *Gay Times* Nov. 1990, p. 71

fatwa /'fætwɑː/, /'fatwa/ *noun* Also written **Fatwa** or **fatwah**

A legal decision or ruling given by an Islamic religious leader.

A direct borrowing from Arabic; the root in the original language is the same verb *fatā* (to instruct by a legal decision) from which we get the word *Mufti*, a Muslim legal expert or teacher.

Actually an old borrowing from Arabic (in the form *fetfa* or *fetwa* it has been in use in English since the seventeenth century), the *fatwa* acquired a new currency in the English-language media in February 1989, when Iran's Ayatollah Khomeini issued a *fatwa* sentencing the British writer Salman Rushdie to death for publishing *The Satanic Verses* (1988), a book which many Muslims considered blasphemous and highly offensive. *Fatwa* is a generic term for any legal decision made by a Mufti or other Islamic religious authority, but, because of the particular context in which the West became familiar with the word, it is sometimes erroneously thought to mean 'a death sentence'.

The ... International Committee ... have capitalized on the outrage felt at the notorious *fatwa* to drive forward with new confidence the long-nurtured campaign for total abolition of blasphemy laws in this country.   *Bookseller* 29 Sept. 1989, p. 1068

This Fatwa ... was written and signed by the Grand Ayatollah of Shia in Iraq, explaining his position regarding the executions of 16 Kuwaiti Pilgrims after the Saudi media quoted his name.

Independent 27 Oct. 1989, p. 10

[He] ... rejected the findings of a BBC opinion poll which claimed that only 42 per cent of Muslims in Britain supported the fatwah.   *Independent* 16 July 1990, p. 5

fax¹ /fæks/ *noun* and *verb* 🖳

noun: Facsimile telegraphy (a system allowing documents to be scanned, digitized, and transmitted to a remote destination using the telephone network); a copy of a document transmitted in this way; a machine capable of performing facsimile telegraphy (known more fully as a **fax machine**).

transitive verb: To transmit (a document) by fax.

An abbreviated and respelt form of *facsimile*; sometimes popularly associated with the respelt form of *facts* in the next entry.

Experiments in different methods of facsimile transmission began in the late nineteenth century; the first successful transmission of a document took place in 1925. *Fax* technology was first written about using this name in the forties, describing a method of transmitting newspaper text by radio rather than by telephone; this was the result of research and development work carried out by the American electrical engineer and inventor John V. L. Hogan during the late twenties and thirties. In 1944, after contributing to military use of facsimile during the Second World War, he was instrumental in forming Broadcasters' Faximile Analysis, a research project linking broadcasters and newspaper publishers in the US, but their plans to provide a facsimile news service in individual homes failed because of licensing difficulties. Legal restrictions on the use of telephone equipment which did not belong to the telephone company also stood in the way of widespread application of telephone *fax*, and the word *fax* remained in the technical jargon of telegraphy until these restrictions were lifted and the machines became widely affordable for business use in the early eighties. By the middle of the eighties, it had already developed the three distinct uses mentioned above as well as being widely used as a verb, and it was commonplace for company notepaper to carry a firm's **fax number** (the telephone number to be dialled to enable the firm to receive a **faxed** document) as well as standard telephone and telex numbers. Derivatives include **faxable** (capable of being *faxed*), **faxee** (a person to whom a *fax* is sent), **faxer** (a sender of *faxes*), **faxham** (a person who uses the *fax* as a radio ham uses short-wave radio to contact unknown enthusiasts), and **faxing** (the sending of *faxes*).

> As the technology improved, fax became faster and cheaper. *Daily Telegraph* 21 Nov. 1986, p. 16

> In a five-storey office building, there may be a fax on each floor.
> *Observer Magazine* 19 June 1988, p. vi

> NFUC sent out several thousand faxes urging the faxees to refax the fax to the fax machines in the governor's office. *Washington Post* 23 May 1989, section C, p. 5

> He had not faxed me specifically, he continued, since he did not know me from Adam—the faxham simply tapped arbitrarily into the void ... hoping sometime, somewhere, to encounter responsive life.
> *The Times* 20 Mar. 1990, p. 14

fax² /fæks/ *plural noun*

Colloquially, facts, information, 'gen'.

A playful respelling of *facts* (compare *sox* for *socks*), in this case reflecting the lack of a *t* sound in most people's casual pronunciation of the word.

This spelling of *facts* was devised by Thackeray in his *Yellowplush correspondence: Fashnable fax and polite annygoats*, first published in 1837. It has been common in popular magazines and newspapers using normal modern orthography since about the 1970s and had formed the second element of trade marks (see CEEFAX and FILOFAX) for decades before that. However, it was only when the Filofax and facsimile (FAX¹) became fashionable in the eighties that *fax* really acquired any popular currency as a word in its own right; the increasing emphasis on information as a commodity in eighties culture has helped it to establish a place in the language that is not simply a newspaper editor's pun.

> Eco-fax. These pages are designed for you to fill in the address and/or telephone numbers you may need. John Button *How to be Green* (1989), p. 230

fax-napping ✂ see FILOFAX

FF ✂ see FUNCTIONAL FOOD

fibre /'faɪbə(r)/ *noun* ⊗ ✂

Food material such as bran and cellulose that is not broken down by the process of digestion; roughage. Often in the fuller form **dietary fibre**; occasionally abbreviated to **F**, especially in the US trade mark **F Plan Diet** (or **F-Plan**), a weight-reducing diet based on a high fibre intake to provide bulk without calories.

A specialized use of *fibre* in its collective sense of 'matter consisting of animal or vegetable fibres'.

Scientists have written about *fibre* in this sense since the early years of this century; what brought it into the more popular domain and made it a fashionable subject was the discovery in the seventies that a *high-fibre* diet could help to prevent certain digestive illnesses, including cancers of the colon, diverticular disease, and irritable bowel syndrome. In the eighties, the GREEN movement added impetus to this by stressing the need to concentrate on natural, un-processed foods (the highly refined foods which most people in developed countries normally eat contain relatively little *fibre*). The *F-Plan* diet (the book of which was published in 1982) is one of many diets put forward in the eighties which emphasize the need for *fibre*, and the word now seems to have taken over from the more old-fashioned *roughage* in popular usage.

> The newly promoted F plan diet, which underlined the nutritional value of beans, fortuitously coin-cided with the Heinz campaign message. 'They were talking fibre; we were talking goodness.'
>
> *Financial Times* 18 Aug. 1983, p. 9

> Bran is one type of fibre, nature's own 'filler' that is present only in plant foods and is essential for proper digestion. *Here's Health* Apr. 1986, p. 127

> Get into a wholefood diet routine, sticking to high-fibre low fat foods, plenty of salads, fresh fruit and vegetables. *Health Shopper* Jan./Feb. 1990, p. 9

Filofax /'faɪləfæks/, /'faɪləʊfæks/ *noun* 〰 ✂

The trade mark of a type of loose-leaf portable filing system; a PERSONAL ORGANIZER.

A respelling of *file of facts* which is meant to reflect colloquial pronunciation.

The *Filofax* has been made for several decades (the trade mark was first registered in the early thirties), but the name was not widely known until the early eighties, when it suddenly became fashionable (especially for business people) to carry a *Filofax*. These small loose-leaf folders usually contain a diary and other personal documentation such as an address book, planner, note section, maps, etc., as well as a wallet with spaces for a pen, credit cards, and other small non-paper items. In the mid eighties the *Filofax* was associated particularly with the yuppie set—the word was even used attributively in the sense 'yuppie'. By the end of the decade all sorts of people could be seen with *Filofaxes*—or with one of the numerous imitations of the *Filofax* proper—and a growing market developed for different types of **filofax insert**. So popular were they that variations on the theme started to appear—notably **Filofiction**, novels pro-duced on hole-punched sheets to fit a *Filofax*. (Some other examples of the birth of *filo-* as a combining form are given in the quotations below.) **Filofax** is even occasionally used as a verb, meaning 'to steal a *Filofax* from (someone) in order to demand a ransom for its return'—a crime apparently known colloquially as **filo-napping** or **fax-napping**.

> The Digger guide to Metropolitan Manners No 1: Yup and Non-Yup by Ivor Pawsh (Advice: consult filonotes when reading this). *Digger* 9 Oct. 1987, p. 26

> Small neat people tend to go for the small neat organizers while fatsos nearly always buy large Filo-faxes and stuff them fit to burst. *The Times* 10 June 1988, p. 27

> An advertisement in last week's *Bookseller* for Filofiction—or what the publishers describe as 'publishing's brightest new idea'. *New Scientist* 28 July 1988, p. 72

Taxpak '89 is a new filofax insert detailing the Budget changes, enabling you to check your income tax allowance. *Investors Chronicle* 17–23 Mar. 1989, p. 35

One of the more Americanised [pop groups] of England's filofax funksters. *Listener* 4 May 1989, p. 36

The filoflask . . . a normal personal organiser but with a hip flask fitted inside, is being marketed. *The Times* 14 June 1990, p. 27

finger-dry /ˈfɪŋɡəˌdraɪ/ *transitive verb* ✖️

To style and dry (the hair) by running one's fingers through it to lift it and give it body while it dries naturally in the warmth of the air. Also as an adjective **finger-dried** and action noun **finger-drying**.

A transparent combination of *finger* and *dry*; the warmth from the fingers apparently also helps to dry the hair.

Hair has no doubt been *finger-dried* since the beginning of time; the technique was only graced with the fashion term *finger-drying* at the beginning of the eighties, when hairdressers sought a more natural look than could be achieved with the *blow-dried* styles of the seventies.

Howard layered Jocelyn's hair, and finger-drying brought out its natural movement. *Woman's Realm* 10 May 1986, p. 29

An advance on the razor is the new texturising technique which forms a feathery, textured look and is ideal for finger-dried styles. *Cornishman* 5 June 1986, p. 8

flak /flæk/ *noun* 〰️ 📷

In business and political jargon, short for **flak-catcher**: a person employed by an individual or institution to deal with all adverse comment, questions, etc. from the public, thereby shielding the employer from unfavourable publicity.

Formed by a combination of semantic change and abbreviation. *Flak* was originally borrowed into English from the German initials of a compound word meaning 'pilot defence gun' in the Second World War, for an anti-aircraft gun and (by extension) anti-aircraft fire; by the late sixties it was being used figuratively to mean 'a barrage of criticism or abuse'. The sense under discussion here arose by shortening the compound *flak-catcher* to *flak* again, perhaps involving some confusion with the word *flack*, an established US term for a press agent which was allegedly coined quite independently by the entertainment paper *Variety* in the late thirties. *Variety* claimed that this word for a press agent was the surname of Gene Flack, a well-known movie agent.

An example of a well-established Americanism that has only gained a place in British English in the past few years. The term *flak-catcher* was popularized at the beginning of the seventies in the US (by the writer Tom Wolfe in *Mau-Mauing the Flak Catchers*); the name was apt enough to stick in US English, and to be applied in British English as well during the seventies to those slick spokesmen who can turn any question to the advantage of the government or organization whose image they are employed to protect. The abbreviation to *flak* belongs to the late seventies in the US and the eighties in the UK. The form **flak-catching** (as an adjective or noun) also occurs.

Spitting Image . . . has firmly established itself as TV's première flak-catching slot. *Listener* 7 Mar. 1985, p. 29

The tone is world-weary, that of the flakcatcher for whom life has become an arduous process of warding off, out-manoeuvring, beating down. *Times Literary Supplement* 31 Oct. 1986, p. 1210

Most U.S. companies employ spokespeople who are paid to parrot the company line . . . To reporters they are derisively known as 'flaks' whose main duties consist of peddling press releases. Bryan Burrough & John Helyar *Barbarians at the Gate* (1990), p. 293

flake /fleɪk/ *noun*

In US slang: an eccentric, dim, or unreliable person, a 'screwball'.

A back-formation from the adjective *flaky*, which in US slang has been used in the sense 'odd, eccentric, unpredictable' since the mid sixties.

Flake was first used in US baseball slang and in college slang generally in the sixties; during the seventies it passed into general slang use in the US, and by the early eighties was becoming more widely known still through its use in political contexts (compare WIMP¹).

> Out in California, Gov. Jerry Brown—often called a *flake*—was campaigning against San Diego Mayor Pete Wilson ... Larry Liebert ... quoted an anonymous Brown aide as asking 'Why trade a flake for a wimp.' *New York Times Magazine* 24 Oct. 1982, p. 16

flashy see GLITZY

flavour of the month /ˌfleɪvər əv ðə ˈmʌnθ/ *noun phrase*

The current fashion; something that (or someone who) is especially popular at a given time. Also with variations, such as **flavour of the week, year**, etc.

A figurative application of a phrase that began as a marketing ploy in US ice-cream parlours in the forties, when a particular ice-cream flavour would be singled out for the month or week for special promotion.

Flavour of the month started to be used figuratively in the news media in the late seventies, and for a while in the early eighties the phrase itself appeared to be *flavour of the month* with journalists. There is often a note of cynicism in its use, implying that the thing or person described as *flavour of the month* is but a passing fashion or whim that will soon be replaced by the next one. It is also sometimes applied to something which is not really subject to fashions, but is especially common or widely reported at a given time.

> In many ways the question of authority in the Church is the theological flavour of the year in Anglican circles. *Church Times* 15 May 1987, p. 7

> Readership surveys were flavour of the month in that sector so he wanted one.
> *Media Week* 2 Sept. 1988, p. 14

> Currently the England dressing room resembles a MASH unit, with finger and hand injuries the flavour of the month. *Guardian* 2 Apr. 1990, p. 15

fly-tipping /ˈflaɪˌtɪpɪŋ/ *noun* Also written **fly tipping** or **flytipping**

In the UK: unauthorized dumping of rubbish on the streets or on unoccupied ground.

Formed by compounding. The *fly-* part is probably ultimately derived from the verb *to fly* (the culprits *tip* and *fly*); it is the equivalent of *fly-posting* (a term which dates back to the early years of this century) except that it involves dumping rubbish rather than putting up posters. Since the thirties, street salesmen have called their unlicensed pitches *fly-pitches*, but this name is probably derived from the adjective *fly*, 'clever'.

The term *fly-tipping* has been used in technical sources to do with waste disposal since at least the late sixties. A topical problem in the Britain of the eighties, *fly-tipping* was the subject of tighter legislation in 1989 to try to tidy up city streets and give the UK a greener image. The term *fly-tipping* has also been applied to the DUMPING of toxic waste in other countries. **Fly-tip** has been back-formed as the verb corresponding to the noun *fly-tipping*; individuals or bodies who do it are **fly-tippers**.

> The LIFT ... Report divides the people who fly tip into four categories: the 'organised criminal', the 'commercial', the 'domestic' and the 'traveller'. The organised criminal fly tipper operates to make money through illegal deposition of wastes.
> *Managing Waste* (Report of the Royal Commission on Environmental Pollution, 1985), p. 71

The Control Of Pollution (Amendment) Bill, to tighten up the law against fly-tippers and stop illegal dumping of builders' rubble, was given an unopposed third reading in the Lords.

The Times 5 July 1989, p. 13

There was the visible evidence of fly-tipping. A mound of rubbish all but obscured an electrical sub-station on which two local hospitals depended. *Independent* 23 Aug. 1988, p. 17

fontware 🖳 see -WARE

food additive ♀ ✖ see ADDITIVE

foodie /'fuːdɪ/ *noun* ✖ 🍴

In colloquial use, a person whose hobby or main interest is food; a gourmet.

Formed by adding the suffix *-ie* (as in *groupie*, etc.) to *food*; one of a succession of such formations during the eighties for people who are fans of, or heavily 'into', a particular thing or activity.

Although gourmets have been around for a long time, the *foodie* is an invention of the early eighties, encouraged by the food and wine pages of the colour supplements and the growth of a magazine industry for which food is a central interest. The *foodie* is interested not just in eating good food, but in preparing it, reading about it, and talking about it as well, especially if the food in question is a new 'eating experience'. An *Official Foodie Handbook* was published in 1984.

He told me about the foodie who sat next to him in a Chinese restaurant and went into transports of enthusiastic analysis about the way in which the chicken had been cooked.

Listener 27 Sept. 1984, p. 19

The oriental chopper . . . —a perfect gift for your favourite foodie, particularly if that happens to be you. *Good Food* Jan./Feb. 1990, p. 11

food irradiation ⊗ ✖ see IRRADIATION

footprint /'fʊtprɪnt/ *noun* 🖳

In computing jargon, the surface area taken up by a computer on a desk or other surface.

A figurative use of *footprint*; the latest in a succession of technical uses employing this metaphor. In the mid sixties, *footprint* had been proposed as the name for the landing area of a spacecraft; from the early seventies onwards it was used for the ground area affected by noise, pressure, etc. from a vehicle or aircraft (an aeroplane's *noise footprint* is the restricted area on the ground below in which noise exceeds a specified level, and the *footprint* of a tyre is the area of contact between it and the ground); it is also used for the area within which a satellite signal can be received.

Interest in the *footprint* of computer hardware began in the early eighties, with the widespread sale and use of PCs and other microcomputers which had to compete for space on people's desks with books, papers, and simply room in which to work. A small *footprint* soon became a selling-point for a microcomputer. In the era of hacking (see HACK), there is some evidence that *footprint* also came to be used figuratively in computing to mean a visible sign left in a file to show that it had been hacked into (the machine-readable equivalent of 'I woz 'ere').

With features like a . . . memory mapper and a footprint of only 12.6 inches by 15.7 inches, it's a difficult micro to fault. advertisement in *Mail on Sunday* 9 Aug. 1987, p. 39

Footsie /'fʊtsɪ/ *acronym* Also written **footsie** or **FT-SE** 〰

In the colloquial language of the Stock Exchange, the Financial Times–Stock Exchange 100 share index, an index based on the share values of Britain's one hundred largest public companies. Also known more fully as the **Footsie index**.

A respelling of *FT-SE* (itself the initial letters of *Financial Times–Stock Exchange*), intended to represent the sounds produced when you try to pronounce the initials as a word.

The *FT-SE index* was set up in January 1984 and almost immediately came to be known affectionately as *Footsie*, perhaps because *FT-SE* /ɛftiːɛsˈiː/ is such a mouthful. Within a few months, traded options and futures which were linked to the index became available and these were described as *Footsie options* etc. (even without a capital initial) almost as though *Footsie* were an adjective. *Footsie* is used with or without *the* to refer to the index; the *100* part of the index's name sometimes follows *Footsie*, especially when the official form, *FT-SE 100 index*, is used.

> The FT-SE 100 (Footsie) Index has already fallen from a peak 1717 early in April to 1565, but if you think calamity lies ahead, it is not too late to buy Footsie Put Options. *Daily Mail* 17 May 1986, p. 30

> With Congress and Administration still deadlocked over the US Budget, the most anodyne political remark is quite capable of shifting Footsie 50 points. *Investors Chronicle* 20 Nov. 1987, p. 29

forty-three ◖◗ see RULE 43

F-plan ⊗ ✖ see FIBRE

-free /friː/ *combining form* ♣ ⊗ ✖

As the second element in a hyphenated adjective: not containing or involving the (usually undesirable) ingredient, factor, etc. named in the word before the hyphen.

A largely contextual development in the use of what is an ancient combining form in English: originally it meant 'exempt from the tax or charge named before the hyphen' (as in *tax-free*, *toll-free*, etc.) and this developed through the figurative sense 'not hampered by the trouble etc. named in the first word' (as in *carefree* and *trouble-free*) to the present use, in which ingredients or processes, often ones formerly thought desirable in the production of something, have been found to be unwanted by some section of the public, and the product is therefore advertised as being free from them.

The sense of *-free* defined here has become particularly fashionable since the late seventies, especially through its use by advertisers (who possibly see it as a positive alternative—with connotations of liberation and cleanness— to the rather negative suffix *-less*). The uses fall into a number of different groups, including those to do with special diets (**alcohol-free, cholesterol-free, corn-free, dairy-free** (an odd term out with **animal-free** in naming the generic source rather than the substance as the first word), **gluten-free, meat-free, milk-free, sugar-free, wheat-free**, and many others), those to do with pollutants or additives (*additive-free* (see ADDITIVE), *Alar-free* (see ALAR), *CFC-free* (see CFC), *e-free* (see E NUMBER), **lead-free**, etc.), those in which an undesirable process or activity is named first (CRUELTY-FREE, **nuclear-free**), and those with the name of an illness or infection as the first element (**BSE-free, salmonella-free**). Occasionally advertisers omit the hyphen, with unintentional comical effect: during the scare about salmonella in eggs in the UK in 1989, for example, some shops displayed posters advertising 'Fresh farm eggs—salmonella free'.

> The Saudis have oil, which the world wants. Now C. Schmidt & Sons, a Philadelphia brewery, has something the Saudis want—alcohol-free beer. *Washington Post* 23 June 1979, section D, p. 9

> Special dishes which are gluten-free, dairy-free and meat-free. *Hampstead & Highgate Express* 7 Feb. 1986, p. 90

> These contain a complex of high potency, dairy-free lactobacilli, good bacteria that help the body to maintain a positive balance. *Health Shopper* Jan./Feb. 1990, p. 4

> The advice of the National Eczema Society is to use either liquids (none of which contains bleaches) or enzyme-free 'non-biological' detergents. *Which?* Apr. 1990, p. 190

> We all feel virtuous because we have gone lead-free; but this is a separate issue from the greenhouse effect. *Good Housekeeping* May 1990, p. 17

> They say they can deliver BSE-free embryos, but no one can guarantee that. *Independent on Sunday* 29 July 1990, Sunday Review section, p. 13

freebase /ˈfriːbeɪs/ *noun* and *verb* Also written **free base** or **free-base**

noun: A purified form of cocaine made by heating it with ether, and taken (illegally) by inhaling the fumes or smoking the residue.

intransitive or *transitive verb*: To make a freebase of cocaine or smoke it as a drug; to smoke (freebase). Also as a verbal noun **freebasing**; agent noun **freebaser**.

Formed by compounding; the *base*, or most important ingredient in cocaine, is *freed* by the process of heating.

The term has been in use in the drugs subculture since the seventies (there are reports of people who claim to have been using *freebase* since 1978, for example), but it was not taken up by the media until 1980, when American comedian Richard Pryor was badly burned while freebasing. It then became clear that *freebase* was a favourite form of cocaine among the Hollywood set, since smoking it was more congenial than 'snorting' cocaine. The cheaper crystalline cocaine, CRACK, was at first also known as *freebase*. The noun and verb appeared simultaneously in printed sources, but it is likely that the noun preceded the verb in colloquial use.

> A police lieutenant said Mr. Pryor had told a doctor the accident happened while he was trying to make 'free base', a cocaine derivative produced with the help of ether.
> *New York Times* 15 June 1980, p. 15

> She recalled that her seven-year-old daughter used to follow her around the house with a deodorant spray because she could not stand the smell of freebasing. *Daily Telegraph* 30 June 1981, p. 15

> A society drugs scandal is introduced as the freebasers start brewing up in their alembics.
> *Times Literary Supplement* 14 Aug. 1987, p. 872

free from artificial additives 🌳 ✕ see ADDITIVE

free radical /ˌfriː ˈrædɪk(ə)l/ *noun* ✕

An atom or group of atoms in which there is one or more unpaired electrons; an unstable element in the human body which, it is thought, can be overproduced as a result of chemical pollution and may then cause cell damage.

Formed by compounding; *free* in its chemical sense means 'uncombined' and *radical* denotes an atom which would normally form part of a compound.

As a chemical term, *free radical* has existed since the beginning of this century. What has brought it into the public eye in the past few years is the interest shown by the alternative health movement and environmentalists in *free radicals* as the apparent link between pollution and late twentieth-century health problems such as cancer and Alzheimer's disease.

> Vincent Lord knew that many drugs, when in action in the human body and as part of their metabolism, generated 'free radicals'. Arthur Hailey *Strong Medicine* (1984), p. 159

> Increasingly essential are the anti-oxidants—vitamins A, C, E and the mineral selenium, which bolster the body's natural defence against disruptive free radicals. Generated in the body as a result of radiation, chemical pollutants, medicinal drugs and stress, free radicals can damage cells and tissues bringing about premature ageing. *Harpers & Queen* Apr. 1990, p. 143

freestyle BMX ✕ 🎮 see BMX

freeware 💻 see -WARE

freeze-frame /ˈfriːzfreɪm/ *noun* and *verb* Also written **freeze frame** ✕ 💻

noun: A still picture forming part of a motion sequence; a facility on video recorders allowing one to stop the action and view the picture currently on the screen as a still.

intransitive or *transitive verb*: To use the freeze-frame facility; to pause (action or a picture) in this way.

Formed by compounding; *freeze-frame* is effectively a contraction of the technical phrase *freeze the frame* as used in cinematography.

Freeze-frame was first used as a noun in cinematography in the early sixties; at that time, before the advent of home videos, the effect was achieved by printing the same frame repeatedly rather than actually stopping on a particular frame, and was also known simply as a *freeze*. The word *freeze-frame* became popularized in the early eighties by the appearance on the general market of video recorders which had the facility; most manufacturers chose to label the control *freeze-frame*, and so it was a natural step to the development of a verb in this form to replace the more cumbersome phrase *freeze the frame*.

> You can freeze-frame sequences for close analysis. *Listener* 12 May 1983, p. 2

> Don't use 'freeze frame' ... for longer than necessary—it increases tape and head wear.
> *Which?* June 1984, p. 250

fresh /frɛʃ/ *adjective*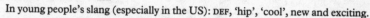

In young people's slang (especially in the US): DEF, 'hip', 'cool', new and exciting.

A sense shift which is perhaps influenced by the pun with *cool*; as a word of approbation in young people's slang it has its roots in RAP talk and ultimately in the street language of HIP HOP.

This is a usage which only began to appear in print in the second half of the eighties, as part of the crop of new slang expressions popularized by the spread of hip-hop culture. A number of rappers used the word in their pseudonyms, and a US sitcom which was centred on hip hop and shown on UK television as well had as its title *The Fresh Prince of Bel Air*.

> Run DMC, the rap group, told it to the audience straighter than most. The other groups at the Fresh Festival, a compendium of rappers and break dancers, had visited Hollywood.
> *Chicago Tribune* 7 July 1985 (Final edition), section 3, p. 5

> According to Freddy, street talkers and rappers long ago abandoned *bad* for such alternatives as *fresh*, *def* and *chillin'*. *Los Angeles Times* 29 Aug. 1988, section 6, p. 2

friendly /ˈfrɛndlɪ/ *adjective*

Of troops, equipment, etc.: belonging to one's own side in a conflict; in specific phrases (such as **friendly fire, friendly bombing**, etc.): coming from one's own side; especially, causing accidental damage to one's own personnel or equipment.

A specialized and slightly elliptical use of the adjective *friendly* in the sense 'not hostile'.

This sense of *friendly* has been in use in military jargon since at least the Second World War (and may go back even further as a noun meaning 'a member of one's own or one's allies' forces'); in the earlier uses, though, *friendly* tended to be followed by *aircraft*, *ships*, etc. The euphemistic phrase *friendly fire* had been used in the Vietnam War (it was chosen in the seventies as the title of a book and film about the parents of a soldier killed by his own side in Vietnam), but was brought to prominence in the Gulf War of 1991, when the majority of fatal casualties among allied troops were attributed to it.

> 'There will be other occurrences of some of our troops potentially being a victim of "friendly fire"', Marine Corps Maj. Gen. Robert B. Johnston, the Central Command's chief of staff, told reporters on Feb. 2. *National Journal* 9 Feb. 1991, p. 335

> Since the war began, more American troops are thought to have been killed by 'friendly fire' than by the Iraqis, most by air-launched missiles. *Independent* 22 Feb. 1991, p. 3

-friendly /ˈfrɛndlɪ/ *combining form*

As the second word in a hyphenated adjective: *either* adapted, designed, or made suitable for the person or thing named in the first word *or* safe for, not harmful to what is named before the hyphen. Hence as a free-standing adjective (often qualified by an adverb): accessible *or* harmless, non-polluting.

Formed on the adjective *friendly*, after the model of USER-FRIENDLY in computing.

One of the most popular ways of forming a new adjective in the late eighties, especially in consumer advertising and writing on environmental issues, *-friendly* has its roots in the extremely successful late-seventies coinage USER-FRIENDLY (the history of which is described under that heading). By the early eighties the computing metaphor was being extended to users of other types of product, sometimes simply as an extension of *user-friendly* itself, but sometimes substituting a new first word (**reader-friendly**, **listener-friendly**, etc.); the gobbledygook of legal drafting was replaced in some legislation by clear, understandable language and this was described as **citizen-friendly**. It was also in the early eighties that the second branch of meaning started to develop, with the appearance on the scene of **environment-friendly** (causing little harm to the environment, ecologically sound); this also gave rise to a stream of imitative formations, notably *ozone-friendly* (see OZONE), **Earth-friendly**, *eco-friendly* (see ECO-), and **planet-friendly**. In the second half of the eighties both branches of meaning grew steadily and became somewhat confused, as new formations arose which did not follow the original pattern. In the sense to do with accessibility and ease of use, for example, the term **computer-friendly** (used of a person, a synonym for *computerate* or *computent* (see the entry for COMPUTERATE) with a nuance of willingness as well as ability to use computers) seemed to turn the tables: the person was now friendly to the computer, rather than the other way round. On the environmental side there were formations like **greenhouse-friendly**, in which the basic meaning 'not harmful to' had been extended into 'not contributing to the harmful effects of' in a potentially confusing way. The fashion for formations in *-friendly* has also led to the use of hyphenated adjectives in which the *-friendly* part means no more than 'friendly' in its usual sense (see the example for *Thatcher-friendly* in the quotations).

There were also grammatical confusions when *-friendly* started to be used as a free-standing adjective. From the late seventies, *friendly* was used as a free-standing word in computing as a synonym for *user-friendly*. As *-friendly* became more and more popular, some sources started to print the compounds with no hyphen between the two words; what is essentially an abbreviated dative phrase 'friendly to . . .' was then interpreted as an adjective qualified by a noun, and this was 'corrected' to an adverb, giving forms such as *environmentally friendly* (see ENVIRONMENTALLY). There were even some examples in which two adjectives were used together, in **environmental friendly** etc. (presumably transferring the adjective from **environmental friendliness**). **Friendliness**, with a preceding noun, and with or without a hyphen, can be used to form noun counterparts for most of these adjectives, but *environmental friendliness* co-exists with **environment-friendliness**.

Companies' requirements for computer-friendly personnel fluctuate dramatically.
The Times 3 Mar. 1987, p. 21

Non-food products such as 'environment-friendly' detergents . . . may not be as widely available.
Which? Jan. 1989, p. 27

Listener-friendly tunes . . . take him close to Michael Jackson in tone and delivery.
Guitar Player Mar. 1989, p. 12

Mitsubishi mixes high performance and environmental friendliness in its new Starion 2.6-litre turbo coupé.
Financial Times 4 Mar. 1989, Weekend FT, p. xxiv

Young people are displaying a lot of behaviour and some attitudes which are Thatcher-friendly.
Listener 4 May 1989, p. 4

It argued that nuclear power had a role to play in a 'greenhouse friendly' electricity supply industry but that this role should not be exaggerated.
Financial Times 18 July 1989, p. 18

Nearly 4,000 products are being analysed according to user- and environment-friendliness in a study sponsored by property developers Rosehaugh.
Sunday Telegraph 13 Aug. 1989, p. 2

On the grocery shelves, garbage and trash bags of all sizes, once the scourge of the environment, now come with planet-friendly certification.
Los Angeles Times 4 Feb. 1990, section E, p. 1

Another well-advanced initiative . . . involves the production of a sterilized sewage and straw compost, a process which disposes of two major pollutants at once, turning them into earth-friendly products which are good growing materials.
The Times 24 Mar. 1990, p. 45

fromage frais /ˌfrɒmɑːʒ ˈfreɪ/, in French /frɔmaʒ frɛ/ *noun* 🔀

A smooth white curd cheese or quark, originally from France; now also any of a number of low-fat dairy desserts based on curd cheese with fruit, sugar, etc. added.

Borrowed from French; literally 'fresh cheese'. This kind of cheese is normally known as *petit suisse* in France, however.

Fromage frais is a product which was introduced to British supermarkets in the early eighties and to American ones a few years later as a way of extending the dairy dessert market in which yogurts were becoming very popular. *Fromage frais* has proved extremely successful as the basis for a whole range of desserts.

> Tell us the fat content of Sainsbury's virtually fat-free fromage frais and you might win a white porcelain gratin dish.   *Good Housekeeping* May 1990, p. 42

> Remove and discard pods, herbs, carrot and celery. Process until smooth with the yogurt or fromage frais, adding a little extra water or skimmed milk to desired consistency.   *She* Aug. 1990, p. 128

front-ending /ˌfrʌntˈɛndɪŋ/ *noun* 💻

In media jargon, direct input of newspaper text by journalists at their own terminals, cutting out the traditional typesetting stage.

Formed by adding the action or process suffix *-ing* to *front end* (the part of a computer system that a user deals with directly, especially a terminal that routes input to a central computer); the term *front end* is used attributively (in *front-end system* etc.), for the 'new technology' which allowed journalists to set their own copy.

Computer scientists used the term *front-ending* from the early seventies to refer to ways of using mini- and microcomputers in networks attached to a single central computer. In the context of newspaper production, the term came into the news in the mid eighties, when the introduction of the system in the UK (especially by the *News International* group producing *The Times*, *The Sunday Times*, *Sun*, and *News of the World*) gave rise to mass picketing by print union representatives who were angry about their members' loss of jobs in typesetting.

> I intend to negotiate the introduction of front-ending and . . . a modern web-offset printing plant.
> *The Times* 10 July 1986, p. 21

fudge and mudge /ˌfʌdʒ ənd ˈmʌdʒ/ *verbal phrase* 🏛

As a political catch-phrase: to evade comment or avoid making a decision on an issue by waffling; to apply facile, ill-conceived solutions to problems while trying to appear resolved.

The verb *fudge* has been used since the seventeenth century in the sense 'to patch up, to make (something) look legitimate or properly done when in fact it is dishonestly touched up'; *mudge* here is probably chosen for its rhyme with *fudge* and influenced by *smudge* or *muddle*, although it might be taken from *hudge-mudge*, a Scottish form of *hugger-mugger*, a noun meaning 'disorder, confusion' but also used as an adjective in the sense 'makeshift'.

The catch-phrase was coined by the British politician David Owen in a speech to his supporters at the Labour Party conference in 1980. In a direct attack on the leadership of James Callaghan, he said:

> We are fed up with fudging and mudging, with mush and slush. We need courage, conviction, and hard work.

Since then it has been used in a number of political contexts, both as a verbal phrase and as a noun phrase for the policy or practice of *fudging and mudging*.

> A short term victory must poison the atmosphere in which much-needed, long-term reforms of pay bargaining are examined. There are occasions on which it is right to fudge and mudge at the margins.   *Guardian Weekly* 14 June 1981, p. 10

Since the Prime Minister has a well-known abhorrence for fudge and mudge, it must be assumed that she agreed to this next step [in joining the European Monetary System] because she intended to take it.
Guardian 28 July 1989, p. 22

full-blown Aids ⊗ see AIDS

functional food / ˌfʌŋkʃən(ə)l ˈfuːd/ *noun* ✖

A foodstuff which contains additives specifically designed to promote health and longevity. Sometimes abbreviated to **FF**.

A translation of Japanese *kinoseishokuhin*.

Functional foods were originally a Japanese idea and by 1990 had an eight per cent share of the Japanese food market. They cleverly turn round the negative connotations of food additives by fortifying foods with enzymes to aid digestion, anti-cholesterol agents, added fibre, etc. and by marketing the foods as beneficial to health—much the same idea as the familiar breakfast cereals fortified with vitamins and iron, but taken a stage further. *Functional foods* have yet to be tested on Western markets.

Unless food manufacturers outside Japan wake up to the market potential of functional foods, a new Japanese invasion of protein-enhanced Yorkshire pudding, high-fibre spotted dick and vitamin-boosted toad-in-the-hole is likely . . . Mr Potter, a food scientist and technologist, explained: 'FF ingredients are products known to have positive health benefits like lowering cholesterol levels, lowering blood sugar, preventing calcium loss from the bone, lowering incidences of heart disease.'
Independent 28 Apr. 1990, p. 3

fundie / ˈfʌndɪ/ *noun* Also written **fundy** or (in discussions of German Green Party politics) **Fundi** /ˈfʊndɪ/ 🌳 🏠

In colloquial use: a fundamentalist; especially *either* a religious fundamentalist *or* a member of a radical branch of the GREEN movement, a 'deep' green.

Formed by adding the suffix *-ie* to the first four letters of *fundamentalist*; the spelling *Fundi* reflects borrowing from the German slang name of the radical wing of the German Green Party.

A nickname which belongs to the political debates of the early eighties, when the Moral Majority and other fundamentalist Christian groups in the US and the Greens in Germany became a political force to be reckoned with. In the green sense, *fundie* has its origins in the arguments from 1985 onwards between the German Greens' *realo* wing, who were prepared to take a normal co-operative approach to parliamentary life, and the more radical fundamentalists, who did not wish to co-operate with other parties and favoured extreme measures to solve environmental problems.

The Fundies are not a serious political force and their current hero is not a serious political candidate.
New York Times 7 Mar. 1988, section A, p. 19

The fundies are the purists who believe the only way to save the Earth is to dismantle industry.
Daily Telegraph 20 Sept. 1989, p. 15

funk /fʌŋk/ *noun* 🎵 ☠

In recent use in popular music, a style that draws upon Black cultural roots and includes bluesy or soulful elements, especially syncopated rhythms and chord progressions including sevenths and ninths; often as the second word in combinations (see below).

In US English the word *funk* originally meant 'a bad smell' but a new sense was back-formed from the slang adjective *funky* in the fifties to refer to the fashion then for down-to-earth bluesy music; *funky* also meant 'swinging' or 'fashionable'. (There is no connection with the British English word *funk* meaning 'a state of fear'.) In the latest development of its meaning, *Funk* has been extended outside the styles traditionally thought of as *funky*, tending to become a catch-all tag for whatever is fashionable in a particular area of popular music.

As mentioned above, *funk* has existed since the fifties, but has acquired a broader meaning recently. The first CROSSOVERS between *funk* and other styles came in the seventies with **disco-funk**, a funky (that is, fast and rootsy) style of disco music. This was followed in the eighties by *electrofunk* (see ELECTRO), **jazz-funk** (which, it has more than once been claimed, is neither jazz nor *funk*), **p-funk** (a style developed by George Clinton of Parliament/Funkadelic), **slack-funk**, **slow-funk**, and *techno-funk* (see TECHNO), to name only a few of the styles which claimed to include *funk* elements. A leading and influential practitioner of *funk* proper is James Brown. Often the *funk* tag signifies no more than an attempt to incorporate Black musical traditions and jagged rhythms, funky chord progressions, or soulful lyrics into the White music style: *funk* has been widely played by White musicians since the mid seventies. Derivatives formed on *funk* have also been common in the eighties: **funker** and **funkster** extended their meaning to cover the broader sense of *funk*, and there were other, one-off formations along the lines of **funkadelic** (originally a proper name but also adopted as a common noun or adjective), **funkateer**, **funkathon**, and **funketize**.

> We scored No 1 disco albums with legendary jazz-funk duo Morrissey Mullen.
> *Music Week* 2 Feb. 1985, Advertisement pullout, p. i

> If old bubblegum music is on I sing at the top of my lungs, and if new funkadelic is on I bop in my seat.
> *New York Times* 14 May 1986, section C, p. 1

> If you've never fancied this kind of frantic funk try this for size. Blackman's wild and witty lyrical style combines macho street level cliche with sharp social awareness. *Hi-Fi Answers* Dec. 1986, p. 78

> These 10 songs demonstrate that all it takes is a good kick in the pants, a bottleneck slide guitar, and a feel for Muscle Shoals slow-funk to make a boy want to whoop and holler all night long.
> *Dirty Linen* Spring 1989, p. 56

> The second track on the album, 'Have a Talk with God' is a simple message to people with problems ... backed with a slack-funk beat. *Shades* No. 1 1990, p. 19

fun run /ˈfʌn ˌrʌn/ *noun* ⊗ �

An organized long-distance run in which amateur athletes take part for fun or to raise money for charity rather than competitively.

A transparent compound of *fun* and *run*, exploiting the rhyme.

The first *fun runs* took place in the US in the mid seventies as a way of bringing together people who had taken up jogging or long-distance running recreationally. The idea was introduced into the UK in the late seventies, and by the mid eighties the *fun run* was an established part of many Western countries' culture, with large races such as the annual London Marathon attracting thousands of participants. Often the **fun runners**, who are only competing for the enjoyment of running or so as to raise money for charity from sponsors, run alongside serious international athletes in the same race.

> Thousands of fun runners and disabled competitors pounded the same rain-soaked course as the stars. *New York Times* 21 Apr. 1986, section C, p. 6

> A fun run over 8km was held at the Phobians Athletics Club. *South African Panorama* Jan. 1988, p. 50

> Before the main race, limited to 150 runners, there will also be a charity one-mile Family Fun Run.
> *Northern Runner* Apr./May 1988, p. 6

futon /ˈfuːtɒn/ *noun* �

A low-slung Japanese-style bed or mattress.

A direct borrowing from Japanese, in which it traditionally refers to a bed-quilt or thin cotton mattress which is laid on a mat on the floor overnight, and may be rolled up and put away during the day.

The word has been used in descriptions of Japanese culture since the end of the last century, but the present Western application dates from the early 1980s. The *futon* as marketed in the West may include a slatted wooden base which stands only a few inches from the floor, is often capable of conversion into a sofa for day-time use, and usually includes a stuffed cotton mattress similar to the Japanese version.

> They fall onto the stripped-pine futon. *Artseen* Dec. 1986, p. 19

> Slatted bases are often used in traditional bedstead designs and low line beds such as futons.
> *Daily Mail DIY Home Interiors* 1988, p. 112

fuzzword /ˈfʌzwɜːd/ *noun*

A deliberately confusing, euphemistic, or imprecise piece of jargon, used more to impress than to inform.

Formed by compounding and abbreviation: a *word* that is *fuzzy* in its twentieth-century sense 'imprecisely defined, confused, vague'. It is also a deliberate alteration of *buzzword* (a fashionable but often meaningless piece of jargon, a vogue word), which has been in use since the late sixties.

Fuzzword was coined by the *Washington Post* in 1983 and is still principally a US usage.

> In the often emotional arms control debate, there may be no more common fuzzword than 'verification'. *National Journal* 14 Apr. 1984, p. 730

• •

G

gag me with a spoon 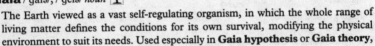 see VALSPEAK

Gaia /ˈgaɪə/, /ˈgeɪə/ *noun*

The Earth viewed as a vast self-regulating organism, in which the whole range of living matter defines the conditions for its own survival, modifying the physical environment to suit its needs. Used especially in **Gaia hypothesis** or **Gaia theory**, the theory that this is how the global ecosystem functions.

Named after *Gaia*, the Earth goddess in Greek mythology (the daughter of Chaos).

The term was coined by the British scientist James Lovelock, who first put forward the hypothesis at a scientific meeting about the origins of life on Earth in 1969; the suggestion that it should be named after the goddess Gaia had come from William Golding. Although not especially well received by the scientific community, the theory reached a wider audience in the eighties and early nineties and proved very attractive both to environmentalists and to the NEW AGE movement, with its emphasis on holistic concepts and an Earth Mother. *Gaia* is used as a proper name for the hypothetical organism itself, and also as a shorthand way of referring to the *Gaia hypothesis*. **Gaian** (as an adjective and noun) and **Gaiaist** (as an adjective) have been derived from it.

> 'The Biosphere Catalogue' expresses a kind of spirituality in science, a metaphysical belief in the biosphere as an entity which has been dubbed 'Gaia', as if to acknowledge its divine qualities.
> *Los Angeles Times* 15 Dec. 1985, p. 12

> Gaians (to use an abbreviation popular at the meeting) argue that this state of affairs is indeed evidence of the interconnectedness of life on Earth, and that it would be foolish to expect to find a series of isolated and independent mechanisms. *Nature* 7 Apr. 1988, p. 483

> Will tomorrow bring hordes of militant Gaiaist activists enforcing some pseudoscientific idiocy on the community? *New Scientist* 7 Apr. 1988, p. 60

> It is at the core of the current debate over the 'Gaia hypothesis', which holds that the planet is one huge organism in which everything interacts to sustain and maintain life on Earth.
> *Christian Science Monitor* 30 Jan. 1990, p. 12

Understanding Gaia means understanding that the survival of the plants, trees and wildlife which live on this planet with us is crucial to our own survival.

Debbie Silver & Bernadette Vallely *The Young Person's Guide to Saving the Planet* (1990), p. 52

galleria /gælə'riːə/ *noun* 〰️ ✖️

In marketing and planning jargon, a collection of small shops under a single roof, either in an arcade or as concessions in a large store.

A direct borrowing from Italian *galleria* 'arcade'.

Architects in English-speaking countries were first inspired by the idea of the Italian *galleria* in the sixties and began to design shopping arcades on the same model, but it was not until the early eighties that the word *galleria* suddenly came into vogue as a fashionable way of saying 'arcade'. The vogue was continued by the application of the term to shops-within-a-shop as well.

Burton and Habitat intend to create a new format at Debenhams with the 'Galleria concept'—an integrated collection of highly-focused speciality stores under one roof.

Yorkshire Post 23 May 1985, p. 4

The winning scheme . . . incorporated the inevitable 'galleria'. *The Times* 17 Feb. 1990, p. 10

Johnson took over eleven floors in an unremarkable glass tower at a suburban shopping center named The Galleria. Bryan Burrough & John Helyar *Barbarians at the Gate* (1990), p. 85

gamete intra-fallopian transfer ✖️ 💻 see GIFT

gaming ✖️ see ROLE-PLAYING GAME

garage /'gærɑːʒ/ *noun* Also written **Garage** 🎵 🎮

A variety of HOUSE music from New York which incorporates elements of soul music, especially in its vocals.

Probably named after the *Paradise Garage*, the former nightclub in New York where this style of music was first played; there may also be some influence from the term *garage band*, which has been applied since the late sixties to groups (originally amateurs who practised in empty garages and other disused buildings) with a loud, energetic, and unpolished sound which is also sometimes known as *garage* or *garage punk*.

New York *garage* developed in the early eighties (principally at the Paradise Garage but later also at other New York clubs), but only came to be called *garage*—or by the fuller name **garage house**—in the second half of the decade. The founding influence on the style was the New York group *The Peech Boys*. In its later manifestations *garage* is very closely related to *deep house* (see HOUSE)—indeed some consider *deep house* to be simply the Chicago version of *garage*, incorporating the lyrical and vocal traditions of American soul into the fast, synthesized dance music which is typical of house.

The void left in trendier clubs following the over-commercialisation and subsequent ridiculing of 'acieed!' . . . is being filled by 'garage' and 'deep house'. *Music Week* 10 Dec. 1988, p. 14

The records will be anything dance-orientated: 'Rap, reggae, hip hop, house, jazz, garage or soul,' says Anita Mackie . . . 'What is garage?' I ask. She consults a colleague and they decide on 'Soulful house'. I decline to ask them what 'house' is. *The Times* 25 July 1990, p. 17

garbage in, garbage out 💻 see EXPERT SYSTEM

gas-permeable ✖️ see LENS

-gate /geɪt/ *combining form* 📓

Part of the name *Watergate*, widely used in compounds to form names for actual or alleged scandals (usually also involving an attempted cover-up), comparable in some way to the Watergate scandal of 1972.

Formed by abbreviating *Watergate*, treating the *-gate* part as a word-forming element in its own right.

Before the Watergate scandal and the ensuing hearings were even fully over, journalists began to use *-gate* allusively to form names for other (major or minor) scandals, turning it into one of the most productive word-final combining forms of the seventies and eighties. In August 1973, for example, the US satirical paper *National Lampoon* wrote of persistent rumours in Russia of a vast scandal, and nicknamed this **Volgagate**; in 1975 the financial paper *Wall Street Journal* called a fraud inquiry at General Motors **Motorgate**, and in 1978 *Time* magazine wrote of an **Oilgate** concerning British North Sea oil. The suffix was used in a variety of ways: tacked on to the name of the place where the scandal occurred (as in the original *Watergate*), to the name of the person or organization at the centre of the scandal (for example **Billygate** or **Carter-gate** for the scandal over the Libyan connections of Billy Carter, brother of US President Jimmy Carter, in 1980), or to the commodity or activity involved (for example **Altergate** for allegations that transcripts of official hearings in the US had been altered in 1983). It was principally a feature of US English until 1978, when the South African **Muldergate** scandal brought it wider publicity.

Perhaps surprisingly, the productivity of *-gate* did not really wane in the eighties: in the US it was kept in the public eye principally because of the Iran–contra affair of 1986 (see CONTRA), immediately nicknamed **Contragate** or **Irangate** (and still sometimes referred to by these names into the nineties) and by scandals over frauds allegedly perpetrated by TELEVANGELISTS, including the punningly named **Pearlygate**; in the UK there was **Westlandgate** in 1985 (involving Cabinet members in conflict over plans to bail out the helicopter company Westland), **Stalkergate** in 1986 (named after the Deputy Chief Constable of Greater Manchester police, John Stalker, who was invited to chair an inquiry into allegations of an RUC 'shoot-to-kill' policy in Northern Ireland and was then removed from this inquiry for several months while allegations of his own improper association with a known criminal were considered and rejected), and **Lawsongate** in 1988, involving allegations that the Chancellor of the Exchequer, Nigel Lawson, had deliberately deceived the public about the economy, to mention but a few.

It suits the White House to flatter Mrs Thatcher's diplomatic pretensions, just as it suits it to deflate those of the Labour leader, Mr Neil Kinnock. But it is a long way from 'Kinnockgate' for the good reason that the Americans are barely aware of the 'Neil-snubs-Ron-snubs-Maggie-snubs-Neil' row they are embroiled in. *Guardian* 30 Mar. 1984, p. 6

The current deterioration of the Ulster environment will continue unabated . . . if future developments significantly touch the RUC ('Stalkergate') or the judiciary. *Marxism Today* Sept. 1986, p. 41

Europeans . . . are not going to stomach the star-spangled strain of bible-thumping religiosity peddled by smooth-talking American preachers like Jerry Falwell, Pat Robertson and Jim Bakker (he of the 'Pearlygate' sex and corruption scandal). *Observer Magazine* 22 Nov. 1987, p. 50

From the 'Lawsongate' headline . . . through to the . . . allegation of a 'cover-up' . . . newspapers were unanimous in their belief that it was Nigel Lawson who had misled people.
Independent 14 Nov. 1988, p. 2

In those days . . . the Higher Skepticism had not yet appeared, fueled by the assassinations of the Kennedys and Martin Luther King and the others and by the Vietnam war and by Watergate . . . and by Irangate, etc. Paul Fussell *Wartime* (1989), p. 167

Blue Heat promisingly pits Brian Dennehy's blue-collar cop against Contragate corruption in high places. *The Face* Oct. 1990, p. 21

gay plague, gay-related immune disease ⊗ see AIDS

gazunder /gəˈzʌndə(r)/ *transitive* or *intransitive verb* ✂

In UK slang, of a house buyer: to reduce the price offered to (the seller of a property) at a late stage in the proceedings, usually immediately before contracts are due to be exchanged; to behave in this way over a house purchase. Also as an action noun **gazundering**; agent noun **gazunderer**.

Formed by altering the word *gazump* 'to swindle, especially in the sale of a house, by raising the asking price'; in the case of *gazunder*, the tables are turned so that it is the buyer rather than the seller who is in a position to do the swindling. Since the buyer comes in with a price *under* the one previously offered, the word *under* replaces the *-ump* part of *gazump*.

It was the slowing down and eventual fall of house prices in the UK in the late eighties, after the boom of the rest of the decade, that turned the housing market into a buyers' market in which the phenomenon of *gazundering* could arise. No doubt the practice existed without a name for a time; the first mentions of *gazunder*, *gazunderers*, and *gazundering* in the press, though, date from late 1988, cropping up first in the tabloid press and later in the 'quality' papers as well.

> The gazunderer goes along with the asking price until days or even hours before contracts are due to be exchanged. Then he threatens to withdraw. *Daily Mirror* 18 Nov. 1988, p. 4

> Media executive Matthew Levin, 44, and his psychotherapist wife Vivienne have just been gazundered in Hampstead. *Daily Telegraph* 6 Jan. 1989, p. 11

> In the heat of the house-price boom I hummed and hawed about protests over gazumping, suggesting that many victims would 'gazunder' their way to a quick buck given half a chance. *Weekend Guardian* 13 Aug. 1989, p. 29

gel /dʒɛl/ *noun* 🗶

A jelly-like substance used for cosmetic preparations of various kinds, especially for setting hair and as a semi-liquid soap for use in showers.

A specialized application of *gel* in its established chemical sense 'a semi-solid colloidal system consisting of a solid dispersed in a liquid'.

The first *gel* for setting and styling hair was developed for salon use as long ago as the late fifties in the US, but this was a setting gel applied before rolling and setting the hair in the traditional way. The *gel* only really came into its own as a product on general sale and in widespread use with the swept-up hair fashions of the punk era (from the late seventies onwards). These preparations could be applied to wet hair before blow-drying, used to 'glue' the hair in place while it dried naturally, or even to fix dry hair into a style. When used on dry hair it produced a glistening, still-wet look that duly resulted in a new hair fashion in the eighties. The *gel* form proved useful for other preparations, too—notably as a shower soap—because it does not run off the hand like a liquid or slip like bar soap.

> Nowadays people are using superglue, lacquer, gel, oils and even soap and water to make their hair stand up. *Telegraph* (Brisbane) 7 Oct. 1985, p. 8

> A luxurious exfoliating gel has been launched by Christian Dior. *Sunday Express Magazine* 17 Sept. 1989, p. 3

> Don't use harsh soaps and shower gels on winter skin—use a cleansing bar. *Health Shopper* Jan./Feb. 1990, p. 4

genco /'dʒɛnkəʊ/ *noun* 〰

A power-generating company; especially, either of the two electricity-generating wholesalers set up to sell electricity in England and Wales.

Formed by combining the first syllable of *generating* with *co* (the abbreviated form of *company*), as in DISCO.

The first *gencos* were set up in the US in the early eighties. The idea of splitting the electricity industry in the UK into generation and supply is a central tenet of the privatization strategy worked out by the government in the closing years of the eighties; the two English *gencos*, National Power and Powergen, are meant to introduce competition into power generation and were privatized in 1991.

If regulators approve the move, the utility would be the first to split into two independent electric-power subsidiaries: a wholesale power generating unit ('genco') that could sell any surplus power it produces to users outside its current turf, and a retail distribution unit ('disco') that would own the power lines and move the product. *Financial World* 5 Jan. 1988, p. 48

gene therapy /'dʒiːn ˌθɛrəpɪ/ *noun* Ⓧ 🖳

The technique or process of introducing normal genes into cells in place of defective or missing ones in order to correct genetic disorders.

Formed by compounding: *therapy* which takes place at the level of the *gene*.

Researchers in medical genetics have been working on the idea of *gene therapy* since the early seventies and during the eighties were approaching a point where their techniques could be applied to human subjects, although most sources spoke of *gene therapy* very much as a hope for the future rather than a practical reality. Since all forms of transgenic research and GENETIC ENGINEERING raise serious ethical issues which have had to be considered by the courts, *gene therapy* could not develop as fast as its inventors would like. Approval for the first real *gene therapy* on human subjects was given in the US in 1990.

Researchers were predicting that common disorders of the red blood cells, such as thalassaemia, would be the first diseases cured by gene therapy. *Listener* 9 May 1985, p. 7

This sort of research, which critics describe as 'playing God', gets even more morally knotty when it comes to gene therapy, with its potential for monitoring and altering human genes to check for and eliminate hereditary diseases. *The Face* June 1990, p. 111

genetic engineering /dʒɪˈnɛtɪk ˌɛndʒɪˈnɪərɪŋ/ *noun* Ⓧ 🖳

The deliberate modification of a living thing by manipulation of its DNA.

A straightforward combination of *genetic* with *engineering* in its more general sense of 'the application of science to design etc.'.

The techniques of *genetic engineering* were developed during the late sixties and seventies and contributed significantly to the boom in BIOTECHNOLOGY during the eighties when applied to industrial processes. There was concern about the possible ecological effects of releasing **genetically engineered** organisms (such as plants resistant to crop diseases, frost damage, etc.) into the environment, but this was allowed under licence in the UK from 1989 onwards. Applications of *genetic engineering* to human DNA have proved even more problematical because of the ethical implications of altering genetic make-up; in the UK, measures to control experiments involving *genetic engineering* on human tissue were added to the Health and Safety Act in 1989.

We are in the process now of bioengineering the world's agroscape. This means moving around the players as well as making new ones through genetic engineering. *Conservation Biology* Dec. 1988, p. 309

Genetic engineering is often presented as producing unnatural hybrids which have no counterparts in the wild. It feeds on people's notions that there is a harmony or wisdom in nature with which we tamper at our peril, even though alongside that people want their videos and their modern medicines and all the other things that science brings by tampering with nature. *Guardian* 6 July 1989, p. 19

genetic fingerprinting /dʒɪˌnɛtɪk ˈfɪŋɡəprɪntɪŋ/ *noun* 🎔 🖳

The analysis of genetic information from a blood sample or other small piece of human material as an aid to the identification of a person.

Formed by combining *genetic* with *fingerprinting* in a figurative sense; the *genetic fingerprint* produced by this technique is as accurate in uniquely identifying a person as an actual fingerprint would be.

Genetic fingerprinting was developed in the late seventies and early eighties and was first widely publicized in the mid eighties. The technique (also known as *DNA fingerprinting*) has a number of applications: it has revolutionized forensic science in the eighties, for example. A sample of blood, semen, etc. or a few flakes of skin left at the scene of a crime can be analysed for the unique pattern of repeated DNA sequences that it displays (its **genetic fingerprint**) and this can be matched with blood samples taken from suspects. The first murder case to be decided on the basis of *genetic fingerprinting* was heard in 1987, but in 1989 a number of cases cast doubt on the reliability of forensic evidence based entirely on this kind of DNA testing. Another quite separate application of *genetic fingerprinting* is in the matching of blood samples in paternity suits or cases of 'disappeared' children (see DESAPARECIDO), since the *genetic fingerprint* can be used to establish whether two people could be related to one another. A slightly more refined process, known as **genetic profiling**, provides a **genetic profile**, or list of all of a person's genetic characteristics.

> Forensic scientists can also use genetic traits found in blood and other tissues to identify bodies. Sometimes known as genetic fingerprints, these include about 70 inherited enzymes that can be used in a form of extraordinarily detailed blood typing.　　*New York Times* 8 July 1985, section A, p. 3

> Genetic profiles are much more sensitive than genetic fingerprints because they give accurate answers based on much smaller samples.　　*Observer* 26 Feb. 1989, p. 8

> Now the baby has been born and blood tests and 'genetic fingerprinting' have proved conclusively that Howitt was not the father.　　*Private Eye* 1 Sept. 1989, p. 6

gentrification /ˌdʒɛntrɪfɪˈkeɪʃ(ə)n/ *noun* 🔀

The conversion of something with humble origins (especially a housing area) into something respectable or middle-class; taking up-market.

Formed by adding the process suffix *-ification* to *gentry*; although in fact it is the professional middle class, rather than the gentry, who have taken over the working-class areas.

Gentrification was first used by town planners in the early seventies to describe the migration of professional, middle-class people back into the inner cities; once there, they began renovating and altering to their own tastes what had been built as artisans' cottages and terraces for the workers originally brought to towns by the Industrial Revolution. As this process became more and more noticeable through the eighties and whole areas of large cities completely changed their character, *gentrification* moved out of the jargon of sociologists and planners and was widely used in the press, often with pejorative meaning. At this stage it also came to be applied to anything which could be moved up-market; in stock-market jargon, even to bonds. The associated verb is **gentrify**; the adjective to describe anything which has undergone this process is **gentrified**.

> Though the area ... is being gentrified, the pub itself has not gone posh.
> 　　*Sunday Times* 30 Jan. 1983, p. 16

> Further down, the first signs of gentrification appear—a renovated colonial house, a vegetarian health food store, and an upmarket boutique. This is ... the vanguard of the yuppie invasion.
> 　　*Courier-Mail* (Brisbane) 6 July 1988, p. 9

> His uncle's place had been gentrified on the outside, presumably to placate the new yuppie neighbors.　　Alice Walker *Temple of My Familiar* (1989), p. 29

ghetto blaster /ˈgɛtəʊ ˌblɑːstə(r)/ *noun* 🔀 🎵

In slang, a large, portable stereo radio (sometimes incorporating a cassette player), especially one on which popular music is played loudly in the street. (Considered by some to be racially offensive.)

Formed by compounding. The music supposedly *blasts* the neighbourhood with its exaggerated volume; this is associated mostly with Black and ethnic-minority areas, which explains the reference to the *ghetto*.

The term originated in the US in about 1980, and was perhaps the most graphic of all the slang names for these outsize portable stereos which, it seems, can only be played at full volume. Other names for the same thing included (in the US) BEAT BOX, BOOM BOX, and the mixed **ghetto box**; MINORITY BRIEFCASE and (in the UK) BRIXTON BRIEFCASE alluded to their having become part of the expected street uniform of HIP HOP and its followers. Despite its rather racialist connotations, *ghetto blaster* proved humorous enough to spread round the world to nearly every English-speaking country where hip hop and BREAK-DANCING became popular: groups of youngsters gathering in the street for break-dancing needed a *ghetto blaster* to provide the accompanying beat. A White American rhythm-and-blues sextet from the Deep South even called themselves *The Ghetto Blasters* in the early eighties. A back-formed verb **ghetto-blast** has also developed, with an action noun **ghetto-blasting** and an adjective **ghetto-blasted** to go along with it.

> Brisbane's breakdancers ... attracted a bigger crowd than the officially-approved buskers; but retribution wasn't long in following. The police came down, the ghetto blasters were turned off and the kids left. *Sunday Mail* (Brisbane) 25 May 1986, p. 3

> Waterproof Sports models have helped restore silence to ghetto-blasted beaches.
> *Q* Oct. 1987, p. 69

GIFT /gɪft/ *acronym* ⊗

Short for **gamete intra-fallopian transfer**, a technique for helping infertile couples to conceive, in which eggs and sperm from the couple are inserted into one of the woman's Fallopian tubes ready for fertilization.

The initial letters of *Gamete Intra-Fallopian Transfer*; a *gamete* is a mature cell able to unite with another in reproduction. Like many recent acronyms, this one seems to be chosen for the significance of the resulting 'word': the technique presents the infertile couple with the much-wanted *gift* of a child.

The technique was developed in the US during the mid eighties as a more 'natural' alternative to *in vitro* fertilization. Since, using this technique, it is possible for fertilization to occur within the human body, *GIFT* has proved more acceptable on moral and religious grounds than IVF, the technique which produces 'test-tube babies'. *GIFT* as a term is often used attributively, in *GIFT technique*, *GIFT delivery*, etc.

> GIFT, which is operating in several non-Catholic hospitals, has a success rate of about 20 per cent.
> *Courier-Mail* (Brisbane) 5 Apr. 1988, p. 17

> They thought that GIFT ... treatment would give them a much-wanted baby.
> *New Statesman & Society* 15 Dec. 1989, p. 22

See also ZIFT

gigaflop 💾 see MEGAFLOP

GIGO 💾 see EXPERT SYSTEM

giro /'dʒaɪərəʊ/ *noun* 🔳

In colloquial use in the UK: a cheque or money order issued through the giro system; specifically, a girocheque in payment of social security benefit.

Shortened from *girocheque*; the word *giro* itself, which originally referred to the system for transferring money between banks, post offices, etc., was borrowed from Italian *giro* 'circulation, tour' in the late nineteenth century.

The colloquial form has been in use since the late seventies or early eighties. The erosion of benefits during the eighties meant that the arrival of the weekly *giro* became a more crucial event than ever for many CLAIMANTs, a fact that has apparently led to the formation of a derivative **girocracy** for the under-class of people who depend on their *giro* for survival, although there is little sign that this derivative will become established.

> 'That my lager?' he inquired, feeling mean even as he uttered the question. 'Yeah, d'you mind?' said Raymond. 'I'll replace it when I get me next giro.' David Lodge *Nice Work* (1988), p. 117

G-Jo ⊗ see ACUPRESSURE

glam¹ ✖ see GLITZY

GLAM² ▮▮ see WOOPIE

glasnost /ˈglæznɒst/, in Russian /ˈglasnəstj/ *noun* ▣

A policy of freedom of information and publicly accountable, consultative government introduced in the Soviet Union in 1985.

A direct borrowing from Russian *glasnost'*, literally 'publicness', which in turn is formed from *glasnyy* 'public, open' (of courts, proceedings, etc.) and *-nost'* '-ness'.

The word has been used in Russian for several centuries, but only acquired its more specialized political meaning towards the end of the Soviet period. It was used in the context of freedom of information by Lenin, and by the dissident writer Solzhenitsyn in an open letter to the Writers' Union in November 1969. *Glasnost* did not become the subject of serious public debate even within the former Soviet Union until January 1985, when an editorial in the state newspaper *Izvestiya* requested letters on the subject. Many were published, most lamenting the lack of basic information—from bus timetables to the reasons for bureaucratic actions—in Soviet society.

When Mikhail Gorbachev used the word in his speech accepting the post of General Secretary of the Communist Party in March 1985, *glasnost* became one of the keywords taken up by the international press to describe his reforming regime. He said

> We are committed to expand *glasnost* in the work of Party, Soviet, State, and public organizations. V. I. Lenin said that the State is made strong through the awareness of the masses; our practice has fully confirmed this conclusion.

At first, journalists attempted to translate the Russian word, using 'publicity' or 'openness'. Soon, though, it became clear that no single English word could sum up the full significance of the Russian meaning, and the Russian word itself became one of the most-used political words of 1986–7. It was not long before it came to be applied to public accountability in general and to the relaxation of political regimes in other parts of the world, acquiring in English a rather broader meaning than in its original language, where the emphasis was very much on the 'right to know' of the Soviet public. It has quickly established its place in English, generating a number of derivatives, some jocular (**glasnostrum, glasnostalgia**), some more serious (**glasnostian, glasnostic, glasnostified**), while others remain true to its Russian roots (**glasnostnik**).

> Exposes of corruption, shortages and economic problems appear virtually daily in the [Soviet] press. It is a change that became evident after Mikhail S. Gorbachev came to office last March and called for more 'glasnost', or openness, in covering domestic affairs.
>
> *New York Times* 22 Feb. 1986, section 1, p. 2

> Life is still hard under glasnost, Vietnamese-style.
>
> headline in *Los Angeles Times* 30 May 1987, section 1, p. 4

> Such recognition of an author [Alexander Solzhenitsyn] once officially scorned as an enemy of the people is a significant marker of the glasnostian literary thaw.   *Daily Telegraph* 4 Aug. 1988, p. 1

See also PERESTROIKA

gleaming the cube ✖ 🐱 see SKATEBOARDING

glitch /glɪtʃ/ *noun* and *verb* ▣

In slang (originally in the US):

noun: A snag, a hitch or hold-up; a technical error.

intransitive verb: To malfunction or go wrong; to suffer a 'hiccup'.

A figurative use of a word that originally (in the early sixties) meant 'a surge of current'—an occurrence which could lead to unpredictable behaviour from electronic instruments or even complete crashes of computer systems. The word's ultimate origins are rather obscure: it has been claimed that it is borrowed from Yiddish *glitsch*, which means 'a slip' in its literal sense of losing one's footing, but this theory has been discredited.

As mentioned above, *glitch* was first used in the early sixties, mainly in the slang of people involved in the US space programme. From there it was taken into computing slang, and by the early eighties had become a fashionable word in the general press for any kind of snag or hold-up, as well as developing more specialized meanings in astronomy and audio recording. It is now used freely in the media in the UK as well as the US, but is still regarded as an Americanism by many British readers. *Glitch* has a derived adjective **glitchy** which can be used of programs, systems, etc. that are particularly prone to malfunction.

> Elsewhere, equipment glitches in the Iranian desert force American commandos to abort the mission to rescue 53 hostages in Tehran. *Life* Fall 1989, p. 15

> The only glitch in the whole Ararat countdown was the failure to get the Project recognized as a charitable institution. Julian Barnes *A History of the World in 10½ Chapters* (1989), p. 267

> No matter how carefully I set the unit up it always glitched a little, especially when using the Diatonic Shift. *Music Technology* Apr. 1990, p. 42

glitterati /ˌglɪtəˈrɑːtɪ/ *plural noun* 🔲

In media slang (originally in the US): the celebrities or 'glittering stars' of fashionable society, especially those from the world of literature and entertainment.

Formed by telescoping *glitter* and *literati* (the people who form the literate, educated élite) into a blend.

A name for the group once known as the *beautiful people* or *jet set*, *glitterati* became a popular term in the media in the late seventies and early eighties, when conspicuous glitter especially characterized the stars of show business (see GLITZY below). The punning name *glitterati* had in fact been coined in *Time* magazine as long ago as 1956, in an article about a party for publicity-conscious editors:

> Bobbing and weaving about the premises are a passel of New York glitterati. There is a highbrow editor of a popular magazine who is keen on starting a new literary journal and wants Tom to round up a staff of 'topnotchers' and decorated veterans from the little magazine wars.

In the late eighties and early nineties it was used for famous or successful people in any field of public interest, from business and politics to pop music and sport.

> In the first two episodes, the mix also runs to Thatcherite glitterati (nesting in their Thameside lofts) and disco gays. *Listener* 30 May 1985, p. 34

> In a Lions tour of Australia that has been desperately short of glitterati England's blind-side flanker has emerged as a player of top quality. *Guardian* 15 July 1989, p. 19

glitzy /ˈglɪtsɪ/ *adjective* 🔲

In show-business slang (originally in the US): full of cheap glitter, extravagantly showy, ostentatious, flashy (often with the implication that there is little of substance under the glitter); tawdry or gaudy.

Probably related to German *glitzerig* or *glitzig* 'glittering' and its Yiddish equivalents, but perhaps influenced by *glitter* and *ritzy*.

The word was first used in American show-business circles in the mid sixties, but it was in the late seventies and eighties that it suddenly became one of the most fashionable reviewers' buzzwords and started to reach a wider audience. This sudden vogue coincided with a particularly showy phase in television entertainment, with the conspicuous wealth and glamour of such upmarket soap operas as *Dallas* and *Dynasty* attracting large audiences in all parts of the English-speaking world. Its new popularity was reflected in a number of derivatives which appeared in the late seventies and early eighties: the nouns **glitziness** and **glitz** (extravagant but superficial display, show-business glamour), from which a verb **glitz** (**up**) was later formed; the adverb **glitzily**; and a number of humorous one-off formations such as **glitzerati** (see GLITTERATI), **glitznost** (the repackaging of the Labour Party: see GLASNOST), **glitzville**, and **Glitzkrieg**. *Glitz* often appears in the same sentence as *glam* (short for glamour) or *hype* to refer to the superficially glamorous and publicity-seeking world of entertainment, or indeed to anything that tries too hard to 'sell itself'. All of these words are usually at least partly pejorative, corresponding to the more established British English word *flashy* (and its derivatives *flashiness* etc.) and serving as an antonym for *classy* (*classiness* etc.).

> The British Film Institute glitzed up its 1985 Awards bash last week . . . by getting an impressive line-up of screen talent to announce the shortlists. *Listener* 9 May 1985, p. 31

> The phrase 'mini-series' brings visions of melodramatic plots, beautiful women, dastardly men, elaborate costumes, sex, death, mystery and Joan Collins . . . But with the four-part series, *In Between*, . . . there is no glam, no glitz and no Joan Collins. *Daily Sun* (Brisbane) 5 Mar. 1987, p. 17

> Nice women grow old and glum, cynical too, in all this glitz of fur, silk, leather, cosmetics, et cetera, of the glamour trades. Saul Bellow *A Theft* (1988), p. 49

> The conventions have become glitzy coronations instead of fiercely-fought inside battles. *Independent* 16 July 1988, p. 6

> Most of the pictures used only impress the British professional because of their earning ability—often they're glitzy superficial rubbish produced to a formula. *Photopro* Spring 1990, p. 4

See also TACK

global /ˈgləʊb(ə)l/ *adjective* 🌳

In environmental jargon: relating to or affecting the Earth as an ecological unit. Used especially in:

global consciousness, receptiveness to (and understanding of) cultures other than one's own, often as part of an appreciation of world socio-economic and ecological issues;

global warming, a long-term gradual increase in the average temperature in climate systems throughout the world as a result of the GREENHOUSE effect.

Both these phrases use *global* in its dominant modern sense of 'worldwide', and are influenced by Marshall McLuhan's famous concept of the *global village* (coined in *Explorations in Communication*, 1960), which recognized the way in which technology and communications allow everyone to experience world events simultaneously and so effectively 'shrink' world societies to the level of a single village or tribe. *Global consciousness* also draws on the fashion for *consciousness-raising* in the sixties.

Global consciousness is originally a US term which arose during the seventies, but became commoner as a catch-phrase (expressing the basis of the 'we are the world' culture) once the GREEN movement gained widespread popular support during the second half of the eighties. It was also during the eighties that *global warming* entered popular usage, although scientists had begun to use the term in the late seventies, as research began to show that increased carbon dioxide emissions in industrialized countries burning large quantities of fossil fuels would almost certainly contribute to the greenhouse effect to such an extent as to affect worldwide climate. The repercussions of even a small increase in world temperatures could be far-reaching, including a rise in sea level and widespread flooding or permanent submersion of land; this is one reason why governments started to treat the problem as a serious one requiring prompt preventive action.

> One of the least pleasant characteristics of our era must surely be its transformation of global consciousness into a sales item. *Nation* 17 Apr. 1989, p. 529

> After the Prime Minister's Downing Street seminar on global warming last year, 'government sources' were quoted as saying that nuclear power had a major part to play. *Which?* Apr. 1990, p. 222

global double zero ☆ see ZERO

glocal /'gləʊk(ə)l/ *adjective* ∿

In business jargon: simultaneously global and local; taking a global view of the market, but adjusted to local considerations. Also as a verb **glocalize**, to organize one's business on a global scale while taking account of local considerations and conditions; process noun **glocalization**.

Formed by telescoping *global* and *local* to make a blend; the idea is modelled on Japanese *dochakuka* (derived from *dochaku* 'living on one's own land'), originally the agricultural principle of adapting one's farming techniques to local conditions, but also adopted in Japanese business for *global localization*, a global outlook adapted to local conditions.

The idea of going for the world market (*global marketing*) was a feature of business thinking in the early eighties. By the late eighties and early nineties Western companies had observed the success of Japanese firms in doing this while at the same time exploiting the local conditions as well; this came to be called *global localization* (or, at first, *dochakuka*), soon abbreviated to *glocalization*. It proved to be one of the main marketing buzzwords of the beginning of the nineties.

> 'Glocalize,' as the Japanese call it. *Fortune* 28 Aug. 1989, p. 76

> We've witnessed what you might have heard called 'glocalization': making a global product fit the local market. To do that effectively, you've got to have individuals who understand what makes that particular market tick. *Advertising Age* 8 Jan. 1990, p. 16

gloom and doom /ˌgluːm ənd 'duːm/ *noun phrase* Also in the form **doom and gloom** ∿ ☆

A feeling or expression of despondency about the future; a grim prospect, especially in political or financial affairs.

A quotation from the musical *Finian's Rainbow* (1947, turned into a film in 1968), in which Og the pessimistic leprechaun uses the rhyming phrase as a repeated exclamation:

> Doom and gloom . . . D-o-o-m and gl-o-o-m . . . I told you that gold could only bring you doom and gloom, gloom and doom.

This allusive phrase was first picked up by US political commentators in the sixties (perhaps as a result of the popularity of *Finian's Rainbow* as a film) and was being used as an attributive phrase to describe any worrying or negative forecast by the seventies. In the early eighties it was perhaps particularly associated with economic forecasting and with the disarmament debate; the emphasis shifted in the second half of the eighties to the pessimistic forecasts of some environmentalists about the future of the planet. Both the nuclear and environmental uses influenced the formation of the word *doomwatch* (originally the name of a BBC television series) for any systematic observation of the planet designed to help avert its destruction. A person who makes a forecast of *gloom and doom* is a **gloom-and-doomster**.

> Amongst all the recent talk of doom and gloom one thing has been largely overlooked. *Daily Telegraph* 7 Nov. 1987, p. 18

> When the grass isn't always greener: gloom and doom that foreign companies are getting ahead in IT is not only a British disease. headline in *Guardian* 17 Aug. 1989, p. 29

gluten-free ⊗ ✖ see -FREE

go /gəʊ/ *verb* 🔲

In young people's speech: to say, to pronounce (usually in the present tense, reporting speech in the past).

An extension of the use of *go* to report a non-verbal sound of some kind expressed as an onomatopoeic word or phrase, as in 'the bell went ding-dong' or 'the gun went bang', perhaps with some influence from nursery talk (as in 'ducks go quack, cows go moo').

This has been used in young people's speech for some time, but was only recently taken up by writers for use in print. Typically the narrative part of the sentence is in the past tense, but *go* is in the historic present, as, for example 'I bashed him on the head, so he goes "What d'you want to do that for?"'

> He liked that very much. So he goes: 'More. Sing it again.'
> Michael Rosen *Quick Let's Get Out of Here* (1983), p. 67

> I go, 'You don't understand how I felt, do you?'
> Elmore Leonard *Bandits* (1987), p. 19

gobsmacked /'gɒbsmækt/ *adjective* Also written gob-smacked

In British slang: astounded, flabbergasted; speechless or incoherent with amazement; overawed.

Formed from *gob* (slang for the mouth) and *smacked*; the image is that of clapping a hand over the mouth, a stock theatrical gesture of surprise also widely used in cartoon strips.

Although it had been in spoken use for several decades (especially in Northern dialects), *gobsmacked* did not start to appear in print until the middle of the eighties. Surprisingly it was the 'quality' newspapers which particularly took it up—perhaps to show their familiarity with the current idiom of young people—although it also appeared in the tabloids, along with a synonym **gobstruck**. A verb **gobsmack** was back-formed from the adjective in the late eighties.

> It's this act...with which she has been gobsmacking the punters in a recent cluster of Personal Appearances in gay clubs, straight clubs, and 'kids clubs'.
> *Melody Maker* 24 Oct. 1987, p. 18

> In short, his work leaves me gobstruck—or would have done, had not a reader written to chide me for using what he calls 'this mean and ugly little word'.
> Godfrey Smith in *Sunday Times* 3 Sept. 1989, section B, p. 3

> When told the price, between 10 and five times over estimate, he was 'gobsmacked'.
> *Daily Telegraph* 21 Sept. 1989, p. 3

go-go /'gəʊgəʊ/ *noun* Also written GoGo

A style of popular music (originating in the Black communities of Washington DC) characterized by an energetic soul sound and an incessant funk-style beat, and using a mixture of acoustic and electronic instruments; a gathering at which this music is played; also, the street subculture surrounding it.

Probably a specialized development of *go-go* as used of discos, their music, and disco-dancing in the sixties. One of the founders of the subculture, Chuck Brown, claims that the name arose when he asked an audience 'What time is it?' and they shouted back 'Time to go-go!'

Go-go is the Washington equivalent of New York's HIP HOP; its musical roots are in the late sixties, when the principle of a continuous beat and the call-and-response style of lyric that characterizes the music were first developed. It remained limited to its Washington audience until the late seventies, when its first big record hits were released, but from the mid eighties onwards was widely promoted outside Washington and became popular in the UK as well. The word *go-go* is often used attributively, especially in **go-go music**.

> Go-go is aggressively live, drawing anywhere from 5,000 to 20,000 people a night to go-gos scattered throughout the city. It is the live performance that defines go-go and denotes its champions.
> *Washington Post* 19 May 1985, section G, p. 4

> Chuck Brown and the Soul Searchers who spearheaded the Go-Go attack in 1986 play three nights at The Town & Country Club in Kentish Town...as part of the Camden Festival.
> *Blues & Soul* 3–16 Feb. 1987, p. 9

gold card /ˌgəʊld 'kɑːd/ *noun*

A preferential charge card (usually coloured gold), which is issued only to people with a high credit rating and entitles them to a range of benefits and financial services not offered to holders of the standard card; hence, a preferential or exclusive membership of any organization.

Named after its colour, which was no doubt chosen for its connotations of wealth, security, and quality.

A preferential credit card was first issued by American Express in the US in the mid sixties, but this did not become known as a *gold card* until the seventies; various other charge-card companies then followed suit. *Gold cards* became available in the UK in the early eighties; here, as in the US, possession of one is regarded as an important status symbol (since high income is a condition of issue, a fee is payable for membership, and they open the door to a better service than other PLASTIC money). A sign of their reputation for exclusivity is the fact that *gold card* has already started to be used figuratively and in an allusive attributive phrase, rather like *Rolls-Royce*, to mean 'expensive' or 'for the élite'.

> Gold cards these days come with a battery of useful services. In the case of NatWest there is Freefone Brokerline for share dealing, plus free personal accident insurance and an investment and tax advisory service. NatWest customers will have to pay £50 a year for their new gold card service on renewal. *The Times* 21 June 1986, p. 27

> Beverly and Elliot Mantle—the film's twin brothers, partners in gold card gynaecology.
> *The Face* Jan. 1989, p. 65

> On offer also is a Gold Membership. Those who hold a Gold Card may enjoy full use of the gymnasium, squash courts, sauna, snooker, pool, darts and the club lounge, which is equipped with hi-fi sound and video. *Oxford Mail* 19 Mar. 1990, p. 26

golden /ˈgəʊld(ə)n/ *adjective*

In business jargon: involving the payment of a large sum of money or other gifts to an employee. Used in a number of phrases humorously modelled on **golden handshake** (a sum of money paid to an employee on retirement or redundancy), including:

golden handcuffs, benefits provided by an employer to make it difficult or unattractive for the employee to leave and work elsewhere;

golden hello, a substantial lump sum over and above the salary package, offered by a prospective employer to a senior executive as an inducement to accept a post;

golden parachute, a clause in an executive's contract guaranteeing a substantial sum on termination of the contract, even if the employee has not performed well;

golden retriever, a sum of money paid to a person who has already left an employer's staff in order to persuade him or her to return.

All of these phrases rely on the association of *gold* with riches; *golden handcuffs, golden hello*, and *golden parachute* consciously alter the earlier *golden handshake*, while *golden retriever* also relies for its humorous effect on the pun with the breed of dog of the same name.

The phrase *golden handshake* dates from the early sixties, but it was not until the late seventies and eighties that the humorous variations on the theme started to be invented: *golden handcuffs* came first in the second half of the seventies, followed by the *golden hello* in the early eighties and the *golden parachute* and *golden retriever* in the late eighties. The theme of *gold* is continued in other areas of business and marketing in the eighties, for example in the expression **golden bullet** for a product that is extremely successful and **golden share**, a controlling interest in a company (especially one which has recently been privatized), allowing the **golden shareholder** (usually the government) to veto undesirable policies.

> Managers . . . have private health insurance, a better than average pension scheme, a car, and perhaps help with independent school fees from the company. These 'golden handcuffs' are a hangover from the days of labour shortages and income policies and higher tax rates.
> *The Times* 4 Apr. 1985, p. 30

It wasn't long before most of RJR Nabisco's top executives 'pulled the rip cords on their golden parachutes' . . . Mr. Johnson's alone was worth $53 million.

New York Times Book Review 21 Jan. 1990, p. 7

Hordes of graduate recruitment managers would appear on one's doorstep clambering and pushing to make the best golden hello/salary/benefits offer. *World Outside: Career Guide* 1990, p. 6

goldmail *see* GREENMAIL

goon *see* ANGEL DUST

Gorby /'gɔːbɪ/ *noun*

A Western nickname for Mikhail Sergeevich Gorbachev, General Secretary of the Communist Party of the Soviet Union 1985–91 and President of the Soviet Union 1990–1; used in compounds and blends including **Gorbymania**, widespread public enthusiasm in the West for Mr Gorbachev and his liberalizing policies.

Formed by adding the diminutive suffix *-y* to the first syllable of *Gorbachev*.

The nickname became widely known throughout the English-speaking world in 1987, when Mr Gorbachev was enthusiastically greeted with cries of *Gorby* from large crowds of people both in Western Europe and in Warsaw Pact countries on trips outside the Soviet Union. His ability to communicate with Western leaders (summed up by Margaret Thatcher's famous phrase 'This is a man we can do business with') as well as his determination to tackle the Soviet economy through PERESTROIKA made him appear to many people in the West as the embodiment of a new order in world politics (even though he could not command the same popularity inside the Soviet Union), and certainly contributed to the disappearance of the Iron Curtain in 1989. The most fevered period of *Gorbymania* (also sometimes written **Gorbamania** or **Gorbomania**) came in 1987–9; it was also called **Gorby fever** in the press. So great was the enthusiasm for *Gorby* that, at the time of the signing of the INF treaty in December 1987, one US commentator sarcastically dubbed it a **Gorbasm**: this word, too, was taken up enthusiastically by journalists (who did not always use it with the critical connotations of William Bennett's remark, quoted below).

He had that smile, he had those surprises, he had the INF Treaty. Gorbachic! Gorbymania! Or, as Secretary of Education William Bennett said, warning of overenthusiasm, 'Gorbasms!'

Washington Post 11 Dec. 1987, section C, p. 13

Gorbymania grips Bonn . . . Mikhail Gorbachev stepped out on to the balcony . . . and appeared overwhelmed by the thousands of Germans cheering his name in a euphoric welcome. 'Gorby! Gorby! Gorby!' they shouted. *Sydney Morning Herald* 15 June 1989, p. 15

In the midst of his country's bout of Gorbymania, the fact that George Bush is . . . cautious . . . may have obscured his own little Gorbasm. Within days of the opening of the Berlin Wall, the defense secretary . . . was asking the services to find 180 billion dollars of cuts over three years.

Spectator 9 Dec. 1989, p. 9

goth /gɒθ/ *noun*

A style of rock music characterized by an intense or droning blend of guitar, bass, and drums, often with mystical or apocalyptic lyrics. Also, a performer or follower of this music or the youth subculture which surrounds it, favouring a white-faced appearance with heavy black make-up and predominantly black clothing.

A back-formation from the adjective *Gothic*; the style of dress and some elements of the lyrics evoke the style of Gothic fantasy.

Goth grew out of the punk movement in the late seventies, with bands like Siouxsie and the Banshees making the transition from punk; by the mid eighties it had attracted large numbers of British youngsters to its subculture. One of the most noticeable things about the *goth* look is its elaborate dress code, including black leather, crushed velvet, heavy silver jewellery, and its elaborate dress code, including black leather, crushed velvet, heavy silver jewellery, and pointed boots, combined with long hair, white-painted faces, and heavy black eyeliner. Although this gives a rather gloomy appearance, most *goths* are actually peace-loving vegetarians who see themselves as the heirs to the hippie movement of the sixties. The leading performers of the music (also known as **goth rock** or even **goth punk**) include Sisters Of Mercy, whose leader Andrew Eldritch reportedly chose his pseudonym from the *Oxford English Dictionary*, where the adjective *eldritch* is defined as 'weird, ghostly, unnatural, frightful, hideous'. A more middle-class and tame version of the *goth* subculture, based on INDIE music and ethnic clothes, is dismissively known as **diddy goth** among young *goths*.

> Siouxsie Sioux is the godmother of goth-punk, and her Banshees' brew hasn't been reformulated in years.
>
> *Washington Post* 14 Oct. 1988, section N, p. 22

> Justin, 22, a computer operator from Southend, explains he's a 'total' goth and fan of SOM, though he does have a surprisingly catholic taste in music ... 'The way I look at it, goth is being into alternative music. We're a mixture of the punk and hippie things. We're into black and the occult.'
>
> *Evening Standard* 22 Mar. 1989, p. 42

gotta lotta bottle see BOTTLE

graphic novel /ˌgræfɪk ˈnɒv(ə)l/ *noun* 🔲

A full-length story (especially science fiction or fantasy) in comic-strip format, published in book form for the adult or teenage market.

Formed by compounding: a *novel* in *graphic* form (that is, told in pictures rather than continuous text).

Graphic novels and comic-books generally have been popular in Japan (where they are known as *manga* 'exciting pictures') since about the sixties, and represent an important section of the publishing industry there. For as long as ten years there has been a cult following among adults in the West for 'adult comics' and for certain comic strips (such as the *Tin-Tin* and *Asterix* stories) in book form; the popularity of this format for science fiction and fantasy, together with the increasing popularity of fantasy in general in the eighties, led to the promotion of *graphic novels* as a distinct section of the publishing market from about 1982—a policy which by the end of the decade had proved a great commercial success.

> By November of this year [they] will be publishing 10 monthlies and will have 11 graphic novels in print.
>
> *Chicago Tribune* 28 Aug. 1986, section 5, p. 1

> There is far more to the graphic novel than recording the exploits of Donatello and his ninja friends.
>
> *Times Educational Supplement* 2 Nov. 1990, Review section, p. 1

See also PHOTONOVEL

graphics card 🖥 see CARD²

gray economy 〰 see GREY ECONOMY

graymail 〰 see GREENMAIL

graze /greɪz/ *intransitive verb* 🔲 🔧

To perform an action in a casual or perfunctory manner; to sample or browse. More specifically, *either* to eat snacks or small meals throughout the day in preference to full meals at regular times; also, to consume unpurchased foodstuffs while shopping (or working) in a supermarket, *or* to flick rapidly between television channels, to ZAP.

These are transferred and figurative uses of the verb *graze* 'to feed', which is normally only used of cattle or other animals.

Although there are much earlier isolated examples of *graze* used with reference to people (for example, Shakespeare's Juliet is told to 'graze where thou wilt'), the new senses defined here first appeared in the US in the early eighties, and focus on the metaphorical similarities of behaviour between human **grazers** and their animal counterparts. Whereas *snacking* has been current since the late fifties, the term **grazing** became most popular in the America of the mid eighties, where it seemed to have become part of the mythology both of the YUPPIE and of the COUCH POTATO: the former too busy to eat proper meals, the latter too preoccupied with the 'tube' to prepare them at home.

The phenomenon of supermarket shoppers (and staff) eating produce straight from the shelves could in part be attributed to larger stores (which are harder to supervise) and consequently longer shopping excursions, but it seems more likely that the problem existed earlier, only becoming a trend when given a name. Technically theft, *grazing* became for some the acceptable (and ingenious) face of shoplifting, perhaps because of its euphemistic name and the fact that the goods are consumed on the premises rather than being taken away.

Only in the late eighties did television become a successful *grazing* ground. Two factors were particularly significant: the growth of CABLE TELEVISION in the US, with the proliferation of channels to graze among, and the popularity of remote control devices (or *zappers*: see ZAP).

The grazer, feeling hunger pangs, drives to the Chinese restaurant and orders a couple of dozen jiaozi . . . This is consumed in the car, using chopsticks kept permanently in the glove compartment.
Observer Magazine 19 May 1985, p. 45

Yuppies do not eat. They socialize, they network, they graze or troll. *New York* 17 June 1985, p. 43

It's thousands of bits from TV shows within one TV show—a grazer's paradise.
USA Today 27 Feb. 1989, section D, p. 3

Brian Finn wandered from room to room, grazing on sandwiches and answering questions.
Bryan Burrough & John Helyar *Barbarians at the Gate* (1990), p. 448

green /griːn/ *adjective, noun* and *verb*

adjective: Supporting or concerned with the conservation of the environment (see ENVIRONMENT'), especially as a political issue; environmentalist, ECOLOGICAL. Hence also (of a product, a process, etc.) not harmful to the environment; environment-friendly.

noun: A person who supports the Green Party or an environmentalist political cause.

transitive verb: To make (people, a society, etc.) aware of ecological issues or able to act on ecological principles; to change the policies of (a party, a government, etc.) so as to minimize harm to the environment.

In this sense, the adjective is really a translation of German *grün*; the whole association of the colour *green* with the environmental lobby goes back to the West German ecological movements of the early seventies, notably the *Grüne Aktion Zukunft* (Green Campaign for the Future) and the *grüne Listen* (green lists—lists of ecological candidates standing for election). There were, of course, antecedents even within English, in which *green* has a centuries-old association with pastoralism and nature: the most obvious, perhaps, is the *green belt*. The noun and verb have arisen through conversion of *green* in its ecological sense to new grammatical uses.

The West German *green* movement grew out of widespread public opposition to the use of nuclear power in the late sixties and early seventies and soon became an important force in West German politics. At about the same time, an international organization campaigning for peace and environmental responsibility was formed; originally operating from Canada, this organization soon became known as **Greenpeace**. These were the two main influences on the adoption of *green* as the keyword for all environmental issues in English and the subsequent explosion of uses of *green* and its derivatives. The transition did not take place until about the middle of the eighties in British English, though. (*Green* was used both as an adjective and a noun to describe West German political developments, but in general the movement was known here as the ECOLOGY movement, and that was also the official title of the party now known as the **Green Party**.) Since that time, the adoption of a *green* stance by nearly all political parties and the re-education of the general public to be environmentally AWARE (the **greening** of country and politics) has led some people to speak of a **green revolution** not just in the UK but throughout the industrialized world (the term had in fact been used in the US before Britons started to use *green* in its ecological sense at all widely).

As *green* became one of the most popular adjectives in the media in the late eighties, its use was extended to policies designed to stop the destruction of the environment (**green labelling**, the same thing as ECO- or ENVIRONMENTAL labelling, **green tax**, etc.), and then to products and activities considered from the viewpoint of their impact on the environment (compare ECOLOGICAL and ENVIRONMENTAL).

Green as a noun was first applied to the West German campaigners, who became known as 'the Greens', but once the adjective became established in the mid eighties, the noun was extended to members of other environmentalist parties and organizations as well, and eventually to anyone who favoured conservation. Colloquially, such a person became a **greenie** or **greenster**; different hues of **greenness** (or **greenism**, or even **greenery**) also began to be recognized—someone who was in favour of very extreme environmentalist measures became a **dark green** or **deep green**, for example.

As political parties began to realize the need to adopt *green* policies in the face of what promised to be the *green* decade of the nineties, it was natural that the word should also come to be used as a verb; *greening* as a 'verbal' noun had already existed for more than a decade in this sense (for example, in the book title *The Greening of America*, 1970). A Centre for Policy Studies report on Conservative Party involvement in *green* issues, written in 1985, was called *Greening the Tories*, turning this round into a transitive verb, and since then the verb has become quite common.

> Mr Cramond said that the Highlands welcomed people from outside with knowledge and expertise who were willing to make things work, but there was no room for green settlers who hoped to live on 'free-range carrots'.
> *Aberdeen Press & Journal* 17 June 1986, p. 9

> While socialists tend to emphasise the liberation of women, greens wish equally to liberate men.
> *Green Line* Oct. 1988, p. 17

> Despite winning 14 per cent of the European vote in Britain, British greens will have no seats at the European Parliament.
> *Nature* 22 June 1989, p. 565

> Labour . . . accused the Government of spending taxpayers' money . . . by agreeing to an unprecedented £1bn 'green dowry' for environmental schemes in the water industry.
> *Independent* 3 Aug. 1989, p. 1

> It may be that 'green' products biodegrade more quickly and thoroughly, since they tend to use surfactants based on vegetable oils rather than petro-chemicals.
> *Which?* Sept. 1989, p. 431

> Vegetarians and the more self-denying Greenies may find themselves in an awkward moral dilemma.
> *Guardian* 23 Feb. 1990, p. 29

> Although 'deep greens' only account for a small percentage of the population, they are becoming more influential.
> *The Times* 28 Mar. 1990, p. 21

> British Gas has been quick to seek to capitalise on worries about the effect of energy consumption on the environment. It has advertised the 'greenness' of its main product—natural gas—in comparison with other hydrocarbons.
> *Financial Times* 20 Apr. 1990, section 5, p. 1

Greenham wimmin 🏠 ⚔ see WIMMIN

greenhouse /ˈgriːnhaʊs/ *noun* 🌳

In environmental jargon, the Earth's atmosphere regarded as acting like a greenhouse, as pollutants (especially carbon dioxide) build up in it, allowing through more heat from the sun than reflected heat rising from the Earth's surface, so that heat in the lower atmosphere is unable to escape and GLOBAL warming occurs; mostly used attributively, especially in:

greenhouse effect, the trapping of the sun's warmth in the lower atmosphere because of this process;

greenhouse gas, any of the various gases that contribute to the greenhouse effect (especially carbon dioxide).

A figurative use of *greenhouse*; in a real greenhouse, the air temperature can be kept high because the glass allows sunlight through but prevents the warmed air from escaping.

The concept of the *greenhouse effect* was first worked on by meteorologists in the late nineteenth century, but it was not given this name until the 1920s. Public interest in the effect, and in the problem of *global warming* generally, has grown steadily since the beginning of the eighties, allowing the term to pass from specialist use in meteorology into a more widespread currency. During the eighties, attributive uses of *greenhouse* multiplied, as *greenhouse* became a shorthand way of saying 'greenhouse effect', and anything which contributed to this could then be described as 'greenhouse *x*'. By far the commonest of these shorthand terms is *greenhouse gas*, but there have also been *greenhouse-friendly* (see -FRIENDLY), **greenhouse pollutant**, **greenhouse potential** (the potential of a substance to contribute to the *greenhouse effect*), **greenhouse tax** (a tax on *greenhouse gases*, also known as *carbon tax*: here *greenhouse* means 'designed to combat the *greenhouse effect*'), and **greenhouse warming** (another name for *global warming*).

> The Greenhouse melted the poles and the glaciers, and those won't reform overnight.
> George Turner *The Sea & Summer* (1987), p. 12

> We calculate that the solar flux necessary to trigger a runaway greenhouse is about 1.4 times the amount of sunlight that currently impinges on the earth. *Scientific American* Feb. 1988, p. 52

> HCFC 142b ... has 40 per cent of the so-called 'greenhouse potential' of CFC 11.
> *New Scientist* 13 May 1989, p. 26

> The criticism was especially pointed in light of Bush's campaign rhetoric promising to tackle the problem of greenhouse warming. *Nature* 18 May 1989, p. 168

> The destruction of the tropical rain-forest is also contributing to the greenhouse effect, since forests help to regulate the amount of carbon dioxide in the atmosphere. *Which?* Sept. 1989, p. 431

greenmail /'griːnmeɪl/ *noun*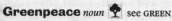

In financial jargon, the practice of buying up enough stock in a company to threaten a HOSTILE take-over, thereby forcing the company's management to buy the shares back at an inflated price if they are to retain control of the business.

Formed by substituting *green* for the *black* of *blackmail*; unlike *blackmail*, *greenmail* remains within the law, and it is backed by dollars ('greens'). This is not the first such alteration of the word *blackmail*: in the seventies there were a number of court cases in the US in which the defence threatened to expose government secrets unless charges were dropped, and these became known as *greymail* (or, in the US, *graymail*) cases.

Greenmail was one of many financial manoeuvres surrounding take-over bids that developed, principally in the US, during the first half of the eighties. In the UK the practice was limited by the Takeover Panel. By the middle of the decade the word had also started to be used as a verb, and an agent noun **greenmailer** had been derived from this. It has been claimed that, when the deal is worth more than a certain sum of money, it becomes known as *goldmail*.

> She went into hostile corporate takeovers, the money being made ... in greenmail and arbitrage.
> Saul Bellow *More Die of Heartbreak* (1987), p. 79

> His clients were little-known 'wanna-be' raiders, third-tier greenmailers such as ... Herbert Haft, the pompadoured scourge of the retail industry.
> Bryan Burrough & John Helyar *Barbarians at the Gate* (1990), p. 157

Greenpeace *noun* 🌳 see GREEN

green PEP ∕ see PEP

grey 🌱 see UNGREEN

grey economy /ˌɡreɪ ɪˈkɒnəmɪ/ *noun* Written **gray economy** in the US ∕

In financial jargon, the consumption, income, earnings, etc. generated by or relating to commercial activity which is unaccounted for in official statistics.

Formed by applying the *grey* of *grey market* to the *economy* as a whole (see below); a lesser version of the BLACK ECONOMY.

The term *grey economy* first appeared in the early eighties; the term *grey market* from which it derives can be traced back to post-war America, where it described the unscrupulous selling of scarce or rationed goods at inflated prices (a lesser *black market*). As the phrase *grey economy* became established its meaning was extended to cover any unorthodox or unofficial trading which is conducted in the wide grey area between official indicators of economic growth and the black market. In specific applications the term has been used with reference to any unwaged but significant activity (such as housework); to the earnings of those who 'moonlight' by taking a second job, often under an assumed name; to the makeshift system of bartering, exchange of goods, etc. which co-exists with the State economy, especially in the countries of the old Eastern bloc; and to the growing practice among small independent retailers in Britain of importing a product direct from its manufacturer or a foreign supplier in order to retail it at a price lower than that of its official distributor. The steady emergence of this last phenomenon during the eighties is in part explained by the strong encouragement given to small businesses in the ENTERPRISE CULTURE.

> Street vendors . . . have sprouted lately as an above-ground grey economy. Their goods—clothes, watches, jewellery—are not stolen, but bought wholesale. *Economist* 2 Apr. 1983, p. 70

> Italy, too, has a thriving entrepreneurial sector, but it is largely part of the 'gray' economy and so does not appear in the figures of tax collectors or government statisticians.
> *Harvard Business Review* Jan.–Feb. 1984, p. 60

greymail ∕ see GREENMAIL

GRID ⊗ see AIDS

grody /ˈɡrəʊdɪ/ *adjective* Also written **groady** 🕷

In the slang of US teenagers: vile, revolting, grotty. Especially in the phrase **grody to the max** (i.e. maximum: see MAX), unspeakably awful, 'the pits'.

This is generally thought to be a clipped form of *grotesque*, like the more familiar *grotty*, but it could perhaps be a diminutive of *gross*, which has been a favourite term of disgust among American youngsters in recent decades (compare *scuzzy* for 'disgusting': see SCUZZ).

Grody has been in spoken use since the late sixties but became fashionable through the spread of VALSPEAK in the early eighties (especially in the phrase *grody to the max*). It was widely popularized by a Moon Unit Zappa record of 1983, in which Moon Unit is heard to say:

> Like my mother makes me do all the dishes. It's like so *gross* like all the stuff sticks to the plates . . . It's like grody, grody to the max.

By 1985 a new noun had appeared: the **grodies** were the bag people, the homeless tramps who slept rough in the streets. *Grody* is not yet used in British English except in conscious imitation of American Valspeak.

> Omigod, Mom, like that's totally beige . . . I mean grody to the max, just gruesome. Gimme a royal break. *New York Times* 12 Dec. 1982 (Connecticut Weekly), p. 4

gross 🔖 see GRODY

groupware 💻 see -WARE

guestage /ˈgɛstɪdʒ/ *noun* 📁

A foreign national held as a hostage (but called a 'guest') in Iraq or Kuwait during the period following Iraq's invasion of Kuwait on 2 August 1990.

Formed by telescoping *guest* and *hostage* to make a blend.

This is a name which the hostages themselves invented in about September 1990. It remained in use until after they were allowed to return home in December 1990, but did not gain the enthusiastic support from the media that such words might usually enjoy, and is unlikely to survive in the language (except, perhaps, in historical accounts of the Gulf War) now that the motivation for it no longer exists.

> In his second television appearance with the 'guestages', as they had come to be known, he [Saddam Hussein] had not bargained for a forthright English woman.　*Independent* 3 Sept. 1990, p. 5

guppie /ˈgʌpɪ/ *noun* Sometimes written **Guppie** or **guppy** 🌱 🍸

Either (mostly in the US) a gay YUPPIE *or* (mostly in the UK) a GREEN yuppie: a yuppie who is concerned about the environment and green issues generally.

Formed by substituting the initial letter of *gay* or *green* for the *y-* of *yuppie* (see YUPPIE).

The word *guppie* was invented by the media in 1984 as one of the many variations on the theme of *yuppie* that arose in the mid eighties (including BUPPIE and others mentioned at YUPPIE). Since it has always had several possible interpretations (apart from those mentioned above, one newspaper even used it for *greedy yuppie*), most sources have needed to expand or explain it, and it has never gained any real foothold in the language despite fairly frequent use in journalism. It has been described as a journalists' 'stunt word', saying more about the influence of *yuppie* than anything else; this may well prove to be true, although with the importance of green issues in the late eighties and early nineties, it could still become established in its own right in the sense of an ecologically aware middle-class person and lose some of its associations with *yuppie*.

> There is one group that is totally universal: 'Guppies'—Gay Urban Professionals ... The so-called 'pink economy' (Guppies' lack of family commitments means money to burn) enables them to acquire possessions and indulge in activities that make straight Yuppies green with envy.
> Russell Ash, Marissa Piesman, & Marilee Hartley *The Official British Yuppie Handbook* (1984), p. 16

> On Wednesdays at midnight, Razor Sharp [a drag queen] appears with her Go-Go Boys at this upper West Side Guppie hangout.　*Newsday* 3 Feb. 1989, section 2, p. 3

> Far from building bridges between environmentalists and big business ... green yuppies or 'guppies' have 'delivered the green movement into the lap of the industrialist'.
> *Daily Telegraph* 20 Sept. 1989, p. 15

gutted /ˈgʌtɪd/ *adjective* 🔖

In British slang: utterly exhausted or fed up, devastated, 'shattered'.

A figurative use of the adjective *gutted*, graphically describing the feeling of having lost all one's 'guts'. An earlier sense in underground slang (current in the nineteenth century) was 'penniless'.

Although probably in spoken use for some time (it has been claimed that it is originally from prison slang), this sense of *gutted* did not start to appear in print until the mid eighties, when it suddenly became a favourite with journalists (especially the tabloid press). People interviewed after disappointments or scandals were often quoted as saying that they were *gutted*, although it was often difficult to be sure whether this was really the interviewee's word or the journalist's.

Seb must be gutted. Pulling out of the 1500m . . . must have been an agonising decision.

Sunday Mirror 4 Feb. 1990, p. 42

I've heard nothing for four months. I'm gutted because I still love him.   *Sun* 6 Feb. 1991, p. 22

●●

H

hack /hæk/ *verb* and *noun*

In computing slang,

transitive or *intransitive verb*: To gain unauthorized access to (a computer system or electronic data); to engage in computing as an end in itself, especially when this involves 'outwitting' the system (an activity known as **hacking**).

noun: A person (also known as a **hacker**) who enjoys using computing as an end in itself, especially when it involves trying to break into other people's systems. Also, an attempt to break into a system; a spell of *hacking*.

In both parts of speech, this is a specialized sense development relying on more than one existing sense. The verb probably arises from a US slang sense of *hack* meaning 'to manage, accomplish, comprehend' (usually in the phrase *to hack it*), since it first appeared in computing slang to describe enthusiastic use of computers, without any connotation of looking at other people's data; as a word for breaking into other computer systems, though, it must also be influenced by the original sense of the verb, 'to cut with heavy blows'. The noun was probably back-formed from *hacking*, but in the sense of an attempt to break into a computer system it has links with a more general US sense, 'a try, attempt'.

Computing enthusiasts first used this group of words in print to refer to enthusiastic (if not obsessive) use of computers in the mid seventies, although they were almost certainly using them in speech before that. By the early eighties, the 'sport' of breaking into computer systems, whether purely for pleasure, to expose some form of corruption, or as part of a more complex crime, had begun to be reported in the media, and soon appeared to be reaching epidemic proportions. Certainly it is the unauthorized type of *hacking* that has received greater media exposure, and therefore this set of meanings that has become widely popularized rather than the earlier ones (which nevertheless remain in use among enthusiasts, who still call themselves *hacks* or *hackers*). The verb is used either transitively (one can *hack* a system) or intransitively, often followed by the adverb *in* or the preposition *into*. With the almost universal use of computers in the business world and in defence planning and research in the late eighties, the activities of *hackers* could prove expensive or dangerous to their targets and various measures were taken to make systems **hacker-proof** or to provide an electronic **hacker watch** to catch the culprits red-handed. In the UK the Computer Misuse Act (1990) was a formal attempt to limit the damage. The jargon of *hackers* (enthusiasts or criminals) has been called **hacker-speak**. A specialized form of *hacking* practised by youngsters involves breaking the software protection on computer games; this is also known as *cracking*.

If you want to keep your street cred in the hacking fraternity, you've got to have an introduction screen with stunning graphics, a message to all the other hacking groups saying 'Hi guys. We did it first,' and comments on how good the software protection was.   *Guardian* 27 July 1989, p. 25

Hacking uncovers design flaws and security deficiencies . . . We must rise to defend those endangered by the hacker witch-hunts.   *Harper's Magazine* Sept. 1989, p. 26

1988: Hacker Robert Morris releases a software virus that kayos 6,000 computer systems.

Life Fall 1989, p. 30

The cost of restoring a computer system which is hacked into can run into hundreds and thousands of pounds for investigating and rebuilding the system.   *The Times* 11 Oct. 1989, p. 2

hack-and-slash /ˌhækənd'slæʃ/ *adjectival phrase* Also written
hack'n'slash ⊠

Of entertainment, especially role-playing and computer games: having combat and violence as its central theme, rather than logical thinking or problem-solving.

So named because the idea is to *hack* and *slash* one's way to a successful conclusion.

A term from *Dungeons and Dragons* (where it originally occurred in the form **hack-and-slay**). A game based on the idea of killing the enemy, or a person who likes this kind of game, is known as a **hack-and-slasher**. Perhaps under the influence of the computer-game use, a film or video whose main theme is gratuitous violence may be called a *hack-and-slash* film or a *hack-and-slasher* (compare SLASHER).

> Added another player: 'This is no hack-and-slash game. You win by creativity.'
> *Christian Science Monitor* 9 Feb. 1981, p. 15

> A pseudo-educational game ... One for the kids, rather than the hack'n'slashers, wethinks.
> *CU Amiga* Apr. 1990, p. 5

hackette /hæ'kɛt/ *noun* ⊠

In media slang, a female journalist. (Dismissive unless used by a fellow journalist.)

Formed by adding the feminine suffix *-ette* (as in *usherette*, but which also often has patronizing or pejorative connotations) to *hack*. As well as being a pejorative word for a writer (implying poor-quality writing produced to a deadline), *hack* is used among journalists as a positive term of solidarity for all those who work in in-house journalism.

A term coined by the British satirical paper *Private Eye*, apparently to describe Emma Soames, *hackette* remains a word particularly favoured by this source, although it has also appeared in a number of the more serious newspapers and has already found its way into fiction. It is principally a British usage, but began to appear in US sources as well from about the middle of the eighties.

> There are distinguished female professors ..., television speakerenes, Fleet Street hackettes, and publishers.
> Tim Heald *Networks* (1983), p. 167

> One hackette ... was ordered to ring up travel writer Bruce Chatwin ... and interrogate him.
> *Private Eye* 3 Apr. 1987, p. 8

> The worlds of newspapers and publishing are unbuttoned, and hackettes can wear pretty well anything.
> *The Times* 11 May 1987, p. 12

half shell ⊠ 🐢 see TURTLE

handbagging /'hænd,bægɪŋ/ *noun* 📇

In media slang, a forthright verbal attack or volley of strident criticism, usually delivered by a female politician (especially Margaret Thatcher, British Prime Minister 1979–90).

Formed on the noun *handbag*; the metaphor intended is that of a verbal battering likened to being bashed about the head by Mrs Thatcher's handbag. This picks up the imagery of comic strips, in which cantankerous women are sometimes shown beating another person (usually a young man) about the head with a handbag. There is also possibly an intentional pun on *sandbagging*, a term used figuratively for political bullying or criticism since the seventies.

The word arises from a remark made by a Conservative back-bencher in 1982. This was reported in the *Economist* as follows:

> One of her less reverent backbenchers said of Mrs Thatcher recently that 'she can't look at a British institution without hitting it with her handbag'. Treasury figures published last week show how good she has proved at handbagging the civil service.

The word became especially popular in the British press in the middle of the eighties—after Mrs Thatcher's often strident protests at EC gatherings and several disagreements with

Cabinet ministers had gained her a reputation for such verbal batterings—and is presumably a temporary term in the language, unless it comes to be applied widely to other female politicians. The verb **handbag** (from which the nound had arisen) and the adjective **handbagging** (describing this style of persuasion) also enjoyed a brief popularity in the media.

No one crosses Margaret Thatcher and gets away with it. And no one is too grand to escape the process of 'handbagging', which has been refined to an art under her premiership.
Independent 11 May 1987, p. 17

In the past, Neil Kinnock has been hand-bagged unmercifully, but he is now beginning to bowl her length.
Observer 22 Oct. 1989, p. 15

Mrs Thatcher has a 'handbagging attitude to German reunification.'
Daily Telegraph 27 Feb. 1990, p. 16

hands-on /ˌhændz'ɒn/ *adjective* 〰️ 🖥️

Involving direct participation; practical rather than theoretical. Also used of a person: having or willing to gain practical experience.

Formed on the verbal phrase *to get one's hands on* (*something*) 'to touch or get involved in' and influenced by the exclamation *hands off!* 'do not touch or interfere!'

Hands-on was first used as an adjective in relation to computer training in the late sixties, when opportunities to learn computing by sitting down at the keyboard and actually using the computer were described as *hands-on* experience. Throughout the seventies this was the dominant sense of the adjective, although towards the end of the decade a number of new applications were beginning to develop: people who had practical experience, or jobs which required it, could now be described as *hands-on*, and the metaphor was taken up in a more literal way by museums devoted to experiential learning, where visitors were encouraged to handle and use the exhibits. It was also at the end of the seventies that *hands-on* came to be used figuratively in **hands-on management**, a style of management in which executives are expected to get involved in the business at all levels, including the production process itself. (The opposite policy, in which managers interfere as little as possible and give their subordinates maximum room for manoeuvre, is called **hands-off management**.) During the eighties *hands-on* has been applied in a wide variety of different contexts to direct, practical participation.

The sucessful candidate will have a solid record of achievement in 'hands-on' management established over several years experience. *Wanganui Chronicle* (New Zealand) 19 Feb. 1986, p. 10

Reactor operators are denied hands-on control until they have proved their competence in a simulator. Just as pilots make their first mistakes firmly fixed to the ground, reactor staff are brought up to standard without the risk of accidentally plunging the world into Armageddon.
Guardian 3 Aug. 1989, p. 27

Zapata, who has been working in the business since she was a teenager, is the hands-on administrator of operations at Dawn. *Delaware Today* July 1990, p. 56

happening /'hæpənɪŋ/ *adjective* 🎇

In young people's slang: trendy, up-to-the-minute, 'hip', that is 'where the action is'.

Formed by shortening the phrase *what's happening* or *where it's* (*all*) *happening* and treating *happening* as an adjective. During the teenage revolution of the sixties, the noun *happening* was widely used to mean any fashionable event, especially a pop gathering, and *happenings* is a slang name for narcotics; the phrase *what's happening?* is a popular street greeting among US teenagers, perhaps originating in the language of jazz.

One of the *happening* words of the late eighties, *happening* as an adjective started in California in the late seventies; in her pastiche of Californian life *The Serial* (1977), American writer Cyra McFadden makes one of her characters say:

Who could live anywhere else? Marin's this whole high-energy trip with all these happening people . . . Can you imagine spending your life out there in the wasteland someplace?

The word then became enshrined in VALSPEAK in the early eighties, and eventually emerged in the pop and rock music world generally around the middle of the decade. In the UK it is still used mainly in writing for young people, but has also started to crop up in fashionable magazines and newspaper colour supplements.

> 'Me and George Michael,' she adds, lapsing into pop-speak, 'may turn out to be a pretty happening scene.' *Sunday Express Magazine* 1 Feb. 1987, p. 13

> Nothing looks sadder than a man wearing voluminous, 'happening' dungarees but with a bemoussed hairstyle that is pure Bros. *Weekend Guardian* 21 Apr. 1990, p. 25

> Manchester is this year's happening place. *Sunday Times Magazine* 6 May 1990, p. 36

hard card 🖳 see CARD²

hard lens 🗶 🖳 see LENS

headbanger¹ /ˈhɛdbæŋə(r)/ *noun* Also written **head banger** or **head-banger** 🎵 😈

In rock music slang: a follower of HEAVY METAL rock music; a person who enjoys a style of dancing to rock music involving head-shaking and rapid bending movements (known as **headbanging**).

Formed on *headbanging*, which in turn is a descriptive name for the dance; the rapid bending and head-shaking look rather like a mime of banging one's head against a hard surface (and in fact there is some suggestion that the early followers of HEAVY METAL actually did bang their heads against the amplifiers). There is also some confusion with the *headbanging* of the mentally disturbed: see HEADBANGER² below.

The term arose in the rock music context in the second half of the seventies, when HEAVY METAL first attracted a large following. Although originally a dismissive nickname, *headbanger* has been adopted by some of the fans themselves, who use *headbanging* to refer to listening to live rock music generally. *Headbanging* is also occasionally used as an adjective.

> Head bangers can find companionship in the mass suppression of individuality that is a heavy metal concert. *Independent* 28 Nov. 1988, p. 14

> Only head-banging heavy metal groups such as Metallica and Guns'n'Roses serve the primary function of rock. *Globe & Mail* (Toronto) 27 May 1989, section D, p. 5

> Headbangers get a chance. We have a fantastic competition for all you heavy metal fans out there ... Ten lucky readers will win a double pass to see *Skid Row*. *Sun* (Brisbane) 23 Apr. 1990, p. 4

headbanger² /ˈhɛdbæŋə(r)/ *noun* Also written **head banger** or **head-banger** 🏳 😈

In young people's slang: a deranged or stupid person; a lunatic or idiot. Hence in political contexts: a person with very extreme political views; someone whose ideas and policies seem 'over the top' (see OTT).

Adopted from psychological jargon, in which a *headbanger* is a child who engages in rhythmic rocking and banging its head against the cot or walls as a comfort mechanism (often as a sign of boredom, neglect, or stress), or an adult who is severely disturbed and shows stress by engaging in similar activity. As a young people's term of abuse it relies more on stereotyped notions of the behaviour of 'lunatics' than on knowledge of psychology.

Long in spoken use (especially, it seems, in Glasgow) as a general term of abuse, *headbanger* has acquired a wider currency in the late seventies and eighties as a result of its use in the newspapers to refer to extremist politicians of the Left and the Right. **Headbanging** in this sense means any militant political extremism.

> If he was to resign from Monday morning's interview ... It was a while since he had been carpeted ... Old Milne was a bit of a headbanger but apart from that. James Kelman *Disaffection* (1989), p. 84

Other drivers spoke about a 'headbanger' and the driving as 'absolute madness'.

The Times 6 Feb. 1989, p. 43

The Tories were always disliked by Christian Democrats for their selfishness and their mindless complacency. In the European Parliament, they sit alone with a few Spanish and Danish head-bangers, while the main conservative grouping excludes them.   *Observer* 19 Feb. 1989, p. 13

headhunt /'hɛdhʌnt/ *transitive verb* Also written head-hunt ⌇

To approach (a manager or other skilled employee who already has a job) with a view to persuading him or her to join another company in which a vacancy has arisen, especially when this approach is made by an agent or agency (a **headhunter**) specifically employed for this purpose by the company seeking staff. Also as an intransitive verb: to act as a headhunter; to engage in the process of executive recruitment known as **headhunting**.

The verb is back-formed from the action noun *headhunting*; this in turn is a case of a derisive nickname for the practice (also labelled *body-snatching* or *poaching*) which eventually became a semi-official term in business circles, losing even its metaphorical association with primitive peoples and the taking of heads as trophies.

Headhunting originated in the US (the practice in the fifties, the name in the second half of the sixties), but was not at all widespread in the UK until the eighties, the term *headhunter* remaining a derisive slang term until then. *Headhunt* as a verb has a similar history—first used in the sixties, but entering a rather different register of usage after the early eighties. During the eighties it became common for senior executives who were unhappy in their jobs to offer their services to *headhunters*, so that the agency's job included finding jobs for individuals as well as individuals for jobs.

He interviewed several people for the position but he did not find anyone suitable. Head-hunting seemed to be the next move.   Jeffrey Archer *First Among Equals* (1984), p. 223

At 45, Peter Birch brought the average age of building society chiefs down by a good few years. Worse, he had not been born and bred in the 'movement', but was headhunted from outside.

Money & Family Wealth Mar. 1989, p. 25

I can't afford an unemployed husband, and there isn't a headhunter in New York who'll talk to Wilder after one look at his curriculum vitae and his job record.   Saul Bellow *A Theft* (1989), p. 6

hearing-impaired ⊗ ❪❫ see DEAFENED

heavy metal /ˌhɛvɪ ˈmɛt(ə)l/ *noun* and *adjective* ♪ ☻

A style of loud, vigorous rock music characterized by the use of heavily amplified instruments (typically guitar, bass, and drums), a strong (usually fast) beat, intense or spectacular performance, and often a clashing, harsh musical style; a later development of 'hard' rock. Often used as an adjectival phrase to describe music of this kind. Sometimes abbreviated to **HM** or **metal**.

Both *metal* and *heavy metal* were used in William Burroughs's novel *Nova Express* in 1964:

At this point we got a real break in the form of a defector from The Nova Mob: Uranian Willy The Heavy Metal Kid.

The phrase was probably more influential when used again in Steppenwolf's record *Born to be Wild* in 1968, referring to the culture of the biker:

I like smoke and lightning, Heavy metal thunder.

In addition to the conscious quotation from these sources, the name may well be influenced by the harsh, metallic sound of the music and its heavy beat, or even by the leather gear with metal studs typically worn by *heavy metal* bands and their followers.

The term *heavy metal* was first used to refer to rock music by the music press of the mid seventies, seeking a dismissive label for what was otherwise known as hard rock. Gradually, though, *heavy metal* acquired a respectable status as a neutral term and came to be applied retrospectively to some of the groups formerly classified as hard rock (notably Led Zeppelin, who have come to be thought of as the founders of *heavy metal*). In the eighties the term was increasingly used adjectively, and *heavy metal* proved to be one of the major strains of White pop music running alongside Black-inspired styles such as HIP HOP.

> The names of Heavy Metal groups like Deep Purple and Motorhead are inscribed on the back of his leather jacket.
> *Daily Mirror* 10 Apr. 1980, p. 12

> New deal and line-up may give Girlschool new impetus in forest of macho HM bands.
> *Rock Handbook* (1986), p. 96

> Heavy Metal band *Skid Row* will be performing at Brisbane's Festival Hall . . . *Skid Row* was voted best new band in the 1989 Hot Metal reader's poll and has worked with metal giants *Bon Jovi, Aerosmith* and *Motley Crue*.
> *Sun* (Brisbane) 23 Apr. 1990, p. 4

See also SPEED and THRASH

helpline see -LINE

heritage /ˈhɛrɪtɪdʒ/ *noun*

In environmental jargon: the sum of the natural and constructed surroundings which a nation can pass on to future generations (especially areas of outstanding natural beauty, architectural monuments, and sites of historical interest). Often used attributively, especially in:

heritage centre, a multi-media museum celebrating local history and traditions;

heritage coast, a stretch of coastline whose natural features are protected by law from destruction;

heritage trail, an organized walk or tour which takes in sites of historical or natural interest, often on a specific theme.

A straightforward sense development from the original sense of *heritage*, 'that which is or may be inherited'.

The word has been used officially, in **national heritage**, to refer to architectural monuments (and especially 'stately homes' with their collections of art, antiques, etc.) since about the beginning of the seventies; *heritage coasts* were also first defined at about that time. It was not until the middle of the eighties, though—in the UK perhaps partly as a result of the creation in 1984 of English Heritage, a new Historic Buildings and Monuments Commission for England—that *heritage* began to be packaged and marketed as a commodity, a development which led to the name **heritage industry** for this aspect of tourism. At about the same time, renewed interest in the natural environment and GREEN issues generally led to a greater emphasis on this aspect of *heritage*. Some writers add an adjective to make their intentions clear— *cultural* or *architectural heritage* for buildings, *natural* or *green heritage* for nature—but often both are implied, and a preceding adjective is not possible when *heritage* is used attributively.

> What significance does the renewed interest in a 'national', 'local' or 'industrial' past packaged as intrinsically 'British' by the relentless 'heritage' machine, have at such a moment? . . . Heritage may indeed be a growth industry.
> *Art* Feb. 1988, p. 28

> The site will become an increasingly popular open air museum and a model of heritage interpretation.
> *British Archaeology* May/June 1989, p. 12

hero in a half shell see TURTLE

herstory /ˈhɜːstərɪ/ *noun*

In feminist jargon, history emphasizing the role of women or told from a woman's point of view (so as to provide a counterbalance to the traditional view, regarded as being male-dominated); also, a piece of historical writing by or about women.

A punning coinage, formed by reinterpreting the word *history* (actually from Latin and Greek *historia* 'narrative') as though it were made up of the masculine possessive pronoun *his* and *story*, and substituting the feminine possessive pronoun *her* for *his*.

The word was coined in the early seventies by militant feminists in the US, who had joined together to form an organization known as *WITCH*. In *Sisterhood is Powerful* (1970), feminist writer Robin Morgan wrote of the expansion of this acronym:

> The fluidity and wit of the witches is evident in the ever-changing acronym: the basic, original title was Women's International Terrorist Conspiracy from Hell ... —and the latest heard at this writing is Women Inspired to Commit Herstory.

Herstory remained effectively limited to feminist writing for some time, but during the eighties acquired a higher profile in general journalism. It is a word which has tended to annoy linguistic purists, who see it as an example of deliberate disregard for the rules of etymology; in a sense, though, this was the reason for its coinage—like WIMMIN, it was intended to shock people into thinking more carefully about male-dominated views of culture. A writer of *herstory* is sometimes called a **herstorian**.

> I have tried to write a herstory of the inner psychic meaning of the ancient religion.
> *Peace News* 2 Oct. 1981, p. 15

> The television cameras overlooked the ... herstorians ... To the eye of the TV camera, the parade was a group of provocatively dressed gays.
> *New Yorker* 13 July 1987, p. 17

> In a series of hot back-flashes we get the 'herstory' so far. As luck would have it, the dead woman was a writer and reader of modern herstory.
> *Sunday Times* 24 Jan. 1988, section G, p. 5

heterosexism /ˌhɛtərəʊˈsɛksɪz(ə)m/ *noun* 🔲

Discrimination or prejudice in favour of heterosexuals (and, by implication, against homosexuals); the view that heterosexuality is the only acceptable sexual orientation.

Formed by adding the suffix *-ism* (as in AGEISM, *racism*, etc.) to the stem of *heterosexuality*, after the model of *sexism*.

The word *heterosexism* was coined at the very end of the seventies in educational circles, when feminism and the gay liberation movement had succeeded in raising public consciousness about attitudes to sexuality enough to make some educators question the traditional assumptions passed on to children through the educational system. The adjective and noun **heterosexist** were coined at the same time. In a paper at the National Council of Teachers of English convention in San Francisco in November 1979, Julia Penelope summed up the feminist viewpoint:

> Heterosexist language, like so many of the social diseases that require radical treatment, must be understood to be, in and of itself, one of the few manifest symptoms of a thorough-going systemic corruption of human intelligence ... Heterosexism ... prescribes that the proper conduct for wimmin is passivity, servility, domesticity ... heterosexuality as the only 'natural' sexual interest.

By the middle of the eighties there was a lively public debate about the issues involved (both in education and in the general area of discrimination on grounds of sexuality), and it was even possible to attend **heterosexism awareness** training. The linking of the Aids risk with gay sex added fuel to this debate: see AIDS and HOMOPHOBIA. It is important to note that *heterosexism* does not always imply discrimination *against* homosexuals; often it is simply the assumption (regarded by many as justified) that heterosexuality is the natural state of affairs and the model on which a society should build.

> Even a non-sexist history may be heterosexist ... in its unquestioned, underlying assumptions; for example, that all women are motivated by an innate desire for men and marriage.
> Lisa Tuttle *Encyclopedia of Feminism* (1986), p. 143

> The branch [of the NUT] also calls on the union to train members not to adopt 'heterosexism' that discriminates against homosexuals.
> *The Times* 1 Feb. 1990, p. 4

hidden agenda /ˌhɪd(ə)n əˈdʒɛndə/ *noun* 📓

A secret motivation or bias behind a statement, policy, etc.; an ulterior motive.

Formed by combining *hidden* in its principal figurative sense of 'secret' with *agenda*, a word which is increasingly used as a countable singular noun meaning 'a list of things to be discussed at a meeting' and hence also 'an individual issue needing discussion or action'.

Like heterosexism, *hidden agenda* derives from the discussion of social issues in education; particularly during the late sixties and seventies there was much discussion of the concept of a **hidden curriculum** in schools, whereby pupils acquired a sense of social value or disadvantage from the prevailing attitudes rather than the subjects that were taught. This concept was translated into that of the *hidden agenda* in political contexts, international relations, labour relations, etc. during the late seventies and eighties and this became a favourite phrase among journalists in the second half of the eighties. *Hidden Agenda* was even the title of a controversial British film dealing with the question of a 'shoot-to-kill' policy in Northern Ireland (see *Stalkergate* in the entry for -GATE).

> There's family politics, sure, but our jobs are not being threatened ... So when we get into disagreements there's no hidden agenda.
> *Cambridge Chronicle* (Massachusetts) 6 Mar. 1986, p. 13

> Barrell's general programme is to point out the presence of a hidden political agenda in the strategies of a poem.
> *Essays in Criticism* Apr. 1990, p. 161

high-fibre ⊗ ❎ see FIBRE

high-five /ˌhaɪˈfaɪv/ *noun and verb* 🎮

In US slang,

noun: A celebratory gesture (originally used in basketball and baseball) in which two people slap their right hands together high over their heads; often in the phrase **to lay down** or **slap high-fives**. Hence also figuratively: celebration, jubilation.

intransitive verb: To lay down high-fives in celebration of something or as a greeting; to celebrate.

Formed by compounding: a *five* (that is, a hand-slap; compare British slang *bunch of fives* for a hand or fist) that is performed *high* over the head.

The *high-five* was originally a gesture developed for use in basketball, where it first appeared among the University of Louisville team in the 1979–80 season; Louisville player Derek Smith claims to have coined the name. By 1980 it was also being used widely in baseball, especially to welcome a player to the plate after a home run (and in this respect is similar to the hugs and other celebratory gestures used by British football players). Television exposure soon made it a fashionable gesture among young people generally; what ensured its eventual importation to the UK was its adoption by the Teenage Mutant TURTLES (in the form **high-three**, since Turtles do not have fingers) as a jubilant greeting.

> All that touched off a wild celebration of hugs, high-fives and champagne spraying.
> *USA Today* 14 Oct. 1987, p. 1

> A month has passed since the election and still Republicans and Democrats are high-fiving.
> *Maclean's* 2 Apr. 1990, p. 11

> So with a flying leap and a double high-five the two teammates celebrated the start of a new season.
> *Sports Illustrated* Dec. 1990, p. 16

high ground /ˌhaɪ ˈɡraʊnd/ *noun* 📓

A position of superiority or advantage (especially one which is likely to accord with public opinion) in a debate, conflict, election campaign, etc.

A metaphorical use of a military phrase whose literal meaning is 'a naturally elevated area providing a strategic advantage to the side which occupies it in a battle'.

The American writer Tom Wolfe attributes this figurative use to Lyndon Johnson in a speech about the US space programme in the late fifties, in which he supposedly said punningly that whoever controlled the high ground of space would control the world; however, although this was certainly the sentiment of his speech, it is not clear whether he actually used the phrase *high ground*. *High ground* really only became a popular political catch-phrase in the eighties; it is used mainly by journalists to describe a position which gives an individual or party the greatest visibility or appearance of right-mindedness in a debate—a position which might or might not accord with any absolute notions of rightness. As such, it seems to fit in well with the excessively opinion-conscious politics of the eighties. Often it is preceded by an explanatory adjective such as *moral*, *intellectual*, or *electoral*.

Her [Nancy Reagan's] seizure of the high ground in the fight against drug abuse has done much to reverse her immense unpopularity.   *The Times* 9 Jan. 1987, p. 7

Why didn't he take the high ground, and argue in favour of universal state benefits and services as ends in themselves?   *Sunday Telegraph* 30 Oct. 1988, p. 24

highlighter /ˈhaɪlaɪtə(r)/ *noun* 🔲

A marker which overlays a printed or written word with a semi-transparent, usually fluorescent, line of colour, leaving it legible but emphasized in the text.

Derived from the verb *to highlight* in the sense 'to make prominent, to draw attention to', by adding the agent suffix *-er*. Originally the word was in the respelt form *Hi-liter*, a US trade mark.

The trade mark was registered in the mid sixties in the US, and by the mid seventies the word in its standard spelling was catching on as a generic term. *Highlighters* in a very wide range of fluorescent colours became available and proved popular for all sorts of business uses from marking important activities and engagements in one's FILOFAX to picking out new words and senses in printed sources for lexicographers. The verb **highlight** was reinvented as a back-formation in the sense 'to mark with a highlighter'; other derivatives include **highlighting** as a verbal noun.

Simply find the hidden words . . . and then circle or highlight them.   *Country Walking* Jan. 1990, p. 16

'Bring me,' she cried, 'a highlighter.' She tinted the discrepancies between her text and the solicitor's in feverish, fluorescent yellow.   *Observer Magazine* 25 Mar. 1990, p. 42

high-tack 🔲 see TACK

high-tech /ˌhaɪˈtɛk/ *adjective* and *noun* Also written **hi-tech** 🔲 🖳

adjective: Making use of or provided with technological innovations, especially microelectronics or computers; automated, advanced.

noun: Technological hardware, automation; also, a style of sparse, functional design that embodies the modern technological ethos.

Abbreviated forms of *high technology*.

The phrase started to be used as an adjective in the early seventies, when electronics began to affect consumer goods and the design of homes, taking over from the phrase *with all mod cons* (that is, modern conveniences). As a name for a style of design, *high-tech* only remained in fashion for a relatively short time; the adjective, though, and the associated noun in the sense of 'technological gadgetry' have remained very common throughout the eighties. So popular was the term in the early eighties that some considered it to have become more or less meaningless; it was also at this time that it acquired a jocular opposite, **low-tech** (which usually implied complete absence of technology).

High-tech laid low: A ruptured $900 gasket dooms *Challenger* . . ., while a Soviet nuclear reactor at Chernobyl melts down.   *Life* Fall 1989, p. 26

The natural childbirth movement attempts to redress the 'high-tech' approach to childbirth.
Dorothy Judd *Give Sorrow Words* (1989), p. 9

Among the hi-tech companies to have prospered is Microvitec, whose technological prowess enabled it to take off with the home and education computing boom for a placing on the USM.

Intercity Apr. 1990, p. 35

Textbooks are unglamorous, low-tech.　　　*Times Educational Supplement* 14 Sept. 1990, p. 19

himbo /'hɪmbəʊ/ *noun*

In media slang, a young man whose main asset is good looks, but who lacks depth and intelligence; the male equivalent of a BIMBO.

Punningly formed on *bimbo*, by replacing the first syllable with the rhyming syllable *him* (the accusative form of the masculine personal pronoun *he*).

A journalistic creation of the late eighties which probably has less chance of surviving in the language than *bimbo*, but is given motivation by the fact that *bimbo* is now overwhelmingly applied to women. (Compare *bimboy* at BIMBO.)

Sex was commonplace, from a Melanie Griffith look-alike stuffed into her gown like salami in spandex to the macho himbo who strutted the Croisette wearing a 16-foot python like a stole around his shoulders.　　　*Washington Post* 29 May 1988, section F, p. 1

The recent spate of kiss-and-tell memoirs by various bimbos and their male counterparts, himbos, throws even more doubt upon the matter.　　　*The Times* 17 Oct. 1988, p. 21

hip hop /ˌhɪp 'hɒp/ *noun, adjective,* and *verb* Sometimes written **hip-hop** or **Hip-Hop**

noun: A street subculture (originally among urban teenagers in the US) which combines RAP music, graffiti art, and BREAK-DANCING with distinctive codes of dress and speech; more specifically, the dance music of this subculture, which features rap (frequently on political themes) delivered above spare electronic backing, and harsh rhythm tracks.

adjective: Belonging to hip-hop culture or its music.

intransitive verb: To dance to hip-hop music.

Formed by combining the adjective *hip* in its slang sense 'cool' with the noun *hop*, which also had a well-established slang sense 'dance'; *hip-hop* had existed as an adverb meaning 'with hopping movements' since the seventeenth century, but *hip hop* as a noun was a quite separate development. Its adoption as the name of the subculture and its music may have been influenced by the rap-funk catch-phrase *hip hop, be bop*, chanted by the disc jockey and rapper Lovebug Starsky in the form 'to the hip hop, hip hop, don't stop that body rock'.

Hip hop originated among young Blacks and Hispanics in New York in the second half of the seventies but was first widely publicized at about the same time as break-dancing in 1982 or 1983. At first the name was used to refer to the assertive and showy culture as a whole, with its visible and flamboyant street manifestations; it was the music which was imported to other cultures, though, and in the UK the word has been used mainly to refer specifically to this since it became popular in British clubs in about 1986. Its popularity as a dance music has led to the development of the verb *hip hop* and the action noun **hip hopping**; someone who listens or dances to the music or follows the culture in general is a **hip hopper**.

Like breakdancing, rap and hip hop in general flourished at street level despite overexposure in too many 'breaksploitation' films and a virtual end to exposure in the media.

Washington Post 30 Dec. 1984, section K, p. 5

Those hip to the beat cats down at Streetsounds bring you the biggest and freshest names in American hip hop.　　　*City Limits* 12 June 1986, p. 89

The look is squeaky clean. In its simplest form, the hip-hopper's kit consists of a hooded baggy top, tracksuit pants and training shoes.　　　*Observer* 24 Sept. 1989, p. 37

hip house 🎵 🎬 see HOUSE

hi-tack ✂ see TACK

HIV /eɪtʃaɪ'vi:/ *abbreviation* Ⓧ

Short for **human immunodeficiency virus**, a name for either one of two retro-viruses (properly called **HIV-1** and **HIV-2**) which cause a breakdown of the body's immune system, leading in some cases to the development of AIDS.

The initial letters of *Human Inmmunodeficiency Virus*.

HIV became the official name for the Aids retroviruses in 1986, after an international commit-tee had looked into the proliferation of names resulting from research in different parts of the world (previously, the same retroviruses had been known variously as *ARV*: Aids-related virus, *HTLV-III* (or *HTLV-3*): human T-cell lymphotropic or lymphocyte virus 3, and *LAV-1* and *LAV-2*: lymphadenopathy-associated virus 1 and 2). The US Center for Disease Control used *HIV* attributively in three of the six stages that it identified: the base state, **HIV antibody seronegativity**, involves no sign in the blood of exposure to *HIV*; **HIV antibody seroposi-tivity** identifies the presence of antibodies; and **HIV asymptomaticity** refers to infection with the virus which has not produced any signs of illness. (For the full list of stages, see AIDS.) Colloquially, *HIV* is sometimes called the **HIV virus**, effectively repeating the word *virus* (but showing that many people are not aware of the expansion of the abbreviation), and **HIV-positive** is used as an alternative for ANTIBODY-POSITIVE (similarly **HIV-negative**). In the late eighties, confusion over the terminology of Aids (and in particular frequent reference to people who actually had only a positive report of HIV infection as 'having Aids') led to the development of the term **HIV disease** for the earlier stages.

> Most people with HIV infection feel entirely well and may remain so for years . . . Some may feel ill . . . at the time they 'seroconvert' (i.e. become HIV antibody positive).
> Allegra Taylor *Acquainted with the Night* (1989), p. 82

> People with haemophilia who are HIV-negative should be able to get life insurance (though it may cost more). *Which?* Sept. 1989, p. 454

> Channel 4's recent Dispatches programme, which repeated the arguments of (among others) molecular biologist Peter Duesberg to suggest that the HIV virus can't cause Aids, has caused out-rage and concern among Aids specialists in Britain. *Guardian* 29 June 1990, p. 38

HM 🎵 🎬 see HEAVY METAL

hog 💉 see ANGEL DUST

homeboy /'həʊmbɔɪ/ *noun* Also written **home boy** or **home-boy** 🎬

In young people's slang (especially in the US): a friend or peer, a member of one's own gang or set; hence (in the usage of adult outsiders) a street kid, a member of a teenage gang.

This is an example of the spread of common Black English expressions into White vocabulary, largely through the medium of RAP (see also BAD, DEF, DISS, FRESH, and RARE). In Black English (especially among youngsters from the Deep South), *homeboy* was an established expression for 'a person from one's home town' and this was extended in Black college slang to anyone from one's own peer group or gang before being taken up by White youngsters as well, from rap lyrics and rap talk generally.

The original use of *homeboy* for a person from one's own home town dates back to at least the late sixties, but this does not seem to have been extended to members of a peer group or gang until the development of the street culture of the late seventies which gave rise to BREAK-DANCING and HIP HOP. Interestingly it is also attested among Black youngsters in South Africa. The spread of the hip-hop culture to White youngsters in the US and the UK during the mid and late eighties ensured that *homeboy* became one of the more prominent 'new' American words of the second half of the decade. The female equivalent is a **homegirl**; in slang use, *homeboy* or *homegirl* can be abbreviated and altered, to **home** or **homes** (and even *Sherlock*, after *Sherlock Holmes*), **homeslice**, etc.

> It's sprayed on walls . . . by some of the 30,000 'home boys', or gang members of the 400 gangs who roam, pretty much at will in LA county.   *Listener* 16 June 1983, p. 14

> Having restrained my homeboys we walked away with dignity, but the whole posse was quite visibly in tears.   *City Limits* 9 Oct. 1986, p. 52

> Just when all my homeboys is just kickin' it, like we all go somewhere.   *Spectator* 28 May 1988, p. 11

> Who cares about its symbolism, homeboy and homegirl has one, why can't I?
>   *Vindicator* (Cleveland State University) 10–24 May 1989, p. 2

> The perfect person to speak to their largely minority audience would be . . . a hip homeboy whose insecurities about making it in an Anglo-dominated world match their own.   *LA Style* Mar. 1990, p. 116

homophobia /ˌhəʊməʊˈfəʊbɪə/, /ˌhɒməˈfəʊbɪə/ *noun* 🔳

Fear or dislike of homosexuals and homosexuality.

Formed by adding the Greek suffix *-phobia* (meaning 'fear' or 'dislike') to the first part of *homosexual*. The formation is objected to by some people on the grounds that *homo-* as a combining form would normally mean 'the same' (as it does in *homosexual*) or that the word was already in use in the sense 'fear of men' (see below).

Homophobia was originally coined in the twenties in the sense 'fear or dislike of men', but as a hybrid formation mixing Latin and Greek elements (Latin *homo* 'man' and Greek *-phobia*) it did not really catch on. The impetus for a completely separate word based on *homosexual* rather than Latin *homo* and meaning 'fear or dislike of homosexuals' came from the gay liberation movement in the US in the late sixties, when consciousness of gay issues among the general public was being 'raised'. The term was popularized by American writer George Weinberg in articles published throughout the seventies, but did not reach a wide audience until the advent of Aids turned the phenomenon it described into a growing reality. A person who fears or dislikes homosexuals is called a **homophobe**; the adjective **homophobic** was derived from *homophobia* in the mid seventies.

> Some [homosexuals] even alleged darkly that a supposedly homophobic Reagan administration was deliberately withholding money so that the 'gay plague' would wipe them out.
>   *The Times* 12 Oct. 1985, p. 8

> Each Wednesday night they attended the Gay Homeowners' Association meeting at the Unitarian church, and the pastor . . . asked, 'Has anyone experienced any homophobia this week?'
>   Don Leavitt *Equal Affections* (1989), p. 24

> 'What part of your life would you recycle into another life?' 'Most of it, but not rottweilers, winebars, racists or homophobes.'   George Melly in *Marxism Today* June 1990, p. 56

Hooray Henry 🔳 see SLOANE RANGER

hopefully see BASICALLY

hospice /ˈhɒspɪs/ *noun* ⊗

A nursing-home dedicated to the care of the dying and the incurably ill.

A specialization of the word *hospice*, which originally referred to a house of rest for pilgrims etc., usually run by a religious order; by the end of the nineteenth century the word was used for any home for the destitute. The early hospices for the dying were mostly set up by religious orders too.

The word *hospice* has actually been in use for a home for the terminally ill since the turn of the century, but did not become widely known in this sense until the rise of the **hospice movement** of the late seventies and early eighties, which led to the setting up of hospices in many countries as places where people could be given a caring environment in which to spend their last days.

> Mother Frances is best known as the founder ..., fundraiser and administrator of Helen House, in Oxford, England, probably the world's first hospice for dying or acutely afflicted children.
> *Washington Post* 30 Aug. 1985, section B, p. 1

> He pays full tribute to his inspirer, Dame Cicely Saunders, who pioneered the hospice movement.
> *Church Times* 8 Aug. 1986, p. 7

hostile /'hɒstaɪl/, in the US /'hɑstəl/ *adjective* 〰️

Of a take-over bid or proposed merger: against the wishes of the target company's management; predatory, contested.

A specialized sense of *hostile* in its figurative use, with an admixture of the literal meaning 'involving hostilities'.

The term arose in the financial markets of the US in the mid seventies. It was the sharp increase in *hostile* bids in the first half of the eighties that led to the growth of devices such as the BUYOUT, the *Pac-Man defence* (see PAC-MAN²), and the POISON PILL.

> Greycoat Group ... is making a hostile £108 million offer for Property Holding and Investment Trust.
> *The Times* 26 Aug. 1986, p. 15

> Mr. Segal insists that hostile takeovers, leveraged buyouts and forced restructurings—which he bundles together under the ... label 'corporate makeovers'—are 'symptoms, not the disease'.
> *New York Times Book Review* 29 Oct. 1989, p. 32

-hostile ▦ see UNFRIENDLY²

host surrogacy ⊗ ⬛ see SURROGACY

hot button /'hɒt ˌbʌt(ə)n/ *noun* ✖ ⬛

A central issue, concern, or characteristic that motivates people to make a particular choice (among consumer goods, political candidates, social structures, etc.).

Formed by compounding: the imagery is that of a particular spot or *button* that must be found and pressed to trigger off the desired responses in the people one wants to influence (an image that had existed before in the figurative sense of *panic button*, used in the phrase *hit the panic button*); *hot* here is used in the combined senses of 'current or fashionable', as in *hot news* and *hot fashions*, and 'tricky', as in *hot potato*. It has been suggested that the term might also refer to the physical buttons on interactive television controls, with which viewers can vote, for example to register their support for an entertainment act or for one of the sides in a debate.

The expression *hot button* originated in the world of marketing in the US in the late seventies, when it was used to refer to the 'upcoming' desires of the buying public that the market would need to satisfy. It acquired a much wider currency when it started to be used in political contexts, though: before the end of the seventies it had been used as a synonym for *hot-spot* (describing Washington and Los Angeles as political *hot buttons*), but it was not widely applied to political issues of current concern (what the British might have called political *hot potatoes*) until the US presidential campaigns of 1984 and 1988. Since then *hot button* has become a political buzzword in the US, developing an attributive use as well (in *hot-button issue* etc.) in which it means 'central, influential, crucial'.

> The news-magazine [*Newsweek*], in the forefront of popularizers of this phrase, listed Republican *hot buttons* as the American Civil Liberties Union, abortion and guns.
> *New York Times Magazine* 6 Nov. 1988, p. 22

> Randall Lewis ... discussed the 'hot buttons' essential to catering to baby boom families.
> *New York Times* 25 Jan. 1990, section C, p. 6

> In the recent Congressional elections, Senator Helms tried to make homosexuality the 'hot button' of his campaign. *Gay Times* Dec. 1990, p. 11

hothousing /ˈhɒthaʊzɪŋ/ *noun* Also written **hot-housing** ⚒ ▮

The policy or practice of artificially accelerating the intellectual development of a child by intensive teaching from babyhood.

A figurative use of the verbal noun *hothousing*. Literally, the verb means 'to cultivate in a hot-house'; in educational *hothousing* the children are treated as hothouse plants which can be 'brought on' by intensive education.

The idea of *hothousing* in education is not especially new: in the early sixties A. S. Neill lamented the fact 'every child has been hothoused into an adult long before he has reached adulthood', and schools for gifted children which concentrated their education in the child's area of excellence were known as **hothouse schools** before the idea of intensively educating babies had been tried. The type of *hothousing* defined above, though, became fashionable in the US in the late seventies and eighties. The underlying principle was that any child could develop into a genius if only all the available time were used for education; using all the available time meant starting intensive training with flash-cards long before the child could talk or understand in the conventional sense what was being taught. The children subjected to this approach were called **hothouse children**.

> Their father . . . wanted to test the hot-housing theory; that if you subject a normally intelligent child to intensive, specialised training in a particular discipline at a very early age, you will produce excellence.　　　　*Observer* 30 Oct. 1988, p. 4

hotline ▮ see -LINE

house /haʊs/ *noun* Also written **House** ♪ ▨

A style of popular music typically featuring the use of drum machines, sequencers, sampled sound effects, and prominent synthesized bass lines, in combination with sparse, repetitive vocals and a fast beat; called more fully **house music**.

An abbreviated form of *Warehouse*, the name of a nightclub in Chicago where music of this kind was first played (see also WAREHOUSE).

House was the creation of disc jockeys at the Warehouse in Chicago and was first played in 1985. It is designed for dancing, and so does away with meaningful lyrics in favour of complicated mixtures of synthesized sounds and a repetitive beat. For these purposes it proved very popular with club-goers and at warehouse parties when introduced in the UK in the late eighties, giving rise to large numbers of sub-genres mixing the features of *house* music with existing sounds: during 1987–9, following on from ACID HOUSE, there was **deep house** (*house* with more emphasis on lyrics and showing the influence of soul music), **hip house** (mixing HIP HOP with *house*), **ska house** (*house* with Jamaican influences), and even **Dutch house** and **Italian house**. As a result of this, the term *house* has come to be used to refer generically to a whole range of sounds which share the characteristics mentioned in the definition above. *House* also contributed its own vocabulary to the language—for example, the verb *jack* in the sense 'move', as in the song titles *Jack Your Body*, *Jack It All Night Long*, etc.

> House is the mystifying music they call the key . . . House is *meta-music*, always referring outwards to other sounds, past and present.　　　　record sleeve of *The House Sound of Chicago* (1986)

> It's huge . . . and last week it became official: The Gallup Top 40 showed that House or House-derived music is occupying the whole Top 5.　　　　*Guardian* 19 Oct. 1989, p. 26

HRT /eɪtʃɑːˈtiː/ *abbreviation* ⊗

Short for **hormone replacement therapy**, a technique designed to relieve some of the unpleasant symptoms suffered by women during and after the menopause, by boosting oestrogen levels artificially.

The initial letters of *Hormone Replacement Therapy*.

The treatment first became available in the late sixties and to begin with was usually known by its full name *hormone replacement therapy*; by the mid eighties it had proved very popular as a safe, long-term treatment for the worst effects of the menopause (in particular brittle bone disease), was widely promoted by famous or successful women who had benefited from it, and was generally known by the abbreviation *HRT*.

> Oestrogen therapy (HRT) for women is increasingly prescribed to stave off post-menopausal symptoms such as brittle bones, thinning and wrinkled skin, falling hair, loss of libido and energy.
> *Sunday Express Magazine* 11 Feb. 1990, p. 45

> No one knows for sure which women should receive hormone replacement therapy. The official line is that it is necessary only for women who are at special risk of the bone-thinning disease osteoporosis. But no one knows exactly who these high-risk people are, so many women play safe and opt for HRT anyway. *Practical Health* Spring 1990, p. 11

HTLV, human immunodeficiency virus, human T-cell lymphocyte virus ⊗ see HIV

human shield /ˌhjuːmən ˈʃiːld/ *noun* 🗎 🎇

A person or group of people placed in the line of fire so as to fend off any kind of attack.

Formed by compounding: a *shield* made up of a *human* or *humans*.

The idea of the *human shield* has been known for some time, and the phrase itself had appeared in print before the end of the seventies. In the late eighties, there was a concentration of uses in connection with the situation in Lebanon. The greatest concentration of all, though, came in 1990–1 with President Saddam Hussein's holding of Western citizens in Kuwait and Iraq, after Iraq's invasion of Kuwait on 2 August 1990; some of these people were transferred to military and industrial installations in order to dissuade Western forces from attacking. The *human shield* policy in Iraq was reversed in December 1990 and most of the hostages were allowed to return to their own countries, but the term *human shield* was by that time very familiar both in the UK and in the US, and continued to be used in news reports in relation to the holding of prisoners-of-war in the Gulf, and in other contexts. For example, when the Red Army arrived in Lithuania in mid January 1991 to seek out draft-dodgers there and take control of strategic buildings in Vilnius, Lithuanians were described as forming a *human shield* to defend those buildings. There is some variation in usage as regards whether it is the whole group of people who are thought of as forming a single *human shield*, or whether each individual person is regarded as a *human shield* (in which case the term can be used in the plural).

> Thirty-nine right-wing French MPs arrived yesterday from Paris to join the 'human shield' around Gen Aoun, who also received the unexpected 11th-hour support of 6,000 'Lebanese forces', or Phalange militiamen. *Financial Times* 30 Nov. 1989, section 1, p. 4

> Forty-one Britons and a number of other Europeans in Kuwait have been rounded up by the Iraqis, apparently as the first of the thousands of foreigners who were waiting last night to be made a human shield for military and other installations. *Daily Telegraph* 20 Aug. 1990, p. 1

> Americans . . . reportedly were taken from the Mansour-Melia Hotel in Baghdad on the night of Oct. 29 and are now presumed to be 'human shields' at an undisclosed strategic site in Iraq.
> *Washington Post* 1 Nov. 1990, section A, p. 1

See also GUESTAGE

human wave 🎇 see MEXICAN WAVE

hunk /hʌŋk/ *noun* 🍴

In media and young people's slang: a sexually attractive, ruggedly masculine young man; a male pin-up.

A figurative sense development of the noun *hunk*, literally 'a large piece cut off from something (especially food)'; in this case, the development arises from an assessment of the man in question entirely from the point of view of physique (as though he were a piece of meat), in response to the plethora of such words used by men about women. An earlier slang sense was 'a large (and clumsy or unattractive) person', but this sense is now normally covered by *hulk*.

First used by jazz musicians in the forties and popular with college students in the US in the late sixties, *hunk* had spread to various other parts of the English-speaking world (including the UK, Australia, and South Africa) by the end of the seventies. During the eighties it enjoyed a fashion among tabloid journalists, along with the adjectives **hunky** and **hunksome**.

> Jumping on the hunk of the month bandwagon is photographer Herb Klein with a 1985 calendar that gives you a different man every month. *Fair Lady* (South Africa) 26 Dec. 1984, p. 11

> Michael Patton pranced his hunky bod around. *Village Voice* (New York) 30 Jan. 1990, p. 83

> Girl fans will be seeing more of the hunk ... in the top ... soap. *News of the World* 11 Feb. 1990, p. 5

hunt sab ✄ ❙ see SAB

hype ✄ see GLITZY

hyper- /ˈhaɪpə(r)/ *prefix* 💻

In computing jargon: involving complex organization of text or other machine-readable media so that disparate sources are linked together and may be accessed simultaneously. Used especially in:

hypermedia, a method of structuring information in different media (text, graphics, sound, etc.) for presentation to an individual user in such a way that related items of information are connected and presented together;

hypertext, machine-readable text that does not form a single sequence or come from a single source, but is so structured that related pieces of text can be displayed together.

The Greek prefix *huper-* 'above, beyond'; these approaches to machine-readable media go *beyond* the concept of searchability to present the user with a highly structured and interconnected resource.

Hypertext and *hypermedia* are concepts which computer scientists have been working on since the sixties, but which were perhaps too far ahead of their time to gain much popular currency until the eighties. Then, with the general public becoming increasingly computer-literate and demanding ever more sophisticated sources of information, and the necessary hardware becoming ever cheaper to produce, *hypertext* and *hypermedia* (sometimes called *multimedia*) were presented very much as the next step after the database and the personal computer, CD player, etc.

> Because different types of data ... can be tied together, hypertext and hypermedia are important in multimedia systems, where they can provide an innovative way to navigate the different data on a multimedia system. *Daily Telegraph* 9 Apr. 1990, p. 29

> Two aspects of the Active Book transcend the most useful Filofax: hyperlink and multimedia. *Independent* 9 Apr. 1990, p. 18

• •

ice /aɪs/ *noun* 💉

In the slang of drug users, a crystalline form of the drug methylamphetamine or 'speed', smoked (illegally) for its stimulant effects.

The name arises from the drug's almost colourless, crystalline appearance during the manufacturing process, like crushed ice. As one Australian newspaper has pointed out, the once in-nocent question 'Would you like some ice?', asked at a party, has taken on an entirely new meaning. In its prepared form, *ice* may be white, yellow, or even brown.

The drug first appeared with this name in Hawaii, and by 1989 had spread to the mainland US. Like the smokable cocaine derivative CRACK, it produces a sustained 'high', is extremely addic-tive, and has a considerable street value. It is smoked through a glass pipe called an *incense burner*, but unlike incense it is almost odourless, and so can be smoked in public with little risk of detection. Older names in the US for essentially the same drug include *glass* and *crystal* or *crystal meth*.

> Like those smoking crack, ice users initially suffer weight loss and insomnia because of the stimula-tion effects. *Daily Telegraph* 3 Oct. 1989, p. 11

> The ice problem is so bad that crack cocaine pales by comparison. *The Times* 7 Nov. 1989, p. 8

> 'However shit your life is, ice, at first, makes things better ...' is how one addict of the new American horror drug ice, describes its effects. *Sky Magazine* Apr. 1990, p. 91

icon /ˈaɪkən/, /ˈaɪkɒn/ *noun* 🖳

In computing jargon, a small symbolic picture on a computer screen, especially one that represents an option or function that can be selected by moving the pointer and clicking (see CLICK) on the icon.

A specialization of sense: in its original sense an *icon* is any representation or picture of some-thing (from Greek *eikon* 'likeness')—probably the best known examples are the religious pic-tures used in the Eastern Orthodox churches.

The *icon* first started to appear widely in the early eighties, when computer manufacturers were trying to make computer screens more USER-FRIENDLY to maximize on the rapid growth of the personal-computer market. The first *icons* typically allowed the computer screen to appear like a familiar desk-top, with the various files and tools available set out upon it in the form of small symbols (for example, a pile of index cards bearing a filename for each of the files which could be opened, a pencil or paintbrush for a program which could be used to 'paint' on the screen, etc.). The processes of computing were thus made to appear as similar as possible to the phys-ical use of files, pencils, etc. and the need to use an unfamiliar command language was mini-mized. As the use of windows (see WINDOW¹) developed during the eighties, whole windows of text could be 'shrunk' to the size of an *icon* so as to make room on the screen for other windows: the verb **iconify** and the adjective **iconified** were derived from *icon* to refer to this facility. In the late eighties, a series of sound equivalents for the *icon* was tried, with different audio mes-sages representing different functions and operations. This concept was punningly named the **earcon** (reinterpreting *icon* as *eye-con*).

> Newwave software, shown here, is one of several that use icons ... to represent different applica-tions. *The Times* 8 Dec. 1987, p. 31

> These 'earcons', a sound equivalent of icons, would tell the user how much memory is left, which task it is performing and how close it is to finishing. *New Scientist* 23 June 1988, p. 46

IKBS 🖳 see INTELLIGENT¹

immune /ɪˈmjuːn/ *adjective* 🖳

Of a computer system: protected against hacking or against destructive software de-vices such as the VIRUS and WORM.

A transferred sense of *immune*, which is normally used of a living thing in the sense 'able to re-sist infection'; compare INFECT.

> The Prolok system is actually a mixture of hardware and software protection. It is immune to the fiendish bit copiers. *Economist* 10 Sept. 1983, p. 71

immuno- /ɪmjʊnəʊ/ *combining form* ⊗

The combining form of the adjective *immune*, used in a wide variety of medical terms associated with the immune system, especially:

immunocompetence, the capacity for a normal immune response; also as an adjective **immunocompetent**;

immunocompromised, having an impaired immune system, especially as a result of illness;

immunodeficiency, immunodepression, a state of reduced immune defences in the body; also as adjectives **immunodeficient, immunodepressed**.

All of these terms have existed in the medical literature for some time; all came to prominence in less technical sources as a result of the growth of AIDS during the eighties and the attendant spurt of interest in the workings of the immune system. *Immunodeficiency* is most familiar to non-specialists as part of the name of *human immunodeficiency virus* (see HIV), the virus which has been associated with the development of Aids.

> They were further down the road than Phylly was. They weren't as tough or as immunocompetent.
>
> Michael Bishop *Unicorn Mountain* (1988; 1989 ed.), p. 310

> The categories of those who most need to take care—infants, the pregnant, etc—now include 'the immuno-compromised'.
>
> *Guardian* 13 July 1989, p. 23

impro /ˈɪmprəʊ/ *noun* ✂

A form of live entertainment based on improvisation and interaction with the audience.

Formed by abbreviating *improvisation* to its first two syllables.

Impro has been a colloquial abbreviation of *improvisation* among actors for some time, but it was only after the publication in 1979 of Keith Johnstone's book *Impro: Improvisation and the Theatre* that *impro* as a basis for live entertainment was developed into a theatrical genre in its own right. In the second half of the eighties it became a popular form of fringe entertainment, allowing the audience to dictate the course of events by suggesting themes, developments, etc., and this idea was even incorporated into television shows.

> 'Impro' stands for 'improvisation' and 'impro' audiences stand for an awful lot.
>
> *Independent* 20 Dec. 1989, p. 25

> The craze of 'impro' is spreading from the TV out into the public domain with the Canal Cafe Theatre putting on Improfest all this week.　*Evening Standard* 21 May 1990, p. 38

incendiary device ✺ see DEVICE

incense burner ✒ see ICE

inclusive /ɪnˈkluːsɪv/ *adjective* 🄸

Of language: non-sexist; deliberately phrased so as to include both women and men explicitly rather than using masculine forms to cover both.

A specialization of sense from the original and dominant use, 'having the character of including'.

The arguments for non-sexist language are as old as the feminist movement, but the name *inclusive* language became fashionable in the late seventies in the US and in the mid eighties in the UK. It has been used particularly in relation to the language of the Bible and of Christian worship, in which much of the imagery is masculine. In *The Word for Us: the Gospels of John and Mark, Epistles to the Romans and the Galatians restated in Inclusive Language* (1977), Joann Haugerud prepared the ground for *inclusive* language in Bible translations, expressing the hope that 'a taste of wholeness will encourage others to work toward providing a whole Bible in inclusive language', and an *Inclusive Language Lectionary* was published in the US from 1983. Although many churches have now adopted a policy of using *inclusive* language wherever possible, the move has not been well received by all members of congregations, especially when it means altering familiar words in the liturgy, hymns, etc.

As in the first edition of An Inclusive Language Lectionary, the word 'God' is often used where the pronouns 'He' and 'Him' appeared before.   *US News & World Report* 17 Dec. 1984, p. 70

'Inclusive language' does not have to mean replacing 'Almighty Father' with an (equally problematic) 'Almighty Mother'.   Janet Morley *All Desires Known* (1988), p. 5

incremental /ɪnkrə'mɛnt(ə)l/ *adjective* and *noun* 🗲

In the UK,

adjective: Of an independent local radio station: additional to the quota of broadspectrum stations; belonging to a set of extra stations designed to provide for a small community or specialized audience.

noun: One of these extra, specialist stations.

An *increment* is an increase or addition; the IBA chose to describe these planned stations as *incremental* in its report of 1988 (see below) because they were to operate in areas where a local radio service already existed, but provide increased minority-interest or specialist coverage, filling in the gaps in what was already available.

The term was first used officially in proposals set out by the Independent Broadcasting Authority in December 1988, when the Home Office authorized the licensing of the first twenty such stations. Typically the *incremental* stations cater for a very local community, an ethnic minority within the community, or a special-interest group (such as devotees of a particular style of music), but all sorts of ideas have come out of the move, including a station broadcasting only travel and flight information from Heathrow and Gatwick airports.

Baldwin suggests a doubling or slightly more of the current 75 franchises (52 stations and 23 incrementals, not all on the air yet) to 150–200.   *Management Today* Dec. 1989, p. 59

Only in 1988 did the IBA bow to the pressure of unsatisfied groups of listeners and allow 20 'incremental' stations to form. KISSFM, the last of these to go on air, opens next month, offering dance music.   *Daily Telegraph* 8 Aug. 1990, p. 28

indie /'ɪndɪ/ *adjective* and *noun* 🎵 💀

adjective: (Of a group or label) independent, not belonging to one of the 'major' companies in the popular-music industry; (of their music) unsophisticated, enthusiastically alternative in style.

noun: An independent artist, group, or label; the style of music typically put out by independents.

An abbreviated form of *independent*. The word was first so abbreviated in the slang of the US film industry in the forties to refer to independent film producers; the world of pop music has simply adopted the word from there.

Although the word was used in the popular-music industry during the sixties, it was not until the eighties that the contribution of independents was recognized as having led to a distinct style of music with its own charts (the **indie charts**). This was also the point at which the word started to be used to refer to the character of the music rather than simply its mode of production. Once the status of *indie* was formalized in this way, though, the character of the music became more static and conventional. By definition, *indie* music is intended to have a minority appeal. Its followers have also sometimes been called *indies* or **indie-kids**.

They're the only one of those indie-type bands that are trying to do something a bit unusual.

Q Mar. 1989, p. 19

From their indie pop beginnings ... The House of Love have ... managed to transform their ... critical acclaim into national popularity.   *Sky Magazine* Apr. 1990, p. 28

Wed Hosted by Dave Booth, a mix of indie (Happy Mondays, Stone Roses) and jazz.
Independent 23 May 1990, p. 31

INF /aɪɛnˈɛf/ *abbreviation* 📷

Short for **intermediate-range nuclear forces**; used especially in **INF treaty**, an agreement on the limitation of intermediate-range nuclear weapons, concluded between the US and the Soviet Union in 1987.

The initial letters of *Intermediate-range Nuclear Forces*.

INF became the preferred US term for theatre nuclear weapons (previously known as *TNF*) in the early eighties, and the abbreviation soon began to crop up frequently in reports of disarmament talks. It was the *INF treaty* of 1987 which resulted in the removal of US cruise missiles from British bases such as Greenham Common, and which heralded the beginning of a new era in East–West relations in the late eighties. The abbreviation is sometimes preceded by a further qualification of the weapons' range: **LRINF**, longer-range *INF*; **SRINF**, shorter-range *INF*.

A Soviet team touched down at Greenham Common yesterday to make a cruise missile inspection under the terms of the INF treaty. *Guardian* 17 Aug. 1989, p. 4

If the success of the INF negotiations can be carried into other areas of the nuclear armoury, then the INF Treaty will be seen as an important milestone.
Steve Elsworth *A Dictionary of the Environment* (1990), p. 326

infect /ɪnˈfɛkt/ *transitive verb* 💻

Of a computer VIRUS or other malicious software: to enter (a computer system, memory, etc.); to contaminate the memory or data of (a computer).

A transferred sense of *infect* which extends the metaphor of the computer *virus* as a contagious 'disease' capable of replicating itself within an organism.

The metaphor of *infecting* a computer system dates from the beginning of the eighties in the US, but became considerably more common in the second half of the decade, after the introduction of computer security hazards such as the VIRUS and the WORM. Systems which have had a virus inadvertently loaded into their memory (usually from a floppy disc), or the affected discs themselves, are described as **infected**; the noun **infection** exists for the process or result of loading, and also as a synonym for *virus*. Like a viral infection in living organisms, the computer virus may lie undetected in its host for some time, silently corrupting data in a succession of files before its effects become apparent.

Viruses usually infect personal computers, spreading through floppy disks and copied programs.
Clifford Stoll *The Cuckoo's Egg* (1989), p. 315

'It's pretty nasty', said Bill Cheswick, a computer science researcher at Bell Labs, who 'dissected' a version of the virus after obtaining it from the infected disk of a co-worker.
Newark Star-Ledger (New Jersey) 13 Oct 1989, p. 14

The problem is heightened by the emergence of 'infections' which, for the first time, have been tracked to virus writers in the Eastern Bloc. *The Times* 1 May 1990, p. 3

info- /ɪnfəʊ/ *combining form* 💻

A shortened form of *information*, widely used in compounds and blends such as:

infobit, a discrete piece of information or data;

infomania, a preoccupation with or uncontrolled desire for information; the amassing of facts for their own sake;

infomercial, a television or video commercial presented in the form of a short, informative documentary (the television equivalent of the newspaper's ADVERTORIAL);

infopreneur, a business person in information technology or the information industry; also as an adjective **infopreneurial**;

infosphere, the area of activity concerned with the dissemination, retrieval, or processing of information, often by computer; the information industry;

infotainment, a form of television entertainment which seeks to present factual material in a lively and entertaining way; docutainment (see DOC, DOCU-);

infotech, information technology.

Info has been a popular colloquial abbreviation of *information* for most of this century, but it was only with the advent of information technology, increasingly influential through the seventies and eighties, that the combining form began to appear. All of the formations mentioned above except *infotech* are American in origin, and all except *infosphere* have entered the language only in the eighties. The *infomercial* is allowed only on cable and satellite television in the UK, and so is still relatively unknown. *Info-* (or *infotech*) is increasingly used in forming the proper names or trade marks of organizations, products, or services, as well as in one-off headings for newspaper columns and advertising copy (in which it competes with FAX²): so we have **infofile**, **infoline**, **infopack**, etc.

I am much impressed by the . . . old-fashioned qualities of greed and mendacity the world of 'infotech' displays.
Listener 18 Aug. 1983, p. 34

American makers have used their knowhow to better commercial ends . . . Other countries—Britain and West Germany particularly—have been inexplicably making life as difficult as possible for their own infopreneurs.
Economist (High Technology Survey) 23 Aug. 1986, p. 15

The myriad factoids and ephemera and random infobits that are the common coin of daily business.
New York Times 6 Dec. 1987, section C, p. 12

Both shows are halfway between hard news and current affairs, being more in the lifestyle/'infotainment' mould. Will this 'infotainment' train ever run out of steam?
Courier-Mail (Brisbane) 23 Sept. 1988, p. 26

Now, in greater numbers than ever on independent stations and cable, comes . . . the half hour or hour that looks like a program . . . but isn't a program. Now comes the infomercial.
Los Angeles Times 12 Mar. 1990, section F, p. 1

Inkatha /ıŋˈkɑːtə/ *noun*

A Black political organization in South Africa, originally formed as a cultural organization in 1928 and revived as a Black liberation movement in 1975 under the Zulu Chief Mangosuthu Gatsha Buthelezi.

From the Zulu word *inkatha*, a sacred head-ring and tribal emblem which is believed to ensure solidarity and loyalty in the tribe. The name is intended to symbolize cultural unity.

Since its revival in 1975 as a Black national movement in South Africa, *Inkatha* has been open to all Blacks, although its following remains predominantly Zulu. It has featured increasingly in the news outside South Africa during the late eighties and early nineties, especially in relation to fighting among rival liberation movements there.

Fighting in Natal between sympathisers of the UDF and its ally, the Congress of South African Trade Unions, and Inkatha loyalists has cost more than 1,000 lives in the past three years, and is inimical to black unity.
Guardian 17 Aug. 1989, p. 10

Local supporters of the ANC have been almost unanimous in calling for more rather than fewer troops as the local police force is seen as being biased in favour of the ANC's opponents, the Zulu Inkatha movement headed by Chief Mangosuthu Buthelezi.
Financial Times 3 Apr. 1990, p. 22

INSET /'ɪnsɛt/ *noun* 🔲

Short for **in-service training**: term-time training for teachers in the state schools of the UK, statutorily provided for in teachers' conditions of service. Often used attributively (with a following noun), especially in **INSET course** and **INSET day**.

An acronym formed by combining letters from *In-SErvice Training*.

The acronym was first used in discussion documents on teacher training written in the mid seventies. Provision for compulsory in-service training for teachers was officially made in the Teachers' Conditions of Service 1987, which stipulated that teachers were to be available for work on 195 days during the year, but that no more than 190 should be spent in teaching classes. The remaining days were to be *INSET days* (or *non-contact days*), during which training could be given. With the introduction of the Education Reform Act of 1988 and the NATIONAL CURRICULUM, *INSET days* were partly used as a way of introducing teachers to the new methods and procedures involved—these days became known colloquially as BAKER DAYs—but they also introduced the acronym *INSET* to a wider audience.

> At the moment, in-service training is a voluntary activity . . . but soon five days of INSET will be a statutory obligation. *Times Educational Supplement* 19 June 1987, p. 18

insider dealing /ˌɪnsaɪdə 'diːlɪŋ/ *noun* 〰

The illicit use of confidential information as a basis for share dealing on the stock market; also known as **insider trading**.

Formed by compounding. In stock-market jargon, an *insider* is a person who is privy to information about a firm which would not be made available to the general public; *insider dealing* or *trading* is trading which is based on the confidential knowledge of *insiders* and is therefore one step ahead of the market.

The term has been used in stock-market jargon since at least the sixties (and the practice for several decades before that). The debate on the moral issues involved and the need to make the practice a punishable offence became quite intense in the UK during the seventies, and the issue reached a considerably wider audience in the eighties as a result of the exposure and prosecution of a number of prominent individuals for *insider dealing*, both in the US and in the UK.

> A quick check shows that if you are caught for insider dealing in France, you are likely to get off more lightly than in Britain. So if anyone is accused of insider trading in Eurotunnel shares (which seems pretty unlikely on past performance), it will clearly pay to make clear that all the action took place on the other side of the Channel. *Guardian* 4 Aug. 1989, p. 14

> Much energy . . . is spent these days on the criminal or near-criminal aspects of the decade's chicanery: . . . the insider trading of Boesky, Milken and others; the cowboy banking habits of Don Dixon. *Nation* 24 Dec. 1990, p. 818

intelligent[1] /ɪn'tɛlɪdʒənt/ *adjective* 🖳

Of a machine: able to respond to different circumstances, developments, etc. or to 'learn' from past experience and apply this knowledge in new situations. Used especially of a computer or other electronic equipment: containing its own microprocessor, SMART.

A transferred sense of *intelligent*, influenced by the term *artificial intelligence* (see AI); unlike the *dumb* machine which can only pass messages to and from a more powerful host and respond to specific instructions, the *intelligent* one can adjust its responses according to circumstance.

The word has been used in computing since the late sixties, although Joseph Conrad had anticipated the concept as long ago as 1907 in his book *The Secret Agent*:

> I am trying to invent a detonator that would adjust itself to all conditions of action, and even to unexpected changes of conditions. A variable and yet perfectly precise mechanism. A really intelligent detonator.

During the seventies and early eighties microelectronics began to be incorporated into a wide variety of consumer goods, bringing this concept of the *intelligent* machine into the public eye and giving the word a wide currency. Software systems can also be described as *intelligent*: an **intelligent knowledge-based system** (or **IKBS**) is similar to an EXPERT SYSTEM in that it stores the decision-making capability of human experts and can act on different data and developments on this basis, but it takes the principle of artificial intelligence one step further.

> The Japanese Fifth Generation computer project aimed at stimulating the development of the next generation of intelligent and powerful computer systems, has laid great emphasis on the importance of Intelligent Knowledge-based Systems (IKBS).
> *Australian Personal Computer* June 1985, p. 101

> An intelligent masterkeyboard ... allows control, via MIDI, of up to eight synthesizers in all registrations.
> *Keyboard Player* Apr. 1986, p. 27

> Gerald Ratner suggests that intelligent tills will generate up to 30 p.c. more profit at the Salisburys shops he bought recently from Next.
> *Daily Telegraph* 6 Feb. 1989, p. 22

> It is an 'intelligent' scanner in that it learns the shape of letters in the text, and can recognise up to ten different type faces per text.
> *English Today* July 1989, p. 49

See also ACTIVE

intelligent² /ɪnˈtɛlɪdʒənt/ *adjective*

Of an office or other building: containing a full set of integrated services such as heating, lighting, electronic office equipment, etc., all controlled by a central computer system which is capable of ensuring the most efficient and sound use of the environment's resources.

A further development from the sense defined in the entry above: the environment is controlled by an *intelligent* computer system, but when this runs all services within the building, it is the building itself that comes to be described as *intelligent*.

The first *intelligent* office buildings were built in the US in 1983 and by the middle of the eighties *intelligent* had become one of the buzzwords of office design both in the US and in the UK. It is difficult to say whether this further development of the adjective will survive in the language, but it certainly seems to express a design concept which is in keeping with the prevailing concern for integrated and efficient use of resources.

> One of Britain's most advanced high tech 'intelligent' office developments, Northgate is nearing completion.
> *Glaswegian* Dec. 1986, p. 12

> To a practitioner in the field of energy, 'intelligent buildings' involve energy engineering and building services, and suggest buildings whose facades, fabric and services combine (passively where possible) to optimise the environment and the consumption of energy.
> *Architech* June 1989, p. 43

intermediate-range nuclear forces 📇 see INF

intifada /ɪntɪˈfɑːdə/, /ˌɪntəˈfɑːdə/ *noun* Also written intifadah 📇

An Arab uprising; more specifically, the uprising and unrest led by Palestinians in the Israeli-occupied area of the West Bank and Gaza Strip, beginning in late 1987.

A direct borrowing from Arabic *intifāḍa*, which literally means 'shake' or 'shudder': the metaphor is that of shaking off the yoke of an oppressor, a concept with a long tradition in Islam.

The word *intifada* had been in use among Islamic groups (in the Lebanon, for example) before the Palestinian uprising of December 1987, but rarely appeared in English-language reports of events. After the beginning of the West Bank *intifada*, though, the word began to appear frequently and soon came to be used without a translation in some newspapers.

> The Palestinians have succeeded for the first time in bringing the *intifada* in the occupied territories within Israel's pre-1967 boundaries.
> *Independent* 14 June 1988, p. 12

Since the beginning of the so-called 'intifada', Israel has spared no effort to control and appease that uprising, with as little loss of life and injury as possible. *Harper's Magazine* Sept. 1989, p. 71

The *intifada* in Gaza and the West Bank is in its third year. Now that we have started, we can go on for three years as well if we have to. *The Times* 22 May 1990, p. 9

intrapreneur /ˌɪntrəprəˈnɜː(r)/ *noun* ᐁ

A business person who uses entrepreneurial skills from within a large corporation to revitalize and diversify its business, rather than setting up competing small businesses.

Punningly formed on *entrepreneur* by substituting the Latin prefix *intra-* in the sense 'within, on the inside' for its first element *entre-* (or by clipping out the middle part of *intra-corporate entrepreneurship*: see below). The result is a hybrid word made up of Latin and French elements, which many people would consider an ugly formation.

The idea of **intrapreneuring** or **intrapreneurship** came from US management consultant Gifford Pinchot in the late seventies. At first he named the concept *intra-corporate entrepreneurship*, but by the mid eighties the shorter form was becoming established. The corresponding adjective is **intrapreneurial**; the view that employees of large corporations should be encouraged to use their skills in this way has been called **intrapreneurialism**. All of these words are still predominantly used in American sources, although the concepts have been tried in many developed countries.

The belief that Japan is lacking entrepreneurs is wrong. 'If you want to set up your own business or go into a partnership, your path is blocked. So an entrepreneur becomes an 'intrapreneur' . . . Intrapreneurs set up the new business ventures. If a venture is a success, the company spins it off as a subsidiary. *Business Review Weekly* Oct. 1987, p. 158

A one day briefing on intrapreneurship: developing entrepreneurs inside Australian organisations. *Courier-Mail* (Brisbane) 21 May 1988, p. 27

Not surprisingly, other parts of the IBM empire reacted jealously against the PC team and the kind of threatening 'intrapreneurial' behaviour that they were encouraged to adopt. *Independent* 21 Mar. 1989, p. 19

investigative /ɪnˈvɛstɪɡətɪv/ *adjective* ▓

Of a style of reporting used especially in television and radio (and also of those who use it): actively seeking to expose malpractice, injustice, or any other activity deemed to be against the public good; penetrative, delving.

A specialized use of *investigative*, which in its most general sense means 'characterized by or inclined to investigation'.

The principle of *investigative* newspaper reporting, which would be so penetrative as to force public officeholders to take account of public indignation at any malpractice, was first established in the US by Basil Walters as long ago as the early fifties. However, **investigative reporting** only really came into its own in the US in the seventies (in connection with the Watergate scandal). In the UK, **investigative journalism** has been associated particularly with television and radio, with a whole genre of 'watchdog' programmes using the technique by the middle of the eighties in fields as diverse as consumerism and foreign aid.

Amateurs and intellectuals should not play at the hard and dirty business of investigative journalism. Philip Howard *We Thundered Out* (1985), p. 66

It may be that . . . the contemporary 'investigative reporter', in contemporary myth, and even by his own account, is inevitably a sort of scoundrel. *New Yorker* 23 June 1986, p. 53

Quality programmes such as drama and plays are expensive to produce, as is investigative journalism and high-standard current affairs and documentaries. *Which?* Feb. 1990, p. 84

See also PILGER

in vitro fertilization ⊗ 🖳 see IVF

Iran-contra 🗋 see CONTRA

Irangate 🗋 see -GATE

irradiation /ɪˌreɪdɪˈeɪʃ(ə)n/ *noun* 🌱 ⊗ ✖

The treatment of food with a small dose of radiation (in the form of gamma rays) as a means of arresting the development of bacteria and so extending the food's shelf-life. (Frequently in the longer form **food irradiation**.)

A specialized application of the standard sense of *irradiation*, 'the process of irradiating'.

The technique of *irradiation* for preserving food is not new (it was discovered in the fifties), but the sale of irradiated food was the subject of considerable debate in the second half of the eighties, bringing the already emotionally loaded words *irradiation* and *irradiated* into the public eye.

> 'Now we've got irradiation to worry about, too,' points out Francesca Annis, shaking her head in disbelief that later this year it will become legal to 'zap' food with radiation, to kill off bacteria and prolong its safe shelf life. 'But nobody knows what the long term risks of eating irradiated food will be.'
> *She* Oct. 1989, p. 18

See also DUTCHING

Italian house 🎵 🎮 see HOUSE

it's more than my job's worth 📇 see JOBSWORTH

IVF /aɪviːˈɛf/ *abbreviation* ⊗ 🖳

Short for **in vitro fertilization**, a technique for helping infertile couples to conceive, in which eggs taken from the woman are fertilized with her partner's sperm in a laboratory and some are then re-implanted in the womb. (Known colloquially as the *test-tube baby* technique.)

The initial letters of *In Vitro Fertilization*; *in vitro* is Latin for 'in glass' (i.e. the laboratory 'test-tube'—although it is actually a small dish that is used).

The technique was pioneered in the late seventies by British obstetrician Mr Patrick Steptoe. During the eighties it became available to larger numbers of women as one of the two principal means of helping infertile couples to have a child (the other being GIFT). *IVF* has been criticized on moral grounds because fertilized eggs (held by some to be living beings from the moment of fertilization) are necessarily wasted in the process, and also because of the high incidence of multiple births resulting from the technique.

> The Hammersmith technique is one of several new off-shoots of IVF, originally designed for the one-in-10 couples who are infertile and of whom an estimated 25 per cent may benefit from IVF techniques.
> *Guardian* 19 July 1989, p. 27

> Clinics are monitored by an interim licensing authority, which is concerned about the number of multiple births and says the Government is throwing away an opportunity to reduce the IVF death rate.
> *Sunday Correspondent* 6 May 1990, p. 3

See also ZIFT

J

jack ♪ 💀 see HOUSE

jack up 🔧 see CRANK

jam ♪ 💀 see DEF

Jazzercise /'dʒæzəsaɪz/ *noun* ⊗ ✕

The trade mark of a physical exercise programme normally carried out in a class to the accompaniment of jazz music.

Formed by telescoping *jazz* and *exercise* to make a blend, after the model of *dancercise* (a similar American invention of the sixties).

Jazzercise originated in the US, where the trade mark was first registered in 1977, claiming a first use in 1974. The programme was invented in 1969 by Judi Sheppard Misset, an American jazz-dance instructor, but only named *Jazzercise* some years later. *Jazzercise* was one of many physical exercise programmes competing for coverage in the fitness-conscious eighties: compare AEROBICS, AQUAROBICS, and CALLANETICS. Although protected by trade mark registration for Misset's programme of exercises, the word is sometimes used without a capital initial in the more general sense of any exercise done to jazz music.

> She wanted to know whether in the jazzercise routine done to the words 'I want a man with a slow hand' your hips bumped left or right on 'hand'.
> *New Yorker* 27 Aug. 1984, p. 36

> Jazzercise, the keep-fit regimen for women of the '80s, should not be overdone ... 'Jazzercise is not a gruelling thing but it does provide the basis for a good fitness program.'
> *Sun* (Brisbane) 21 Sept. 1988, p. 17

jazz-funk ♪ 💀 see FUNK

job-sharing /'dʒɒb,ʃeərɪŋ/ *noun* Also written **jobsharing** or **job sharing** 〰 ⓘ

A working arrangement in which two or more people share the hours of work, duties, and pay of a single post.

Formed by compounding: the *sharing* of a *job*.

The idea of *job-sharing* has been discussed since the early seventies, but was rarely put into practice before the early eighties. In the campaign to attract more women back into the job market, *job-sharing* offers greater flexibility than the traditional approach of one person, one job, but it requires considerable co-operation between the job-holders (or **job-sharers**). The verb **job-share** has been back-formed from *job-sharing*, and **job-share** is also used as a noun, for the post affected by *job-sharing*, in attributive phrases such as **job-share scheme**, or as a synonym for *job-sharing* itself. In the UK a programme of **job-splitting** (in which employers were given incentives for splitting full-time posts into two or more part-time ones) was tried in the mid eighties.

> John Lee ... said at Jobshare's national launch in Manchester ... the job-splitting scheme ... had not been a big success.
> *Independent* 7 Apr. 1987, p. 5

> Many are women who left teaching to have a family and have not returned. To attract them back there will need more flexible working hours (both job share and part-time), refresher courses and priority in the queue for nursery school places.
> *Guardian* 18 July 1989, p. 22

jobsworth /'dʒɒbzwəθ/, /'dʒɒbzwɜːθ/ *noun*

An employee or official who upholds petty rules and bureaucracy for their own sake.

A contraction of the phrase 'it's more than my *job's worth* (*not*) *to*'—the supposed justification that such a person would give for petty insistence on the rule.

A peculiarly British word, *jobsworth* has been in colloquial use since the early seventies. It was brought to greater prominence from the early eighties by television comedians; when, in September 1982, the well-known television consumer programme *That's Life* invented a *jobsworth* award (in the form of a gaudy commissionaire's hat) for the official who insisted on the silliest rule, its place in the language was assured. Introducing the award, Esther Rantzen said it was for 'the stupidest rule and the official who stamps on the most toes to uphold it', and Jeremy Taylor sang a song entitled *Jobsworth*—actually composed some years earlier for a revue—in honour of its first presentation, to a council which would not allow a woman to erect a white marble headstone on her husband's grave.

> Andropov turned out to have learned nothing at all since, as the imperial governor-general in Hungary in 1956, he carried out the crushing of the Revolution; a bureaucratic jobsworth, his reign was as useless as it was mercifully brief. *The Times* 9 Mar. 1987, p. 12

> Now, we all know park-keepers—'jobsworths' to the man. ('It's more than my job's worth to let you in here/play ball/walk on the grass/film my ducks.') *Punch* 20 May 1987, p. 47

> I was suddenly accosted by a Jobsworth who uttered the classic words, 'You can't do that in here.' *Personal Computer World* Dec. 1989, p. 122

jojoba /həʊ'həʊbə/, in Spanish /xoˈxoba/ *noun*

A desert shrub belonging to the box family, whose seeds contain an oil which is used as a lubricant and in cosmetics. Also, the oil which comes from these seeds.

The Mexican Spanish common name of the shrub *Simmondsia chinensis*.

The word is not new to American English, but only became current among British English speakers as a result of a flurry of interest in *jojoba* oil from the mid seventies onwards, first as a substitute for sperm whale oil and later as an ingredient of soaps and cosmetics. The first cosmetics containing *jojoba* were marketed in the early eighties.

> The Renewer Lotion contains collagen, jojoba oil and a special firming ingredient to smooth and soften the skin and increase cell renewal. *Look Now* Oct. 1986, p. 68

journo /'dʒɜːnəʊ/ *noun*

In media slang (originally in Australia): a journalist.

Formed by abbreviating *journalist* and adding the colloquial suffix *-o* (as in *milko* for *milkman*, etc.). This suffix is particularly popular in forming Australian nicknames and colloquialisms: see also MUSO.

In use for several decades in Australia, *journo* was popularized in the British newspapers from the mid eighties onwards, especially by the columnist Philip Howard. The word's popularity in the late eighties perhaps reflects the fashion for things Australian in the entertainment world generally; in particular, the ownership of many British newspapers by Australian tycoon Rupert Murdoch, and the fashion for Australian soap operas and television series, which have brought Australian forms of speech into prominence.

> You meet a better class of person there [at a girl's school] than egocentric journos. *The Times* 20 July 1984, p. 10

> Compared to the excesses for which Fleet Street journos are traditionally noted, chocolate addiction seems positively virtuous. *She* Aug. 1990, p. 69

jukebox /'dʒuːkˌbɒks/ noun Also written juke-box 🖳

In computing jargon, an optical storage device containing a number of CDs and a mechanism for loading each one as required for the retrieval of data.

A figurative use of *jukebox*; like the musical version, the computer *jukebox* has a number of discs which the user can select and load at will.

The technology for exchanging discs in a computer data store has been referred to in computing literature as the *jukebox* principle since the early sixties. However, it was the development of the optical disc as a storage medium in the eighties that made the *jukebox* a realistic possibility for ordinary businesses. The storage capacity is vastly greater than any other medium yet made available, and the *jukebox* mechanism makes for speed of access as well.

> One-and-a-half juke-boxes could store the names and addresses of every person in the world.
>
> *Daily Telegraph* 21 Nov. 1986, p. 15

> A CD-ROM jukebox, about the size of a suitcase . . . holds up to 270 CD-ROM discs—the equivalent of 72 million pages of text. *The Times* 2 Mar. 1989, p. 36

> Reflection Systems, formed in Cambridge last year, offers a deskside optical juke-box with two drives for users who need 47 gigabytes of data storage. *Guardian* 28 June 1990, p. 29

junk bond /'dʒʌŋkˌbɒnd/ noun 〰️

In financial jargon (especially in the US): a bond bearing high interest but deemed to be a very risky investment, issued by a company seeking to raise a large amount of capital quickly (for example, in order to finance a take-over); a type of MEZZANINE finance.

Formed by compounding: the *bond* is dismissively called *junk* ('rubbish') because of doubt over the issuing company's ability to pay the interest from income generated by the assets purchased.

The concept of the *junk bond* arose in the US in the mid seventies. It became a particularly prominent feature of corporate finance there from the early eighties, associated especially with Michael Milken of investment bankers Drexel Burnham Lambert and with the whole financial ethos of leveraged buyouts (see LEVERAGE and BUYOUT), mezzanine finance, and corporate 'raiders'. Debt incurred through the issuing of *junk bonds* is known as **junk debt**; finance based on them is **junk finance**.

> Mr. Milken told them it was time for some companies to de-leverage, urging many companies to swap their junk debt for a combination of equity and higher-grade debt.
>
> *Wall Street Journal* 18 Sept. 1989, p. 1

> As Drexel Burnham fell, two warring junk-bond titans scrambled for their payoffs.
>
> *Vanity Fair* May 1990, p. 50

> To Giuliani, the junk-bond monger's offense was to undermine the apparent 'integrity of the marketplace'. If people don't believe in this integrity, Giuliani said, they won't participate in the 'capital-formation system'. *Nation* 17 Dec. 1990, p. 755

junk food /'dʒʌŋkˌfuːd/ noun 🚫 ❌

Food such as confectionery, potato chips, and 'instant' meals that appeals to popular taste (especially among young people) and provides calories fast, but has little lasting nutritional value.

Formed by compounding: *food* that is *junk* from a nutritional point of view.

The term *junk food* arose in the US in the mid seventies, when it became clear that young people in particular ate a high proportion of instant foods containing much carbohydrate (often in the form of refined sugars), and were not getting the balanced diet needed for proper nutrition. This proved to be true of eating habits in other countries, too; the peak of concern about *junk foods* occurred in the late seventies and early eighties, before the health-and-fitness revolution of the eighties had started to affect people's diets, but both the phenomenon and the name have survived into the nineties. The term is sometimes used figuratively (compare FAST-FOOD).

Blyton may be junk food but it's not addictive. *The Times* 12 Aug. 1982, p. 6

He's a pretty average kid . . . Likes junk food, noneducational TV, and playing with guns.
Perri Klass *Other Women's Children* (1990), p. 5

With the demise of the traditional school dinner, more and more pupils are turning to junk food at lunch-times and unhealthy snacks at breaks. *Health Guardian* Nov.–Dec. 1990, p. 13

juppie see BUPPIE

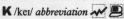

K

K /keɪ/ *abbreviation*

One thousand (widely used as an abbreviation in computing and hence also in financial contexts, newspaper advertisements, tables, etc.).

The initial letter of *kilo-*, the combining form used to denote a factor of 1,000 in metric measurements such as *kilogram*, *kilometre*, etc. and to represent either 1,000 or 1,024 in computing, as in *kilobyte* etc.

The abbreviation *K* has been used in computing since the early sixties, especially to denote a *kilobyte* (1,024 bytes) of memory. Although, for technical reasons, *K* does not represent exactly 1,000 in this context, it was the computing use that brought the abbreviation to public notice during the seventies and early eighties (as computers became commonplace in most people's working lives in industrialized countries) and, at least in popular usage, established its meaning as '1,000'. In the late sixties, job advertisements for computing personnel would sometimes give the salary offered as '$ . . . K' or '£ . . . K', meaning '. . . thousands of dollars or pounds sterling'. By the early eighties this practice had been picked up in job advertisements outside computing as well; *K* also began to be used in place of the three zeros in prices of houses offered for sale etc. It was even possible to hear *K* in spoken use (unusual for an initial-letter abbreviation); this was associated particularly with the 'yuppiespeak' (see YUPPIE) of the mid eighties.

Financial administrator, Thames Valley, from £12k.
advertisement in *Daily Telegraph* 26 Feb. 1986, p. 25

Alfa-Romeo—'84 . . . Perf. cond. 23k ml.
advertisement in *Washington Post* 31 Aug. 1986, section K, p. 24

I told him I had been approached by a cash purchaser with thirty-five k.
Andrew Davies *Getting Hurt* (1989), p. 95

karaoke /ˌkærəˈəʊkeɪ/, /ˌkærɪˈəʊkɪ/ *noun*

A sound system with a pre-recorded soundtrack of popular music from which the vocal part has been erased so as to allow an individual to sing along with it, often recording his or her performance on tape or video. Also, the pastime of singing to this kind of system.

A Japanese compound word which literally means 'empty orchestra'. The coincidence of two vowels which results from joining *kara* and *oke* makes the Japanese word even more difficult than most for English speakers to pronounce; some solve the problem by changing the first of these two vowels to /ɪ/.

Karaoke was invented in Japan and is extremely popular with Japanese business people visiting bars and clubs on the way home from work. It has a Western precedent in 'Music minus One', the recordings of classical concertos with the solo part missing which have been available for some years, and *karaoke* itself was successfully introduced both in the US and in the UK during the eighties (although not taken up with such popular enthusiasm as in Japan). The word is often used attributively, especially in **karaoke bar** or **karaoke club** (where *karaoke* is the main form of entertainment, with the customers themselves providing the cabaret) and in **karaoke machine**, the jukebox on which the accompaniments are recorded.

> The hotel people had provided a *karaoke* kit: a microphone and amplifier with backing tapes for amateur songsters.　　　　　　　　　　　　　　　James Melville *Go Gently Gaijin* (1986), p. 16

> Karaoke nights ... on Fifth Avenue ... are the hippest events in the entire city ... A natural extension of the No Entiendes theme, which encouraged anyone with enough bottle to get up and perform, karaoke has attracted the cream of Gotham.　　　　　　　　　　*Arena* Autumn/Winter 1988, p. 183

> The *karaoke*, or singing bar, is a few yards off Shaftesbury Avenue ... The idea of the karaoke bar is very simple. You get roaring drunk, chat up the bar girls and sing maudlin popular songs, dreadfully out of tune.　　　　　　　　　　　　　　　　　　　　*Daily Telegraph* 19 May 1989, p. 15

> They improve on the usual rugby songs by putting a lot of effort into the singing, aided and abetted once a week by a karaoke machine.　　　　　　　　　　　　*Evening Standard* 19 Apr. 1990, p. 19

keyboard /'ki:bɔːd/ *noun* 🎵 🖳

An electronic musical instrument with keys arranged as on a piano, and usually a number of pre-programmed or programmable electronic effects such as drum rhythms, different 'voices', etc.; known more fully as an **electronic keyboard**.

Formed by dropping the word *electronic* from the more formal name *electronic keyboard*. The word *keyboard* originally meant 'the row of keys on musical instruments such as the organ and piano'; the modern *keyboard* looks like a section of piano keyboard in a flat plastic casing.

Although electronic keyboard instruments of one kind and another have been in existence since the early years of this century, the type now known as an *electronic keyboard* or simply a *keyboard* did not become available until the late seventies. Much more compact than the earlier electronic organ, the *keyboard* (which is really little larger than the depth and width of the set of keys) relies on microchip technology to produce a wide range of sounds and effects. *Keyboards* became popular and versatile instruments for pop and rock music during the eighties, especially with the development of MIDI, allowing several to be linked together. They were also heavily marketed as ideal instruments for home entertainment. A player of a *keyboard* is known as a **keyboardist**.

> Combine this with a virtuoso stick player and MIDI keyboards and you get organs, guitars, synthesizers, and lots of other different sounds.　　　　　　　　　　　　*Dirty Linen* Spring 1989, p. 15

> Let's play keyboard video and the complete keyboard player book. Takes you through the initial learning exercises to the complete keyboard player.
> *Family Album Home Shopping Catalogue* Spring and Summer 1990, p. 959

keyboarder /'ki:bɔːdə(r)/ *noun* 🖳

A person who enters text at a keyboard, especially in typesetting or data CAPTURE.

Formed by adding the agent suffix *-er* to the verb *keyboard*, which was adopted in computer technology from well-established use in typesetting terminology.

A word which has been used in the printing industry for some decades, but which has acquired a much wider currency with the spread of computer technology during the eighties. The word is now sometimes applied to anyone who works at a keyboard, whether or not this is part of a programme of data capture, and might eventually take over from *typist* as the typewriter gives way to the computer keyboard.

> Much of this work is performed by keyboarders who don't understand English.
> *Fortune* 4 Feb. 1985, p. 51

> The standard of accuracy achieved by the keyboarders is outstanding.
> *Review of English Studies* Feb. 1990, p. 77

keyhole surgery /ˌkiːhəʊl ˈsɜːdʒərɪ/ noun ⊗ 🖳

Colloquially, minimally invasive surgery, carried out through a very small incision, using fibre-optic tubes for investigation and as a means of passing tiny instruments into the tissue.

Formed by compounding: *surgery* done through a hole which is so small that it is likened to a *keyhole*.

Keyhole surgery, a technique that is dependent upon advances in fibre optics in the seventies and eighties, has been practised for about a decade, but the colloquial nickname belongs to the second half of the eighties, when it became possible to carry out what would otherwise have been major operations using the technique.

Never an admirer of 'keyhole' surgery, I decided on liberal exposure of the problem.

Sunday Mail (Brisbane) 1 May 1988, p. 28

The first operation in Britain to remove a kidney . . . by minimal invasive surgery, or 'keyhole' surgery in popular jargon, was carried out in Portsmouth. *The Times* 17 May 1990, p. 20

keypad /ˈkiːpæd/ noun Also written key pad 🖳

A small panel (either hand-held or attached to a larger keyboard) with an array of push-buttons which can be used to control an electronic machine such as a television, video recorder, calculator, or telephone.

Formed by compounding: *keys* arranged on a plastic *pad* (smaller than the *board* of *keyboard*).

The word was introduced in the mid seventies in connection with teletext systems, and was soon also being used for TV remote-control monitors and the push-button controls which replaced dials on telephones. Many computer keyboards have a separate **numeric keypad** which can be used as a calculator, and may also have separate groupings of keys which act as *keypads* for selecting functions, moving the cursor, etc.

Pressing the mute button on the keypad temporarily cuts off your caller.

Sunday Times Magazine 28 Oct. 1984, p. 118

This new terminal has . . . a numeric keypad, a function keypad and a tamper-resistant pinpad.

Computer Bulletin June 1986, p. 3

kidflation /kɪdˈfleɪʃ(ə)n/ noun 〰

Humorously, economic inflation as it affects the price of children's toys and activities.

Formed by substituting the word *kid* 'child' for the first syllable of *inflation*.

A humorous example of the inventive ways in which *-flation* has been tacked on to words as though it were a combining form since the late seventies; more serious examples included *oilflation* and *taxflation* (inflation caused by increases in oil prices and taxes respectively).

The record and confection industries are among several that believe they have lost sales at the hands of 'kidflation'. When the recording industry, for example, fell into a slump in 1979, some industry officials said part of the reason was that the rising cost of albums was pushing them beyond the financial reach of young people. *Wall Street Journal* 2 Mar. 1981, p. 12

kidult /ˈkɪdʌlt/ adjective and noun 🎬 📺

In US media slang,

adjective: Of a television programme or other piece of entertainment: designed to appeal to all age groups; intended as 'family viewing'.

noun: A piece of entertainment designed to appeal to children and adults equally. Also, a person who likes this kind of entertainment; an adult with immature tastes and interests.

Formed by telescoping *kid* and *adult* to make a blend.

The word was coined in the US as long ago as the late fifties to refer to the kind of adventure series that naturally appeals to a young audience but can be so designed as to attract a cult following among older viewers, too. The adjective remained popular with US television reviewers throughout the sixties and seventies (often with the implication that the programme so described was truly appealing to neither group, but fell between two stools), but only acquired any currency outside the US towards the end of the seventies. During the late eighties the noun acquired the secondary sense of the 'typical' viewer of *kidult* entertainment.

> Not a film for either children or adults, but for 'that new, true-blue American of the electronic age, the kidult, who may be 8, 18, 38 or 80'. *New York Times* 29 Jan. 1989, section 2, p. 30

kidvid /'kɪdvɪd/ *noun* 🔲

In media slang (originally in the US): children's television or video; a children's programme or videotape.

A clipped compound, formed by combining the rhyming parts of *kids'* and *video*.

Kidvid has been an established slang name for children's TV in the US for more than two decades (it first appeared in a new words dictionary in the US in 1955 and is typical of the abbreviated nicknames created by the entertainment paper *Variety*), but has recently acquired a new lease of life in British use with the explosion of the UK video market during the eighties. In American English it is often used attributively (with a following noun), in *kidvid programming*, etc. An alternative form **kideo** (for children's video, often used in trade marks) only recently started to catch on outside the US, while in Australia another variation on the theme, **kidflick** (a children's film), was more successful.

> At the network he moved from the kidvids, those barely animated cartoons he is said to really love, to the grown-up stuff. *Listener* 26 Jan. 1984, p. 11

> Kids Vid, as the trade calls it, has suddenly become Big Business. *The Times* 27 Jan. 1986, p. 9

> With the summer holidays in full swing there are plenty of 'kideo' videos available. *Daily Express* 20 Aug. 1986, p. 21

> Ever since the early days of movies, the burning question has always been 'Is there a life after "kid-flicks"?' *Sunday Mail* (Brisbane) 31 Jan. 1988, p. 24

> The second Mom and Dad disappear, it's—click—on to the sugar-coated treats of commercial kid-vid. *New Age Journal* July–Aug. 1990, p. 12

krytron /'kraɪtrɒn/ *noun* 🔲 🔥

A kind of high-speed, solid-state switching device that is used in the detonation of nuclear weapons.

The derivation of the word is uncertain: the *-tron* element is almost certainly taken from *electronic*; the *kry-* could be a partial respelling of CRYO-, or part of the word *krypton*.

The *krytron* first appeared in technical literature in the early seventies and would no doubt have remained limited to technical use but for an incident in early 1990, when it appeared that American-made *krytrons* had been obtained by President Saddam Hussein of Iraq and a political scandal ensued. For a short time the word was prominent in the media.

> Some forms of krytron can be bought commercially...The order aroused CSI's suspicions because it required krytrons of a specification which could only have a military use. *The Times* 30 Mar. 1990, p. 9

L

lab ⚒ see NAB

lager lout /'lɑːgə ˌlaʊt/ *noun* ⚒

In the UK: a young (usually affluent) man who typically spends leisure time drinking large quantities of lager or other beer as one of a group in a pub, and takes part in rowdy, aggressive, or boorish group behaviour.

Formed by compounding: a *lager*-drinking *lout*. This form takes advantage of the alliterative effect of two words beginning with *l*—a factor which gives it more popular appeal than the original coinage *lager culture* (see below).

The idea originated with a speech by John Patten MP, then Home Office Minister of State responsible for crime prevention, in September 1988. Lamenting the increase in violence, especially in country towns which had formerly been thought of as quiet and peaceful, Mr Patten put the blame on affluent young men who would normally act respectably but had nothing better to do with their leisure time than drink too much beer. He described this as a **lager culture** and asked responsible citizens to help the police stop what he called 'lager culture punch-ups'. The form *lager lout* started to crop up in the newspapers about a fortnight after Mr Patten's speech; *Sun* journalist Simon Walters claims to have been the first to make the transformation, although *lager lout* itself is often attributed to Mr Patten. The form *lager culture* has since died out, but *lager lout* continues to be used and has even been used figuratively and as the basis for an adjective, **lager-loutish**.

> Lager louts . . . may be educated into drinking at a much earlier age than executives in the alcohol industry believe. *Independent* 13 Dec. 1988, p. 17

> I would ask you to dismiss the idea that this was lager-loutish behaviour. *The Times* 27 June 1989, p. 3

> Having produced so many phoney dummies, the editor of the new lager-lout among 'quality' newspapers has only himself to blame. *Private Eye* 15 Sept. 1989, p. 6

lambada /læm'bɑːdə/ *noun* ⚒

A fast and erotic dance of Brazilian origin, in which couples dance with their stomachs touching each other; also, the rhythmic music to which it is danced.

A Brazilian Portuguese word which literally means 'a beating, a lashing'.

The *lambada* has been danced in Brazil for many years, but was suddenly taken up as a fashion in North and Central America in the late eighties, perhaps in response to the craze for 'dirty dancing' (after the film of the same title, 1987). *Lambada* became the focus of considerable media hype during 1989 and 1990, and was included in the title of a number of films and of a disc which reached the top of the charts. This media interest caused it to be popularized in the UK and Australia as well. A verb **lambada** also exists; so striking was the promotion and 'packaging' of the dance for the Western market that the whole process of taking world or ETHNIC culture and marketing it in the West has been referred to as **lambadazation**.

> We were dancing the lambada face to face and sort of going up and down against each other. *Sun* 11 Apr. 1990, p. 3

> First it was disco, then dirty, then lambada—whatever way you want to kick up your heels. *Delaware Today* July 1990, p. 48

LAN /læn/ *acronym* 🖳

Short for **local area network**, a computer network (see NETWORK²) in which computers in close proximity to each other are enabled to communicate and share resources.

The initial letters of *Local Area Network*.

The first *local area networks* were developed in the late seventies; by the early eighties, the acronym *LAN* was being used as a pronounceable noun in its own right. The *LAN* is most useful for inter-communication within a single business or department, giving a higher quality of service than the wider networks (see WAN) and at the same time enabling groups of computer users to share resources. *LANs* were therefore in extremely widespread use throughout the computerized world by the end of the eighties, sometimes linking electronic audio or visual equipment as well as text-handling computers.

> We've installed and continue to support a number of varied network environments—from LANS to WANS.   *New York Times* 17 Oct. 1989, section C, p. 13

> ETHERNET and Novell NetWare still dominate the local area network market. It seems IBM's Token Ring and Microsoft's OS/2-based LAN Manager have made little headway outside those bits of the corporate market with Big Blue-tinted glasses.   *Guardian* 28 June 1990, p. 29

Lance /lɑːns/ *noun* 🎖

A short-range surface-to-surface ballistic missile system designed to be used mainly with nuclear warheads; also, a missile used by this system.

A figurative application of a historic weapon-name.

The *Lance* missile system was developed in the US in the sixties, for use by the US army. What brought it into the news in the eighties was controversy over its replacement in NATO after the conclusion of the INF treaty of 1987, which removed intermediate-range nuclear weapons from the European NATO armoury. The programme to develop a successor was written about as the **follow-on to Lance** programme and the weapon itself as the **Lance replacement** or **Son-of-Lance**. The cause of the controversy was the proposal to give this new weapon a longer range, bringing it near in range to the intermediate-range Soviet weapons then being destroyed as a result of the INF treaty. In May 1990 the US announced its decision not to modernize the NATO *Lance*, after coming under pressure from Germany (where many of the old *Lance* missiles are based) to cancel the development plans.

> There is no intention of extending the range so as to run foul of the INF treaty. But the Soviet Defence Minister blurred this distinction by describing the Lance replacement as having a range of 'up to 500 kilometres', and being 'similar to the SS-23'. Should the Soviet Union go on destroying its SS-23s when Lance was being modernised, he asked rhetorically.   *Guardian* 29 July 1989, p. 9

> Better even than the B-2 as a symbol, the committee halted work on two mischievous missiles—the SRAM-T (air-to-surface) and the Son-of-Lance (surface-to-surface). Each of these was designed to fall barely beneath the distance ceilings of the 1987 Intermediate Nuclear Forces Treaty.
> *Boston Globe* 3 Aug. 1990, p. 11

landfill /ˈlændfɪl/ *noun*

In full **landfill site**: a place where rubbish is disposed of by burying it under layers of earth.

Formed by abbreviating *landfill site*; the term *landfill* had been in use since the forties in the US for this method of disposing of rubbish, and since the sixties for the rubbish buried in this way.

Landfill has been used as a method of waste disposal in developed countries for several decades; *landfill site* was first abbreviated to *landfill* during the seventies. In the mid eighties, the subject of *landfills* came into the news in connection with growing concern for the environment, especially when it was revealed that hazardous wastes had been buried in them, and that the land had in some cases been re-used for residential sites: see DUMPING.

> Manila's huge landfill at Tondo receives garbage from nearly two million people every day.
> *Listener* 12 July 1984, p. 16

> Truck carrying 1,800 gallons of waste oil believed to contain cancer-causing PCBs was held at landfill pending tests.   *USA Today* 18 Oct. 1985, section A, p. 5

landside ✖ see AIRSIDE

laptop /'læptɒp/ *adjective* and *noun* Also written **lap-top** 🖳

adjective: Of a computer: small, light, and usually not dependent on a mains power supply, so that it can be used on a person's lap.

noun: A portable microcomputer designed to be used on a person's lap. (Short for **laptop computer** or **laptop portable**.)

Formed by compounding, after the model of DESK-TOP; normally one would not speak of the *top* of the *lap*. As ever-smaller computers were invented, the terminology was changed to keep up with them: successors to the *laptop* have included the *lunchbox*, the *notebook*, and even the *palm-top*.

The *laptop* micro was first marketed in the US in the early eighties, and by the middle of the decade accounted for a sizeable proportion of microcomputer sales worldwide. Most models work on rechargeable batteries and are no larger than a small briefcase; one of their main advantages is that they can be used anywhere, whether there is a mains power supply available or not. By the second half of the eighties it was commonplace to see business people using them in a variety of public places, including trains, cars, and aircraft. **Lap-portable** is sometimes used as an alternative term for *laptop*.

> The Z-181 and Convertible are aimed at the real lap-portable market of journalists, academics, travelling salespersons and suchlike. *Practical Computing* Oct. 1986, p. 63

> You don't have to be a genius to know that a laptop usually costs more than its equivalent desktop. *Intercity* Apr. 1990, p. 4

See also LUGGABLE

laser angioplasty ⊗ 🖳 see ANGIOPLASTY

laserdisc /'leizədɪsk/ *noun* Also written **laser disc, laser disk**, or (as a trade mark) **LaserDisc** 🖳

A disc on which signals or data are recorded digitally as a series of pits and bumps under a protective coating, and which is 'read' optically by a laser beam reflected from the surface; also called an *optical disc* or CD. In the form **LaserDisc**: the trade mark of software developed for the Philips LaserVision system.

Formed by compounding: a *disc* which is both written and read by *laser*.

The technology associated with the *laserdisc* was developed by Philips in the second half of the seventies (see CD and LASERVISION). The name *laserdisc* started to be used more generally from the beginning of the eighties, contributing to the vogue for any new technology to contain the word *laser* in its name at this time.

> Any videocassette or laserdisc featuring the Premiere Recommends seal in its advertising has been approved by our editors with your home-viewing satisfaction in mind. *Premiere* June 1990, p. 142

> A laser disk player, together with a computer, a monitor, and probably a printer, adds up. *Smithsonian* Feb. 1991, p. 24

LaserVision /'leizəvɪʒ(ə)n/ *noun* Often written **Laservision** 🖳

The trade mark of a video system in which the signal is recorded as a series of pits and bumps on an optical disc and 'read' by laser; a type of CD video (see CD).

Formed by compounding, after the model of *television* and *Cablevision* (see CABLE TELEVISION): *vision* made possible by *laser* technology.

Laservision was developed by Philips during the seventies and first made commercially available in the early eighties as one of a number of videodisc formats competing for the CD video market. The quality of reproduction from the digital recording on compact discs is much higher than can be achieved using videotape; Philips went on to develop an interactive version (*CDI*: see under CD) which is designed to make this system more versatile in the age of MULTI-MEDIA.

> The CD-I Enabling Initiative will provide software tools and a manual to help designers to transfer programmes from Laservision and computer format to CD-I, thus broadening the choice of course-ware and helping to reduce its cost.
> *Guardian* 20 July 1989, p. 29

> When I saw my first LaserVision demo, it was, in the immortal words of Yogi Berra, 'deja vu all over again'. The picture was sharp.
> *Stereo Review* Dec. 1989, p. 94

LAV ⊗ see HIV

Lawsongate ☐ see –GATE

LBO 〜 see BUYOUT

leaderene /ˌliːdəˈriːn/ *noun* ☐

In the UK, a female leader.

Formed by adding to *leader* the otherwise unknown suffix *-ene*, possibly under the influence of the French feminine suffix *-ine* as used in the very similar Franglais word *speakerine* (for a female TV presenter), a word which caused heated discussion among French purists during the sixties and seventies. (Franglais also boasts *le leader* and *le leadership* among its political borrowings, but not *leaderine*.)

The word was coined by Norman St John Stevas, then MP for Chelmsford, as a humorous nickname for Margaret Thatcher when she was Leader of the Opposition in the late seventies. The nickname proved very successful and continued to be used of her, usually with a capital initial, throughout her period as Prime Minister (1979–90); it was a particular favourite of the satirical paper *Private Eye*. The usage also spread beyond its original limited context, and by the mid eighties was often used as a humorous word for any female leader, especially if she shared some characteristic with Mrs Thatcher. It will be interesting to see whether this extended use survives the end of Mrs Thatcher's leadership career.

> The British security services seem to be the out-and-out villains under their new leaderene, a Thatcher-like figure of absurd proportions.
> *Listener* 26 Apr. 1984, p. 33

> In Finchley Central, part of the glorious leaderene's own constituency, there is only one policeman on patrol during the wee small hours.
> *Private Eye* 29 May 1987, p. 8

lead-free ☑ see –FREE

leading edge /ˌliːdɪŋ ˈɛdʒ/ *noun* and *adjective* Usually written **leading-edge** when used as an adjective 🎌 🖥

noun: The forefront of progress or development, especially in technology; the 'state of the art'.

adjective: Representing the most advanced technology; state-of-the-art.

A figurative application of a term that originally belonged (as a noun) to aerodynamics and aeronautics, where it was used of the forward edge of a moving object such as an aircraft's wing; the imagery here is of technology as a body moving constantly forwards, but with some aspects and designs further advanced than others and acting as a vanguard for future developments.

The figurative use arose in the world of computer technology in the second half of the seventies, and during the eighties was enthusiastically taken up by advertisers as a fashionable way of claiming their products to be in the forefront of design. In the UK the term *leading edge* was even chosen as the name for a chain of shops selling technological gadgetry and new design 'concepts'. An alternative term for the same idea, also popular with advertisers, is **cutting edge**.

Three choices from the Burton Group's spring ranges. Sophisticated style from Principles . . . Leading-edge young fashion from Top Shop . . . Mainstream young fashion from Dorothy Perkins.

Daily Telegraph 26 Feb. 1986, p. 13

The information systems available in the dealing room are quite astonishing for someone whose idea of leading-edge technology is teletext. *Meridian* (Midland Group) Spring 1990, p. 15

The company also puts out *Gorgon*, on horror movies, and *Impact*, on cutting-edge pop culture.

Premiere May 1990, p. 96

lemon law /ˈlɛmən ˌlɔː/ *noun* 〽

In the US, a law designed to provide some redress for buyers of faulty or substandard cars.

Formed by compounding; in US slang, a *lemon* is anything that is faulty or undesirable.

The first *lemon laws* were passed in the US (as individual State Laws) in the early eighties, after much public discussion during the seventies of the high proportion of *lemons* among new and second-hand cars, and the impossibility of doing anything about their poor quality. The different laws passed for different States vary in their provisions, but all give the buyer of a substandard car some redress from the manufacturer or salesperson.

There are now at least 42 variations on the three basic types of 'lemon laws' among the states. To say the least, most manufacturers do not find such variation among the states encouraging.

Legal Times 11 Apr. 1988, p. 19

Mr Forth, American Consumer Affairs Minister, has rejected demands from consumer organisations to adopt American-style 'lemon laws' for purchasers of cars. *Daily Telegraph* 24 Jan. 1989, p. 4

lens /lɛnz/ *noun* 🅧

Short for **contact lens**: a small, very thin piece of plastic which can be worn inside the eyelid, in contact with the eyeball, to correct faulty vision; often in the plural **lenses**.

An abbreviated form of *contact lens*.

Contact lenses were invented by Dr A. E. Fick of Zurich as long ago as the 1880s (when they were made of glass), but did not become available to the general public until the forties, and have only been widely worn from about the sixties onwards. The full term *contact lens* had been abbreviated to *contact* by the early sixties and to *lens* by the seventies; by the eighties it was nearly always abbreviated in colloquial use, although the full term remained in use among opticians. The technology has developed during the seventies and eighties to make several types available: **hard lenses**, the original type available to the public, are made of rigid plastic; **soft lenses**, made of a hydrophilic gel which is soft to the touch and moulds itself to the shape of the eye, were introduced in the sixties as less harmful to the cornea; **gas-permeable lenses**, which are more rigid but allow the passage of oxygen to the eye, were developed soon afterwards and became widely available in the eighties. The fact that *contact lens* became the slang name for a mixture of hallucinogenic drugs in the eighties is an indication that *lenses* are considered commonplace in modern society.

Although many astigmatics can wear lenses successfully, prescribing and fitting them can be complex. *Which?* June 1987, p. 272

These are extended-wear lenses . . . and people should be aware that they run a 20 per cent higher risk of bacterial infection. *Woman's Journal* Mar. 1990, p. 155

leverage /ˈlɛvərɪdʒ/ *intransitive verb* 〽

To speculate financially (or cause someone else to do so), using borrowed capital and relying on the profits made being greater than the interest payable.

The verb is formed on the noun *leverage*, which originally meant the action or power of a lever, but acquired a figurative use in the nineteenth century. In the 1930s a specialized meaning developed in US financial circles: the ratio of a company's debt to its equity, which could be used to maximize returns on an investment. Although *leverage* is normally pronounced /'liːvərɪdʒ/ in British English, the verb reflects in its pronunciation the specialized American sense of the noun from which it derives.

Leverage was first used in US financial writing in the thirties, but remained limited to the technical vocabulary of finance for several decades. The increasing involvement of ordinary people in the stock market, as well as the adventurousness of investment generally, brought it into the public eye in the eighties, but it remains principally an American word. The verbal noun **leveraging** is used for the practice of speculating in this way; the adjective **leveraged** is applied to companies and transactions based on borrowed capital (see also BUYOUT). In the late eighties, after a decade of **leveraging**, there was a widespread move to **deleverage** in the US and UK markets.

> The corporation discovered that the more it borrowed, the higher the earnings and the higher the stock, so it began to leverage. *'A. Smith' Supermoney* (1972), p. 209

> Safeway's announcement that it intends to deleverage itself via a $160 million public share issue was heralded as the start of a trend. *Observer* 18 Feb. 1990, p. 53

leveraged buyout 〰 see BUYOUT

lifestyle /'laɪfstaɪl/ *noun and adjective*. Also written **life-style** 〰 ▓

In marketing jargon:

noun: The sum total of the likes and dislikes of particular customers or a section of the market, as expressed in the products that they would buy to fit their self-image and way of life; a marketing strategy based on the idea of appealing to this sense of self-image and way of life.

adjective: Using or belonging to this strategy of marketing; (of a product) fitting into or conceived as part of such a strategy, appealing to a customer's sense of lifestyle.

A specialized use of the compound noun *lifestyle* in the sense 'way of life', itself a concept of the sixties.

The concept of *lifestyle* merchandising goes back to at least the beginning of the eighties, but was particularly in evidence in the second half of the decade, as advertisers attempted to cash in on and shape the demand for fashion goods, interior decorations, foods, and sports equipment that expressed the new awareness of *lifestyle*. In consequence *lifestyle* came to be used over-freely and imprecisely in marketing, sometimes ending up as an almost meaningless adjectival 'filler'. At the same time a movement in the very opposite direction, away from conspicuous consumption and consumerism, was also under way; this movement, influenced by A. H. Dammers' book *Lifestyle*, urged a simpler and greener *lifestyle* on Western societies. Both the consumers of YUPPIE *lifestyle* products and the followers of this movement towards simplicity have been called **lifestylers**.

> Being a meat-free lifestyler on Gozo is no problem. *Lean Living* Feb.–Mar. 1987, p. 4

> Creative talents in marketing have grasped the concept of lifestyle so insistently that it is changing the face of the high street, the commercials break, even the media. *Creative Review* Jan. 1988, p. 14

> B & Q is targeting the 'lifestyle' market with . . . quick-drying acrylic paints . . . in tins featuring illustrations of country house interiors. *Design Week* 26 May 1989, p. 6

> Swissair has gone life-style with its series of 'customer portraits' (would you buy a second-hand seat from this man?). *International Management* Mar. 1990, p. 60

lig /lɪg/ *intransitive verb* ▓ ▓

In media and youth slang: to sponge or freeload; to gatecrash parties.

Lig was originally a dialect word corresponding to standard English *lie*, mainly in Scottish, Northern Irish, and Northern English dialects. It entered standard English in the early sixties in the general sense 'to idle or lie about' and was then adopted by media people in the more specialized meaning given above.

This is a usage which arose in the late seventies, especially among journalists and entertainers, whose lifestyle involves accepting free hospitality of one kind and another. The word was popularized by media people themselves during the mid eighties. The corresponding action noun is **ligging**; the word for a freeloader is **ligger**.

> [I] suddenly twigged what ligging was all about when I got my first job as a researcher on *Aquarius* I found . . . I could get free tickets for everything, everywhere.　　　　　　*Radio Times* 6 Apr. 1985, p. 16

> A penniless young man who begins in Trafalgar Square with nothing but a pair of underpants and ligs his way onward and upward with clean-cut charm.　　　　　　*The Times* 9 Apr. 1985, p. 8

> Once the last lingering ligger has been escorted out, Dylan and his three piece band . . . shamble through on to the dimly lit stage.　　　　　　*Q* Dec. 1989, p. 64

light /laɪt/ *adjective* Often written **lite** in brand names ▨

Of foods and drinks: containing few calories; especially, low in fat or cholesterol.

A specialization of sense arising almost entirely from the use of the word in advertising and brand names; the current use when applied to food and drink deliberately combines elements of a number of well-established senses. On the one hand, it is the food that is being described as *light* (in the same sense as one might speak of a *light meal*, or think of lager as *light* compared with bitter); on the other, it is the effect on the consumer that is at issue (implying that *light* foods and drinks will not make you fat and heavy). *Light* has been used of drinks (especially beer), as in *light ale*, to mean 'not strong' since the late nineteenth century (and in this sense is the opposite of *stout*), but in the 1980s this development moved one step further. The spelling *lite* in brand names reflects the same process as the one which produced *nite* from *night*.

This is a usage which has become especially common as a result of the prevailing fashion in the eighties for a low-fat, high-fibre diet and the consequent marketing of foodstuffs, drinks, and prepared meals specifically to take advantage of this. The first beer to carry the brand name *Lite* was launched in the late sixties by Meister Brau in the US; this became *Miller Lite* in the seventies and started to become very popular in the second half of that decade. Now, the word *light* (or *lite*) is often part of the name of a product, following a proper noun (as in the trade marks *Meadow Lea Lite* and *Vitaquell Light* margarines, *Budweiser Lite* beer, etc.)—a departure from the normal pattern of usage in English, where adjectives would normally precede the nouns they qualify, but consistent with a trend in the naming of products. In the US the word has also been applied to other consumables, such as cigarettes with a low tar content.

> Its idea of what makes a light beer light is that it contains 100 calories or less in a 12-oz serving.　　　　　　*Marketing Week* 29 Aug. 1986, p. 16

> Polyunsaturated Meadow Lea Lite and Mrs McGregors Lite are reduced fat spreads with only half the fat and half the kilojoules of regular margarine and butter.　　　　　　*Courier-Mail* (Brisbane) 28 June 1989, p. 29

line /laɪn/ *noun* ✎

In the slang of drug users: a dose of a powdered drug (especially cocaine).

So named because the powder is formed into a long trail like a *line* on a shiny surface, ready for 'snorting' through a straw or tube. An earlier use of the word in drugs slang was as an abbreviation of *mainline*, a main artery into which drugs such as heroin could be injected.

A term of the late seventies and eighties, this word is rarely found in print but is apparently in common spoken use among drug users.

> Graffiti recently collected at the University of North Carolina (Chapel Hill) include: . . . Cocaine is like a good joke. You can't wait for the next line.　　　　　　*Maledicta* Winter 1979, p. 276

> [She] produced a six-inch ivory tube, sank to her knees and greedily did her lines, sniffing angel dust into each nostril.　　　　　　Roger Busby *The Snow Man* (1987), p. 21

-line /laɪn/ *combining form* ✖ ◖

A telephone service. (Usually as the second element of a compound name, the first part of which describes the purpose or target of the service.)

From the noun *line* in the sense 'telephone connection', perhaps with some conscious alteration of *hotline* (see below).

A well-known early example of this use was the so-called **hotline**, or emergency telephone link, set up between the US and the Soviet Union in the early sixties. During the seventies some organizations offering help or advice, especially in emergencies, would call the service a *hotline*, but from the beginning of the eighties the first part started to be replaced by some other word describing the service. Any service that offered help and advice to people in difficulty was named a **helpline**, with *hotline* now reserved for matters of extreme urgency (although this apparently includes 'rushing' orders to mail-order companies!). *Helplines* devoted to particular types of advice are sometimes named accordingly—for example **Aidsline** for people with Aids, **Childline** for children in trouble or danger (especially as a result of CHILD ABUSE), **Parentline** for parents who need advice about their children. The *helpline* which simply gives the caller a chance to talk over the problem with an anonymous helper is also often called a **talkline**. In the second half of the eighties there was public consternation over the high telephone bills run up by teenagers using a service called a **chatline**, which allowed them to take part in a conference call with other youngsters who just wanted a chat. In the UK, the familiar speaking clock has been renamed **Timeline**, and a service allowing a business to pay for the calls made direct to it by prospective customers is known as **Linkline**. Many formations using *-line* are trade marks and are therefore written with a capital initial.

> Although Jenni seems to have the only official help-line in the country for battered husbands, there are other places where men can go for help. *Woman* 20 Feb. 1988, p. 13

> The controversial telephone chatlines, withdrawn earlier this year after complaints about exorbitant bills, are likely to be allowed to resume in the near future. *The Times* 28 July 1989, p. 3

> Since the beginning of 1988, 13 volunteers have run a 'telephone friendline' for latchkey children—youngsters who return to empty homes after school—in La Verne and San Dimas. *Los Angeles Times* 7 Sept. 1989, section 9, p. 8

> The Wellington Parentline, a telephone advice service, has received 32 calls reporting violence from children towards parents. *Independent* 29 Jan. 1990, p. 8

linkage /'lɪŋkɪdʒ/ *noun* ▣

The linking together of quite different political issues in international negotiations by declaring that progress on one front is relevant and necessary to progress on other fronts.

A specialized use of *linkage* in the sense 'connection, the act or process of linking together'.

Linkage emerged in the US in the context of US–Soviet relations in the mid and late sixties, when it was used by senior White House officials in order to establish a link between nuclear arms control and general East–West political relations; in practice, it became associated with the way that Cold War tensions were eased by a bargaining process in which one side made concessions in a given area in return for a promise on arms control or other concessions in a different area. *Linkage* remained an important concept in the seventies and eighties—as, for example, the US demand in 1987 for progress on arms control in return for Soviet movement on human rights and withdrawal from Afghanistan—but it acquired an especial currency after Iraq's invasion of Kuwait in August 1990, when Saddam Hussein and his allies sought unsuccessfully to place the Palestinian question firmly on the agenda for any negotiations about Iraq's withdrawal.

> Mr. Kissinger's version of détente included a strategy of 'linkage' designed to deter the Russians from misbehaving. The idea was that Moscow would not risk the loss of favorable arms agreements ... by engaging in risky adventures around the world. *US News & World Report* 29 Mar. 1976, p. 17

Many speculate that the message carried by Hussein will only be a repeat of Saddam's call for Israel to withdraw from the occupied territories and Syria to leave Lebanon. The State Department has dismissed this proposal out of hand, calling it 'false linkage'. *USA Today* 16 Aug. 1990, section A, p. 2

liposuction /'laɪpəʊˌsʌkʃ(ə)n/ *noun* ⊗ ✖

A technique used in cosmetic surgery in which particles of excess fat beneath the skin are loosened and then sucked out with a vacuum pump through a tube or cannula inserted into a small incision.

Formed from *lipo-*, the combining form of Greek *lipos* 'fat', and *suction*.

The technique of *liposuction* was developed in the early eighties, principally as a means of removing unwanted fat which is resistant to dieting and exercise. Not surprisingly, though, it was hailed by the media and the public-at-large as the long-awaited end to all dieting for those with a weight problem and little will-power.

She says he recommended a tummy tuck for her overhanging stomach and liposuction for her legs, bra line and chin. *New Age* (Melbourne) 16 Aug. 1986, p. 25

The liposuction that promises to suck bodies into shape carries the risks of all general anesthesia. *Philadelphia Inquirer* 20 Sept. 1989, section A, p. 17

For a consultation on . . . spot fat reduction (Liposuction) call us on the number below. *Vogue* Sept. 1990, p. 432

listener-friendly ✖ see FRIENDLY

little devil ✐ see BASUCO

liveware ▣ see -WARE

living will /ˌlɪvɪŋ 'wɪl/ *noun* ▮▮

A document written by a person while still legally fit to do so, requesting that he or she should be allowed to die rather than be kept alive by artificial means if subsequently severely disabled or suffering from a terminal illness; a request for euthanasia.

Formed by compounding: a kind of *will* dealing specifically with an individual's understanding of what constitutes worthwhile *living*.

The concept of the *living will* was first discussed in legal circles in the US in the late sixties; the coinage is claimed by an American lawyer, Luis Kutner. The documents themselves acquired legal status in several States during the seventies, and by the end of the eighties most States in the US recognized them. In the UK there was little mention of the *living will* until the end of the eighties and the legal force of these documents has not yet been fully tested in the courts.

Henry Campbell discovered he had Aids in 1984. That year, after two major bouts of pneumonia, he drew up a living will. *Independent* 18 May 1990, p. 19

LMS /ɛlɛm'ɛs/ *abbreviation* ▮▮

Short for **local management of schools**, a system set up by the Education Reform Act of 1988, providing for a large proportion of the financial and administrative management of state schools in the UK to become the responsibility of the governors and head teacher respectively.

The initial letters of *Local Management of Schools*.

The Act set out the two basic principles of applying formula funding to all primary and secondary schools, based on the need to spend, and of handing over budgetary control to the governors of schools over a certain size; funding was to be linked to pupil numbers, giving schools an incentive to attract and retain pupils. It did not, however, introduce the terms *local management of schools* or *LMS*—these terms came in a Coopers & Lybrand report on the scheme, published in January 1988:

> The changes require a new culture and philosophy of the organisation of education at the school level. They are more than purely financial; they need a general shift in management. We use the term 'Local Management of Schools' (LMS).

From here the phrase was taken up in a Department of Education and Science circular, and soon became institutionalized. The idea had its origins in an experiment carried out in a village school in Cambridgeshire in the early eighties; at that time the scheme was known as *Local Financial Management* (*LFM*). The main consequence of *LMS* itself was that, for the first time, many schools' budgets would be controlled by the governors, who would also become the employer of all the school staff. The role of the head teacher centred on the day-to-day management of the school. Each Local Education Authority had to devise and submit its own scheme for approval; most had done this by 1991, but the Inner London schemes were left for approval and implementation later.

> The key to future waves of opting out . . . lies in the Act's provisions for local management of schools (LMS) . . . Heads and governors operating LMS will control 90 per cent of their budgets, increase their funds on the basis of the number of pupils they attract and have power to hire and fire staff.
> *Daily Telegraph* 23 Feb. 1989, p. 15

lock ⚒ 💀 see BREAK-DANCING

logic bomb /'lɒdʒɪk ˌbɒm/ *noun* 💻

A set of instructions surreptitiously included in a computer program such that if a particular set of conditions ever occurs, the instructions will be put into operation (usually with disastrous results).

Formed by compounding: the equivalent of a *time bomb*, metaphorically speaking, except that it is a particular set of circumstances built into the *logic* of the program, rather than the passage of time, that will set it off. A similar set of instructions designed to be implemented on a given *date* is in fact called a *time bomb* in computing but the distinction between the two terms is not always clearly made.

The *logic bomb* is one of a number of malicious or even criminal uses of computing know-how that have been invented since computers became widely accessible and affordable in the second half of the seventies. It has been used as a way of destroying evidence of a computer fraud as soon as information which might lead to the culprits is accessed, as the basis for blackmail, and as a way for a programmer to take revenge on an employer by causing the system to crash mysteriously.

> If you damage someone's computer—whether by attacking it with a hammer or crippling the program with a logic bomb—it's . . . a crime.
> *Independent* 21 Sept. 1988, p. 2

> Slip a logic bomb into the development software; it'll be copied along with the valid programs and shipped to the rest of the country.
> Clifford Stoll *The Cuckoo's Egg* (1989), p. 232

See also TROJAN, VIRUS, and WORM

loopy dust 💉 see ANGEL DUST

lose one's bottle see BOTTLE

low-alcohol beer ⚒ see NAB

low observable technology 🌿 see STEALTH

low-tech 💻 see HIGH-TECH

LRINF see INF

luggable /'lʌgəb(ə)l/ *adjective* and *noun* 📖

adjective: Of a computer: rather larger than a portable; light and small enough to be carried short distances with some effort.

noun: A computer which is not quite small enough to be easily portable.

Formed by adding the suffix *-able* to the verb *lug* 'carry (something heavy)', after the model of *portable*.

One of a series of terms for different sizes of personal computer which came into the language during the first half of the eighties. *Luggable* was originally used to refer to the PC which had been made rather lighter than usual to allow it to be moved about from one location to another; as such, it was still in a distinct category from the portable LAPTOP (which had an LCD screen and was not dependent on mains power). With the development of ever smaller computers in the second half of the eighties (see the examples listed under LAPTOP) came smaller and lighter *luggables*—of about twenty rather than thirty pounds—without which the maufacturers would have been unable to compete successfully in the microcomputer market.

> The success of these 30lb 'luggables', as they are more appropriately known, owes more to their wide range of software . . . than to their ease of carting about. *Sunday Times* 26 Aug. 1984, p. 49

> At a time when portables are getting smaller and lighter, IBM has come up with a mains luggable the size of a small suitcase and weighing some 20lb. *PC Magazine* July 1989, p. 46

lunchbox 📖 see LAPTOP

Lyme disease /'laɪm dɪˌziːz/ *noun* ⓧ

A form of arthritis which mainly affects the large joints, is preceded by a rash, and is thought to be transmitted by a bacterium carried by deer ticks.

Formed from the name of the town of *Lyme*, Connecticut (where the first outbreak occurred in 1975) and *disease*.

Lyme disease, at first called **Lyme arthritis** in the medical literature, caused much concern in the US during the late seventies and eighties and was identified in British patients as well in the mid eighties.

> The ticks feed on small mammals and birds, and in their adult stage, on deer, but not all deer ticks are infected with Lyme disease. In order to become carriers of Lyme disease, they must first feed on an animal which already has the spirochete. *Madison Eagle* (New Jersey) 3 May 1990, p. 5

lymphadenopathy syndrome ⓧ see AIDS

• •

M

McGuffin /mə'gʌfɪn/ *noun* Also written **MacGuffin** ⚒

A device used in a film or work of fiction whereby some fact or activity seems all-important to the characters involved while actually only providing an excuse for the plot as a whole; the thing which absorbs the characters and misleads the audience in this way.

The word was invented by the film director Alfred Hitchcock in the thirties in relation to the film *The Thirty-Nine Steps*; when interviewed by François Truffaut in the sixties, he claimed that he always liked to use a *McGuffin* in his films:

> The theft of secret documents was the original MacGuffin. So the 'MacGuffin' is the term we use to cover all that sort of thing: to steal plans or documents, or discover a secret, it doesn't matter what it is. And the logicians are wrong in trying to figure out the truth of a MacGuffin, since it's beside the point. The only thing that really matters is that in the picture the plans, documents, or secrets must seem to be of vital importance to the characters. To me, the narrator, they're of no importance whatsoever.

The word itself may be derived from *guff*; it was apparently borrowed from a Scottish joke involving a man carrying a mysterious parcel on a train; but the joke may also be a *McGuffin* in its own right.

Although Hitchcock had been using the word for several decades, *McGuffin* did not start to appear more widely in film criticism until the early eighties, when it suddenly acquired a more general currency, and was used to refer to the underlying impetus for the plot of novels and television series as well as horror films.

> There's a funny scene in which Wilder, looking for a gold coin—the film's McGuffin—ventures into the bathroom of a beautiful woman villain and encounters her in the shower.
>
> *Sydney Morning Herald* 27 July 1989, p. 14

> Maddeningly, neither the deal nor its unmaking are anything but McGuffins in this misfiring comedy.
>
> *Los Angeles Times* 22 June 1990, section F, p. 6

McKenzie /məˈkɛnzɪ/ *noun* 🏛

In the UK, a person who attends a court of law to help and advise one of the parties to the case. Often used attributively, especially in **McKenzie friend** or **McKenzie man**.

Named after the case of McKenzie v. McKenzie (1970), in which the precedent was set for a non-professional helper to be allowed in court.

According to the Law Reports on the case of McKenzie v. McKenzie,

> Any person, whether he be a professional man or not, may attend a trial as a friend of either party, may take notes, and may quietly make suggestions and give advice to that party.

During the seventies these people were generally called *McKenzies* or *McKenzie men* in legal journals and the like, but the term had little currency outside legal sources. In the early eighties greater use was made of the precedent by people who wanted to do without legal representation or who could not afford it, and the terms started to appear in the newspapers; by the end of the decade the preferred form in this more popular usage was clearly *McKenzie friend*.

> Mr Dave Nellist, MP for Coventry South-East, said he intended to appear before Coventry magistrates as a 'McKenzie friend'.
>
> *Daily Telegraph* 24 July 1990, p. 2

mad cow disease /ˌmæd ˈkaʊ dɪˌziːz/ *noun* ⊗

Colloquially, BSE.

So nicknamed because the disease affects the brain and central nervous system of the infected cows, causing them to stagger, fall down, or generally behave as though deranged.

For history, see BSE. Although only a popular nickname for the disease (originally popularized by journalists), *mad cow disease* came to be used in a number of reputable sources without inverted commas. It caught the popular imagination to such an extent that a number of humorous variations were coined during 1989 and 1990; most were one-off instances like the examples printed below, but **mad bull disease** (making use of the pun with the stock-market concept of *bullishness*) cropped up quite frequently in financial reports. *Mad cow disease* itself is sometimes shortened to **mad cow**.

> Fresh call for bigger 'mad cow' payouts.
>
> headline in *The Times* 6 Feb. 1990, p. 6

The process could be accelerated ... with salmonella infection on the increase and the frightening spectre of mad cow disease crossing the species barrier. *Health Guardian* May/June 1990, p. 1

Fears are growing that the continuing—perhaps worsening—problems associated with mad cow disease could accelerate what many regard as an alarming drift from the land.

Guardian 9 June 1990, p. 4

School BSE, or mad classroom disease, exists largely as a result of the ridiculous notion that a teacher's primary duty is to make lessons interesting. *Daily Telegraph* 21 June 1990, p. 14

What we have here is a bompin' stompin' monsta groova, a toe tanglin', heart manglin', floor fanglin' 125 bananas per minute of sheer joy—mad fruit disease in the area. *Sounds* 28 July 1990, p. 24

Madrid conditions /məˌdrɪd kənˈdɪʃ(ə)nz/ *noun* ⚡ 🏛

The set of conditions (laid down by UK Prime Minister Margaret Thatcher at the European summit held in Madrid in June 1989) for the entry of the UK into full participation in EMS.

Formed by compounding: *conditions* laid down at *Madrid*.

Mrs Thatcher had claimed for some considerable time before the Madrid summit of June 1989 that the pound would join the ERM (the exchange-rate mechanism at the heart of EMS: see the entry for EMS) 'when the time is ripe'. It was in the *Madrid conditions* that she first stated explicitly when she thought that would be. The conditions covered five areas, the most important of which was that UK inflation must first be brought down to a level near to the average in other EC countries. In fact, when her Chancellor, John Major, took the UK into the ERM in October 1990, this condition had not been met—a circumstance which gave rise to much discussion of the *Madrid conditions* in the media. The other four conditions were that France and Italy should abolish exchange controls, that the single internal market of the EC should first be completed, that there should be progress towards a free market in financial services, and that competition policy should be reinforced.

Last week the Chancellor, more cautious than the Foreign Secretary, but working with him, set out his stall. He stressed the importance of completing the 1992 single market and other Madrid conditions. *Guardian* 19 June 1990, p. 6

magalog /ˈmægəlɒg/ *noun* Also written **magalogue** ⚡

A marketing publication issued periodically and combining features of the glossy magazine with characteristics of a mail-order catalogue.

Formed by telescoping *magazine* and *catalogue* (or, in the US, *catalog*) to make a blend. The same principle was followed in the formation of *Specialog(ue)*, the trade mark of a type of specialized catalogue.

The *magalog* was an invention of US advertisers in the second half of the seventies which caught on in many other affluent countries during the eighties. Typically, the 'magazine' is issued free of charge to a limited number of people (cardmembers of a particular credit card, users of a mail-order house, etc.) or given away in another publication; the content is a mixture of editorial, ADVERTORIAL, and straightforward advertising. Many *magalogs* are issued at regular monthly or quarterly intervals and are difficult to distinguish visually from a magazine (except, perhaps, for the absence of a price from the cover).

GUS, the market leader in traditional mail order, is also responding to the new challenge. Next month sees the launch of Complete KIT, a fashion magalogue (its word), through W H Smith and associated newsagents. *Daily Telegraph* 18 Feb. 1988, p. 17

The products include bulletin boards, early learning books, post-it notes and reading aids. The Kids' Stuff magalog also contains editorial pages and teaching tips. It is mailed twice a year.

DM News 15 Apr. 1988, p. 74

magnetic resonance imaging ⊗ 🖳 see MRI

mainline 💉 see LINE

makeover /ˈmeɪkəʊvə(r)/ *noun* Also written **make-over** 〜 ▓

A complete transformation or remodelling; specifically, the remodelling of a person's appearance (or some aspect of it, such as hairstyle), especially when this is carried out by a professional.

Formed by turning the verbal phrase *to make over* ('to refashion') into a compound noun.

The noun *makeover* was first used in the late sixties and by the seventies was not unusual in professional hairstylists' and beauticians' publications. It remained in relatively limited use until the end of the seventies, when it started to appear in magazines aimed at a wider audience; by the mid eighties it had become a part of the stock vocabulary of women's magazines, especially those which featured an opportunity for an ordinary reader to have her whole appearance and image rethought by experts, with markedly different 'before' and 'after' photographs. This was extended to all kinds of remodelling (for example, of interior decoration, houses, etc.) from the early eighties. The word was also taken up in the business world in a figurative sense from about the mid eighties: when a company is restructured by a new management, this is described as a *makeover* or **corporate makeover**, especially if the results seem only cosmetic.

> Mr Segal insists that hostile takeovers, leveraged buyouts and forced restructurings—which he bundles together under the . . . label 'corporate makeovers'—are 'symptoms, not the disease'.
> *New York Times Book Review* 29 Oct. 1989, p. 32

> The make-over of California Cosmetics has worked. Although sales slipped . . . last year, . . . the company is now more profitable than ever. *Financial Review* (Sydney) 23 Feb. 1990, p. 48

> We did this make-over for six ladies in the region. You know the sort of thing—you get an expert in to show them what they should wear. *She* Oct. 1990, p. 9

mall /mɔːl/ *noun* ▓

A covered shopping precinct, usually situated outside a town and provided with car-parking facilities and other amenities.

A *mall* has meant 'a covered or sheltered walk' since the eighteenth century; some towns have the evidence of this historical usage in the name of a particular street or promenade, but this is usually pronounced /mæl/. The shopping *mall* is a specialized use of this sense.

A well-established concept in North America (where they were first written about in the late sixties), *malls* were tried in the UK during the seventies, but with little success. In the eighties, however, increasing traffic congestion and parking problems in large towns, as well as the changeover to the MEGASTORE approach to shopping, meant that the *mall* became increasingly popular. In the UK the longer term **shopping mall** is still commoner than *mall* alone.

> Most striking is the way individually-designed shop fronts spill over into the malls themselves.
> *Which?* Aug. 1989, p. 406

> The downtown Los Angeles car wash used in the original [film] was recently torn down and replaced by a mini-mall. *People* 19 Feb. 1990, p. 51

> Telecommuting will also be promoted, along with no-go zones for cars, pedestrian shopping malls and park-and-ride schemes. *BBC Wildlife* July 1990, p. 456

management buyout 〜 see BUYOUT

marginalize /ˈmɑːdʒɪnəlaɪz/ *transitive verb* Also written **marginalise** 📋 〔〔

To treat (a person or group of people) as marginal and therefore unimportant; to push from the centre or mainstream towards the periphery of one's interests, of society, etc. Also as an adjective **marginalized**; adjective and noun **marginalizing**; process noun **marginalization**.

Formed by adding the verbal suffix *-ize* to *marginal*; the verb was originally formed in the nineteenth century in the sense 'to make marginal notes (on)'.

Marginalization was originally a sociologists' term, in use from about the early 1970s. It was during the mid to late seventies that a number of interest groups and liberation movements (including feminism, Black power, and gay rights groups) took up the term to focus public attention on their causes, eventually turning it into one of the main social buzzwords of the eighties.

> Society, taking its lead from the media and its politicians, begins to reject a whole class and marginalizes them in the job market.   Caryl Phillips *The European Tribe* (1987), p. 123

> One of the many tales that we have been told is that there was once a homogenous national culture which is now under threat from multiculturalism, as if there was, is, or is ever likely to be, *one* tradition within England—not to mention the traditions within each of the marginalised nations in the United Kingdom.   *New Statesman* 17 June 1988, p. 46

> Although the curve of decline has been flattening gradually, it is not yet clear that the church's long years of marginalisation in our national life have been ended.   *Independent* 29 July 1990, p. 20

market maker /ˈmɑːkɪt ˌmeɪkə(r)/ *noun* Also written **market-maker** or marketmaker ⩘

In the jargon of the Stock Exchange after BIG BANG, a broker-dealer who deals in wholesale buying and selling, guaranteeing to make a market in a given stock; essentially the same thing as a stock-jobber before Big Bang.

Formed by compounding; the one who *makes* a *market*. The phrase *make a market* has been in use on the London Stock Exchange since the turn of the century; the form *market maker* also already existed before the big bang, but was not an official term and was used pejoratively (see below).

The word *market maker* is not new, but it has been used in a new sense in the Stock Exchange since the deregulation of 1986. Whereas the *market maker* of the turn of the century specialized in making a market by dealing in a stock to drum up interest in it, today's *market maker* simply guarantees to buy and sell a specified stock and so make the market available. The main business of a *market maker* consists in buying stock wholesale and then selling it on at a profit; this is essentially what stock-jobbers did before the distinction between brokers and jobbers was abolished in 1986. The activity of a market maker is **market making**; occasionally the intransitive verb **market-make** is also used.

> After last week's hefty fall on Wall Street there must be many in the City wondering if the London equity market will suffer bouts of guruitis . . . when the American market makers begin to extend their influence.   *Sunday Telegraph* 13 July 1986, p. 23

> Marketmakers are obliged to deal at the price shown on their screens.   *The Times* 20 Oct. 1986, p. 25

mascarpone /ˌmæskɑːˈpəʊnɪ/, in Italian /maskarˈpone/ *noun* ▨

A soft, mild cream cheese from Lombardy in Italy.

A direct borrowing from the Italian name of the cheese *mascarpone* or *mascherpone*.

Mascarpone, which is a relative of the better-known *ricotta*, has been written about in English since at least the thirties; for some reason it became a fashionable food in the mid and late eighties, cropping up frequently in writing for and by FOODIEs.

> Tiramisù, which means 'pick-me-up', consists of layers of espresso-soaked spongecake or ladyfingers, sprinkled with rum and slathered with sweetened mascarpone cheese.
> *New York Times* 8 Mar. 1989, section C, p. 3

> Chef Leigh correctly detected a touch of horse-radish in the cream topping . . . but affected not to have heard of the other principal ingredient, *mascarpone*.   *The Times* 17 Feb. 1990, p. 36

masculist /ˈmæskjʊlɪst/ *noun* and *adjective* 🔩

noun: A person who upholds the rights of men in the same way as a feminist upholds those of women; also, a person who opposes feminism.

adjective: Representing or upholding men's rights or masculine attitudes in general.

Formed by adding the suffix *-ist* to the stem of *masculine*, after the model of *feminist*. The word *masculinist* had already been coined in the same meaning by Virginia Woolf in 1918, and is also in current use (although rare).

The word was coined at the beginning of the eighties, after the feminist movement had radically altered the position of women in Western societies. The term **masculism** is also sometimes used for the men's rights movement or the attitudes that it enshrines, but it is considerably less common than *masculist*.

> What is claimed to be the first ever European petition for men's rights is to be handed in to the European Parliament by a new 'masculist' group . . . There are already some 20,000 militant masculists in Europe. *The Times* 20 Mar. 1984, p. 6

> It does not matter if the cartoon is insulting to men. The number of such cartoons is so small that, set against the insults to women broadcast by every newsagent and television channel, only a loony masculist would object to them. *Guardian* 23 Nov. 1989, p. 38

> Phoebe thought that science in general was a crude product of masculist thinking, designed to separate knowledge and experience. Sara Maitland *Three Times Table* (1990), p. 93

massage /ˈmæsɑːʒ/ *verb* and *noun* 〰️

transitive verb: To manipulate (figures, computer data, etc.) so as to give them a more acceptable or desirable appearance.

noun: The action of manipulating figures or data in this way.

A figurative application of *massage*, which had already been used metaphorically in the sixties to refer to the 'touching up' of written material such as an official report.

The business use of the word dates from the mid seventies, when the widespread application of computing to business statistics made **data massage** possible. During the eighties, the verb in particular became increasingly common, and it is now usually printed without inverted commas. In most cases, the activity is not actually fraudulent, but takes place on the fringes of legality and propriety as a way of putting the desired 'spin' on the data. Figures which have been manipulated in this way are described by the adjective **massaged**.

> He . . . uses the manipulated data to prove the link between money and prices . . . Professor Hendry's feat, however, is to take this heavily massaged data and show that not even such distortion can save the empirical support for Friedman's theory. *Guardian Weekly* 25 Dec. 1983, p. 9

> The headline writers will be wondering endlessly about Mrs Thatcher's choice of an election date; with the drear descant that, if she delays, the figures for the following year will have to be massaged all over again. *Guardian* 20 July 1989, p. 22

> Numbers can be massaged by putting them in different places in the accounts . . . but it is difficult to manipulate them over several years. *Business* Apr. 1990, p. 59

See also CREATIVE

max /mæks/ *noun* and *verb* 🎰

noun: In the US slang phrase **to the max**, totally, completely, to the highest degree.

transitive or *intransitive verb*: In US slang, to do (something) to the limit; to excel, to perform to maximum ability or capacity, to peak. (Often as a phrasal verb **max out**.)

Max has been an abbreviated colloquial form of *maximum* since the middle of the nineteenth century, and there is some evidence that it was also occasionally used as a verb at that time. Both the phrasal uses result from the tendency for 'in' expressions to become fixed phrases among a particular group of people and then be picked up as phrases by outsiders. *Out* can be added to almost any verb in US slang: compare PIG OUT and MELLOW OUT.

The phrase *to the max* may have originated in US prep school slang in the late seventies, but is now particularly associated with the speech of young Californians. In the late eighties it started to appear in British sources as well, but is still a conscious Americanism. The verb *max out* has its roots in US prison slang, where it has been used in the sense 'to complete a maximum prison sentence' since at least the mid seventies. In the eighties, it was used in a wide variety of different contexts, including the financial (giving or spending to the limit of one's resources), the physical (for example, exercising to the limit of one's endurance), and cases in which it simply means 'to peak'. The phrasal verb is the foundation for an adjective **maxed out**, at the limit of one's abilities, endurance, etc.

> In the past three years, 81 percent of those who've 'maxed out' on psychiatry (that is, exceeded the Blues' $50,000 lifetime limit on outpatient bills) have been from Washington.
> *Washington Post Magazine* 22 Nov. 1981, p. 28

> Pop 1987 was choc-a-bloc with 'good songs', was human-all-too-human, warm and fleshy to the max. *New Statesman* 18 Dec. 1987, p. 36

> On stage and in interview, Sandra Bernhard works her sharp tongue to the max.
> *The Face* Jan. 1989, p. 20

> 'We are maxed out. We are practically pushing the walls out', said Jane Marie Schrader, library director. *Newark Star-Ledger* (New Jersey) 14 Jan. 1990, p. 56

See also GRODY

MBO 〰 see BUYOUT

MDMA 💉 see ADAM and ECSTASY

ME /ɛmˈiː/ *abbreviation* ⊗

Short for **myalgic encephalomyelitis**, a benign but debilitating and often long-lasting condition which usually occurs after a viral infection and causes headaches, fever, muscular pain, extreme fatigue, and weakness.

The initial letters of *Myalgic Encephalomyelitis*.

ME, which has also been known as *post-viral fatigue syndrome* or *post-viral syndrome* (because it so often follows a viral infection), or *Royal Free* or *Iceland disease* (after two famous unexplained outbreaks), has been the cause of considerable debate in the medical world since the late seventies. Although there have been documented cases of the symptoms associated with *ME* since the fifties, no definite cause could at first be found (some connection with coxsackieviruses was identified in the late eighties); it is really only during the eighties that *ME* was recognized as anything more than a psychosomatic condition by doctors and public alike. The syndrome tends to attack high achievers with a busy lifestyle, causing them to take months or even years to recover from what at first sight appeared to be no more than an attack of flu—hence the colloquial nicknames which have been applied to it, including YUPPIE FLU. The abbreviation *ME* has been in common use since the early eighties.

> Post-viral syndrome, or Myalgic Encephalomyelitis (ME), is a mysterious illness, a chronic disease a generation of doctors dismissed as 'shirker's sickness'.
> *Woman's Day* (Melbourne) 4 Jan. 1988, p. 29

> Maria-Elsa Bragg, 23, has been battling for more than two years against the mystery disease ME … The illness, full name Myalgic Encephalomyelitis, affects about 150,000 Britons, mostly women.
> *Sunday Mirror* 16 Apr. 1989, p. 9

> My local bookshop has just given 'ME' (myalgic encephalomyelitis) the final seal of approval, its own shelf. *British Medical Journal* 3 June 1989, p. 1532

meat-free ⊗ ✕ see -FREE

mechatronics /ˌmɛkəˈtrɒnɪks/ *noun* ▣

A technology (originally from Japan) which combines mechanical engineering with electronics, mainly so as to increase automation in manufacturing industries.

Formed by putting together the first two syllables of *mechanics* and the last two of *electronics*.

The word first started to appear in English-language sources in the early eighties in descriptions of Japan's pioneering work in the field. Often *mechatronics* involves developing robots to carry out very precise manufacturing tasks, and this is probably what most people in English-speaking countries think of as *mechatronics*, especially in relation to car assembly; however, the word can be applied to many different aspects of the manufacturing process. It is nearly always a way of reducing the human workforce, and is therefore an important economic consideration for any industry.

> Renault's contribution to the new generation of systems now being developed lies in three areas: 'mechatronics', communications and signal processing. Mechatronics embraces the use of the latest combination of electronics, mechanical and electrical engineering and allied technologies to develop new, functional systems for the auto industry. *Scientific American* Dec. 1984, section A, p. 14

> Australia's leading roboticists are gathering in Perth this week . . . Our *Mechatronics* section next week will report on this important meeting. *The Australian* 13 May 1986, p. 23

> An unattended operation requires the construction of a computer control system and the introduction of technology related to mechatronics and robots. *The Times* 20 May 1986, p. 32

mecu 〰 see ECU

meeja /ˈmiːdʒə/ *noun* Also written **meejah** or **meejer** ✕

In humorous or dismissive use in UK slang: the media; journalists and media people collectively.

A respelling of *media*, meant to represent a common colloquial pronunciation of the word.

A form which first cropped up in the early eighties, *meeja* (along with its variants) became increasingly common as the decade progressed. This was perhaps partly a result of public debate about the role of the media (especially the intrusion of journalists from the popular press into people's private lives), and the generally high profile of media 'personalities'.

> The British public, whose contempt for politicians rivals that for the meejer.
> *Spectator* 25 July 1987, p. 7

> We aren't middle-class poor anymore, you know. I am part of the rich *meeja*.
> Janet Neel *Death's Bright Angel* (1988), p. 41

mega /ˈmɛgə/ *adjective* ✕ ▩

Colloquially, very large or important; on a grand scale; great.

From the Greek *megas* 'great'. The adjective was probably formed because the combining form *mega-* (as in MEGASTAR and MEGASTORE) was sometimes written as a free-standing element (*mega star*, etc.), which later came to be interpreted as a word in its own right. This process is not uncommon with Latin and Greek combining forms: see ECO- and EURO-, and compare *pseudo*, which has been used as a free-standing adjective for several decades.

Mega has been in colloquial use, especially in the entertainment industry, since at least the beginning of the eighties. At first it was used mainly in variations on *megastar* and *megastore* (describing a person as a *mega* bore or a development as a *mega* project). By the middle of the decade it had also started to be used predicatively (as in 'that's mega'). In the business world, any transaction involving large sums of money (millions of dollars) can be described as *mega*; **mega bid**, **mega deal**, and **mega merger** are all in use, sometimes written solid (and therefore probably based on the combining form rather than the adjective). By the end of the eighties, *mega* had been taken up as a favourite term of approval among young people, with a weakening of sense to 'very good' (a similar story to that of *great* two decades previously).

> I was mega, but not mega enough for the job. *New Yorker* 25 Mar. 1985, p. 41

> The insurance companies helped promote the industry as a whole with their mega launches and promotions. *Investors Chronicle* 8 Jan. 1988, p. 28

> I got the gabardine there. I must say that I think that it's absolutely mega. I got it in Auntie Hilda's shop—for a quid. I'm afraid she doesn't have much concept of the value of stylish clothes. *Guardian* 3 Aug. 1989, p. 34

megaflop /'mɛgəflɒp/ noun 📖

In computing jargon, a processing speed of a million floating-point operations per second.

Formed from the combining form *mega-* in its usual sense in units of measurement, 'a million times', and a 'singular' form of the acronym *FLOPS*, 'floating-point operations per second' (the *s* being dropped as though it were there to mark the plural form of a regular noun *flop*).

A term which has been used in computing circles since the second half of the seventies, and is now also found in less technical sources. A measure of the speed at which the field develops is that the computing world talks of today's supercomputers' speeds in terms of *gigaflops* (billions of floating-point operations per second), and tomorrow's in *teraflops* (trillions of floating-point operations per second).

> The Cray 2 has busted out of the 'megaflop' realm, where speed is measured in millions of 'flops'—floating-point operations per second. Its peak speed is 1.2 billion flops, or gigaflops. *Business Week* 26 Aug. 1985, p. 92

> The TC2000 can have up to 504 processors, providing 9,576 mips (millions of instructions per second) or 10,080 megaflops (floating-point operations per second). Prices start at $350,000. *Guardian* 27 July 1989, p. 25

megastar /'mɛgəstɑː(r)/ noun ✖

A performer or media 'personality' who has achieved fame and fortune on a very large scale and enjoys the publicity and lavish lifestyle that go with stardom; a star who is considered greater even than a superstar.

Formed from the combining form *mega-* (from Greek *megas* 'great') and *star*.

In the late nineteenth and early twentieth centuries, the entertainment industry produced *stars*; between the twenties and the seventies some were great enough to be called *superstars*; by the late seventies and early eighties, the next step on the ladder of increasing media hype was to call them *megastars*. Some of the ingredients of **megastardom** seem to be international renown, perhaps in more than one medium (especially films and television), great wealth and extravagance of lifestyle, and a vigorous publicity machine to keep the GLITZY image in the public eye. The Australian comedian Barry Humphries, in his role as Dame Edna Everage, has done much to popularize—and at the same time to debunk—the concept of the *megastar* on television.

> Elton—born Reginald Kenneth Dwight—did not, as Jagger and Lennon did, become a tax exile and disappear off into megastardom. *Independent Magazine* 11 Feb. 1989, p. 23

> Sometimes, when I'm doing my shows, I see people in the audience slipping from their seats into a kneeling position and I say, 'Get up! Off your knees! Back into your seat!' After all, I'm just a megastar, no more than that. I'm frail. I have my weaknesses. Above all, I want to show my human side. 'Dame Edna' in *She* Oct. 1990, p. 116

megastore /'mɛgəstɔː(r)/ noun ✖

A very large store, usually situated on the outskirts of a town or city, provided with its own parking facilities, and often selling goods from its own factory direct to the customer.

Formed from the combining form *mega-* (as in the entry above) and *store*.

The original idea of the warehouse-style *megastore* was that people could bring their own transport and buy furniture, do-it-yourself equipment, electrical goods, etc. direct from the manufacturer. This has been practised in the UK since the late sixties or seventies, but many such outlets were at first called *warehouses*. The name *megastore* was popularized throughout the world by Richard Branson's Virgin chain in the mid eighties, but this time it simply referred to a very large retail outlet. In the late eighties, the *megastore* in the US and the UK tended to be a large retail store bringing together many different kinds of goods under one roof.

> Walk into any of the new megastores now sprouting up—themselves a new way of consuming pop, a far cry from the listening booths or record counters of yesteryear—you will find an immense variety of music from the last forty years on offer. *New Statesman* 4 July 1986, p. 26

> Richard Branson . . . will arrive in Sydney tomorrow to open his first Australian 'megastore' next week . . . The store, at Darling Harbour, is billed as Australia's biggest record shop. *Sydney Morning Herald* 28 Apr. 1988, p. 6

mellow out / ˌmɛləʊ ˈaʊt/ *intransitive verb*

In US slang (especially in California): to relax; to release one's tensions and inhibitions; to become 'laid-back'.

Formed by adding *out* to the verb *mellow* in its figurative sense 'to soften, become toned down or subdued'; as is often the case in these US phrasal verbs with *out*, there is strong influence from the slang use of the first word in another part of speech. In this case, *mellow* had been used as a fashionable adjective in Californian slang for several decades in the sense 'feeling good and relaxed after smoking marijuana': to *mellow out* is therefore to reproduce this feeling in oneself (though not necessarily by using drugs).

The phrasal verb has been used in US slang since the mid seventies; during the eighties, American television series made it a familiar expression to viewers in other countries too, although most British English speakers would only use it in parody of Californian speech. The adjective **mellowed out** is also sometimes found. So prevalent is the word *mellow* in its various guises in Californian speech that in the late seventies the cartoonist Garry Trudeau coined the word **mellowspeak** to describe this particular variety of English; the word has survived and extended its meaning to any bland, laid-back, or jargon-ridden language.

> He's getting it all together at last, mellowing out (in the jargon). Susan Trott *When Your Lover Leaves* (1980), p. 75

> 'You told me on the phone that the highest rock climb would be 15 feet.' 'Ah, I did?' he said in his most mellowed-out tones. 'Well, it was no problem, really, eh? You did fine.' *Sports Illustrated* 16 May 1988, p. 12

meltdown /ˈmɛltdaʊn/ *noun* 〰

A disastrous and uncontrolled event with far-reaching repercussions; especially in financial jargon, an uncontrolled rapid fall in share values, a crash.

A figurative application of *meltdown* in its nuclear physics sense, 'the melting of the core of a nuclear reactor'—an event which, once started, cannot easily be controlled, and which causes widespread destruction and contamination.

This figurative sense arose in the US in the mid eighties after the Three Mile Island accident, and was reinforced by the near meltdown of a nuclear reactor at Chernobyl in the Ukraine in 1986. In the financial world, it was applied especially to the stock-market crash of October 1987, when dramatic falls in share values on Wall Street had repercussions in all the world markets. Monday 19 October 1987 was given the nickname **Meltdown Monday** (but see also BLACK MONDAY). *Meltdown* is now used in more trivial contexts as well, with a weakening of meaning to 'slump, failure'.

The rapidly growing international hotels group, Queens Moat Houses, yesterday asked its shareholders to dip into their pockets for the third time since Meltdown Monday, to help pay for further expansion.

Guardian 17 Aug. 1989, p. 12

The Expos ... suffered another meltdown and sank to fourth place.   *New Yorker* 11 Dec. 1989, p. 74

Smarties-to-coffee giant Nestle disappointed chocoholics with a 5% meltdown in its half-way profits.

Today 15 Sept. 1990, p. 35

metal ♪ ☢ see HEAVY METAL

Mexican wave /ˌmɛksɪkən ˈweɪv/ *noun* Sometimes in the form **Mexico wave** 〽 ✕

A rising-and-falling effect which ripples successively across different sections of a crowd; also, a similar effect in the movement of statistics etc.

The effect, which looks like a moving *wave*, was so named because it was first widely publicized by television pictures of sports crowds doing it at the World Cup football competition in *Mexico* City in 1986.

The *Mexican wave* was apparently first practised (under the name *human wave*) by American football crowds in the early eighties; the crowd in the grandstand expresses appreciation of what is happening in the match by standing up one lateral section at a time, raising their arms, and then sitting down again as the next section rises. When this was done at Mexico City, it was seen on television by millions of people and later widely copied. The figurative use of the term is very recent, and perhaps unlikely to survive.

Play was first delayed when another rendition of the Mexican wave, that mental aberration which cricket should long have discouraged, was accompanied by a confetti storm of torn-up paper.

The Times 12 June 1989, p. 46

Unlike the crash in 1987 and the mini crash last October the Mexican wave effect, by which market movements sweep around the globe from Tokyo to Hong Kong to London to Wall Street, has failed to materialise.

Guardian 26 Apr. 1990, p. 11

mezzanine /ˈmɛtsəniːn/ *adjective* 〽

In financial jargon: representing an intermediate form of finance, debt, etc. between two more established or traditional ones. Used especially in:

mezzanine debt, debt consisting of unsecured loans (intermediate between secured loans and equity), usually as a component of a management or leveraged BUY-OUT (compare *junk debt* at JUNK BOND);

mezzanine finance (or **funding**), *either* the financing of a leveraged buyout using subordinated or unsecured debt *or*, in companies financed by VENTURE capital, the final round of funding before the company's public flotation (intermediate in seniority between the venture capital financing and bank financing).

A figurative use of *mezzanine*, which was originally a noun meaning 'a storey of a building between two others', but which was so commonly used attributively (in *mezzanine floor* etc.) that it came to be reinterpreted as an adjective meaning 'intermediate between two floors or levels'.

The fashion for *mezzanine* finance arose in US financial markets in the late seventies or early eighties, and was widely discussed when financier Michael Milken of investment bank Drexel Burnham Lambert persuaded institutional investors to take the risk of JUNK BONDS in return for the high yield that they offered. In 1983 the Charterhouse Group launched a **Mezzanine Fund** specifically to provide the *mezzanine* finance for corporate buyouts. In some of its uses, *mezzanine* is simply a more official synonym for *junk*.

Others, such as Seragen in Hopkinton, Mass., raised seed money easily but now find venture capitalists 'more discriminating' when investing in a 'mezzanine', or third, funding round.

Scientific American June 1988, p. 92

The Citicorp fund will be dollar-based and provide mezzanine debt for deals led by the group both inside and outside the United States.

Daily Telegraph 16 Aug. 1988, p. 21

microwave /ˈmaɪkrəʊweɪv/ *verb and adjective* ✖

transitive or *intransitive verb*: To cook (food) in a microwave oven; to be suitable for or undergo microwave cooking.

adjective: (Of food or food containers) intended for cooking in a microwave oven; **microwavable**.

Formed by changing the grammatical function of *microwave*, originally the name of the type of electromagnetic wave which is passed through the food to cook it; by the mid seventies, though, it was already being used widely as a short name for a *microwave oven*.

Microwave ovens were in widespread use in the US by the late sixties and in the UK by the seventies; the development of a verb meaning 'to cook by microwaves or in a microwave oven' was to be expected as soon as the cooker had become a standard household item, and in fact the earliest uses of the verb date from the mid seventies. The regular adjective for food which has been cooked in this way is **microwaved**. During the early eighties, a number of food and cookware manufacturers started to describe their products as *microwavable* (or *microwaveable*), but in speech most people described them simply as *microwave*; this informal use eventually also found its way into print and is occasionally used as a synonym for *microwaved*, too.

He went to the pub and had a microwave mince and onion pie and crinkle-cut chips.

Sue Townsend *The Growing Pains of Adrian Mole* (1984), p. 59

When cooking or reheating: food should be very hot throughout—when you take it out of a conventional oven, or after standing times when microwaving, it should be too hot to eat immediately.

Which? Apr. 1990, p. 205

It was only last year that the F.D.A. learned that dioxin . . . was migrating from bleached paperboard cartons into milk and fruit juices and from microwave meal packages.

New York Times 7 May 1990, section D, p. 11

middleware 📟 see -WARE

MIDI /ˈmɪdɪ/ *acronym* Also written **Midi** or **midi** ✖ ♫

noun: An interface which allows electronic musical instruments, synthesizers, and computers to be interconnected and used simultaneously.

adjective: Making use of this kind of interface, usually as part of a complete music system.

An acronym, formed on the initial letters of its official name, *Musical Instrument Digital Interface*.

MIDI was invented in the US in the early eighties at a time when increasing use was being made of synthesizers in the world of music, both classical and popular. It was the introduction of this standard means of linking a number of synthesizers with a computer which made possible some of the most characteristic musical developments of the eighties: SEQUENCERS, SAMPLING, and TECHNO music generally depend upon the possibility of recording and remixing sounds and effects from electronic sources. What really brought the word *MIDI* into the high street, though, was the appearance on the mass market in the mid eighties of **MIDI keyboards**, **MIDI sequencers**, and **MIDI software** that could be run on a standard PC.

Some professional musicians already use MIDI connections to play several synthesizers at once from a single keyboard.

Newsweek 28 May 1984, p. 89

Combine this with a virtuoso stick player and MIDI keyboards and you get . . . lots of . . . different sounds.

Dirty Linen Spring 1989, p. 15

milk-free ⊗ ▓ see -FREE

mind-boggling see BOGGLING

minder /'maɪndə(r)/ *noun* ▢ ◖

A person employed to protect a celebrity, politician, etc. from physical harm or from unwanted publicity. Also, a political adviser (especially a senior politician who protects a more inexperienced one from embarrassment or mistakes, for example in an election campaign); anyone whose job is to 'mind' another person and ensure that he or she does not overstep the mark.

A sense which has developed from the use of *minder* in criminals' slang since the twenties. A criminal's bodyguard or assistant was known as a *minder*, and this word has now simply been applied in a wider and more official context, perhaps under the influence of the very successful television series *Minder* (1979–), about a petty criminal and his bodyguard, whom he hires out to 'mind' other people's property.

Extended uses of the slang sense of *minder* started to crop up quite frequently in the press from about the mid eighties, usually with the word *minder* in inverted commas; within a few years the inverted commas had been dropped and *minder* seemed to have moved from slang into the standard language. Pop stars and other celebrities often employ a whole group of *minders*, as much to ward off the unwanted attention of journalists and inquisitive members of the public as to avoid physical harm.

> He goes out alone: unlike fellow multimillionaires like Prince, Madonna and Michael Jackson, he refuses to employ a minder.
> *Today* 10 Nov. 1987, p. 20

> The minder, Mr Simon Burns, Conservative MP for Chelmsford, directed all enquiries about the plans of Mr Nigel Lawson to the press office.
> *The Times* 30 Nov. 1988, p. 7

> Her London lawyer and minder ... had struck a deal with a British newspaper to reveal the secrets she has so far coyly refused to disclose.
> *The Times* 5 Apr. 1989, p. 7

mindset /'maɪndsɛt/ *noun* Also written **mind-set** ◖

In colloquial use: an attitude or frame of mind; an unthinking assumption or opinion.

A weakened sense of *mindset*, which was originally a more precise psychological and sociological term referring to habits of mind which had been formed as a result of previous events or environment and which affected a person's attitudes.

This more general use of *mindset* became a fashionable synonym for *attitude*, starting in the late seventies in American journalistic writing, and spreading to British use as well during the eighties. The vogue made the more precise and original sense difficult to use, since many readers now think of *mindset* as being the same thing as *attitude*, rather than an event or condition imprinted on the psyche in such a way as to inform attitudes.

> The Kemeny report asserted that a change in 'mind-set', or mental attitude, was essential if nuclear safety was to be assured.
> *Scientific American* Mar. 1980, p. 33

> The Western scientists noted the Chernobyl reactor had the best operating record of any in the Soviet Union and said the operators had got into a 'mindset' that nothing could go wrong.
> *Australian Financial Review* 26 Sept. 1986, p. 39

> The mindset of a team ... is ... critical.
> *Toronto Sun* 13 Apr. 1988, p. 32

miniseries /'mɪnɪˌsɪəriːz/ *noun* Also written **mini-series** ▓

A television series, often dramatizing a book or treating a particular theme in a few episodes, and shown on a number of consecutive nights.

Formed by adding the combining form *mini-* 'small' to *series*.

Miniseries originated in the US in the early seventies; by the mid eighties they were being shown in the UK as well and the word had become so common that it seemed any television series could be called a *miniseries* (even *The Forsyte Saga* was once described as one). The difference between a series and a miniseries is partly a matter of length and partly the screening of the *miniseries* in a tight sequence, with more than one episode on the same night or all on consecutive nights (although the usage has not always supported this distinction). It has become a preferred format for television dramatizations of novels and biographies.

> At this stage, a big budget movie rather than a television miniseries was in prospect.
> *Listener* 5 Jan. 1984, p. 10

> The mini-series, which will be screened on Thursday and Friday evenings at 8.30pm, tells the story of Franciscan friar Padre Rufino who saved hundreds of Jews from the Nazis.
> *Telegraph* (Brisbane) 7 Aug. 1986, p. 43

minority briefcase /maɪˌnɒrɪtɪ ˈbriːfkeɪs/ *noun* 🔲 🎵

In dismissive US slang, the same thing as a GHETTO BLASTER.

For etymology and history, see GHETTO BLASTER.

> Maybe one day, just for the hell of it, I'll plug my mini-headphones into my minority briefcase, cruise down the street, and go find myself a watermelon.
> Transcript of *Macneil/Lehrer Newshour*, 28 Aug. 1986

MIRAS /ˈmaɪrəs/, /ˈmaɪræs/ *acronym* 〰️

Short for **mortgage interest relief at source**, a scheme providing for people paying off house-purchase loans in the UK to have the tax relief on their interest repayments paid by the Government direct to the company providing the loan.

The initial letters of *Mortgage Interest Relief At Source*.

The scheme, which was designed to simplify the system of tax relief, was introduced in 1983 to provide direct tax relief on the interest paid on loans of up to £30,000 (or on the first £30,000 of larger loans). At a time when the Government was keen to encourage home ownership, *MIRAS* made possible mortgages on a very high proportion of the purchase price of a house, since it was no longer necessary to find the full repayment and later reclaim the tax relief.

> Most people now get basic tax relief under the system known as MIRAS (Mortgage Interest Relief At Source). Under MIRAS, you pay a reduced amount to the lender and the Government makes up the difference.
> *Which? Tax-Saving Guide* March 1989, p. 26

mirror-shades group 🔲 see CYBERPUNK

moi /mwɑː/, /mwʌ/ *pronoun*

Humorously (especially when feigning pretentiousness or false modesty): me, myself.

French for *me*.

This has become a sort of humorous shorthand for pretentious reference to oneself in the late seventies and eighties, based on the obvious pretension of slipping into a foreign language. It was largely popularized through its use on television, especially by *The Muppets* (a children's puppet show created by Jim Henson), in which it was liberally used by the main female character, Miss Piggy. The theme was also taken up by a number of adult cult shows both in the US and in the UK.

> So Harry says, 'You don't like me any more. Why not?' And he says, 'Because you've got so terribly pretentious.' And Harry says, 'Pretentious? *Moi*?'
> John Cleese and Connie Booth *Fawlty Towers* (1988), p. 190

> I think it's going to be a great advantage for Ventura and for moi . . . A methanol sign on the freeway will lead them to my station. *Los Angeles Times* 30 June 1988 (Ventura County edition), section 9, p. 6

mondo /'mɒndəʊ/ adverb

In young people's slang, originally in the US: utterly, ultimately, extremely.

Formed by interpreting the (originally Italian) word *mondo* 'world' as an adverb, in attributive uses of phrases such as *mondo bizarro* (see below).

In 1961 the Italian film director Gualtiero Jacopetti produced the film *Mondo Cane*, which was released in the English-speaking world in 1963 as *A Dog's Life*. Ostensibly a documentary, it consisted of thirty sequences of such peculiar aspects of human behaviour as cannibalism and a restaurant for dogs, and became wildly popular: the original title became sufficiently well known for other films of an equally anarchic nature to be given similar titles (often with a mock-Italian flavour), such as *Mondo Bizarro* (1966) and *Mondo Trasho* (1970). During the seventies such formations became more common outside the cinema, with the meaning 'the weirder or seedier side of (a particular place, activity, etc.)': **mondo bizarro** began to be used attributively in the sense 'extremely bizarre', and *mondo* began to be reinterpreted as an adverb (and the following word as an adjective). The connotations of seediness or grossness persisted for some time, but by the time it had been absorbed into VALSPEAK in the early eighties it had become a simple intensifier, similar to *serious*—see SERIOUS²—and likewise also sometimes used as an adjective. It was, however, the adoption of *mondo* by the TURTLEs that led to its spreading outside North America, predominantly in expressions of approval like **mondo cool**.

> It was just part of a week in which the news, particularly on ABC, went further and further into the realm of Mondo Bizarro.   *Washington Post* 19 Apr. 1980, section C, p. 1

> Last weekend Mom let me go visit her and stay in the dorm and everything. It was MONDO party time.   Mimi Pond *The Valley Girl's Guide to Life* (1982), p. 49

> Why this fascination with Miller? Because he's so mondo cool, even though he's not British and doesn't have spiked hair!   *Stereo Review* Apr. 1986, p. 53

monergy /'mʌnədʒɪ/ noun

Economical use of energy; fuel conservation leading to greater cost-effectiveness in running one's home. (Originally, money spent on energy costs: see below.)

Formed by telescoping *money* and *energy* to form a blend; the word was apparently invented by the advertising agency Saatchi and Saatchi.

Monergy was originally part of the slogan 'Get more for your monergy'—the catch-phrase of a Government energy-saving campaign in the UK in 1985. The whole campaign soon came to be known by the one word *monergy*, which was widely criticized as an ugly and unnecessary formation. Perhaps unsurprisingly, it is already rarely seen, despite the greater emphasis on energy conservation which has been urged by the green movement in the late eighties and early nineties.

> Efficiency in use also requires conservation, lower energy appliances and domestic insulation, and the government's soft pedalling on its 'monergy' campaign is to be regretted.
> *Planet* 82 Aug./Sept. 1990, p. 60

monetarism /'mʌnɪtərɪz(ə)m/ noun

An economic theory based on the belief that only control of the money supply can successfully bring about changes in the rate of inflation or the level of unemployment.

Formed by adding the suffix *-ism* in the sense of 'a system, belief, or ideological basis' to *monetary* as used in *monetary control* etc.

This is not a particularly new word—the theory was first proposed by David Hume in the eighteenth century and the word has been used in relation to the economic theories of Professor Milton Friedman and his followers since the late 1960s—but it is one which has been used so frequently in the eighties to refer to the economic basis of the political administration both in the UK and in the US that it deserves an entry here for its high profile in recent years.

Monetarism has been the underlying principle for controlling inflation used by the Conservative government in the UK under Mrs Thatcher and Mr Major, and the US Presidential administrations of Ronald Reagan and George Bush, and as such it has affected the lives of millions of British and American citizens. It has been the main opponent of Keynesianism (based on the theories of J. M. Keynes), which puts an obligation on governments to create employment and put money into people's pockets through public spending. A believer in the economic principle of *monetarism* is a **monetarist**; the adjective to describe policies founded on the principle is also **monetarist**.

> Not even the fierce monetarism of the last decade has prevented us from paying ourselves far more in relation to what we produce than any of our major competitors.　*Guardian* 3 July 1989, p. 11

> In the early 1980s the insights of monetarism were dissipated because the claims of the monetarists for control of the money supply as a cure-all were exaggerated.　*Financial Times* 3 Apr. 1990, p. 21

moonwalk ✕ ◑ see BREAK-DANCING

Moral Majority /ˌmɒr(ə)l məˈdʒɒrɪtɪ/ *noun* ▢

In the US, a right-wing political movement emphasizing traditional moral standards in society and drawing support mainly from fundamentalist Christian groups. Hence more generally (as **moral majority**), upholders of traditional right-wing social values.

So named because it claims to represent a *majority* of the American people favouring the re-establishment of *moral* standards.

The *Moral Majority* movement was founded by Revd Jerry Falwell in Washington DC at the end of the seventies, originally as a 'legislative research foundation' to promote conservative Christian viewpoints. During the eighties it attracted considerable support and was able to put its message across through commercial religious broadcasting (the 'electric church'), even putting one of the TELEVANGELISTS up as a possible presidential candidate in the middle of the decade. In 1986 it was renamed the Liberty Federation but by this time the phrase *moral majority* had acquired the more general meaning of the conservative or traditionalist component of society.

> As well as the relentlessly Ann Summers view of sex, metal's other great shock tactic is horror and devil worship imagery. Accusations of satanism have stirred up America's moral majority to call for outright bans, a guarantee for enhanced teen appeal.　*Guardian* 11 Aug. 1989, p. 24

more than my job's worth ❬❬ see JOBSWORTH

mountain bike /ˈmaʊntɪn ˌbaɪk/ *noun* ✕

A bicycle with a sturdy lightweight frame, fat, deep-treaded tyres, and multiple gears, originally designed for riding on mountainous terrain.

Formed by compounding: a *bike* for *mountain* riding.

Although originally designed for hill-riding, the *mountain bike* became the most fashionable and sought after style of bicycle for town and road cycling as well during the late eighties, rising to the height of a status symbol by 1990. The fashion began in the US and Canada in the early eighties and by 1987 had spread to the UK. At first, *mountain bikes* were custom-made in California rather than being mass-produced; the name began as a component of the brand names of these 'designer' bikes, such as the Ritchey Mountain Bike. The *mountain bike* has a distinctive appearance with its thick, heavily treaded tyres and straight handlebars, but the reason for its popularity is more likely to be its versatility and performance, achieved mainly through the wide choice of gears (more than twenty on some models). The sport of hill-riding on a *mountain bike* is known as **mountain biking**; someone who takes part in it is a **mountain biker**. *Mountain bikes* are also sometimes known as *off-roaders* or *all-terrain bikes* (*ATBs*).

Mountain biking demands hill-walking stamina as well as track-riding skills. Initially, choose gentle routes among familiar terrain—or risk prolonged shoulder-carries! *Country Living* Nov. 1987, p. 164

80 per cent of all bikes sold in London are now mountain bikes. *The Face* Jan. 1989, p. 8

Cycling, like walking, is one of the best ways of seeing and enjoying the countryside, and mountain bikes have proved to be the latest and most popular method of 'green' transport: over 1 million of them were sold last year. *National Trust Magazine* Autumn 1990, p. 9

mouse /maʊs/ *noun* 🖱

A computer peripheral consisting of a small plastic box with a number of buttons and a lead, which may be moved about on a desk or tablet to control the position of the cursor on a monitor, and used to enter commands.

A metaphorical use of the animal name, arising from the appearance of the computer device, with its compact body and its trailing flex resembling a tail, as well as its effect of making the cursor 'scamper' across the screen. This is the latest in a long line of technical uses of *mouse* based on physical resemblance to the furry animal: these include a nautical term for a type of knot and a plumber's lead weight on a line.

This kind of *mouse* was invented by English and Engelhardt, computer scientists at Stanford Research Institute in California, and was first named by them in print in 1965. By the seventies the device was produced commercially, but it was only during the eighties that it became widely popularized as WIMPs (see WIMP²) became available to personal computer users. The usage debate has centred on the correct plural form in this sense, with some computer scientists using the regular plural *mice*, others *mouses*; *mice* certainly has the majority. A measure of the popularity of the *mouse* is the number of compounds it has produced, notably **mouse-button** (any of the keys on a *mouse* which allow one to enter commands), and adjectives such as **mouse-controlled** and **mouse-driven**.

Mouse-driven software has caught the imagination of American hardware designers. *Australian Personal Computer* Aug. 1983, p. 60

In a world of two- and three-button mice, why did Apple decide on the ... one-button mouse? *A+* July 1984, p. 35

mousse¹ /muːs/ *noun* and *verb* 🔳

noun: A foamy substance sold as an aerosol or in a pressurized form, usually for applying to the hair to give it body and help to set it in a style.

transitive verb: To apply mousse to (the hair or some other part of the body).

Mousse was originally a French word meaning 'froth'. It has been applied in English cookery to frothy purées using whipped cream or egg since the nineteenth century; the beauty preparation has a similar consistency to an edible *mousse*, but it may represent a fresh borrowing from French (see below).

Hair-styling products in the form of a pressurized foam (for home perming, for example) have been on the market for fifteen years or more, but were not generally known as *mousses*; the impetus to develop a non-sticky setting foam that could be used outside salons came from the increased popularity of blow-dried women's hairstyles in the late seventies. The first *mousse* for the general market was developed at the beginning of the eighties by the French firm l'Oréal; their marketing of the product using the untranslated French word *mousse* was probably the deciding factor in the establishment of *mousse* as the generic term for hair-styling foams. *Mousse* was so popular in the eighties (especially in creating the sculpted, swept-up styles that were fashionable then) that manufacturers of other pressurized beauty products also began using the word *mousse*, and combinations such as **body mousse** started to appear on labels and in advertising.

'People will try to mousse everything,'predicts stylist Louis Licari. *People* 10 Sept. 1984, p. 79

All these looks were created on one permed head and styled using a selection of mousse, gel, and spray. *Hair Flair* Sept. 1986, p. 10

See also GEL

mousse² /muːs/ *noun* 🌳

A frothy mixture of oil and sea-water which may develop after an oil spill and which is very difficult to disperse; known more fully as **chocolate mousse**.

The same word as MOUSSE¹ above; in this case, definitely so named because of its resemblance to the edible *mousse*.

The term was first used (in the fuller form *chocolate mousse*) in relation to the *Torrey Canyon* disaster in 1967, and appears to have taken the unusual route for a technical term of starting in the writing of lay reporters in the press and only later being taken up by specialists as a precise term (a water-in-oil emulsion of 50 to 80 per cent water content). From technical writing in the seventies, it moved back into the popular press each time there was a major oil spill—most recently in relation to the *Exxon Valdez* incident in Alaska in 1989.

The Ixtoc 1 well released oil for 9 months into the open ocean where winds and currents dispersed the floating mousse . . . which had formed at the wellhead. *Nature* 19 Mar. 1981, p. 235

He said the main part of the slick is about 30 miles from shore, half the distance from the ship to the shore, and thin streamers of oil with the consistency of mousse extend another 10 miles toward shore. *New York Times* 15 June 1990, section A, p. 12

MRI /ɛmɑːrˈaɪ/ *abbreviation* ⊗ 🖳

Short for **magnetic resonance imaging**, a technique which provides sectional images of the internal structure of the patient's body by plotting the nuclear magnetic resonance of its atoms and converting the results into graphic form by computer.

Formed by compounding: the *image* is based on the varying *magnetic resonance* of the atoms making up the body.

Like CAT scanning (see CAT¹), *MRI* was developed in the mid seventies as a diagnostic technique which would do away with the need for exploratory surgery. At first it was known as the **nuclear magnetic resonance** (**NMR**) technique or *zeugmatography*, but *magnetic resonance imaging* and the abbreviation *MRI* now seem to be becoming the established terms in popular sources. The technique works by passing low-frequency radiation through the soft tissues of the body in the presence of a strong magnetic field and scanning the temporary magnetic realignment that this produces in the nuclei of the elements; the machinery required to do this (an **MR scanner**) only became commercially available in the UK in the first half of the eighties. *MRI* produces a clear image of soft tissue even if it is obscured by bone, and is likely to become one of the foremost diagnostic techniques of the nineties. The abbreviation *MRI* is also sometimes used for **magnetic resonance imager** (another name for an *MR scanner*).

The company's intensive work on developing semiconductor magnet systems has resulted in today's applications in . . . magnetic resonance imaging (MRI). *Physics Bulletin* Jan. 1987, p. 9

MRIs are like CAT scan machines, but they create images by placing a patient in a strong magnetic field. *Baltimore Sun* 7 Mar. 1990, section C, p. 10

muesli belt /'mjuːzlɪ ˌbɛlt/, /'muːzlɪ ˌbɛlt/ *noun* ⊗

Humorously, an area largely populated by middle-class health-food faddists.

Formed by compounding. *Belt* has long been used in the sense of a zone or region, especially with a preceding word denoting the main characteristic or product (such as *corn belt*, *rust belt*, etc.). *Muesli* is seen as the archetypal health food; in this case there is also some allusion to the *Bible belt*, with the implication that belief in health foods is fundamental to the way of life of this group.

The term arose soon after the middle-class obsession with health foods took hold in the late seventies. A report published in 1986 showed that the children of health-food faddists tend to be undernourished, a fact which gave rise to the term **muesli-belt malnutrition**.

Team vicar required. An attractive post in S.W. London 'Muesli belt'.

advertisement in *Not the Church Times* 22 Sept. 1981, p. 6

muggee /mʌ'giː/ *noun* 🗡

The victim of a mugging; a person who is or has been mugged.

Formed by adding the suffix *-ee*, denoting the person affected by an action, to the verb *mug*, 'to rob violently, especially in a public place'.

The word has been used in US English (which tends to form nouns in *-ee* more freely than UK English) since the early seventies. With the increasing problem of street muggings in the eighties, and the difficulty of finding an alternative word for the victim, it has spread beyond the US to other parts of the English-speaking world.

Have the muggees, the majority of whom are white, no right to be protected against muggers?

Spectator 28 Nov. 1981, p. 4

After proving four were tougher than one the muggers drove off and the muggee went home to bed.

Brisbane Telegraph 9 Apr. 1987, p. 14

Muldergate 📷 see -GATE

multilevel /ˌmʌltɪ'lɛv(ə)l/ *adjective* Also written **multi-level** 〰

In business jargon: operating on a number of levels simultaneously. Used especially in **multilevel marketing** or **multilevel sales**: a selling technique involving direct contact with the customer through a network of independent distributors.

So named because the system makes use of sellers at a number of different *levels* in the organization, each buyer taking on the responsibility of finding further sellers as well as trying to sell the product.

Multilevel marketing originated in the US in the early seventies as a name for a development of the type of marketing operation that is sometimes also called *direct sales* or *pyramid selling* (an earlier term with more critical connotations, dating from the sixties)—the technique best exemplified by Tupperware parties and home shopping representatives. *Multilevel* seemed to become one of the buzzwords of the sales world in the eighties, but the system has been criticized because it tends to exploit those in the middle of the pyramid, putting great pressure on them to find more sales staff.

Merchant Associates said it was working for a California-based organisation selling health products on a multi-level (or pyramid) system. *Daily Telegraph* 4 Feb. 1987, p. 22

To avoid problems, he says, USA Today no longer takes ads for multilevel sales organizations, where you make your biggest money not by selling products but by bringing in new sales people into the game. *Chicago Tribune* 17 Oct. 1988, section 4, p. 7

multimedia 🖳 see CD

muso /'mjuːzəʊ/ *noun* 🎵

In musicians' slang (originally in Australia): a musician, a music fanatic.

Formed by abbreviating *musician* and adding the colloquial suffix *-o*; like JOURNO, a typically Australian slang nickname.

Muso has been used in Australia since the late sixties, and is used there of classical as well as popular musicians. It had started to appear in the popular music press in the UK by the late seventies (and so was probably in spoken use for some time before that), but in British use it seems to be more or less limited to the pop and rock scene.

Since he's also a muso, and has a brother . . . with Whitesnake connections, it seemed like a good idea to turn all the background knowledge of crass horrors into more than a Trapeze reunion, a rockstravaganza called 'Phenomena'.   *Sounds* 27 July 1985, p. 17

It's hard to imagine many people, apart from die-hard musos and dedicated Gabriel fans, would want to listen to this in the comfort of their own home.   *Empire* Sept. 1989, p. 108

myalgic encephalomyelitis see ME

• •

N

nab /næb/ *acronym*

Short for **no-alcohol beer**, a beer from which almost all the alcohol has been removed after brewing.

The initial letters of *No-Alcohol Beer*.

Nabs became increasingly popular in the late eighties as the message of 'don't drink and drive' finally started to sink in and alcohol-free drinks became more widely available in bars and restaurants. The low-alcohol equivalent of a *nab* is a *lab* (*low-alcohol beer*); these too became more popular and widely available during the eighties. In the trade, the two categories are sometimes grouped together as **nablabs**.

Alcohol-free or low-alcohol beers, the so-called Nablabs, are now available in almost every public house in Britain.   *The Times* 2 Dec. 1988, p. 7

Next on the agenda is image. The so-called 'nablab' sector . . . is growing at the rate of 100 per cent each year, 200 per cent in the case of low-alcohol wines.   *Daily Telegraph* 3 Dec. 1988, p. 13

Nabs and labs . . . are brewed as normal beers and then go through a further process to remove or reduce the alcohol.   *Daily Telegraph* 24 Oct. 1990, p. 36

nacho /'nɑ:tʃəʊ/, /'nætʃəʊ/, in Spanish /'natʃo/ *noun*

A tortilla chip, usually served grilled and topped with melted cheese, jalapeño peppers, spices, etc.; often in the plural **nachos**, a 'Tex-Mex' snack with these ingredients.

The word is clearly borrowed from Mexican Spanish, but its further origins have been the subject of some debate. The dish was first served in the late forties, and one attractive theory is that it was named after the chef who first prepared it. *Nacho* is the diminutive form of the Spanish given name *Ignacio*, and one Ignacio 'Nacho' Anaya, a Mexican chef working in the Texan border area of Piedras Negras in the forties, has claimed the honour. The apparent plural form may have originated as a misinterpreted possessive *Nacho's*. Another theory is that the word is borrowed from the Mexican Spanish adjective *nacho*, meaning 'flat-nosed'.

Although first prepared as long ago as the forties, *nachos* did not spread far outside Texas and North Mexico until the seventies, and only became widely known through fast-food chains in the eighties. The original dish consisted of a wedge of tortilla, garnished and toasted, but in Britain the basic ingredient has always been corn chips.

The chain of Mexican fast-food restaurants is busily expanding its product line to include . . . a nacho side dish, and a salad.   *Fortune* 14 Nov. 1983, p. 126

I can tell you what they served. It was guacamole and nachos and there was Gallo jug wine and shrimp dip.   Jonathan Kellerman *Shrunken Heads* (1985), p. 86

naff¹ /næf/ *adjective*

In British slang: unfashionable, lacking in style, vulgar or kitsch; also, useless, dud.

Despite its resemblance to the verb (see NAFF²), the two words do not seem to be etymologically related. The origins of the adjective may lie in English dialects, several of which have similar words of contempt for inept or stupid people: in the North of England, for example, an idiot is a *naffhead*, *naffin*, or *naffy*, and *niffy-naffy* as an adjective (meaning 'stupid') has been recorded since the last century. In Scotland, *nyaff* is a term of contempt for any stupid or objectionable person.

The word was first used in the late sixties, mostly among young people, as a new alternative for 'square'. The rise of social groups such as the Sloane Rangers and the yuppies in the eighties made it socially desirable for people to know how to avoid being *naff* (just as, some decades earlier, the social élite had wanted to know how to be U rather than non-U); and in 1983 a whole book (*The Complete Naff Guide*) was devoted to the subject. Although principally a British word, *naff* has been borrowed into US English. Now overtaken by other words among the really young, it is used by those who want to sound younger than they are. The nouns corresponding to *naff* are **naff** (for the whole style) and **naffness** (for the quality of being *naff*).

No electricity...I think it's just a naff battery connection.   Liza Cody *Bad Company* (1982), p. 13

'I shan't bother with that,' a chap retorted on hearing what preview I had attended. 'One-word title that doesn't make sense—bound to be naff.'   *Daily Mail* 6 Apr. 1985, p. 6

Issues [of the magazine]...embodied even more the spirit of naff than had earlier been the case.
     Harpers & Queen Dec. 1989, p. 235

naff² /næf/ *intransitive verb* 🔲

A slang word used euphemistically to avoid saying 'fuck'; usually in the phrase **naff off**: go away, 'eff off'. Also as an intensifier or empty filler, in the adjectival form **naffing**.

The origin of this word is uncertain; it may be an example of back-slang, reversing the sounds in *fan* (a long-established shortened form of *fanny*). Alternatively it could be connected in some way with the wartime NAAFI: Keith Waterhouse, who was the first to use it in print (in *Billy Liar*, 1959), points out that *naffing* was a general-purpose expletive in the RAF during the Second World War.

Although first used in 1959, *naff* really became popularized by the BBC television series *Porridge* from the mid seventies onwards. When, in 1982, Princess Anne told persistent press photographers to 'naff off', it acquired an unexpected respectability; this was reinforced by its association in some people's minds with the (in fact unrelated) adjective in the entry above. A new phrasal verb **naff about** (to make a fool of oneself) arose from this confusion.

'It's all been arranged, it's all set up, right? So naff off', I said.
     Dick Clement & Ian La Frenais *Porridge* (1975), p. 63

Stealing your tin of naffing pineapple chunks? Not even my favourite fruit.
     Dick Clement & Ian La Frenais *Another Stretch of Porridge* (1976), p. 16

'Salute'...does not mean naffing about in a tutu.   Suzanne Lowry *Young Fogey Handbook* (1985), p. 30

naked /'neɪkɪd/ *adjective* 〰️

In financial jargon, of an option, position, etc.: unhedged, not secured or backed by the underlying stock, and therefore high-risk.

A figurative use of *naked* in the sense 'not covered'; the writer of a *naked* option does not actually own the stock concerned, so in this sense it is 'not covered'.

The practice of writing unhedged or *naked* options was first reported in the US in the early seventies; in the middle of the decade it was the subject of a number of prosecutions for fraud. As high-risk financial deals involving JUNK BONDS and MEZZANINE finance became more common in the eighties, **naked writing** spread to other financial markets and the **naked writer** became a recognized (although still slightly suspect) figure in stock dealing.

Some traders were using more risky index-trading strategies, sources said. One involves writing naked puts—selling someone the right to force you to buy a stock index at a set price in the future.

Newsday 26 Oct. 1989, p. 58

NAM ⊗ ✖ ♪ see NEW AGE

nanny state /ˌnænɪ ˈsteɪt/ *noun* 📛 ⚙

A derogatory nickname for the Welfare State, according to which government institutions are seen as authoritarian and paternalistic, interfering in and controlling people's lives in the same way as a nanny might try to control those of her charges.

Formed by compounding: the *state* perceived as playing the role of *nanny*.

The coinage of the nickname *nanny state* has been attributed to both Bernard Levin and Ian Macleod; certainly it was first applied to the paternalistic British Welfare State, with its insistence on limiting individual's freedoms if this could be argued to be for the individual's own good. Under the Conservative government of Margaret Thatcher in the eighties the term acquired a new emphasis as the ethos of individualism and enterprise was presented as a better alternative to spoon-feeding from the *nanny state*; the government's programme of privatization was one way in which individuals were to be weaned from reliance on such spoon-feeding. However, opponents of the government argued that authoritarianism and paternalism were stronger than ever in other areas, leaving the *nanny state* intact in so far as it affected individual rights and freedoms. From the mid eighties the term was used in Australian politics as well.

The British, we are incessantly told, have now rejected the 'nanny state' and regard the social worker as a boring pest. *Washington Post* 14 Aug. 1983, p. 5

The Nanny State is alive and well Down Under. The immediate target is the cigarette industry and individual smokers, but the drive to purify our lives will not end there.

Weekend Australian (Brisbane) 9–10 Apr. 1988, p. 20

A measure of privatisation of adoption is called for, with a diminution in the powers of ... ideological apartheiders of the nanny State. *The Times* 28 Sept. 1989, p. 17

narcoterrorism /ˌnɑːkəʊˈtɛrərɪz(ə)m/ *noun* Also written narco-terrorism 🗡 ⚙

Violent crime and acts of terrorism carried out as a by-product of the illicit manufacture, trafficking, or sale of drugs, especially against any individual or institution attempting to enforce anti-drugs laws.

Formed by adding *narco-* (the combining form of *narcotic*) to *terrorism*.

Narcoterrorism came into the news in the mid eighties, when it became clear that, in a number of countries where dangerous but highly profitable drugs such as cocaine are produced, the influential producers or 'drug barons' were in alliance with guerrilla and terrorist organizations to defeat any attempts to enforce anti-drugs laws. Alleging government collusion with *narcoterrorism* in a number of Central and South American countries, some US authorities favoured intervention in the affairs of foreign countries to stop the flow of drugs into their own country; in view of the serious and rapidly growing problems of drug abuse and drug-related crime within the US in the second half of the eighties, some argued that to manufacture drugs at all was itself a **narcoterrorist** act. In the late eighties reports of the activities of the **narcoterrorists** centred on the plight of Colombia, where a government determined to stop the drug traffic was the target of repeated attacks in 1989–90.

Mr. Belaunde Terry said the victims [of a raid on an anti-drug team in Peru] were 'heroes' and the killers were 'narco-terrorists'. *New York Times* 19 Nov. 1984, section A, p. 14

Calling cocaine manufacture 'narco-terrorism', as White House spokesman Edward Djerejian did in defense of the raid, the State Department merges its all purpose justification for intervention with the politics of drug warfare. *Nation* 2 Aug. 1986, p. 68

It is the consensus among anti-drug officials here [in Colombia] that those two men are the master-minds of a 'narcoterrorist' campaign that has driven this nation of 32 million people into a state of widespread anguish and fear. *Los Angeles Times* 13 Dec. 1989, section A, p. 6

nasty /'nɑːstɪ/ *noun* ✂ ▓

Colloquially, a horror film, especially one on video; a video film depicting scenes of violence, cruelty, or killing (known more fully as a **video nasty**).

A specialized use of *nasty*, which had existed as a noun meaning 'a nasty person or thing' since the thirties.

The problem of *nasties* (the word is often used in the plural to describe the genre as a whole) was discussed a good deal in the newspapers in the early and mid eighties—at the beginning of the video rental boom in the UK—when large numbers of these films first became widely available and proved worryingly popular. In particular, there was public concern over the potential influence of the more violent *nasties* on the behaviour of those who watched them.

Three videos, part of the current crop of 'nasties' available in thousands of High Street rental shops, have been sent to the DPP. *Sunday Times* 6 June 1982, p. 3

With its tougher law on videocassettes, West Germany hopes to keep its youth away from the nasties. *Christian Science Monitor* 3 May 1985, p. 30

See also SLASHER and SNUFF

national curriculum /'næʃən(ə)l kəˈrɪkjʊləm/ Frequently written **National Curriculum** ▓

In the UK, a programme of study provided for in the Education Reform Act of 1988, to be followed by all pupils in the maintained schools of England and Wales, and comprising core and foundation subjects to which appropriate attainment targets and assessment arrangements are to be applied at specified ages.

Self-explanatory: a *curriculum* to be followed on a *national* basis (though in fact the schools of Scotland are not statutorily included, since education is separately administered there).

As originally proposed, the *national curriculum* was intended to provide higher and more uniform standards of education across the various schools and parts of the country at a time when there was serious public concern over the content and standards of British education. **National Curriculum Councils** were set up for England and Wales to co-ordinate proposals for the content of the curriculum, standards, etc., but the Act gave final responsibility for specifying the attainment targets and programmes of study to the Secretary of State for Education and Science. The early proposals were quite ambitious in their scope and were based on the premise that all pupils should study certain subjects (the 'core' subjects) up to a certain age, their level of attainment in those subjects being assessed by organized testing at the 'key stages' of ages 7, 11, 14, and 16—the testing was to be based on *standard assessment tasks*, or *SATs*. As these proposals were implemented from 1990 onwards, it became clear that the original scope had been over-ambitious, and the number of subjects in which testing was to take place was reduced accordingly.

This autumn, 25 Hampshire schools and colleges will be taking part in trials using CA material for teaching of maths and science under the new National Curriculum. *Which?* Sept. 1989, p. 413

The Department of Education and Science said: 'An increased workload in the short term will bring long-term benefits for teachers and pupils as the national curriculum brings a clearer framework for teaching. The Government is pacing its vital reforms and deferring appraisal to meet concerns about teachers' workload.' *Financial Times* 3 Apr. 1990, p. 12

national heritage 🌳 see HERITAGE

neato /'niːtəʊ/ *adjective* 🎨

In young people's slang, especially in the US: really good, desirable, or successful; extremely 'neat'.

Formed by adding the suffix *-o* (here intensifying the force of the adjective) to *neat* in its colloquial sense 'excellent, desirable'.

Neato was in spoken use in the late sixties, but became a particularly fashionable term of approval among young people in the late seventies and early eighties. It was at this time that it also spread outside the US to other English-speaking countries.

> We would probably never have heard of Peter Wagschal, or of his neato Ouija Board Studies Program, if it hadn't been for one Larry Zenke, a pretty neato guy himself.
>
> *Underground Grammarian* Jan. 1982, p. 1

> Those were the days when Beaver used to . . . have what she calls 'a neato free time'.
>
> *More* (New Zealand) Feb. 1986, p. 49

necklace /'nɛkləs/ *noun and verb* 🔒

In South Africa,

noun: A tyre soaked or filled with petrol, placed round the neck and shoulders of a victim, and set alight, used as a form of unofficial execution. Often attributive, in **necklace killing**, **necklace murder**, etc.

transitive verb: To kill (a person assumed to be a police informer or collaborator) using this method. Also as an action noun **necklacing**.

A figurative use of *necklace*, based on the fact that the tyre is placed round the neck. In the days of hanging, a noose was also sometimes referred to metaphorically as a *necklace*.

It was in the mid eighties that Western newspapers began reporting the use of the *necklace* by South African Black activists on fellow Blacks who were suspected of betraying the Black rights movement. Such reports continued into the early nineties, even after the unbanning of the African National Congress and the move towards greater recognition of Black rights which followed.

> Four more blacks . . . have been killed in 'necklace' murders . . . in South African townships.
>
> *The Times* 22 Apr. 1986, p. 7

> We heard that two nine year olds in that area had been 'necklaced', having rubber tyres filled with petrol put round their necks and set alight. *Tear Times* Summer 1990, p. 6

need not to know 🔒 see DENIABILITY

neighbourhood watch /ˌneɪbəhʊd 'wɒtʃ/ *noun* Written **neighborhood watch** in the US 🎨

An organized programme of vigilance by ordinary citizens in order to help the police combat crime in their neighbourhood; crime prevention achieved by this method.

Formed by compounding: the idea is for ordinary citizens to keep a *watch* on their *neighbourhood*.

The idea of *neighbourhood watch* came from the US, where the first scheme was set up in the early seventies. By the mid eighties it was also catching on in the UK as a popular response to the rising number of burglaries and thefts. The underlying principle is local co-operation: that neighbours should be prepared to watch out for each other's property and welfare and co-operate with the police in ensuring that anything suspicious is reported and investigated.

> Neighbourhood watch schemes are catching on fast. In January a Home Office minister said 8,000 schemes were in operation. *New Socialist* Sept. 1986, p. 5

The words 'neighborhood watch' mean more than just keeping an eye out for suspicious activity. Here . . . some 35 area block clubs' representatives meet regularly to figure out how to make their streets safer and cleaner. *Modern Maturity* Aug.–Sept. 1989, p. 18

neo-con /ˌniːəʊˈkɒn/ *noun* and *adjective* Also written neocon

In North American politics (especially in the US),

noun: A neo-conservative; a member of a political movement known as *neo-conservatism*, which rejects the allegedly utopian values of liberalism but supports democratic capitalism in which there is a measure of social conscience.

adjective: Of or belonging to the neo-conservative movement.

Formed by abbreviating *neo-conservative*.

The *neo-conservative* movement in the US arose in the seventies under the influence of a group of contributors to the journal *The Public Interest*, and by the end of the decade had crystallized its ideas (for example on the place of a welfare state within a conservative society and the need for practical realities rather than utopian dreams) to become the focus of the 'soft' right in US politics. By the end of the seventies *neo-conservative* was being abbreviated to *neo-con*; in the course of the eighties this became a standard way of referring to conservatives of this complexion.

The neo-con intellectuals are privately dismayed at the choice of 'a Kemp without Kemp's baggage'.
New York Times 18 Aug. 1988, section A, p. 27

On the right, the hard-core conservatives and the neocons are left lamenting what they perceive as Reagan's unfortunate drift to détente. *Washington Post* 2 Dec. 1988, p. 27

Neo-Geo /ˌniːəʊˈdʒiːəʊ/ *noun* and *adjective* Also written Neo Geo or neo-geo

noun: An artistic movement characterized by a high degree of geometric abstraction and often by the inclusion of consumer products such as manufactured goods. Also, an artist belonging to this movement.

adjective: Of or belonging to this movement.

Formed by adding the prefix *neo-* 'new' to the abbreviation *geo* (for *geometric*).

Throughout the twentieth century abstract artists have often shown an interest in 'geometric' figures, producing precisely drawn pictures of straight lines and simple shapes: a particularly extreme form of this was the **Neo-Plasticism** of Piet Mondrian and his followers. Consequently, when in the mid eighties a small group of artists in New York's East Village began to exhibit works which showed a similar approach, the supposed 'school' that this represented became known as *Neo-Geo*. The hallmarks of the work of these artists were their interest in mass-production and the idea of creating something which has a suggestion of having been manufactured, interpreted by some as an ironic comment on the technological society. Other proposed labels for the genre include **Neo-Conceptualism**, **Neo-Pop**, and *Smart Art*.

The question of what to call the new thing has not been settled. 'Neo-geo', the catchiest title, may not stick, because it refers only to one ingredient of the package—the geometric abstract painting that mimics and comments on earlier geometric abstract painting. *New Yorker* 24 Nov. 1986, p. 104

Worst of all are the Neo-Geos, who are like children aping their elders. *Art & Design* Oct. 1987, p. 31

nerd /nɜːd/ *noun*

In US slang: a contemptible or boring person, especially one who is studious, conventional, or 'square'; a DWEEB.

Of uncertain origin: possibly a euphemistic alteration of *turd*, but perhaps simply an allusion to a nonsense word used in Dr Seuss's children's book *If I Ran the Zoo* (1950):

And then, just to show them, I'll sail to Ka-Troo And Bring Back an It-Kutch, a Preep and a Proo, a Nerkle, a Nerd, and a Seersucker, too!

Nerd itself has been in use in US slang since the sixties, but enjoyed a fashion in the late seventies and early eighties which led to the development of a number of derivatives and compounds. Notable among these are the adjectives **nerdish**, **nerdlike**, and **nerdy** and the nouns **nerdishness** and **nerdism**. The *nerd* affects a fussy, conventional (and, some would say, pretentious) style of dress and appearance which became known as the **nerd look**; the quintessential characteristic of the *nerd*, a plastic pocket protector worn in the top pocket to prevent pens from soiling the fabric, was nicknamed the **nerd pack**. The word *nerd* had supposedly gone out of fashion by the late eighties in favour of *dweeb* and other synonyms, but it and its derivatives had by then already spread to the UK and continued to appear frequently in print, even in US sources, into the early nineties. A British variation on the same theme is **nerk**, a stupid or objectionable person (probably formed by telescoping *nerd* and *jerk* to make a blend); the corresponding adjective is **nerkish**.

> To make the simplest and most effective statement of your nerdishness, all you need to do is go out and buy a bra. Not the kind associated with women, but the black, oozy, plastic kind that dimwits put on the front of their cars. The auto bra is at its nerdish best when used on cars costing less than $10,000. *Car & Driver* Oct. 1989, p. 3

> Cedrico and Angelita . . . would call them aunt and uncle if they didn't consider such titles nerdy.
> Alice Walker *Temple of My Familiar* (1989), p. 395

> Most people think of BBSs as crude hacker forums where computer nerds trade tips on how to pirate software or break into the Pentagon's computers. *Computer Buyer's Guide* 1990, part 3, p. 34

> Nerdpacks are for engineers and computer programmers who have earned their status as nerds, or compulsive-obsessive gadget freaks. Michael Johnson *Business Buzzwords* (1990), p. 97

net 🖳 see NETWORK² and NEURAL

network¹ /ˈnɛtwɜːk/ *intransitive verb* 〰 ◖

To make use of one's membership of a network, one's contacts, etc. to acquire information or some professional advantage, often while appearing to be engaged only in social activity. Frequently as the verbal noun **networking**, the use of contacts in this way; also as agent noun **networker**, a person who uses this technique.

The verbal noun was formed on the noun *network*, with the simple verb as a later back-formation from it. The verb *to network* in the sense 'to cover with a network' had existed since the late nineteenth century and had developed technical uses in broadcasting and computing in the forties and seventies respectively.

As the feminist movement gathered momentum during the seventies, it was realized that men had always used the *old boy network* to get ahead, and there was no reason why women should not do the same. By the late eighties, particularly as the individualistic ethos of the Thatcher and Reagan economies became evident, *networking* was recognized as an important way of advancing all kinds of interests (not just among women).

> Over a networking lunch of smoked salmon sandwiches . . . she learned all that she needed to know about the status, income and prospects of her Valentine date. *The Times* 9 Feb. 1985, p. 11

> Party delegates are gathering . . . and 'networking'. *Independent* 16 July 1988, p. 6

network² /ˈnɛtwɜːk/ *noun and verb* 🖳

noun: A system of interconnected computers, especially within a business organization etc.; a local area network (see LAN) or wide area network (see WAN). Sometimes abbreviated to **net**.

transitive verb: To link (computers or other electronic equipment) together to form a network, so as to make it possible to transfer data, share resources, or access the system from a number of different locations. Also as an adjective **networked**; action noun **networking**.

A further specialized development of *network* in the sense of 'something which resembles a net in its complex organization and interconnectedness'; earlier examples had included the broadcasting *network*.

The first computer *networks* were set up in the sixties; by the early eighties the word was frequently used as an abbreviation of the longer terms *local area network* and *wide area network*, especially by those who did not feel comfortable with the acronyms *LAN* and *WAN*. The further abbreviated form *net* originated in the jargon of computer scientists in the seventies, but by the mid eighties was beginning to gain a wider currency. The general public perhaps met it most frequently as a suffix for the proper names of large computer *networks* or their components, such as *Ethernet* and *Internet*.

> Extras: . . . ECONET network interface. *Which Micro?* Dec. 1984, p. 20
>
> The net requires you to have intelligence at the terminals but the PCs don't have to be flash and you have to be careful the network will support them. *Today's Computers* Nov. 1985, p. 125
>
> One result of buying different types of equipment has been their lack of compatibility within a network. *Daily Telegraph* 21 Nov. 1986, p. 4
>
> The term 'ION' stands for 'Image Online Network' and means that this camera has the potential to be connected—or 'networked'—to a range of other equipment, such as computers, desktop publishing systems and copiers. *Video Maker* July/Aug. 1990, p. 37

See also NEURAL

neural /ˈnjʊər(ə)l/ *adjective* 🖳

In computing jargon: modelled on the arrangement of neurons in the brain and nervous system; used especially in **neural network** (or **neural net**), a computer system which is designed to simulate the human brain in its ability to 'learn' probabilistically and carry out complex processes simultaneously at a number of different nodes.

A figurative use of the adjective *neural*.

The development of computer *neural networks* was founded on the work of mathematicians studying neurophysiology as a model for the construction of automata from the late forties onwards; it was not until the eighties, though, that computer scientists announced that they had succeeded in building a computer which worked on the *neural* principle. The basic principle underlying the **neural net computer** is that of *connectivity*; essentially this means doing away with a central processor in favour of a number of simple calculating elements which work in parallel and are connected in patterns similar to those of human neurons and synapses. Such a system, unlike the digital computer, can solve problems even when there are minor inaccuracies in the starting data, and can also be 'trained' to use a technique for reaching correct solutions based on trial and error. The *neural net computer* is therefore seen as one of the most promising areas of AI research in the early nineties.

> A number of special neural networks will be designed and interlinked to create a neural computer . . . Research into neural computing is now a multi-million pound scientific endeavour. *The Times* 25 Mar. 1989, p. 5
>
> We're also looking at advanced neural nets and doing quite a lot of work on VLSI (Very Large Scale Integration), to make sure that the memory we develop is properly structured and packaged in a chip. *CU Amiga* Apr. 1990, p. 91
>
> There's something big just below the surface of neural-net technology, something real big. *PC Magazine* June 1990, p. 170

Neuromantic 🔳 see CYBERPUNK

New Age /ˌnjuː ˈeɪdʒ/ *noun and adjective* ⊗ 🔳 🎵

noun: An umbrella term for a cultural movement (known more fully as the **New Age Movement**, abbreviated to **NAM**), covering a broad range of beliefs and activities and characterized by a rejection of (modern) Western-style values and culture and the promotion of a more integrated or 'holistic' approach in areas such as religion, medicine, philosophy, astrology, and the environment.

adjective: Belonging to, characteristic of, or influenced by the New Age approach to health, society, music, etc.

Formed by compounding: an *age* that is *new*. The term may be used to describe any new era or beginning, but, from about the turn of the century, it also became an alternative name in astrology for the *Age of Aquarius*, that part of the zodiacal cycle which the world is due to enter in the late twentieth or early twenty-first century, and which is believed to signal an era of new spiritual awareness and collective consciousness.

Although *New Age* originated in and remained strongly associated with California and the West Coast of the US, its influence spread throughout the US and northern Europe and became established in communities such as Findhorn in Scotland from about the beginning of the seventies. Many of the various components that make up the *New Age Movement*—including the wide range of ALTERNATIVE and COMPLEMENTARY therapies, the practice of Eastern religions, and the fascination with the occult and parapsychology—are of course not 'new'; and moreover, at first sight, they seem to follow directly from aspects of the hippie movement of the sixties. What made *New Age* different (and in this sense 'new') was that, whereas the hippie movement involved mainly young people and tended to operate in opposition to contemporary Western society, *New Age* was by the early eighties attracting not only an older age group but also middle-class people who had both money and status within society. Such people—some of whom were in fact the hippies of the sixties now grown older—not only gave the movement a reputation for being a kind of 'religion for yuppies', they also, by the late eighties, ensured its rapid growth and extraordinary success in commercial terms, whether it was in publishing *New Age* books on organic gardening or astrological charts, or in promoting CRYSTAL HEALING or water-divining. A person involved with *New Age* ideas was soon referred to by the agent noun **New Ager**.

The general theme within the *New Age Movement* was that in the harsh post-industrial world of the late twentieth century, people had somehow become out of BALANCE both with their own spiritual selves and with nature and the environment as a whole; this theme was strongly featured in **New Age music**. From about the middle of the eighties, this term was loosely applied to a particular brand of music that tended to be characterized by light melodic harmonies and improvisation, by the lack of a strong beat or prominent vocals, and by the use of such instruments as the piano, harp, and synthesizer. The idea was to create a relaxing or dream-like atmosphere; sometimes sounds were reproduced from the natural world such as 'planetary' sounds and the calls of dolphins and whales.

Most New Agers favor replacing nuclear and fossil fuels with ecologically sound solar power which represents a kind of marriage between technology and spirit. *Nation* 31 Aug. 1985, p. 146

Most of them listen to New Age music—waves lapping, whales calling, amplified heartbeats and so on. None of them listen to the Beach Boys. *Sunday Express Magazine* 23 Aug. 1987, p. 30

So-called New Age philosophy has much in common with the worldmind and Gaia: the self is subsumed in the larger whole. *Raritan* (1989), volume IX, p. 132

Mrs. Brandon is less furiously New Age; her hair is frosted and shaped into a ladylike little flip. Perri Klass *Other Women's Children* (1990), p. 65

new-collar /ˌnjuːˈkɒlə(r)/ *adjective* and *noun* 🔊

adjective: Belonging to a supposed socio-economic group made up of white-collar workers who are more affluent and better educated than their parents.

noun: A person who belongs to this group.

Formed by compounding: having a *collar* of a *new* kind.

Ralph Whitehead, a Chicago reporter who later became a University professor, was one of many people writing in the seventies about the demographic changes that had taken place in the US since the war. He noticed that as a result of the declining manufacturing sector, large numbers of people from working-class ('blue-collar') backgrounds were moving into new areas of employment, and were as a result beginning to acquire new, supposedly more 'educated' values—and to vote differently. In a series of articles, Whitehead described this subgroup of 'baby boomers' in detail: the idea caught on amongst political commentators, and from about the mid eighties the *new-collar* worker became a stereotype, to be courted by advertisers and politicians like the less numerous (but even more affluent) YUPPIES.

> There has arisen what Whitehead calls the 'new-collar class'. New collars are to the middle class what yuppies are to the upper-middle class . . . New collars earn from $20,000 to $40,000. But what new collars lose in individual wealth when compared to yuppies, they gain back in numbers.
>
> *New Republic* 30 Dec. 1985, p. 20

new heroin ✒ see DESIGNER DRUG

New Wave /ˌnjuː ˈweɪv/ *noun* and *adjective* ♪ 🔊

noun: A style of rock music which grew out of punk rock, but later developed a more restrained character of its own and proved more enduring than punk.

adjective: Belonging to this style of rock.

There had already been a *New Wave* in jazz and a similar movement in French cinema (also known as *nouvelle vague*); the punk rockers simply adopted the term and applied it in a new context.

New Wave developed in the late seventies as a toning-down of some of the more shocking features of punk rock, especially in the US. The angry, socially conscious lyrics of punk remained, but more tunefully and in a more sophisticated minimalist rock framework than before. In practice, nearly all new rock groups of the late seventies and early eighties were described as *New Wave* except those which clearly belonged to HEAVY METAL. A performer of *New Wave* music was sometimes called a **New Waver**.

> [Laurie] Anderson is a borderline New Waver who looks as though she has been out in the rain upside down.
> *Washington Post* 10 June 1982, section D, p. 10

> They refused to conform to the prevailing fashions of the San Francisco new wave/punk scene.
> *Guitar Player* Mar. 1989, p. 41

nibble /ˈnɪb(ə)l/ *noun* Also written **nybble** 🖥

In computing jargon, half a byte (four 'bits') of information.

Formed humorously on *byte*, treating it as the same word as *bite*; something which is only half as big as a *byte*.

Nibble began as a piece of computer programmers' slang in the seventies and soon found its way into print in technical sources. It remains largely an 'in' joke in computing, but sometimes appears in popular magazines for enthusiasts and explanations of computing for the layperson.

> The quarter-frame message breaks down the SMPTE number into 'nibbles', or pieces of bytes (I didn't make this up), and the second byte of each message is one nibble.   *Keyboard* Mar. 1990, p. 94

nicad /ˈnaɪkæd/ *noun* Also written **NiCad** or **ni-cad** ✖ 🖥

A nickel and cadmium battery which, because of its construction, can be recharged frequently and is able to deliver short bursts of high current. Often used attributively, especially in **nicad battery**.

A clipped compound, formed by combining the initial syllables of *nickel* and *cadmium*.

Nicads were first used in the fifties, amongst other uses in experimental electric cars, but proved too expensive to be very successful at that time. During the sixties they were among the types of battery tried out in US spacecraft. What really ensured their success was the search for a lightweight rechargeable battery for the growing market in portable computers in the late seventies and early eighties. As the eighties progressed, public interest in GREEN issues led to a greater demand for rechargeable batteries for all kinds of consumer durables, and the word *nicad* entered the general vocabulary, initially through advertising of these products.

> Ni-cads are better able to provide a sufficient current but, at 1.2 V instead of 1.5 V each, the effect is much the same. *Cycletouring* Jan. 1986, p. 31

> Clock version has high capacity NiCad battery—never needs replacing!
> *Amiga User International* May 1990, p. 99

niche /niːʃ/, /nɪtʃ/ *noun* 〰️

In business jargon, a position from which an entrepreneur is able to exploit a gap in the market; a profitable corner of the market.

A specialized figurative sense of *niche* (literally 'a recess'), similar to *corner* in its business sense.

This sense of *niche* was first used by Frederik Barth in his book *The Role of the Entrepreneur in Social Change in Northern Norway* (1963). In the late seventies and eighties it gave rise to a number of compounds and derivatives, including **niche advertising**, **niche analysis**, **niche business**, and **niche marketing** (all referring to the exploitation of *niches*), **niche player** (a person who exploits a *niche*), **nichemanship** (the practice or technique of exploiting a *niche*), and the verb **niche market**.

> The only sensible strategy for non-bank financial institutions is nichemanship.
> *Business Review Weekly* 29 Aug. 1986, p. 56

> At the very time when Campbell was niche marketing trendy vegetables in its bid to be the 'well-being company', it was embroiled in a messy farm labor dispute.
> Warren Belasco *Appetite for Change* (1989), p. 219

> The pizza chains . . . plug valuable niches in the Retail Division between the Berni and other restaurants at one extreme and the pubs and pub-restaurants at the other. *Intercity* Apr. 1990, p. 17

> But if you had a real niche fund, say a French authorised second section oil fund for instance, then you could raise interest from foreign investors who wanted into that niche.
> *European Investor* May 1990, p. 10

> The areas of assistance available through the program include technology transfers, OEM agreements, distribution networks, market niche analysis for products and technologies, joint ventures, mergers, and acquisitions. *UnixWorld* Jan. 1991, p. 157

Nikkei /'nɪkeɪ/ *noun* 〰️

Used attributively in **Nikkei index**, **Nikkei (stock) average**, etc.: an index of the relative prices of representative shares on the Tokyo Stock Exchange (also known informally as the **Nikkei Dow** (or **Nikkei Dow Jones**) **average**).

A borrowing from Japanese; it is formed from the initial syllables of the first two words of *Nihon Keizai Shimbun* 'Japanese Economic Journal', the title of Japan's main financial newspaper, where the index is compiled and published (compare FOOTSIE).

The Tokyo Stock Exchange calculated its own stock average from 1949; this work was taken over by the *Nihon Keizai Shimbun* in 1974. In the late seventies and eighties Western economic and financial sources started to publish figures from the *Nikkei index* and *Nikkei* was frequently mentioned in television and radio reports, bringing the word into popular use alongside *Footsie* and *Dow*. Like *Dow Jones*, *Nikkei* is sometimes used on its own as a short form of *Nikkei average*, etc.

> A major aim of the $90 million fund is to negotiate the region's sky-high p/e multiples and towering 28,000 Nikkei Dow without giving its investors nosebleeds. *Financial World* 20 Sept. 1988, p. 51

> The Nikkei average plummeted 1,978.38, or 6.6 per cent, to close at its low for the day of 28,002.07—its steepest decline since just after New York's Black Monday crash in October 1987, when the Nikkei dropped 3,936.48 points. *Financial Times* 3 Apr. 1990, p. 41

Nilkie ⚫ see DINK

NIMBY /'nɪmbɪ/ *acronym* Frequently written **Nimby** or **nimby** 🌱 🏠

The initial letters of the slogan 'not in my back yard', expressing objection to the sit-ing of something unpleasant, such as a nuclear waste dump, in one's own locality (al-though, by implication, not minding this elsewhere). Hence as an adjective, having the attitude that such unpleasant developments should not be allowed in one's own neighbourhood; as a noun, a person with this attitude, a protester against local de-velopments.

An acronym, perhaps coined with pronounceability in mind. It very quickly acquired its own grammatical status as an adjective and noun.

The abbreviation originated in the US as a derogatory label for the anti-nuclear movement, and is attributed to Walton Rodger of the American Nuclear Society. In its earliest usage (around 1980), it was simply an abbreviated form of the slogan itself, but it soon came to be used as an adjective (especially in **Nimby syndrome**), to describe an attitude increasingly prevalent both in the US and in the UK. In the UK it was widely used as a noun in connection with reports in 1988 of the then Environment Secretary Nicholas Ridley's opposition to hous-ing developments near his own home. The noun can have the plural *Nimbies* or *Nimbys*, the first attesting to its acceptance as a common noun in the language, subject to the morphological rule that words in *-y* form their plural in *-ies*, the second remaining faithful to the original slogan's initial letters. Derivatives such as **Nimbyism** and **Nimbyness** are sometimes found.

> He simultaneously made clear his belief that all waste disposal options should be properly examined and expressed unalloyed support for the government's nuclear expansion plans. It would be hard to find a more classic and indefensible example of the NIMBY . . . syndrome.
>
> *New Statesman* 7 Mar. 1986, p. 11

> Nicholas Ridley's embarrassment over revelations that he has on several occasions objected to proposed developments . . . near his Cotswolds home shows that there may be a closet Nimby . . . in all of us.
>
> *Independent* 16 June 1988, p. 26

nineteen ninety-two /ˌnaɪntiːn ˌnaɪntɪˈtuː/ *noun* Usually written **1992** 〰 🏠

The date for the completion of a SINGLE MARKET in the EC, often used allusively to refer to the single market itself or to one or more of the characteristics of the Euro-pean economy that would result from it.

The year in which the changes were to be implemented fully; actually, the single market was not to be complete until the end of the year, so 1993 would be the first year in which its full effects would be felt.

For history, see SINGLE MARKET. *1992* was the focus of the British Department of Trade and In-dustry's advertising campaign to prepare businesses and individuals for the single market, and thus became a term with more currency than *single market* itself.

> With 1992 just around the corner, Eisner and the rest of his 'Yo-team-let's-go' management will be eagerly looking to Disneyize Europe and then the rest of the world. *Broadcast* 18 Aug. 1989, p. 10

> Over the past five years there has been a new renaissance, as Eurosclerosis was replaced by the excitement of the 1992 programme. *European* 11–13 May 1990, p. 23

> As 1992 looms closer and cross-border deals become increasingly important, we do have an ace up our sleeve: a knowledgeable European network. *World Outside: Career Guide* 1990, p. 94

ninja /'nɪndʒə/ *noun* and *adjective* Also written **Ninja** ⚔ 🌿

noun: A Japanese warrior trained in **ninjutsu**, the art of stealth or invisibility, which was developed in feudal times in Japan and later practised more widely as a martial art.

adjective: Of, belonging to, or characteristic of the ninjas or their techniques.

A direct borrowing from Japanese, in which it is a compound word meaning 'practitioner of stealth', made up of the elements *nin* 'stealth' and *ja* 'person'.

Ninjutsu is an ancient art in Japan—it was practised by the warriors employed by feudal war lords for espionage and assassination—but the words *ninja* and *ninjutsu* were hardly used in English-language sources before the seventies. A rare use in spy fiction comes in Ian Fleming's *You Only Live Twice* (1964):

> My agents are trained in one of the arts most dreaded in Japan—*ninjutsu*... They are now learning to be *ninja* or 'stealers in'.

The rise of interest in oriental martial arts in the seventies meant that some Westerners became interested in the history of the *ninjas* and started to try to emulate them. *Ninjas* also began to figure in role-playing and fantasy games. What brought the words *ninja* and *ninjutsu* into popular use, though, was the commercial success in the late eighties of the TURTLES (whose full name, in the US at least, was *Teenage Mutant Ninja Turtles*).

> I'm inside a recreated Japanese ninja training hall—on the walls a collection of exotic chains, knives, swords, whips, staffs, and other sadistic tools that would make a hardened dominatrix blush.
> *Omni* Mar. 1990, p. 64

> The first level [in a computer game] starts off with Ninjas suspended from trees.
> *CU Amiga* Apr. 1990, p. 28

> There is far more to the graphic novel than recording the exploits of Donatello and his ninja friends.
> *Times Educational Supplement* 2 Nov. 1990, Review section, p. 1

Ninja Turtle 🏏 🎮 see TURTLE

NIREX /'naɪərɛks/ *acronym* Also written **Nirex** 🔌

Short for **Nuclear Industry Radioactive Waste Executive**, a body set up to oversee the disposal of nuclear waste in the UK.

Formed from letters taken from the name *Nuclear Industry Radioactive waste EXecutive*.

NIREX, a government-sponsored body, was established in 1982 by a group of English and Scottish generating boards and nuclear energy authorities. Its brief includes the development of plans to build a nuclear waste repository for the UK by the year 2005.

> Environmentalists are angry that NIREX has not considered as an option the long-term storage of nuclear waste above ground. *New Scientist* 14 Jan. 1989, p. 30

NMR ⊗ 💻 see MRI

no-alcohol beer 🏏 see NAB

noise footprint 💻 see FOOTPRINT

non-ism /'nɒnɪz(ə)m/ *noun* 🏏 🍴

A policy or lifestyle of avoiding all activities and substances (foods, drink, drugs, etc.) which might be harmful to one's mental or physical health; an extreme form of total abstention.

Formed by combining the prefix *non-* 'not' with the suffix *-ism* to make a word which does not, strictly speaking, contain a root (but perhaps this emphasizes the point: it is a non-word).

The increasing preoccupation in the late eighties with health and fitness on the one hand, and with prevention as preferable to cure on the other, produced a feeling not infrequently expressed that it had become difficult to consume or do anything without worrying about its possibly deleterious effects. *Non-ism* is a name for the most extreme response to the wealth of information on preventive medicine; a person who practises it is a **non-ist**. The word was brought into the news by reports in 1990 of a Boston psychiatrist whose son had given up almost all pleasures; he seemed to typify a growing trend in US society.

His son . . . is stuck in a limbo of non-ism . . . He gave up drinking, drugs and caffeine, meat, sugar, dairy and wheat products, and sex. He is depressed and lethargic. 'He's a pleasure anorexic,' said his father. *New York Times* 27 May 1990, p. 22

The rule . . . for the 1990s . . . is to define yourself through denial . . . This new creed of 'non-ism', as the academics are calling it, draws on the fashion for abstention from drink, tobacco, drugs . . . and all other contaminants. *The Times* 13 June 1990, p. 11

noov /nu:v/ *adjective* Also written **noove**

In slang, a member of the *nouveaux riches*; someone who has recently come into money and thereby moved up to a higher socio-economic bracket.

Formed by abbreviating *nouveau* (itself sometimes used as a short form for *nouveau riche*), re-spelling the resulting word to reflect its anglicized pronunciation; English speakers might be tempted to pronounce *nouv* /naʊv/.

Noov and *nouveau* became popular slang abbreviations of *nouveau riche* in the late seventies or early eighties.

A neighbour of ours . . . A real noove, pretending to be a farmer.
Susan Moody *Penny Post* (1985), p. 31

The pupils: 45 per cent sons of Old Etonians . . . Also largish element of noovs to keep up academic standards and/or provide useful business contacts. *The Times* 7 Oct. 1986, p. 14

notebook see LAPTOP

nouvelle /nu:'vɛl/ *adjective*

Of a restaurant, food, etc.: using or characterized by **nouvelle cuisine**, a style of cooking, originally from France, in which simplicity, freshness, and aesthetically pleasing presentation are emphasized.

Formed by abbreviating *nouvelle cuisine*, literally 'new cooking' to its first word, 'new'.

Nouvelle cuisine became fashionable outside France in the late seventies and early eighties, offering as it did a completely different approach from the elaborate sauces and richness of traditional French cooking. *Nouvelle* also became a fashionable adjective in the second half of the seventies to describe cooking that incorporated any of the principles of *nouvelle cuisine*, such as lightness, short cooking times, artistic presentation (some of the *nouvelle* dishes were likened to works of art, designed only for photographing and not for eating), or small helpings (since the bare surface of the plate had a part to play in framing the artistic arrangement of the food). All of these characteristics were the object of criticism as well as praise, so the adjective *nouvelle* could be either approving or derogatory, depending on the view of its user.

Plates arrive from the kitchen under silver covers that are removed with a flourish to reveal distinctly *nouvelle* still-life-like arrangements on those handsome basket plates popularized by Michel Guérard. *Gourmet* July 1981, p. 90

One establishment we visited served every dish flanked by the same ludicrously inappropriate clutter: a frilly lettuce leaf pinned down by a couple of hefty spring onions, a pallid slice of kiwi fruit and a strawberry. Oh nouvelle cuisine, what have you spawned! *Country Living* Aug. 1990, p. 68

nuclear device see DEVICE

nuclear-free see -FREE

nuclear winter /ˌnju:klɪə 'wɪntə(r)/ *noun*

A prolonged period of extreme cold and darkness which, according to some scientists, would be a global consequence of a nuclear war because a thick layer of smoke and dust particles in the atmosphere would shut out the sun's rays.

Formed by compounding: an artificial *winter* caused by a *nuclear* conflict.

The theory of the *nuclear winter* was formulated by five American scientists, originally for a conference in Washington DC in October–December 1983, and popularized particularly by one of them, Carl Sagan, who attributes the coinage to another, Richard Turco. Writing in the *Washington Post*'s *Parade* magazine at the time of the Conference, Sagan describes their research as follows:

> We considered a war in which a mere 100 megatons were exploded, less than one per-cent of the world arsenals, and only in low-yield airbursts over cities. This scenario, we found, would ignite thousands of fires, and the smoke from these fires alone would be enough to generate an epoch of cold and dark almost as severe as in the 5000-megaton case. The threshold for what Richard Turco has called The Nuclear Winter is very low.

The lowering of temperatures and lack of light caused by radioactive debris in the atmosphere would, according to this theory, destroy the cycles of nature and ruin crop growth, so that any human survivors of a nuclear exchange would soon run out of food. The theory of the *nuclear winter*, which was widely discussed in the mid eighties, had an important influence on the military strategy of the superpowers in the second half of the decade. It possibly contributed to the spirit of disarmament which marked the late eighties and early nineties, since it showed a nuclear first strike to be a potentially suicidal act on the part of any country using it, whether or not it led to a nuclear exchange. As the theory was refined it became clear that the global winter scenario was perhaps an exaggeration, and it was supplemented by the idea of a **nuclear autumn**, in which temperatures would drop significantly, altering the climate with agricultural consequences, but not causing global famine. The underlying principle was raised again in a non-nuclear setting in 1991, when Iraqi troops set light to hundreds of oil wells in Kuwait before leaving at the end of their occupation of the country, and smoke from these oil fires, blocking the sun's rays, had a similar effect on local temperatures and light levels.

> Downwind from Chernobyl, the first faint chill of a nuclear winter has caused . . . shivers of anxiety.
>
> *The Times* 20 May 1986, p. 14

> Calculations that the aftermath of a nuclear war might resemble 'nuclear autumn' rather than 'nuclear winter' are probably wrong.
>
> *New Scientist* 1 July 1989, p. 43

nuke /njuːk/ *transitive verb* ▧

In US slang, to cook or heat (food) in a microwave oven.

A transferred use of the slang verb *nuke*, which since the late sixties has meant 'to attack or destroy with nuclear weapons'. The transfer is explained by the fact that both nuclear bombs and microwave ovens generate electromagnetic radiation (although of very different kinds!).

> 'This potato', he said listlessly, 'is undernuked.' Half a pulse later and it was dropped back onto his plate like a spent cartridge. Now it was overnuked.
>
> Martin Amis *London Fields* (1989; paperback ed. 1990), p. 400

> It was a perfect night to nuke some popcorn and curl up in front of a Duraflame.
>
> *New Yorker* 11 Dec. 1989, p. 14

numeric keypad ▧ see KEYPAD

nyaff see NAFF[1]

nybble ▧ see NIBBLE

O

offender's tag see TAG¹

off-roading /ˌɒfˈrəʊdɪŋ/ noun

Driving on dirt tracks and other unmetalled surfaces as a sport or leisure activity; also known more fully as **off-road racing**.

Formed from the adjective *off-road* (which dates from the early sixties) and the action-noun suffix *-ing*, perhaps by abbreviating *off-road racing*.

Off-roading originated on the West coast of the US in the late sixties, when recreational vehicles such as the *beach buggy* were first in fashion among young people. From California it spread across the US as a more serious sport, and from the late seventies and early eighties was increasingly practised in an organized way outside the US as well. An **off-roader** is both a vehicle used in *off-roading* and a person who takes part in it (but see also MOUNTAIN BIKE). Although *off-roading* began as *off-road racing*, racing is not an essential element of the sport, which focuses more on the enjoyment of driving away from the traffic and pollution of metalled roads.

> A serious off-roader is more interested in what a vehicle can do once its wheels start rolling.
> *Outdoor Life* (Northeast US ed.) Oct. 1980, p. 29

> Unsurfaced roads . . . are becoming muddy death traps for other countryside users as off-roading becomes an increasingly organised leisure activity. *Daily Telegraph* 13 Jan. 1988, p. 25

> The new all-drive platform is aimed at the rustbelt market, not at serious off-roaders, so the MPV 4WD doesn't sit six feet off the ground or ride on giant knobby tires. *Car & Driver* Sept. 1989 p. 131

oilflation ⌁ see KIDFLATION

oink ⧌ see DINK

on-and-on rap ♪ 🔲 see RAP

onsell /ɒnˈsɛl/ transitive verb Also written **on-sell** ⌁

To sell (an asset, especially one recently acquired) to a third party, usually for profit.

Formed from the phrasal verb *sell on*, by converting the adverb *on* into the prefix *on-*. This process of converting a phrasal verb into a prefixed one is quite common in verbs used in business: compare *onlend* (a formation of the seventies), *outplace* (see OUTPLACEMENT), and *outsource*.

This is a piece of financial jargon of the late seventies and eighties that has acquired some limited currency outside the financial markets as well.

> The Euro CP dealers, in bidding for paper, will most likely remain exposed to interest rate movements overnight, since they cannot onsell it until the following morning.
> *Euromoney* (Supplement) Jan. 1986, p. 79

> We will buy some works by contemporary artists this year and may on-sell them if it means we can buy some better examples. *Business Review Weekly* 19 Feb. 1988, p. 98

on your bike see BIKE

optical disc 🔲 see CD

option card¹ ⚡ see CARD¹

option card² 🔲 see CARD²

Oracle /'ɒrək(ə)l/ *noun* ❎ 🔲

In the UK, the trade mark of a teletext system (see TELE-) originally operated by the IBA.

A figurative use of *oracle*, based on the popular transferred sense of the phrase *consult the oracle*, 'to seek information from an authority': the purpose of the service is to provide information on the television screen.

Oracle was introduced in the mid seventies and is now a standard option on most new television sets in the UK. The name has been used in other trade marks, especially in information technology.

> Ceefax and Oracle are both teletext systems. At present teletext is limited to the amount of information that may be transmitted on the two available lines on a television screen, but it is a free service.
> *Bookseller* 29 Mar. 1980, p. 1430

orbital /'ɔːbɪt(ə)l/ *adjective* 🔅

In British youth slang, of a party (especially an acid-house party: see ACID HOUSE): taking place beside or near the M25 London orbital motorway.

The word is taken from the official name of the M25, *London orbital motorway*.

Orbital parties were a phenomenon of 1989–90, taking the place of WAREHOUSE parties in popularity among London's youth. They probably represent a passing fashion.

> If you've been to any of the major house parties, you'd know them by sight, if not by name. Their multiscreen projections of slides and film loops have featured in orbital parties, at the Astoria and Heaven, in Rifat Ozbek's 1988/89 fashion shows, and at Energy's recent Docklands all-dayer.
> *The Face* June 1990, p. 18

organic /ɔː'gænɪk/ *adjective* 🌱 ❎

Of food: produced without the use of chemical fertilizers, pesticides, etc., by adding only organic material to the soil.

Organic in this sense was originally applied to the fertilizers themselves, signifying that they were derived from living matter, unlike the *inorganic* chemical fertilizers. The adjective was then applied to the method of farming in which *organic* fertilizers were used (from about the early forties onwards), and finally to the produce of this method of farming. A term such as *organic vegetables* therefore represents two stages of abbreviation from the more accurate but impossibly cumbersome *vegetables grown using a method of agriculture employing only organic materials*. Such vegetables are *organic* in the sense that they contain no traces of the inorganic chemicals often used in vegetable production, but the term *organic vegetables* rightly strikes some people as a tautology, since all living things are *organic*.

Organic was first applied to the produce of organic farming methods in the seventies, when environmental concerns began to gain a place in the public consciousness. However, *organic* produce was considerably more expensive than that produced by modern methods and for some time it was considered to be the province of health-food freaks (an attitude which had prevailed in developed countries when organic farming was first tried in the forties as well). However, demand for *organic* produce grew markedly in the eighties, as did awareness of the meaning of the term; this was largely because of the success of the GREEN movement and growing public concern about the potentially harmful effects of agricultural chemicals (fed by such scares as the one over ALAR in apples). By the end of the eighties **organically** grown fruit and vegetables were regularly on sale alongside those produced by mainstream farming techniques, and it was even possible to buy *organic* meat (that is, meat from animals that had been fed only on *organic* produce).

> High-tech greens who like the way microwaves cook their organic veg could find the new foodprobe ... worth investigating. *Practical Health* Spring 1990, p. 9

> More recently, the desire for organically grown, pesticide-free produce has created a new kind of city garden where food plants are mixed with flowers. *Garbage* Nov.–Dec. 1990, p. 36

organizer /'ɔːɡənaɪzə(r)/ *noun* ▓

Something which helps a person to organize (objects, appointments, papers, etc.); a container which is arranged in sections or compartments so as to make systematic organization of the contents easier.

A sense shift involving abbreviation of a longer phrase; an *organizer* would normally be a person who organizes, but here it is the object which helps a person to *organize*, that is, a product *for the organizer*. No doubt the manufacturers of these products would be happy for *organizer* in this sense still to be interpreted as though the organization were done for its owner by the product, but as Stephanie Winston has pointed out in her book *Getting Organized* (1978):

> You're bound to be disappointed if you buy lots of boxes, containers, and 'organizers' in the wistful hope that they will somehow *make* you organized. They won't.

Products described as *organizers* (often with a preceding word describing the thing to be organized, as, for example, **desk organizer**) started to appear on the market in the late sixties. The fashion for *organizers* in the office was followed in the late seventies by the idea of the **organizer bag**, a handbag with many different compartments and pockets. In the eighties, when *getting organized* was synonymous with *getting on*, *organizer* was often used as a short form for PERSONAL ORGANIZER, the generic term for sectioned notebooks like the FILOFAX which became so fashionable in the early eighties for organizing one's life. Perhaps trying to jump on the bandwagon, advertisers tended to overwork the word *organizer* in the mid and late eighties: any piece of furniture with shelves or compartments, or even a simple box file was enthusiastically transformed into an essential *organizer* by the copywriters. The word *organizer* is often used attributively in naming these products (following the model of *organizer bag*), in **organizer unit** etc.

> Our gift to you—an organizer unit to store your player and discs. *New Yorker* 4 June 1984, p. 1

> It has one shelf and two small plastic 'organisers' to hold all your baby's toiletries. *Practical Parenting* Apr. 1988, p. 8

> The desk-sized professional organizer now makes up 10 per cent of sales, and a small pocket organizer has been launched. *The Times* 7 Apr. 1989, p. 25

OTE /əʊtiːˈiː/ *abbreviation* 〰

Short for **on-target** (or **on-track**) **earnings**, a level of pay at which a person is earning to full potential by receiving a basic salary and commission representing top performance.

The initial letters of *On-Target* (or *On-Track*) *Earnings*.

OTE began to appear as an abbreviation in job advertisements in the second half of the eighties; it is really a shorter and euphemistic way of saying 'earning potential with commission'. Unlike *profit-* or *performance-related pay* (PRP), it is dependent upon the individual's performance rather than the company's.

> Computers. £30,000 Basic. £60,000 OTE. *Sunday Telegraph* 1 July 1990, section A, p. 16

otherly abled ▐▌ see ABLED

OTT /əʊtiːˈtiː/ *abbreviation*

In slang, short for **over the top**: (especially of a person, or a person's appearance, manner, opinions, etc.) extreme, exaggerated, outrageous; characterized by excess.

The initial letters of *Over The Top*; this phrase began in the sixties as a colloquial verbal phrase *go over the top*, 'to go beyond reasonable limits' and was itself based on the army metaphor of going *over the top* of the trenches and into battle.

Over the top began to be used as an adjectival phrase among young and middle-aged people in the early eighties and was soon being abbreviated to *OTT*, even in print. It is mentioned as a SLOANE RANGER expression in the *Official Sloane Ranger Handbook* (1982), but is just as likely to be found in the popular music papers or youth magazines as in writing for or by the upper classes. Anything that seems overdone or offends a person's sense of proportions and propriety can be described as *OTT*, but it is used especially of people or of things in which a human agent has been at work to stir up (sometimes only mock-serious) outrage.

> I think that's puritanical. It's totally over the top. *Green Magazine* Dec. 1989, p. 38

> Fans will be happy enough to get half a dozen previously unreleased tracks, including a typically OTT Watkins offering. *Folk Roots* Aug. 1990, p. 35

out /aʊt/ *transitive verb*

To expose the homosexuality of (a prominent or famous person); to force (someone) to come 'out of the closet'. Also as an action noun **outing**, the practice or policy of making such a revelation, especially as a political move on the part of gay rights activists; agent noun **outer**.

Formed by turning the adverb *out* (as in the phrase *come out* (that is, *out of the closet*), meaning 'to make public one's homosexuality') into a verb. The transitive verb *out* already existed in a number of more general senses.

The practice of *outing*, also known as *tossing*, was first brought to public attention in the US in early 1990, when public revelations about the sexual orientation of some famous people were used as a political tactic by gay rights activists; they were concerned mainly about lack of support for the victims of Aids, even among those who were closet gays. The word *out* and its derivatives very quickly acquired a currency among gay groups in the UK as well; wherever it was practised, *outing* caused considerable controversy. The New York gay magazine *OutWeek* became particularly associated with *outing*, revealing the homosexuality of a number of prominent film stars and public figures who, it said, were betraying the cause of gay rights by remaining silent.

> Instead of tossing or outing this congressman, I ... called to his attention the hypocrisy that he had been legislating against gays. *Los Angeles Times* 22 Mar. 1990, section E, p. 23

> This [i.e. Aids] is the new factor that gives outing both its awful appeal and its power and, most precisely, exposes the motives of the outers as terrorism. *Sunday Times* 6 May 1990, section C, p. 6

outlaw technologist see CYBERPUNK

outplacement /ˈaʊtˌpleɪsmənt/ *noun* Also written **out-placement**

Assistance in finding a new job after redundancy, given to an employee by the employer making him or her redundant or by a special outside service; hence, euphemistically, the act of making someone redundant, 'dehiring'.

Formed by adding the prefix *out-* to *placement*; *placing* (a person) *out* rather than within one's own staff.

Outplacement has been a standard term in the US business world since the early seventies, but only became current in the UK in the mid eighties. The verb **outplace** has a similar history to *outplacement*; derivatives such as the adjective **outplaced** and the agent noun **outplacer** (a person or firm that does the *outplacement*) arose in the early eighties.

> If you ever do get canned . . . you might count yourself lucky to be placed in the hands of the out-placers. *Forbes* 19 Jan. 1981, p. 77

> Career counselling—or 'outplacement', as the service is called when it is pitched instead at companies that are trying to chop senior executives as mercifully as possible.
> *Sunday Times* 26 July 1987, p. 69

> Up to 150 staff will be 'outplaced', with the group administrative services unit and the professional services unit (lawyers) being hardest hit. *Financial Review* (Sydney) 28 Aug. 1987, p. 18

ozone /ˈəʊzəʊn/ *noun* 🌳

A colourless unstable gas with a pungent smell and powerful oxidizing properties, which makes up the **ozone layer**, a layer of naturally occurring ozone in the earth's upper atmosphere that absorbs most of the sun's harmful ultraviolet radiation. Used especially in compounds to do with environmental concerns about the ozone layer:

ozone depletion, a reduction of ozone concentration in the ozone layer caused by atmospheric pollution and the build-up in the atmosphere of **ozone-depleting** chemicals such as CFCs;

ozone-friendly, of a product, material, etc.: not containing chemicals which harm the ozone layer (see also -FRIENDLY);

ozone hole, an area of the ozone layer in which serious ozone depletion has occurred; also used as a synonym for *ozone depletion*.

Concern about the damaging effects of modern industrial chemicals on the *ozone layer* was expressed by environmentalists as long ago as the seventies, but most of the other terms defined here came to public attention only in the mid eighties, as environmental concerns were in general brought to prominence by the GREEN movement. Public awareness of the potentially damaging effects of creating an *ozone hole* was possibly heightened by the results of research which linked overexposure to ultraviolet radiation with skin cancers, although the environmental effects of a large ozone hole would be so devastating to weather systems, agriculture, and animal life on the planet that some argued that the cancer risk was a minor concern. Other terms using *ozone* in this context include **ozone-benign**, **ozone destroyer** (and **ozone destruction**), **ozone safe**, and *ozone-unfriendly* (see UNFRIENDLY²).

> Scientists expected from some mathematical models that the next very large ozone hole over Antarctica would occur in 1990. *New York Times* 23 Sept. 1989, p. 2

> Many ozone-friendly aerosols use hydrocarbons as the propellant; these have a higher risk of ignition or explosion if misused. *Which?* Sept. 1989, p. 431

> HCFC–123 . . . has the potential to break down some ozone, although its ozone depletion potential (ODP) has been calculated at only 0.02. *New Scientist* 15 Sept. 1990, p. 34

> First of all, polystyrene loose fill is not made with ozone-depleting CFCs or HCFCs, but with hydrocarbons. *Garbage* Nov.–Dec. 1990, p. 73

ozone-unfriendly 🌳 see UNFRIENDLY²

••

P

package /'pækɪdʒ/ *noun*

In computing jargon, a closely related set of programs, usually all designed for the same purpose and sold or used as a unit.

A specialized use of the figurative sense of *package*, 'any related group of objects that is viewed or organized as a unit'.

The word *package* has been used in computing for at least two decades, but it was the appearance on the market in the early eighties of large numbers of commercial software packages for home computers and PCs that brought the word into popular usage. To the lay user, the commercial software *package* can appear to be a single program, since it contains all the software required to carry out a single function (such as word processing or statistical analysis) and there is usually a user interface which draws together the various programs into a single menu of functions.

> The finished animation was then imported into Macromind Director, a 2D moving graphics package, where it was layered over a textured background. *Creative Review* Mar. 1990, p. 52

> It's the first UNIX spreadsheet package to take advantage of windowing, mouse support, dialog boxes, and pulldown menus. *UnixWorld* Apr. 1990, p. 145

Pac-Man¹ /'pækmæn/ *noun* Also written **PacMan** or **Pac-man**

The trade mark of an electronic computer game in which the player guides a voracious blob-shaped character through a maze, gobbling up lines of dots on the way and avoiding being eaten by opposing characters. Also, the name of the central character, represented on the screen as a yellow circle with a section missing for the mouth (similar to a pie-chart from which one 'slice' of the pie has been removed).

Like most trade marks, this one is of uncertain origins; *Pac* is probably a respelling of *pack*, referring to the fact that the little creature's whole object in life is to *pack away* (eat) everything that gets in its way.

Pac-Man appeared on the market in October 1980, at the height of a boom in video games in the US, and proved one of the most successful and popular of the games then available in video arcades. Surprisingly it was not registered as a trade mark in the US until 1983, by which time it was widely available in other countries and the video arcade market was beginning to wane. The *Pac-Man* character had become a well-known symbol in its own right by the mid eighties—giving rise to the figurative sense in PAC-MAN²—and even acquired a family (including **Pac Baby** and a cat) in versions for home video use. The idea of the game was copied in a computer VIRUS in the late eighties (see the *Network World* quotation below).

> Among the viruses now invading or about to invade systems are: The PacMan virus. This one shows up on Apple Computer, Inc. Macintosh systems. The user gets to watch as PacMan eats the file on the screen. *Network World* 6 Feb. 1989, p. 85

> 1981: Joystickmania was led by Pac-Man, which gobbled up nearly $1 billion—25 cents at a time—in a nation suddenly hip-deep in video arcades. *Life* Fall 1989, p. 63

Pac-Man² /'pækmæn/ *noun*

Used attributively (in **Pac-Man defence** or **Pac-Man strategy**) of a company's response to a take-over bid: involving a counter-bid in which the company facing the take-over threatens to take over the 'predator' instead.

A figurative use of PAC-MAN¹: the situation is likened to a game of *Pac-Man*, in which the central character can, in certain circumstances, gobble up the monsters that threaten to devour it.

The *Pac-Man strategy* was first so named in 1982—less than two years after the video game came on to the market—bearing witness to the way in which the little yellow gobbler had caught the imagination of the general public. The name was coined by New York investment bankers and first reported by Deborah A. De Mott in the *Wall Street Journal* in August 1982. By the end of 1982 it had been used in a number of markets outside the US as well.

> Martin Marietta's strong countermove is in line with a budding takeover defense plan that Wall Street arbitragers and investment bankers alike yesterday were calling 'the Pac-Man strategy'. 'That's where my client eats yours before yours eats mine,' a merger specialist at one major investment banking firm said. *Wall Street Journal* 31 Aug. 1982, p. 3

> The board saw the tactic as an ASCAP, an assured second-strike capability; someone else called it a Pac-Man defence, after the video gobblers. *Courier-Mail* (Brisbane) 26 Dec. 1987, p. 16

paintball /'peɪntbɔːl/ *noun* 🎌 🎖

A type of war-game practised as a sport or hobby, in which teams of combatants in military clothing attempt to capture the opposing team's flag, eliminating members of the opposition by firing pellets of brightly coloured paint from a type of airgun; also, the pellet of paint used in this pastime.

Formed by compounding: the bullet is replaced by a *ball* of *paint*, which bursts on impact to stain the clothing of the opponent.

The sport of *paintball* began in the US in the early eighties, but *paintball* did not, it seems, become its established name until about the middle of the decade. In the second half of the eighties it became an increasingly popular leisure activity in the US and the UK, an international association was formed for the sport, and a number of magazines were published on this subject alone. The *paintball* itself, which is fired from a gun using carbon dioxide as a propellant, is a thick-skinned gelatin capsule filled with paint, which may be of any colour; its purpose is to 'tag' a player as having been hit, since it bursts on impact and leaves a bright-coloured stain on the opponent's clothing. Protective eyewear prevents any injury from the *paintball* if it hits the face. Some people saw the rapid growth of interest in *paintball* as a worrying sign of an increasingly violent and militaristic ethos among the young (see RAMBO and SURVIVALISM), but its followers emphasized the fact that it was actually a very safe sport, teaching teamwork and strategic thinking. Use of the guns was nevertheless declared illegal in the UK in October 1991. The word *paintball* is often used attributively, in **paintball combat**, **paintball (war)-game**, and **paintball team**. A player of the sport is sometimes called a **paintballer**.

> Tucker has found a way to shoot people by playing a war game, Paintball, in which he and squads of weekend guerillas stalk each other through the woods with air guns that fire blobs of paint instead of bullets. *Chicago Tribune* 18 Dec. 1987, section 5, p. 3

> Five years since their introduction into Britain, the industry of paintball wargames continues to expand, attracting grown men and women back to a more sophisticated version of the games they once played as children with toy guns in their gardens. *Guardian* 3 July 1989, p. 20

> Paintballers come from all walks of life and we share a love of excitement and the open air. *Paintball Games* Oct. 1989, p. 5

palm-top 💻 see LAPTOP

paper /'peɪpə(r)/ *noun* 🔪

In the slang of drug users, a packet containing a dose of a drug; in recent use, especially a packet of ICE.

A piece of paper folded up as a container or wrapper for something (such as a medicinal powder) has been called a *paper* for many centuries (the earliest examples in English go back to the sixteenth century); it is a logical step—admittedly after a long interval—to this more specialized use, even though in practice the drugs may be in small bags rather than folded pieces of paper.

A folded piece of paper containing some illicit drug has been known as a *paper* since illegal drug-taking first became a problem in the thirties; by the sixties the word was being used for any packet or dose of drugs, whether in a folded paper or not; a heroin pusher was known as a **paper boy**. When the drug *ice* first came on the market in 1989, a one-tenth gram dose immediately became known as a *paper* even though there is no evidence that it was ever distributed in folded paper.

> In Hawaii, one-tenth gram or 'paper' of ice costs $50 and usually produces an eight- to 30-hour high.
> *Boston Globe* 8 Dec. 1989, p. 3

parasailing /'pærəseɪlɪŋ/ *noun* Also written **para-sailing** ⊠

The sport of gliding through the air attached to an open parachute and towed by a speedboat.

Formed by combining the first two syllables of *parachute* with *sailing*, probably after the model of PARASCENDING.

Parasailing developed at the very end of the sixties but did not become established as a sport until the second half of the seventies. Essentially, *parasailing* is an airborne variation on water-skiing; it differs from parascending in that the person being towed remains attached to the tow boat rather than letting go once the right height has been reached. The verb **parasail** has been back-formed from *parasailing* and can be used transitively or intransitively; a person who does this is a **parasailer** or **parasailor** (the spelling variation displaying uncertainty as to whether verbs ending in *-sail* should form their derivatives in the same way as *sail*: compare *boardsailer* and *boardsailor* under BOARDSAILING).

> There are glass-bottomed boats, Canadian canoes, sailboats and windsurfers—you can even go parasailing.
> *Meridian* Spring 1990, p. 42

parascending /'pærəsɛndɪŋ/ *noun* ⊠

A variation on the sport of parachuting, in which participants are first towed by a motor vehicle or speedboat while wearing the open parachute, so as to gain sufficient height from which to descend.

Formed by telescoping *parachute* and *ascending* to make a blend.

Parascending was an earlier innovation than PARASAILING, having developed in the sixties, at first as a safe variation on parachuting which dispensed with the complications of making a parachute jump. By the mid seventies it was becoming established as a sport in its own right, and during the eighties was among the group of fast-growing action sports that managed to increase their popular appeal. The verb **parascend** was back-formed from *parascending*; a person who practises the sport is a **parascender**.

> New amendments to the Air Navigation Order and the revision of CAP 403 'Code of Conduct for Air Displays' now encompass the modern features in aviation, such as microlights and parascending which were not previously mentioned.
> *Air Display* Dec. 1988–Feb. 1989, p. 3

Parentline ⚓ see -LINE

passive smoking /ˌpæsɪv 'sməʊkɪŋ/ *noun* ⊗

Involuntary inhalation of tobacco smoke from smokers in one's immediate vicinity or with whom one shares an environment.

Formed by compounding: *smoking* which is *passive* rather than *active*. The English term may be modelled on the German compound word *Passivrauchen*.

Passive smoking was first recognized and named by medical researchers investigating the health hazards of tobacco smoke in the early seventies. The health risks of smoking became clearer and its popularity waned during the seventies and eighties; at the same time the plight of the **passive smoker**, living or working with a heavy smoker and forced to breathe smoke-filled air, gained ever greater popular awareness and sympathy.

The passive smoker is exposed mainly to 'sidestream' smoke given off directly from a cigarette, pipe or cigar.
Scotsman 16 June 1986, p. 11

In recent years scientists have found that passive smoking is a significant hazard for healthy people too. In 1988 the Froggatt Report, the Fourth Report of the Independent Scientific Committee on Smoking and Health, stated that exposure to tobacco smoke increased the risk of lung cancer in non-smokers by up to 30 per cent and may account for several hundred deaths in Britain each year.
Independent on Sunday 29 July 1990, Sunday Review section, p. 51

Patriot /ˈpeɪtrɪət/, /ˈpætrɪət/ *noun* 🎖️

The name (more fully **Patriot missile system**) of a computerized air-defence missile system developed in the US and designed for early detection and interception of incoming missiles or aircraft; also, a missile deployed as part of this system (known more fully as a **Patriot missile**).

A figurative use of *patriot* 'a person who is devoted to and ready to defend his or her country'; the *Patriot missile* is ready to defend the home country from attack by airborne forces.

The *Patriot* system was developed by Raytheon in the US during the late seventies and early eighties; the first *Patriots* were put into service by NATO in Germany in 1985, as a replacement for the Hawk and Hercules systems. The first *Patriot* ever to be fired operationally, however, was in the Gulf War of 1991, when the system was deployed to great effect by allied forces against Iraqi SCUD missiles. The computerized tracking system of the *Patriot* locates incoming missiles, works out their expected trajectory, and if necessary launches an intercepting *Patriot missile*, which 'locks on' to the incoming missile and destroys it in mid air. The name *Patriot* is sometimes used as a proper name, without a preceding article.

The antimissile era has dawned in thunder and flame as wave after wave of Patriots has knocked Iraqi Scuds out of the sky. But the Patriot is just the beginning.
New York Times 5 Feb. 1991, section C, p. 1

Iraq has fired 68 Scud missiles—35 at Israel, 33 at Saudi Arabia. The allies have launched about 130 Patriots against them.
Independent on Sunday 17 Feb. 1991, p. 2

PC /piːˈsiː/ *abbreviation* 💻

Short for **personal computer**, a microcomputer designed for personal office or home use by a single user at any given time; specifically, such a computer designed and marketed by International Business Machines Corporation and known as the IBM PC.

The initial letters of *Personal Computer*.

From 1982 until it was replaced by the PS2 series at the end of the eighties, the IBM PC was the acknowledged standard among 16-bit microcomputers, with the result that the abbreviation was very often used to refer to this particular model. Other computer manufacturers quickly set about copying the *PC*; such a model became known as a **PC clone** (sometimes simply a CLONE) or a **PC-compatible** (also used as an adjective). By the end of the decade, though, with IBM marketing the PS2, *PC* alone was regularly used again for any *personal computer*. A *personal computer* with a hard disc might be described as a **PC XT** (after the appropriate IBM model) and one with 'advanced technology' (using a more advanced chip) as a **PC AT**, on the same principle.

BGL Technology's LaserLeader line of plotter/printer splits the responsibilities for the front-end work and graphics processing between an embedded PC AT and a graphics processor.
UnixWorld Sept. 1989, p. 137

Choose a PC which has . . . a colour EGA (enhanced graphics adaptor) monitor which will be able to display the games and educational software, and has a resolution high enough for your word processing.
Which? May 1990, p. 271

PCB[1] /piːsiːˈbiː/ *abbreviation* 🌳

Short for **polychlorinated biphenyl**, any of a number of chemical compounds which are obtained by adding chlorine atoms to biphenyl and which cause persistent environmental pollution.

The initial letters of parts of the chemical name *PolyChlorinated Biphenyl*.

PCBs were widely used in old electrical transformers, hydraulic and lubricating oils, paints, lacquers, varnishes, and the plastics industry, until they were recognized as very toxic pollutants in the late sixties. They are difficult to dispose of and have been shown to be carcinogenic in people and animals, with the result that production of them was stopped in the US and the UK during the late seventies. What brought them into the public eye in the eighties was the general upsurge of interest in environmental issues; the persistent problem of disposing of the *PCBs* which were so liberally used in the fifties and sixties, before it was realized that they could be so dangerous, has meant that they remain on the green agenda.

> The emergency meeting of 18 scientists . . . called for every effort to be made to reduce the leakage into the environment of an extremely long-lasting and toxic type of pollutant, polychlorinated biphenyls (PCBs). *Independent* 12 Aug. 1988, p. 1

> The otters take in the PCB from the fish that they eat along with other pollutants. *Earth Matters* Summer 1990, p. 4

PCB[2] /piːsiːˈbiː/ *abbreviation* 💻

Short for **printed circuit board**, a flat sheet carrying the printed circuits and microchips in a microcomputer or other microelectronic device.

The initial letters of *Printed Circuit Board*.

A common abbreviation in writing on computing and electronics since the seventies; it is now sometimes used in less technical sources and is included here to distinguish it from the commoner use above.

> If you look inside its workings, you will find the PCB (printed circuit board), with all the chips or ICs (integrated circuits), neatly plugged into it. *Observer* 3 Oct. 1982, p. 21

PCP[1] /piːsiːˈpiː/ *abbreviation* 💉

In the slang of drug users, the drug **phencyclidine hydrochloride**, taken illegally for its hallucinogenic effects.

The initials are said to come from *PeaCe Pill*, an early street name for the drug, although they could as easily come from *PhenCyclidine Pill*.

The drug was introduced as an anaesthetic in the late fifties, but was soon limited therapeutically to veterinary use. It began to be taken illicitly as a hallucinogen in the psychedelic sixties; in the eighties it enjoyed a revival with the new psychedelia of ACID HOUSE. *PCP* has had over 150 street names, some of which are listed in the entry for ANGEL DUST (the most enduring of all of them).

> In parallel with the rise in gang warfare has been the increasing availability of PCP . . . on the street drug-market. *Listener* 7 June 1984, p. 7

> We talked to kids who got stoned on PCP at eight in the morning, just to start the day. *Girl About Town* 30 Jan. 1989, p. 11

PCP[2] /piːsiːˈpiː/ *abbreviation* ⊗

Short for **pneumocystis carinii pneumonia**, a fatal form of pneumonia caused by infection with the *Pneumocystis carinii* parasite, which especially affects the immuno-compromised (such as people with AIDS).

The initial letters of *Pneumocystis Carinii Pneumonia*.

Pneumocystis carinii pneumonia, in which numerous cysts form inside the lung cavity, was first observed and named in the fifties and commonly abbreviated to *PCP* from the mid seventies. It was its rapid spread among people with Aids in the early and mid eighties that brought the name and the abbreviation out of the specialized domain of medical vocabulary and into widespread public use, especially in the US.

> Three months after we'd moved in together, we learned Keith had [Aids]. The tip-over diagnosis was PCP.   Michael Bishop *Unicorn Mountain* (1988; 1989 ed.), p. 61

peace camp /ˈpiːs ˌkæmp/ *noun* 🗎 💥

A camp set up by peace campaigners, usually outside a military establishment, as a long-term protest against the build-up of weapons.

Formed by compounding: a *camp* for *peace*.

The *peace camp* was a phenomenon of the early eighties, when the campaign against nuclear weapons in particular was at its height and peace campaigners felt that their protests had as yet found little response in the actions and policies of the superpowers. In the UK, the name *peace camp* is particularly associated with the women's camp outside the US airbase at Greenham Common in Berkshire (see WIMMIN), where some campaigners continued to live a decade or more after the camp was set up in 1981.

> Soviet newspapers are full of praise for the anti-nuclear activities of the women's peace camps at Greenham Common in Britain and elsewhere.   *Economist* 15 Mar. 1986, p. 63

peace dividend /ˈpiːs ˌdɪvɪdɛnd/ *noun* 🗎

A saving in public spending on defence, brought about by the end of a conflict or successful disarmament negotiations.

Formed by compounding: a *dividend* for the public purse because of a period of *peace*.

The idea of the *peace dividend* originated in the US in the late sixties as people began to speculate about an end to the Vietnam War. In practice, the expected surplus of public money did not materialize in the mid seventies and talk of a *peace dividend* largely died down until the late eighties. Then it was much discussed as an expected benefit—for the US, other NATO countries, and the former Warsaw Pact—of the ending of the Cold War and the resulting disarmament on both sides. Once again, it largely failed to materialize, this time because of the allied involvement in the Gulf War in 1991.

> Two Senate committees, Budget and Armed Services, have ... already held hearings on what has come to be called the 'peace dividend'. That is the money that will become available as military spending is reduced because of improved relations with the Soviet Union.
> *International Herald Tribune* 21 Dec. 1989, p. 6

> The awful truth may be that the peace dividend, if there is one, will be of less benefit to Europe than to the Americans, who have talked of cutting their defence budget by 25 per cent.
> *Observer* 13 May 1990, p. 16

peace pill 🔪 see PCP¹

peace wimmin 🗎 💥 see WIMMIN

Pearlygate 🎮 see -GATE

PEP /pɛp/ *acronym* Also written **P.E.P.** or **pep** 〜

Short for **personal equity plan**, an investment scheme intended to extend share ownership in the UK, under which investors are allowed to acquire shares up to a given value in UK companies without paying tax on dividends or capital gains.

The initial letters of *Personal Equity Plan*; the acronym might well have been chosen with the resulting 'word' in mind, suggesting that this initiative would *pep* up the market in UK shares.

The *PEP* was an innovation introduced in the mid eighties by the then Chancellor of the Exchequer Nigel Lawson as a deliberate incentive to widespread share ownership in the UK; the scheme coincided with the beginning of the government's privatization programme which, it hoped, would result in a large proportion of the British population owning and controlling their own service industries. The scheme presupposed long-term investment, so the tax advantage could only be earned if the investment remained in the Plan for a minimum period. Many high-street banks and other financial institutions introduced their own *PEPs*, many of which included the services of a **PEP manager** to make the investment decisions if the investor did not wish to manage his or her own portfolio. There was also provision for a particular preference or bias to be put on the investments—the investor might request ETHICAL INVESTMENT or even a **green PEP** (one concentrating on environmentally sound investment), for example.

> PEPS—Personal Equity Plans—are Mr Lawson's subtle persuaders which will, he hopes, turn us into a nation of shareholders.   *Estates Gazette* 9 Aug. 1986, p. 555

> Your mortgage can be repaid by an endowment linked to an Ethical Fund or indeed by a Green P.E.P.
> *Green Magazine* Dec. 1989, p. 55

perestroika /ˌpɛrɪˈstrɔɪkə/ *noun* Also written **perestroyka** 🔼

The 'restructuring' or reform of the economic and political system in the former Soviet Union, first proposed in 1979 and actively promoted under the leadership of Mikhail Gorbachev from 1985 until 1991. Hence any fundamental reorganization or reform, especially of a formerly socialist society.

A direct borrowing from Russian *perestroyka*, literally 'rebuilding, restructuring'. The same Russian word had been used within the Soviet Union to refer to the electrification programme of the twenties.

The policy of *perestroika* in the Soviet Union evolved out of an awareness among the central leadership of the deep economic and social crisis that the country seemed to be facing at the very end of the seventies, with widespread corruption, excessive bureaucracy, and industrial stagnation as some of its principal symptoms. The problem was the subject of a series of decisions of the Central Committee of the CPSU in April 1979; these were reported to the 26th Party Congress by Leonid Brezhnev, who said:

> It is a question of restructuring—yes, this was not a slip of the tongue, I said restructuring—many sectors and areas of ideological work.

Despite this announcement, little actual progress was made towards *perestroika* until 1985, when Mikhail Gorbachev came to power and made it a central tenet (along with GLASNOST) of his policy. The Central Committee considered a detailed programme for *perestroika* in April 1985, based on a careful analysis of the state of the economy. This became the basis for a plan announced by Mikhail Gorbachev at the 27th Party Congress in February–March 1986. This Congress was unique in the history of CPSU Congresses for its open criticism of Soviet industry, bureaucracy, and society, and its call to radical change. Gorbachev himself saw *perestroika* as nothing less than a new revolution; as he wrote in his book *Perestroika* (1987):

> In the spring of 1985, the Party put this task on the agenda. The gravity of accumulated and emerging problems, and the delay in their understanding and solution necessitated acting in a revolutionary way and proclaiming a revolutionary overhaul of society. Perestroika is a revolutionary process for it is a jump forward in the development of socialism.

Perestroika was widely discussed in the West at the time when it was first announced, and was generally seen as a sign of real change in Soviet society, especially since it was to be based on democratization. However, it proved less popular within the Soviet Union, where it seemed to make little difference to the availability of goods. By the early nineties *perestroika* had become the focus for a head-on fight between Mr Gorbachev and Boris Yeltsin, leader of the Russian Federation and subsequently President of Russia. In the end it was Mr Yeltsin's even more radical economic reforms that won approval as the Soviet Union disintegrated. Meanwhile *perestroika* had become a byword in English for any radical reform, especially of a socialist country or system; one sign of the word's acceptance into the language was the fact that it soon acquired the derivative **perestroikan** (an adjective and noun).

> Were Czechoslovakia to catch perestroika fever as strongly as Poland and Hungary, the troika could embark on a path that would seriously threaten Moscow's strategic interests.
> *Guardian* 29 July 1989, p. 8

> Mr Kohl, the clever tactician who substitutes instinct for any lack of intellect is playing a hand of fear: a fear that perestroika could soon be over and with it the Soviet willingness to accept a new order of democracy in Europe.
> *European* 25–27 May 1990, p. 9

> Yesterday's NEC decision to reduce the clout of the union block vote at conference was a valuable if partial and belated contribution. But as Frank Field knows, you can't get perestroika overnight, particularly when your route to reform requires the assent of the very institutions which need reforming.
> *Guardian* 28 June 1990, p. 18

personal computer 🖳 see PC

personal equity plan 〰 see PEP

personal identification number 〰 see PIN

personal organizer /ˌpɜːsən(ə)l ˈɔːɡənaɪzə(r)/ *noun* 🎴

An ORGANIZER for keeping track of one's personal affairs (appointments, commitments, finances, etc.), in paper or electronic form: *either* a loose-leaf notebook with sections for different types of information, pockets for credit cards, pens, etc. (a generic term for FILOFAX) *or* an electronic diary and notebook in the form of a pocket-sized microcomputer or software for a personal computer.

Formed by compounding: an *organizer* for one's *personal* life.

The transformation of the appointments diary into the *personal organizer* took place in the early eighties as the fashion for the Filofax among yuppies encouraged other firms to manufacture similar systems and a name was sought which was not protected as a trade mark. A growing preoccupation with organizing information (especially in the form of electronic data) coincided in the second half of the eighties with the development of ever smaller computers at affordable prices; the term *personal organizer* was not yet so firmly associated in the public mind with loose-leaf notebooks as to preclude its application to these electronic organizers as well, a process which began to take place in the late eighties and early nineties.

> These busy people all rely on personal organizers—compact, three-ring binders designed to keep track of various aspects of one's life.
> *Los Angeles Times* 20 Aug. 1985, section 4, p. 1

> We have given you the chance to get your life back into some sort of shape with the amazing Agenda word processor/personal organizer.
> *CU Amiga* Apr. 1990, p. 71

personkind /ˌpɜːs(ə)nˈkaɪnd/ *noun* 🎴

The human race; humankind. (Invented as a humorous non-sexist substitute for *mankind*.)

Formed by substituting the non-sexist word *person* for *man* in *mankind*.

It was the feminist movement of the seventies that promoted the word *person*—both as a free-standing word and as a word-forming element—as the successor to *man* in its centuries-old broader sense of 'human being'. Many of the formations which resulted, including *chairperson* (see CHAIR) and STATESPERSON, appeared awkward or even comical to those who had grown up with the forms ending in -*man* without ever thinking of them as referring exclusively to males, and the view was not infrequently expressed that the move towards INCLUSIVE language had gone too far too fast. It was in this context that the word *personkind* was coined in the early seventies as a humorous alternative for *mankind*, intended to ridicule the use of *person-* for *man-*. During the eighties, as the feminist view of language became more widely accepted, the word *personkind* retained a place in the vocabulary of English but remained largely tongue-in-cheek in its use.

Sonja fights for her life and the lives of all personkind. *Video Today* Apr. 1986, p. 36

The artificial ring of the new alternatives (like 'personkind') is counterproductive because it is faintly ridiculous to scrupulously avoid all possible references to gender, even when no reference to a particular sex is implied. *Music Technology* Apr. 1990, p. 10

person with Aids ⊗ ⦃ see PWA

p-funk ♪ 📷 see FUNK

phencyclidine 💉 see PCP¹

phonecard /ˈfəʊnkɑːd/ *noun* Also written **phone card** or **phone-card** ▩ ▣

In the UK, a plastic card (see CARD¹) providing a specified number of units of telephone time, which may be bought in advance from any of a number of retail outlets and then used in a special call-box known as a *cardphone* or *phonecard kiosk/phone*, etc.

Formed by compounding: a *card* for the *phone*.

Plans for a *phonecard* system, which would solve many of the problems with theft and vandalism that plagued coin-in-the-slot pay phones, were announced by British Telecom in 1980 (at first using the name **Phonocard**). A public-service trial of the system began in 1981, and within three years it was being expanded to provide several thousand more *cardphone* kiosks. The *phonecard* is inserted into a slot before dialling; a liquid crystal display on the computerized box shows the caller how many units remain to be used and what the computer is deducting for the current call. At first, the kiosks that were fitted to take the credit-card-sized *phonecard* were known as *cardphones*; by the middle of the eighties, though, the logo on the kiosks read *phonecard* and it seemed that British Telecom was trying to simplify things by using a single name for all the parts of the system. Colloquially, though, there is some variation; *cardphone* remains in use, as do synonyms for *phonecard* such as *telephone card*.

There are 700 Phonecard phones in London and these are expected to be increased to around 5,000 by 1987. *Ambit* Sept. 1985, p. 8

Subscribers will be sent a 'smart' card—a bit like a phonecard—which switches on the decoder. *Which?* Sept. 1989, p. 444

He went into an Indian grocery and provided himself with a telephone card and a stack of change. He walked over Putney Bridge and into Fulham, where he found a cardphone box that had to be functioning because it had a long queue. He waited. Two people, a black man and a white woman, exhausted their cards. Antonia Byatt *Possession* (1990), p. 327

photonovel /ˈfəʊtəʊˌnɒv(ə)l/ *noun* Also written **photo-novel** ▩ 📷

A piece of (usually romantic) fiction for young adults, in which the story is told in strip-cartoon fashion as a series of photographs with superimposed speech bubbles (rather than actual cartoons).

Formed by compounding: a *novel* told in a series of *photos*.

The *photonovel*, which is often published in serial form with each individual story sometimes known as a **photonovelette**, was originally a popular form of romantic fiction for European (especially Italian) teenagers, dating from at least the early sixties. In the late seventies the idea was imported to the US with some success, being used among other things for the 'book' publication of a number of films for teenagers such as *Grease* and *Alien*. By the mid eighties, *photonovel* series were available in the UK as well; their popularity among certain groups of young people was seen by some as a symptom of declining literacy.

Photonovels are here ... These photonovels are the American counterparts of magazines that have been raging successes in Europe for decades. *Daily News* (New York) 11 July 1978, p. 40

He ... fronted a rock band, wrote a porno photo-novel, and for a decade worked for the state phone company. *Time* 30 Jan. 1989, p. 68

See also GRAPHIC NOVEL

photo opportunity /ˈfəʊtəʊ ˌɒpəˈtjuːnɪtɪ/ *noun* Also written **photo-opportunity** 🟪 📄

In media jargon (especially in the US): an organized opportunity for press photographers and cameramen to take pictures of a celebrity or group of celebrities.

Formed by compounding: an *opportunity* to take a *photo*.

The idea of the *photo opportunity* originated in the US in the mid seventies, but was turned to advantage particularly by President Ronald Reagan and his administration during the eighties—a technique which other politicians did not fail to note. Journalism developed in such a way during the eighties that a historic agreement or summit of world leaders could be summarized in the results of a *photo opportunity* and perhaps a SOUND BITE of an official statement, and politicians became the ones hounded for a picture, taking the place of the film stars of previous decades. This approach to world events has been called **photo-opportunistic**; a person who takes advantage of it is a **photo opportunist**. By the end of the eighties *photo opportunity* itself was often abbreviated to **photo op**.

They operate in the slick new tradition of political handlers, whose job is to reduce a campaign to photo ops and sound bites. *Time* 21 Nov. 1988, p. 144

We must not be dazzled by these photo-opportunistic images. This modern magical foil for our memory can help us discover anew the luxury of retrospect. *Life* Fall 1989, p. 37

The rebuilding of Eastern Europe offers Tories photo-opportunities galore but confronts the Foreign Office with one of its trickiest tests in years. *Economist* 2 June 1990, p. 29

piece 🟩 see TAG²

pig out /ˌpɪg ˈaʊt/ *intransitive verb* 🟪

In US slang: to overindulge one's appetite, to overeat; to 'make a pig of oneself'.

Formed by adding *out* to the verb *pig* in the sense 'to act or eat like a pig', making a phrasal verb on the same model as MELLOW OUT and *max out* (see MAX).

The expression *pig out* has probably been in spoken use in US English for some considerable time, but in the late seventies and eighties it started to appear in printed sources, often without any indication of its slang origins. Typically, one *pigs out on* a particular food; a binge of overindulgence can be referred to by the noun **pigout**.

Troy and Vanessa ... pig out for days on leftover Halloween candy. Jane Fonda *Jane Fonda's Workout Book* (1981), p. 29

To prevent Americans from pigging out on between-meal snacks, herewith some ... tips. *Time* 11 May 1987, p. 29

pilger /ˈpɪldʒə(r)/ *intransitive verb* 🟪

In British media slang, to treat a subject or present an investigation in a manner supposedly characteristic of the investigative journalist John Pilger, especially when this entails exposing human suffering or drawing conclusions which reflect badly on the actions of a powerful government or institution.

The surname of Australian-born investigative journalist John *Pilger*, treated as a verb.

Pilger was the creation of *Spectator* journalist Auberon Waugh and has remained a favourite word with him and a small group of other journalists since the mid eighties. There is wide variation in the way that it is used, reflecting differing attitudes to John Pilger's own style of reporting. On the one hand (represented by Waugh and friends), it can be a highly critical and negative word, implying that the subject is being treated emotionally and with little regard for factual detail; sometimes, in fact, it is used as though it were only one step removed from outright lying. On the other hand (usually represented by the politically left-wing), there are those who admire Pilger's style and nerve and who use it with implications of compassionate reporting on behalf of powerless victims against the rich and powerful. A plethora of other words based on *pilger* grew up during the eighties, the commonest being the action noun **pilgering** and the adjective **pilgerish**; rarer and less established derivatives include **pilgerism, pilgerist**, and **pilgerization**.

It was a brilliant piece of pilgering to claim that he knew of a miner's family in Durham which possessed only one pair of shoes, although at the time of writing he has not produced so much as a photograph of this model family for us to weep over as John [Pilger] would undoubtedly have done.

Auberon Waugh in *Spectator* 24 Nov. 1984, p. 8

Le pilgerisme. From the English verb 'to pilger', this expresses the continuous action of going on the television and suggesting at length . . . that war, pestilence, governmental corruption in South-east Asia/Central America/the Lebanon etc. are essentially the fault of the Americans in general and the lack of land reform in particular. *Spectator* 24 Mar. 1984, p. 12

J. G. Dudley's question (Letters, 31 January) about the word 'pilgering' and 'pilgerish' is quickly answered. The verb to pilger means to regard with insight, compassion and sympathy.

Spectator 7 Feb. 1987, p. 26

PIN /pɪn/ *acronym* 〽️

Short for **personal identification number**, a confidential code-number allocated to the holder of a cash card or credit card for use when the card is inserted into a CASH DISPENSER or ATM.

The initial letters of *Personal Identification Number*.

The *PIN* (sometimes tautologically called a **PIN code** or **PIN number**) appeared at the beginning of the eighties, when greatly improved machines ensured that public take-up of automatic cash dispensing began to increase, and greater protection against misuse became necessary. The *PIN* is a security measure, designed to render the cards useless to a thief, since the machine will not carry out a transaction until the *PIN* has been keyed in correctly; the *PIN* relating to a particular card must therefore be revealed only to the card-holder, who must keep it secret. This need for secrecy has led to all kinds of mnemonics and means of writing the number down in a way which a thief would not recognize. Self-service machines which allow a customer to pay for goods and services using a credit card and the appropriate *PIN* were introduced in 1984 with the trade mark **Pinpoint**.

Where the card-holder had disclosed his PIN, or recorded the PIN with the card, the card-holder was liable for any unauthorized transactions. *Courier-Mail* (Brisbane) 14 July 1986, p. 25

For motorists . . ., we're installing Pinpoint machines for buying petrol in Shell garages all over the country. *Daily Telegraph* 24 Feb. 1987, p. 5

I reported the missing credit cards . . . but I did not call my bank that evening, trusting that nobody could use that card without the PIN code. *New York Times* 21 Nov. 1989, section A, p. 24

PLA, PLWA ⊗ ▌▌ see PWA

planet-friendly 🌻 see -FRIENDLY

plastic /'plæstɪk/ *noun* 〽️

Colloquially, credit cards, debit cards, and other plastic cards which can be used in place of money to pay for goods and services.

So named because this form of credit is obtained using a piece of *plastic* which serves as a membership card: see CARD¹. Probably abbreviated to *plastic* from the longer (and earlier) *plastic money* (see below).

The explosion of credit facilities and the consequent proliferation of credit cards which people carried in the seventies led to the development of the term **plastic money** in the US in about the middle of the decade; by the beginning of the eighties this was being abbreviated to *plastic* alone, and used colloquially as a collective term for all forms of credit. Thus 'Do you take plastic?' became a common way of asking to pay by credit card.

> It [is] easier than ever to spend money without seeing the real thing. 'The acceptance of plastic has reached an all-time high,' John Bennett, senior vice-president of Visa, said. 'Plastic has become a way of life.'
> *Globe & Mail* (Toronto) 10 Oct. 1985, section B, p. 13

> To use your plastic in a cash machine, you need a personal identification number (PIN).
> *Which?* July 1988, p. 299

plausible deniability 🖻 see DENIABILITY

playing the dozens 🔖 see DISS

pneumocystis carinii pneumonia Ⓧ see PCP²

poaching 〰 see HEADHUNT

poison pill /ˌpɔɪz(ə)n ˈpɪl/ *noun* 〰

Any of a number of ploys (such as a conditional rights issue) which may be adopted by the intended target of an unwelcome take-over bid in order to make itself unattractive to the bidder.

A metaphorical application of a word-combination which is usually used in the context of combat and espionage. Whereas the spy carries a *poison pill* for personal use when cornered, the company facing a hostile bid uses it to give the aggressors a taste of their own medicine.

In its literal sense, *poison pill* has been in use since the Second World War; the figurative use arose in the US financial markets in the early eighties, at first usually in the phrase **poison pill defence** or **poison pill device**, and was allegedly coined by US lawyer Martin Lipman in his defence of El Paso Natural Gas in 1982. It was adopted (as a device and a term) on the British Stock Exchange in the mid eighties. Despite attempts to limit the practice, it remained popular in a number of markets and generated several variants. Another name for essentially the same type of defence is a *shark repellent*.

> Lenox played hard to get ... and implemented a novel anti-takeover devise to discourage Brown-Forman Distillers takeover bid. The move is called the 'Poison Pill defense'.
> *New York Times* 19 June 1983, section 3, p. 14

> An American appeals court judge last week issued an important ruling against the use of ... 'poison pills' ... which embattled corporations are adopting as a defence against hostile takeover bids.
> *Daily Telegraph* 4 Aug. 1986, p. 19

> A 'poison pill', limiting shareholders' voting rights to 5% regardless of the size of their stake, killed Veba's enthusiasm for the merger.
> *EuroBusiness* June 1990, p. 7

poll-capping 〰 see CAP

poll tax /ˈpəʊl ˌtæks/ *noun* 〰 🖻

A colloquial name in the UK for the COMMUNITY CHARGE, used especially by its opponents.

So named because it is a capitation tax, that is a tax levied on every person, or 'head' of population (*poll* being an old word for 'head'); *poll-tax* is an ancient term, first used in England (in this form or as *poll-money*) for the capitation taxes of the fourteenth, seventeenth, and eighteenth centuries.

The community charge was nicknamed *poll tax* by its opponents almost as soon as it was announced in 1985, and this name was soon used as frequently in print as its official counterpart (especially in the popular press). The growing wave of protest which the *poll tax* provoked centred on this derogatory nickname; its historical associations with the oppression of the populace in earlier centuries, when the *poll money*, too, had provoked civil unrest, meant that it offered protesters a considerably more emotive focus than the bland and official name *community charge*. For further history see COMMUNITY CHARGE.

> Militant supporters started to form local anti-poll tax unions or to hijack ones formed by other groups . . . Many of the 'smash the poll tax' leaflets . . . are being printed by Militant—the wealthiest of the Trotskyite groups—at its East London presses.　　　　　　　　　　　*The Times* 8 Mar. 1990, p. 5

> Mrs Thatcher's new communications supremo, Brendan Bruce, was quoted yesterday saying that the handling of the Harrods report was 'a classic cock-up'. How would he describe the handling of the poll tax fiasco?　　　　　　　　　　　*Today* 12 Mar. 1990, p. 6

> Leading poll tax protester Alistair Mitchell admitted organisers had asked European activists to join in.　　　　　　　　　　　*Daily Star* 23 Oct. 1990, p. 2

polychlorinated biphenyl see PCB¹

pop, popping see BODY-POPPING

Popmobility /ˌpɒpməʊˈbɪlɪtɪ/ *noun*

The name of a physical exercise programme designed to be performed to the accompaniment of popular music.

Formed by compounding: *mobility* to the accompaniment of *pop*.

Popmobility classes appear to have been a feature of local authority evening classes in the UK since the late seventies, perhaps providing a British counterpart for JAZZERCISE. During the eighties it had to compete with a large number of other fitness programmes, including AEROBICS, AQUAROBICS, and CARDIOFUNK.

> Reasons for learning specific crafts vary, from taking up woodcarving because the 'Popmobility' classes were full.　　　　　　　　　　　*Artists Newsletter* Nov. 1987, p. 20

posse /ˈpɒsɪ/ *noun*

A gang of Black (especially Jamaican) youths involved in organized or violent (often drug-related) crime in the US. Now more widely in youth slang, one's gang or crowd; a group of friends.

A specialized sense of the existing word, representing a substantial shift of meaning: a *posse* was originally a group of people whose purpose was the enforcement of the law (and in this sense will be familiar to all lovers of Westerns). From here it developed to mean any strong band or company, was taken up in Black street slang (see below), and then came to be used specifically by police and journalists for a forceful band operating on the *wrong* side of the law.

The first reports of the criminal kind of *posse* arose from the spread of the cocaine derivative CRACK in the US, and the associated rise of drug-related crime there in the mid eighties. Originating as it does from Black street slang, where it means no more than 'a gang or crowd' (and has been used since at least the early eighties), the word figured in the names of RAP groups and lyrics and thereby spread to White youngsters as well, so that by the end of the decade it had become a fashionable way to refer to a group of one's friends—the people with whom one 'hangs out'.

> Having restrained my homeboys we walked away with dignity, but the whole posse was quite visibly in tears.　　　　　　　　　　　*City Limits* 9 Oct. 1986, p. 52

> Copeland's people are called the Beboes, a violent Jamaican drug posse operating big time in Queen's and Brooklyn.　　　　　　　　　　　*Newsday* 17 May 1989, p. 3

> You gotta mention my baby daughter AJ and the CIA dance posse.　　　*Sky Magazine* Apr. 1990, p. 18

post-bang _see_ BIG BANG

post-boomer _see_ BOOMER

post-lingually deafened _see_ DEAFENED

post-viral (fatigue) syndrome _see_ ME

pre-Aids _see_ AIDS

pre-lingually deaf _see_ DEAFENED

primeur _see_ BEAUJOLAIS NOUVEAU

privatizer /ˈpraɪvətaɪzə(r)/ _noun_

A person who advocates the transfer of nationalized industries to the private sector; someone who carries out privatization.

Formed by adding the agent suffix -_er_ to the verb _privatize_, which has been used since the early seventies in the sense 'to assign (services, industries, etc.) to private enterprise'.

Privatizer arose at the beginning of the eighties and has been used especially of members of the Conservative government in the UK, with its policy of selling national service industries and encouraging ordinary citizens to own the shares.

> Mr Redwood, the new under secretary, is an evangelical privatiser of similar persuasion and a leading light in the No Turning Back group of radical reformers. _Guardian_ 27 July 1989, p. 18

priviligentsia /prɪvɪlɪˈdʒɛntsɪə/ _noun_ Also written **privilegentsia**

A class of intellectuals and Party bureaucrats in Communist countries who, until the reforms of the late eighties, enjoyed social and economic privileges over ordinary citizens; more widely, any privileged class.

Formed by telescoping _privilege_ and _intelligentsia_ to make a blend.

Priviligentsia was coined, probably by Western observers, as the name for the privileged class of important Party members in the Soviet Union and other Warsaw Pact states as long ago as the fifties, but remained a specialized word used only in academic journals until the early eighties. Then it was taken up by the media as a convenient shorthand for all those who could avoid food shortages by shopping in special shops, speed through the traffic by travelling in specially reserved lanes, get jobs through friends and contacts, and generally lead a life of privilege and luxury which starkly contrasted with the life of ordinary people in the Soviet Union. The _priviligentsia_ was one of the main targets of Mikhail Gorbachev's policy of PERESTROIKA in the second half of the eighties, and the group which had most to lose from the reform programme. By the middle of the decade, the English-language press had already extended the word's use to cover any group of people who either enjoyed or advocated privilege.

> An unholy alliance of Labour 'egalitarians' and the Tory 'priviligentsia'.
> _Daily Telegraph_ 28 Jan. 1985, p. 18

> These bureaucrats get their jobs under the nomenklatura or privilegentsia system, whereby Communist party members nominate their friends in return for kickbacks and privileged access to rationed goods. _Economist_ 30 May 1987, p. 72

> When technology is expanding as fast as it ... is now, freer markets bring gains to everybody except the conservative privilegentsia. _Sunday Telegraph_ 9 Aug. 1987, p. 20

pro- /prəʊ/ _prefix_

In favour of; used in a number of adjectives relating to the abortion debates of the late seventies and eighties, especially:

pro-choice, in favour of a woman's right to choose whether or not to have an abortion;

pro-family, promoting family life and a return to a Christian moral code based on the family unit (and therefore opposed to the legalizing of abortion);

pro-life, in favour of upholding the right to life of the developing foetus (and therefore against abortion).

The Latin prefix *pro-* used in its usual sense 'in favour of, on behalf of'; in all of these formations, whichever side of the issue they represent, there is an attempt to present a positive approach by choosing a term containing this prefix rather than a complementary term containing *anti-*: see the comments at ANTI-CHOICE.

All of these terms arose in the US in the seventies and by the early eighties had become central to an understanding of political debate there and important election issues in many States. *Pro-choice* was first used in the mid seventies, sometimes as a noun (short for **pro-choice movement**) as well as an adjective; by the end of the decade a supporter of this view was regularly known as a **pro-choicer**. *Pro-life* was a more positive adjective which the anti-abortion lobby applied to itself from the late seventies onwards (see the discussion under ANTI-CHOICE); a supporter of this view is a **pro-lifer**. The *pro-family* campaign was a rather broader political issue (also a product of the late seventies), advocating a return to the values of family life and the moral standards of biblical Christianity, but this, of course, also embraced a stand against abortion.

> Some 'pro-family' activists . . . noisily pressed their antiabortion and 'morality' platform.
>
> Bob Frishman *American Families* (1984), p. 15

> Right-to-life groups, re-energized by the ruling, press for new laws limiting abortion, and their pro-choice counterparts rally to protect the gains embodied in Roe v. Wade, the 1973 Supreme Court decision.
>
> *New York Times Magazine* 6 Aug. 1989, p. 18

> Abortion was legalized in 1973, but with 1.5 million women annually opting for the procedure during the '80s, the issue flared anew. Right-to-life advocates fostered shows of civil disobedience while a lunatic fringe bombed clinics. Last July the Supreme Court retreated from its landmark *Roe* v. *Wade* decision by allowing individual states to impose restrictions. [Photo caption] Cleveland: Steven Green, 25, is hauled from the entrance of an abortion clinic that he and other members of Operation Rescue, a national 'pro-life' group, had been blocking.
>
> *Life* Fall 1989, p. 98

See also RIGHT-TO-LIFE

professional carer ⟦ see CARER

program trading /ˈprəʊɡræm ˌtreɪdɪŋ/ *noun* Also written **programme trading** 〰

In financial jargon, trading in a basket of securities rather than single issues; more specifically, a type of arbitrage (see ARB) in which traders take advantage of a difference in market values between a portfolio of securities and stock-index futures on essentially the same stocks, by taking a long or short position in the stocks at the same time as an offsetting position in a futures contract.

Formed by compounding: this form of *trading* is complex and sophisticated, and can only be carried out with the aid of high-powered computer *programs* which show when there is a suitable discrepancy in values for the trader to exploit.

Program trading is a phenomenon of the computerized financial markets of the eighties and arose in the US in the early years of the decade. It is a low-risk form of arbitrage, but one which normally involves very large portfolios of securities and considerable sums of money, and so it is only practised by those with substantial capital behind them. It has been criticized for creating great volatility in the markets, particularly at the times when options are about to expire (see TRIPLE WITCHING HOUR), since a great deal of buying and selling can be sparked off at these

times by *program trading* and the computer-driven nature of these deals means that they are re
garded as less controllable than deals decided upon by human agents.

> The collapse of Wall Street's biggest sustained rally last week sparked new controversy over the
> use of computers by big investors for so-called program trading.
>
> *Courier-Mail* (Brisbane) 27 Jan. 1987, p. 2

> Wall Street is gradually returning to some semblance of stability. This process will greatly be helped
> by the curbs on computerised programme trading announced on Thursday by the New York Stock
> Exchange.   *Financial Times* 4 Nov. 1989, Weekend FT, p.

> If small investors want to end the stock market volatility that is being caused by program trading
> they may have to stop complaining to their congressional representatives and stockbrokers and, in
> stead, send off an angry letter to the guy who watches over their own pension money.
>
> *Washington Post* 5 Nov. 1989, section H, p. 1

PRP 〽 see OTE

psychobabble ✖ see -BABBLE

puff-ball /'pʌfbɔːl/ *noun* Also written **puffball** ✖

A short full skirt which is gathered in at the hemline to produce a soft puffy effect;
balloon skirt. (Usually attributive, in **puff-ball dress** or **puff-ball skirt**.)

So named because the resulting shape of the garment is like that of the *puff-ball* fungus.

The *puff-ball* has been known to fashion designers under this name since the sixties; it enjoyed
a brief fashion in 1986–8 after being promoted by a number of the top Paris designers, and this
brought the word into the news.

> Christian Lacroix, the Paris designer, ... is credited with introducing the *pouffe*, otherwise known a
> the puffball, into the grandest parties.   *The Times* 9 June 1987, p. 2

> She has abandoned skintight leathers and puffball minis, platinum rinses and bootlace ties.
>
> *Sunday Mail* (Brisbane) 16 Oct. 1988, p. 1

pull-by date ⊗ see SELL-BY DATE

puppie ❙❙ see YUPPIE

PWA /piːdʌb(ə)ljuːˈeɪ/ *abbreviation* ⊗ ❙❙

Short for **person with Aids**, an official designation in the US which is also the pre
ferred term for themselves (rather than *Aids patient*, *Aids sufferer*, or—most disliked
of all—*Aids victim*) among those who have AIDS.

The initial letters of *Person With Aids*.

The term *PWA* arose as a direct result of the coming together of people with first-hand experi
ence of Aids at the second Aids forum in the US, held in Denver, Colorado, in December 1983
At this forum a group of people who had Aids or ARC (see AIDS) formed themselves into the
Advisory Committee of People with Aids and issued a statement objecting to some of the other
terms which had been applied to them in the past:

> We condemn attempts to label us as 'victims', which implies defeat, and we are only occasionall
> 'patients', which implies passivity, helplessness, and dependence upon the care of others.

A variation on *PWA* is **PLWA** or **PLA**, both denoting **person living with Aids**. This arose
again among the people most intimately concerned, in the second half of the eighties and wa
designed to counteract the negative responses of the general public by emphasizing the fact o
living with—rather than *dying* from—Aids. Among journalists and others who influence popu
lar usage, however, *PWA* is the only one of these designations which has gained any currency
in the US in particular, it had become a well-known and widely used abbreviation by the early
nineties, although the terms to which *PWAs* most object also remained frequent in the popula

press. Sometimes the apparent sensitivity of the writer to the feelings of *PWAs* is cancelled out by an insensitive reversion to *Aids victim* within a few words.

> He found a place to live thanks to the Shanti Project, a charity subsidised by the municipality to help PWAs. It makes houses available to AIDS victims. *Guardian Weekly* 26 Jan. 1986, p. 12

> He explains that the race and class of most straight PWA's are proof that the 'heterosexual epidemic continued to fail to show up'. *Village Voice* (New York) 30 Jan. 1990, p. 61

•••

Q

qinghaosu /tʃɪŋhaʊˈsuː/ *noun*

A naturally occurring compound (also known as *artemisinin*) which is extracted from the Chinese plant *Artemisia annua* for use in the treatment of malaria.

A direct borrowing from Chinese *qīnghāosū*, itself derived from *qinghao*, the Chinese name for the *Artemisia* plant, and a suffix meaning 'active principle'. The plant (a member of the wormwood family) grows alongside rivers in the North-East and South-West of China and is used as feed for pigs or against mosquitoes.

The Chinese have known about the anti-malarial properties of the *qinghao* for many centuries—the leaves and stems are used in traditional Chinese medicine against fevers—but it was not until the early seventies that these were confirmed by rigorous testing and identification of the active ingredient, *qinghaosu*. News of the discovery was reported in the West in the late seventies and eighties; one reason for excitement over the discovery in medical circles is that this natural drug is effective against some types of malaria that are not treatable with synthetic anti-malarials. During the eighties *qinghaosu* was extracted from *Artemisia* plants cultivated outside China as well.

> One of the plants to come under scrutiny was a weed with a long history of use known in China as *qing hao*…The Chinese named the crystalline compound *qinghaosu*, meaning active principle, and the western version of the name is Artemisinin. *The Times* 22 July 1985, p. 12

quaffable /ˈkwɒfəb(ə)l/ *adjective*

Of a wine: lending itself to being drunk copiously, drinkable.

Formed by adding the suffix *-able* to the verb *quaff* 'to drink (liquor) copiously'.

This is one of the many words on the borderline between wine-lovers' slang and technical terminology that have thrived in the growing literature on wine in the eighties.

> It is an intensively fruity, soft-bodied wine, … charming and eminently quaffable. *Washington Post* 1 Dec. 1982, section E, p. 1

> Were it not for 'a little local difficulty' we would here in Britain already be able to drink the very quaffable wines of Argentina. *Wine Society Annual Review* 21 Apr. 1987, p. 12

quagma /ˈkwægmə/ *noun*

In physics, a hypothetical state or body of matter consisting of free quarks and gluons.

Formed by combining the first three letters of *quark*, the initial letter of *gluon*, and the last two of *plasma* to make an artificial word designed to rhyme with *magma*.

One of the most important areas of development in particle physics in the past two decades arises from M. Gell-Man's theory of sub-atomic particles called *quarks* (after a line in James Joyce's *Finnegans Wake*, 'Three quarks for Muster Mark!', but pronounced /kwɔːk/), first postulated in the mid sixties. These quarks, according to the theory as it developed in the seventies, are bound together by the colour force carried by massless *gluons* (so named because they act as a kind of sub-atomic glue). The idea that under certain conditions the quarks and gluons would become mixed into a kind of plasma, called a *quagma*, was postulated in the mid eighties.

> Theory suggests that when the density of energy in nuclear matter is high enough, the quarks and gluons will no longer remain confined but will form a quagma.   *New Scientist* 3 Mar. 1988, p. 45

quark ▣ see QUAGMA

quilling /ˈkwɪlɪŋ/ *noun* ▨

The art or craft of paper filigree, in which elaborate pictures and designs are built up from curled strips of paper.

Formed by adding the action suffix *-ing* to the verb *quill* 'to form (ribbon, etc.) into small cylindrical plaits or curls'. The word *quilling* has been in use since the eighteenth century in the sense 'a ribbon, strip of lace or other material gathered into small cylindrical folds'.

Quilling is a traditional craft, practised as *paper filigree* in the UK and as *quill work* in parts of the US for decades or even centuries. Like a number of other traditional crafts, though, it began to be promoted outside the small community in which it was traditionally practised during the seventies and benefited from the revival of interest in crafts which took place during the late seventies and eighties. In this revived use, the name given to the craft throughout the English-speaking world was *quilling*, and the word soon passed from technical terminology into more widespread usage. A practitioner of *quilling* is a **quiller**.

> Quillers have used all varieties of paper . . . In modern quilling, the choice of colors is broad.
> Betty Christy & Doris Tracy *Quilling: Paper Art for Everyone* (1974), pp. 34 and 37

quiteron /ˈkwɪtərɒn/ *noun* ▣

An electronic device which operates rather like a transistor in switching and amplifying, but uses superconducting materials rather than semiconductors and needs less power to do its switching.

Formed from letters taken from the full technical name of the effect on which its working depends, *QUasiparticle Injection Tunnelling Effect*, and the last three letters of *-tron*.

The *quiteron* was developed by Sadeq Faris for IBM and patented in the US in 1982. When the invention was first announced to the electronics community in 1983, it was thought that it could eventually replace the principle of the semiconducting transistor; whether it will in fact do so remains to be seen.

> The quiteron is not the first superconducting device that engineers have considered for chips.
> *New Scientist* 10 Feb. 1983, p. 369

Quorn /kwɔːn/ *noun* ▨

The trade mark of a type of textured vegetable protein derived from a small edible fungus and marketed as a vegetarian meat substitute.

This vegetarian product is named after the Leicestershire company which originally made it, itself named after the village of Quorn (now Quorndon); ironically, this is also the name of a famous traditional fox-hunt in the area, an example of the blood sports to which many vegetarians would object on principle.

> Food novelties based on mycoprotein—now trade-named Quorn—should be in the shops during this year.   *Financial Times* 7 Jan. 1987, p. 11

> Where Quorn scores over these other meat alternatives is that its plant fibres are almost identical in size to the fibres in meat, which produces the similar texture and eating quality.
> *Fitness* May 1988, p. 29

R

racquet abuse see ABUSE

rad /ræd/ *adjective*

In young people's slang (especially in the US): really good or exciting; 'cool', 'hip', AWESOME.

Formed by abbreviating *radical*, itself a favourite term of approval among American youngsters in the eighties and originally (like TUBULAR) a word used in Californian surfers' slang. Such slang terms of approval often get abbreviated to a snappy monosyllable—in the UK BRILLIANT became *brill* by the same process.

The longer form **radical** was used from the late sixties by surfers to describe a turn or other manoeuvre that was at the limits of control and safety, presumably by extending the political sense of the adjective 'representing the extreme section of a party'; this specific surfers' use was interpreted as the equivalent of *far out* and, like *far out* itself some time earlier, was soon weakened to express no more than approval and admiration for something. In the early eighties, as Californian surfers' slang became diluted and spread to a generation of young Americans through films and VALSPEAK, *radical* and the abbreviated form *rad* began to crop up frequently as the currently fashionable accolade. By the middle of the decade it had spread outside the US as well; its popularity in the UK, especially among the very young, was fed by American television shows, comics, and the craze for the TURTLEs in the late eighties.

> Kim Robb . . . sat down with a group of Prairie teenagers to discuss things that were 'cool' . . . 'The word now,' says Robb, . . . 'is rad.' *Maclean's* 6 Sept. 1982, p. 48

> The raddest moments on *Louder Than Love* sound like the raddest moments on the Cult's *Sonic Temple*. *Spin* Oct. 1989, p. 99

radical see FREE RADICAL

radical hard SF see CYBERPUNK

radicchio /ræˈdiːkɪəʊ/ *noun*

A type of chicory with reddish-purple white-veined leaves, used as a salad vegetable and as a decorative garnish.

A direct borrowing from Italian *radicchio* 'chicory'; this variety of chicory originally comes from Italy.

The move towards a greater variety of fresh raw vegetables in British and American cooking was one of the beneficial results of the fashion for *nouvelle cuisine* (see NOUVELLE) in the late seventies and early eighties. *Radicchio* satisfied the desire of the health-conscious for more interesting salad vegetables as well as offering colour to those more concerned with the aesthetic quality and presentation of the food; it therefore became a regular feature of restaurant fare and food-market stock by about the middle of the decade. Since Italian spelling conventions are not completely self-explanatory to English speakers, some try to pronounce the word /ræˈdiːtʃɪəʊ/, using an English -*ch*- sound for the last consonant.

> The big public market specializes in . . . sophisticated imports from rice to radicchio. *St Louis Post-Dispatch* 28 May 1986, section D, p. 1

> Superb spring rolls filled with radicchio, mozzarella and salsa. *Vogue* Sept. 1990, p. 192

ragazine /rægə'zi:n/ *noun* 🃏 🎭

In US media slang, a cheaply produced news-sheet or magazine devoted to the dissemination of gossip.

Formed by telescoping *rag* (a contemptuous word for a cheap or worthless newspaper) and *magazine* to make a blend.

The word was coined in relation to a publication called *Hollywood Kids*, designed to spread gossip about who was doing what in Hollywood, which first appeared in the mid eighties.

The Hollywood Kids is a ten-page 'ragazine' which prints tall tales like the ones outlined above.
Empire Sept. 1989, p. 32

You wanna be a gossip columnist? Dish the dirt in your own eponymous, Xeroxed 'ragazine'.
Los Angeles Times 9 Mar. 1990, section E, p. 1

rage /reɪdʒ/ *noun* and *verb* 🃏 🎭

In young people's slang, especially in Australia:

noun: A party, a good time.

intransitive verb: To revel, to have a good time. Also as an agent noun **rager**, a party-goer or reveller.

An extended use of *rage* in its figurative sense 'to be violent or boisterous, to rush', probably passing through an intermediate stage when it meant 'to go on a spree'.

This is an Australian usage which became established in the early eighties; it came to prominence outside Australia as well, largely as a result of the popularity of Australian soap operas and other television series in the UK in the second half of the eighties.

The Roxy churns out an endless stream of disco, dancing, and drinking, tailor-made for young working people who ... are looking for 'a rage'. *Courier-Mail* (Brisbane) 26 June 1986, Supplement, p. 8

'I still go out and rage occasionally,' says the former sidekick to Greg Evans ..., 'but I can't do it like I used to, not five nights a week.' *TV Week* (Melbourne) 28 Mar. 1987, p. 4

rah-rah skirt /'rɑ:rɑ: ˌskɜ:t/ *noun* Also written **ra-ra skirt** 🃏 🎭

A very short flounced skirt, similar in design to the type worn by American cheer-leaders.

So named because it is the type of skirt worn by a *rah-rah girl* or cheer-leader, who is herself named after the chorus of *rah-rah-rah* with which she cheers on her team.

The *rah-rah skirt* came into fashion in 1982 as the first really successful attempt to revive the mini-skirt of the sixties, but its success was largely limited to a restricted clientele of slim teenage girls. The participation of British teams in the World League of American football, complete with their own teams of cheer-leaders, could perpetuate the fashion.

For evening, the bomber jacket was worked in black satin and leather, with floaty chiffon ra-ra skirts.
Daily Telegraph 19 Mar. 1991, p. 2

rai /raɪ/ *noun* 🎵 🎭

A style of popular music, originally from North Africa, which fuses Arabic and Algerian ETHNIC or folk elements with Western styles.

Like ZOUK, *rai* was popularized on the WORLD MUSIC scene in Paris during the second half of the eighties.

Look for Stevie Wonder to introduce America to the latest music rage sweeping northern Africa. Called rai ... the sound is described as space-age Arabic folk music. *People* 24 Feb. 1986, p. 29

rainbow coalition /'reɪnbəʊ kəʊəˌlɪʃ(ə)n/ *noun* 🏛

In political jargon (originally in the US): an alliance of minority peoples and other disadvantaged groups, acting together in an election or political forum so as to gain greater recognition for their cause.

Formed by compounding: a *coalition* of people of many colours (summed up by the image of a *rainbow*).

The idea of the *rainbow coalition* originated in the Southern US in 1982 and was first widely written about in the early eighties, when liberal groups (and in particular the Democratic Party's Jesse Jackson, trying for a presidential nomination) put forward the idea that racial minorities, disadvantaged White groups, and women's interests could be combined to form a potentially powerful political pressure group. By the middle of the decade the imagery, at least, had spread to the UK, where the term was used to refer to possible coalitions of parties of differing political colours (such as the possibility of co-operation between the Liberal–Social Democrat Alliance and Labour).

Jackson's prediction that he would attract a 'rainbow coalition'—of blacks, Hispanics, women, American Indians, peace advocates and others—has not come to pass.

New Yorker 28 May 1984, p. 115

The Alliance's best chance of something spectacular is in Liverpool where they hope to gain minority control by forming a 'rainbow' coalition with Labour opponents of council deputy leader Derek Hatton. *Today* 6 May 1986, p. 16

Rambo /'ræmbəʊ/ *noun* Also written **rambo** ⬚ ⬚

A person who resembles the film character Rambo in attitudes or behaviour; specifically, *either* a macho male type who practises survival techniques and likes to live as a 'loner' *or* a person who advocates or carries out violent retribution.

An allusive use of the name of the hero of David Morrell's novel *First Blood* (1972), a character widely popularized by the films *First Blood* (1982) and *Rambo: First Blood Part II* (1985).

In the novel and films, the Rambo character is a Vietnam veteran who lives as a loner and is bent on violent retribution for the wrongs that he thinks society has done him. By the middle of the eighties the name *Rambo* was being used in a number of transferred contexts, often in derivatives such as the adjectives **Ramboesque** and **Rambo-like**, to refer to things as diverse as international diplomacy and PAINTBALL games, but which all seemed to reflect the world-view of this character. The word was used attributively as well, almost passing into an adjective meaning 'savage': any violent killing, especially when carried out by a person in combat dress, could be described as a **Rambo killing**, and the newspapers nicknamed Michael Ryan, who carried out the Hungerford massacre of 1987 (see SURVIVALISM), the **Rambo killer**.

Given the bomb-'em-kill-'em suggestions pulsing from the typewriters of 100 literate Rambos, a boycott of the airport was the most reasonable act suggested.

Washington Post 6 July 1985, section A, p. 19

To lawyers, as to other Americans, Ronald Reagan apparently has become the stars and stripes for ever. By his own oft-stated, Rambo-like standards, the hostage crisis was a downer. There was none of the threatened 'swift and effective retribution'. *Washington Post* 9 July 1985, section A, p. 2

Sensitive to charges of encouraging a new generation of Rambos, the companies organising the games insist more excitement than aggression is stimulated. *Guardian* 3 July 1989, p. 20

One of the first victims, World War 2 veteran Pat Surgrue, was attacked by a 2.5m 'rambo' roo [kangaroo] on his front lawn. *Australasian Post* 17 Feb. 1990, p. 14

rap /ræp/ *noun* and *verb* ⬚ ⬚

noun: A style of popular music (also known more fully as **rap music**) in which (usually improvised) words are spoken rhythmically, often in rhyming sentences, over an instrumental backing. Also, a song or piece of music which incorporates this technique; an individual 'poem' or refrain in this style.

intransitive verb: To perform rap music; to talk or sing in the style of rap. Also as an action noun **rapping**; agent noun **rapper**.

A specialized development of the US slang sense of the noun and verb *rap* '(to) talk', which it-self dates from the turn of the century. This had already been taken up by US Blacks in the six-ties as a name for the special style of verbal repartee which developed as an important part of their street culture and peer-group behaviour (see DISS); the transformation of *rapping* of this kind into a type of performance poetry which could be associated with a particular style of pop-ular music completed the process of specialization.

Rap, an important element of the youth subculture known as HIP HOP, developed among Black youngsters on the streets of New York during the seventies, but did not become a recognizable genre of popular music known by this name until the early eighties. *Rap* has links with other more formal styles of Black (especially West Indian) performance poetry known as *dub* and *toasting*, which began to reach a wide audience in the seventies as a result of the popularization of West Indian culture through *reggae* and *ska*. At first the New York *raps* themselves were im-provised live over the rhythmic backing of music from a BOOM BOX OR GHETTO BLASTER; in the early eighties the style was taken up by disc jockeys in New York's clubs, and a number of **rap groups** recorded the music and enjoyed great commercial success with it, popularizing *rap* within White youth culture as well as Black and establishing it as one of the most important styles of the eighties. The influence of *rapping* is evident in a number of areas outside Black cul-ture, such as the language and creative writing of young Whites in the UK (words such as BAD, DEF, DISS, FRESH, and RARE might never have spread beyond a quite limited population of young people but for their use in *rap*); another sign of *rap*'s influence is the fact that distinct styles (such as **rude rap** and **on-and-on rap**) are recognized among groups of youngsters far re-moved from *rap*'s New York origins.

Many raps still brag about the rapper's financial success and superior cool but others talk about such topics as friends and basketball. *Wall Street Journal* 4 Dec. 1984, p. 16

But when he realised that black classmates were listening to a different rap group each week he de-cided that rap was much more progressive than rock 'n' roll. *New Musical Express* 9 May 1987, p. 30

Cartel distributors, Revolver, have great hopes for the ... hip-hop EP ... consisting of 'Anyone', 'The Dark' and 2 raps. *Tower Records' Top* Feb. 1988, p. 7

D.J.'s Matt Dike, ... and Mike Ross, ... got Los Angeles rapping. *Interview* Mar. 1990, p. 52

rap and scratch ♪ 🎛 see SCRATCH

rare /rɛə(r)/ *adjective* 🎛

In young people's slang: extremely good or impressive; 'hip', 'cool'.

A revival of a colloquial sense of the adjective *rare* which first developed in the fifteenth century, but was considered archaic in the early twentieth century. The usage probably found its way into young people's slang through US Black street slang and RAP.

'Rare!' is an expression of wonder, gasped rather than spoken. *New Statesman* 16 Feb. 1990, p. 12

rate-capping 〰 🗎 see CAP

reader-friendly ✖ see -FRIENDLY

read my lips /ˌriːd maɪ 'lɪps/ *phrase* 🗎

In US politics, a catch-phrase promoted during the Republican presidential cam-paign of George Bush to emphasize commitment to lower taxes; also sometimes used as an adjectival phrase to refer to the tax policy of his administration or to its policies in general.

The phrase comes from Mr Bush's speech to the Republican Party convention in New Orleans in August 1988:

Congress will push me to raise taxes, and I'll say no, and they'll push, and I'll say no, and they'll push again, and I'll say to them 'Read my lips: no new taxes'.

During the election campaign that followed this was repeated to reporters and questioners as *read my lips* followed by the silently mouthed words 'no new taxes'. The phrase itself is, of course, older than this in other contexts; the imagery is that of someone talking to a deaf person, or of a parent emphasizing something to a child and urging visual as well as aural concentration on what is said, the equivalent of 'I really mean this'—or even the television catch-phrase 'I will say this only once'. There is also sometimes a suggestion that what follows *read my lips* represents a sub-text, a deeper meaning or message that can only be mouthed and not spoken aloud. In these broader uses the phrase *read my lips* was well known by the time Mr Bush used it at the convention (it had even been the title of a rock-music album).

So often did Mr Bush use the technique described above during the election campaign that it became a hallmark of his promised policies, so that the phrase *read my lips* alone became enough to signify a promise of no new taxes during his presidency. It also became a yardstick by which the American public could measure his administration and assess once and for all the reliability of election promises.

It appears the 'read my lips' President is simply giving lip service to his environmental concerns.
Philadelphia Inquirer 20 Sept. 1989, section A, p. 16

Sen. Phil Gramm, . . . aiming to rescue the administration's 'read my lips' strategy, plans an alternative amendment.
Washington Post 1 Oct. 1989, section D, p. 7

Truth caught up with Mr Bush last week when he tiptoed into Congress and agreed, no doubt with everyone reading his lips, to raise $25 bn in new taxes.
Punch 13 July 1990, p. 20

realo  see FUNDIE

recycling /riːˈsaɪklɪŋ/ *noun*

The conversion of waste (such as paper, glass, etc.) into reusable materials.

Formed from the verb *recycle*, literally 'to return to an earlier stage in a cyclic process'; when a waste product is recycled it is returned to its raw-material state so as to be formed into a new product.

The idea of *recycling* paper waste in particular is several decades old, but the whole concept of reusing waste rather than dumping it in the environment gained a new impetus and a more positive public profile as a result of the success of the GREEN movement of the eighties. Whereas it was only a few keen environmentalists who took the trouble to save and reuse domestic waste in the seventies, the eighties saw the development of government-sponsored *recycling* programmes, collection points for **recyclable** containers (such as the BOTTLE BANK and the CAN BANK) appeared in many towns, and in certain areas (including Canada and some States in the US) the division of domestic waste into *recyclable* and *non-recyclable* elements was required by law. The availability of **recycled** products also improved, as did their quality and market image, with advertisers working hard to convince shoppers that they could 'do something for the environment' by choosing *recycled* paper, containers, etc. Manufacturers keen to present themselves as ecologically aware had to consider the **recyclability** of the packaging that they used as well as the possibility of using **recyclables** in the product itself.

Manufacturers have jumped on the bandwagon, slapping 'biodegradable', 'ozone friendly', 'recyclable', and . . . any other environmentally correct slogan . . . on everything from diapers to deodorant.
New Age Journal July–Aug. 1990, p. 10

A pilot Blue Box scheme which covers 3,500 homes in Sheffield—the first recycling city—is proving to be the most successful collection method in the UK.
Earth Matters Summer 1990, p. 4

So far, Canada has accepted seven [EcoLogo] sectors: Zinc-Air Batteries, water based paint, fine recycled paper, miscellaneous recycled paper, recycled newsprint, heat recovery ventilators, and cloth nappies.
Earth Matters Summer 1990, p. 9

Recycling was encouraged by . . . the buy-back value for recyclables (paid out at privately owned drop-off centers).
Garbage Nov.–Dec. 1990, p. 27

red-eye /ˈrɛdaɪ/ noun Also written red eye 🎯

In colloquial use: an overnight flight, especially one on which the traveller crosses one or more time-zones.

So named because the passengers can be expected to arrive *red-eyed* from lack of sleep.

The term *red-eye* (at first in attributive form, as **red-eye flight** or **red-eye special**) has been in colloquial use in the US since at least the late sixties. In the late eighties, with transatlantic commuting a reality, it became a fashionable term among British business executives for the overnight flight from New York to London; arriving at breakfast time on such a flight, the traveller has a full business day ahead and a time difference of five or six hours to cope with.

> Three days ago (is it?) I flew in on a red-eye from New York. I practically had the airplane to myself.
> Martin Amis *London Fields* (1989), p. 1

> Participants . . . were ushered aboard the late night 'red eye' for the non-stop flight to Tokyo.
> *Gramophone* Feb. 1990, p. 1547

red route /ˈrɛd ˌruːt/ noun 🎯 🏠

A proposed expressway (marked by a red line along the edge of the road) designed to ease traffic congestion on certain urban roads and similar in operation to a clearway, except that more severe penalties would be incurred by the driver of any vehicle which stopped or otherwise infringed the rules.

Formed by compounding: a *route* marked by a *red* line.

The idea of the *red route* as a way of easing urban traffic congestion in the UK was devised by a group of Conservative politicians called the **red route group** in the second half of the eighties. Initially intended to solve some of London's traffic problems, their scheme would place tight restrictions on parking, unloading, stopping, and roadworks on the selected routes and would provide for a special force of traffic wardens to impose the steep fines which anyone infringing these restrictions would incur. The proposals did not meet with unqualified enthusiasm from the general public or the government.

> Red routes, designed to speed the flow of vehicles of all kinds indiscriminately, could only make things worse.
> *Independent* 20 Dec. 1989, p. 18

reflagging /riːˈflægɪŋ/ noun Also written re-flagging 🏠

The practice of registering a ship under a new national flag or flag of convenience, especially so as to enable it to qualify for protection in disputed waters.

Formed by adding the prefix *re-* and the suffix *-ing* to *flag*.

Although the word *reflagging* was not new to the language in the eighties (it had been used in specialized sources for some years before that), it was only during the mid and late eighties that the issue was brought into the public eye through widespread reporting of the situation in the Persian Gulf and the word was therefore used frequently in the newpapers. Most of the reports concerned the difficulties experienced by Kuwaiti ships passing through the Straits of Hormuz with cargoes of oil in 1986–7, during the Iran–Iraq war; the question was whether they should be allowed to avail themselves of naval protection from NATO countries or from the Soviet Union after one or other of these countries had offered to **reflag** them under its own national flag. In practice, this was done mainly by the US, whose warships subsequently escorted the **reflagged** Kuwaiti tankers safely through the Straits, and the lead was later followed by the UK, but the rights and wrongs of this approach were hotly disputed both in the US and in the UK.

> Reflagging Kuwait's tankers as 'American' vessels. *US News & World Report* 8 June 1987, p. 20

> Two reflagged Kuwaiti tankers hoisted the Stars and Stripes and signalled to their escort of four American warships that they were ready to sail. *Daily Telegraph* 22 July 1987, p. 1

> We reflagged the tankers because the Kuwaitis were going to ask the Russians to do it.
> *USA Today* 21 Oct. 1987, p. 6

reflexology /ˌriːflɛkˈsɒlədʒɪ/ *noun* ⊗

A COMPLEMENTARY therapy based on the application of pressure to specific points on the feet and hands.

Formed by adding the suffix *-ology* 'subject of study' to *reflex*, the term used for the pressure points on the feet and hands which are used in this technique (because each point has a corresponding effect—a secondary manifestation or *reflex*—on a particular part of the body).

Reflexology is also known as *zone therapy of the hands and feet*; like ACUPRESSURE, it is an ancient oriental therapy whose techniques date back thousands of years, but which has only this century been taken up and widely practised in the West. It was rediscovered in the twenties by William Fitzgerald, an American ear, nose, and throat specialist, popularized in the US by Eunice Ingham, and brought to the UK in the sixties by a student of hers named Doreen Bayly. However, it was only in the eighties, with the growth and success of alternative and complementary therapies, that *reflexology* was taken up by significant numbers of people. The underlying principle is very similar to that of acupressure, except that an entire 'map' of pressure points affecting the whole body is found in the feet, and it is mainly these reflexes (together with occasional use of a corresponding set in the hands) that are worked on to produce an improvement of circulation to the corresponding part of the body, a relaxation of tension there, and eventually a return to BALANCE. A practitioner of *reflexology* is a **reflexologist**.

> For the reflexologist, there are 10 channels, beginning (or ending) in the toes and extending to the fingers and the top of the head. Each channel relates to a zone of the body, to the organs in that zone—the big toe relates to the head, for example. By feeling patients' feet in certain prescribed ways, reflexologists can detect which energy channels are blocked.
>
> Brian Inglis & Ruth West *The Alternative Health Guide* (1983), p. 112

> Apparently, the Princess of Wales, the Duchess of York and others are advised . . . on a form of . . . reflexology, and it keeps those treated healthy, young and beautiful.
>
> *New Scientist* 23 June 1990, p. 112

refusenik /rɪˈfjuːznɪk/ *noun* Also written **refusnik** 🔒 ▮

Colloquially, any person who has been refused official permission to do something or who has refused to follow instructions, especially as a form of protest.

A transferred sense of a word which was originally a partial translation of the Russian word *otkaznik* (itself made up of the stem of the verb *otkazat'* 'to refuse' and *-nik*, the agent suffix used in other English words such as *beatnik* and *peacenik*). When first borrowed into English, *refusenik* was used only in the specific sense of Russian *otkaznik* 'a Soviet Jew who has been refused permission to emigrate to Israel'.

The plight of the Soviet *refuseniks* was first widely reported in the English-language press in the second half of the seventies and by the early eighties the word would have been familiar to the readers of most quality newspapers. By the mid eighties journalists had started to apply it in other contexts (in much the same way as other Russian borrowings such as GLASNOST and PERESTROIKA would later be applied in new and often trivial home contexts); perhaps under the influence of the punning style of newspaper headlines, or possibly just as a result of misunderstanding or forgetting the original import of the word (since many of the original *refuseniks* had been dissidents), they then began to use *refusenik* for the person who does the refusing rather than the one who is its victim, so that it became a milder synonym for *dissident* or *protester*.

> The 30 'refuseniks' who would not go to Wapping have been joined by 50 people.
>
> *City Limits* 10 Apr. 1986, p. 7

> 'Refuseniks' of Voyager lobby Hawke.
>
> *Courier-Mail* (Brisbane) 1987, p. 19

See also RETURNIK

remastered ♪ 💻 see DIGITAL

Restart /'riːstɑːt/ *noun* Also written **restart** 〰

A return to paid employment after a period of absence or unemployment; in the UK, the name of a government programme to facilitate retraining and re-employment.

A specialized use of the existing noun *restart*.

The government's *Restart* scheme began in September 1988, a time when, despite high unemployment, employers complained that they were unable to find suitable staff to fill their vacancies. This situation was particularly acute in inner cities (especially inner London), so the schemes started there and a few months later spread nationwide. The scheme (parts of which, at least, are compulsory after six months' unemployment) includes opportunities for training in interview technique, self-presentation, etc. to help the candidates to 'fit' the employers' requirements.

> If you're still unemployed after six months, you're obliged to attend a Restart interview. This gives you the first opportunity to retrain on a state scheme or join a Jobclub.   *Which?* Aug. 1988, p. 378

restructuring 🏠 see PERESTROIKA

retro /'rɛtrəʊ/ *noun* and *adjective* 🗾

noun: A style or fashion that harks back to the past, a throw-back; a movement to revive past styles.

adjective: Reviving or harking back to the past; nostalgically retrospective.

Although the prefix *retro-* has a long history in English, forming words with the meaning 'backwards-' on Latin roots (such as *retrograde*), it was actually through the French word *rétrograde* that this word reached English. The French began to abbreviate *rétrograde* to *rétro* specifically in relation to fashion in late 1973, when the styles of the thirties were revived by Paris designers. The abbreviation stuck in French, and it was only the abbreviated form that was borrowed into English.

The earliest uses in English closely follow the developments of 1973–4 in France, and use the word both as a noun and as an adjective, as was already the case in French. As nostalgia in a number of cultural areas became increasingly fashionable in the eighties, both the adjective and the noun were used to form compounds such as **retro-culture**, **retrodressing**, **retromania**, **retrophobia**, and **retro-rock**.

> The icy charms of the Group TSE's productions, beginning as far back as 1969, have been in the vanguard of the French vogue for 'retro'.   *Guardian Weekly* 18 May 1974, p. 14

> Kevin was delighted . . . Any guy who wore a retro tux would have to be.
>   Erica Jong *Parachutes & Kisses* (1984), p. 157

> Rebecca is a 19-year-old Retrogirl . . . [She] dresses in semi-hippie garb and offsets this with a studded belt and pointed black boots.   *Courier-Mail* (Brisbane) 27 Sept. 1988, p. 17

retrovirus /'rɛtrəʊˌvaɪərəs/ *noun* Also written **retravirus** ⊗

Any of a group of RNA viruses (including HIV) which form DNA during the replication of their RNA.

Formed by adding the initial two letters of *REverse* and *TRanscriptase* and the combining-form suffix *-o* to *virus*; one of the distinguishing characteristics of a *retrovirus* is the presence of *reverse transcriptase*, the enzyme which acts as a catalyst for the formation of DNA from an RNA template.

The family of *retroviruses* was first given the Latin name *Retroviridae* in the mid seventies; during the late seventies there was increasing scientific interest in them, which was boosted in the early eighties by the race to find the viral cause of (and ultimately a cure for) AIDS. It was this connection with Aids that ensured that the word *retrovirus* became popularized rather than remaining limited to technical literature; however, although the word appeared in popular sources in the eighties it was probably not as widely understood as this popular usage would suggest. The corresponding adjective is **retroviral**.

> It turns out this virus is a retrovirus, and it's a close, kissing cousin of the AIDS virus.
> *USA Today* 29 Oct. 1990, section A, p. 13

returnik /rɪˈtɜːnɪk/ *noun* 🔳 🔳

An émigré from an East European country who has returned home, especially after a change of political regime there.

Formed from the verb *return* and the suffix *-nik*, on the model of REFUSENIK.

This inventive formation gave a new lease of life to the *-nik* suffix in English during the second half of the eighties, when the media began to take an interest in the growing number of émigrés from the Soviet Union and other East European countries who wished to return once a more democratic government was in power. The phenomenon of *returniks* had existed before, however: of the people who successfully emigrated from the Soviet Union, for example, there were always a few who found that their ties to the motherland were so strong that they could not be happy anywhere else and who tried to find some way to return home even without a change of political regime there.

> The Gross family are Returniks—Russians who emigrated to the West and have now decided to return. They . . . swapped one of the most prestigious New York addresses, Waterside Plaza in Manhattan, for two dingy rooms which the Grosses, who have three children, share with her [Olga's] mother.
> *Sunday Mail* (Brisbane) 17 May 1987, p. 21

> Known as the returniks, these natives of Czechoslovakia, Poland, Hungary . . . are helping manufacture consumer goods and build housing.
> *Time* 2 July 1990, p. 48

rhythmic gymnastics /ˌrɪðmɪk dʒɪmˈnæstɪks/ *noun* ⊗ 🔳

A form of gymnastics which emphasizes rhythmic movement and incorporates dance-like routines, performed with ribbons, hoops, or other accessories, used as extensions of the gymnast's body.

Formed by compounding: *gymnastics* based on *rhythmic* movement.

Although the phrase *rhythmic gymnastics* was used as long ago as 1912 to refer to a form of gymnastics based on rhythmic movement, it was not adopted as the official name of a recognized style of gymnastics until the seventies, and this style only became a sport which was popularized through international competition in the eighties.

> Bianca Panova . . . , the Bulgarian champion, practising . . . for the Rhythmic Gymnastics International at Wembley Conference Centre tomorrow.
> *Daily Telegraph* 4 Nov. 1989, p. 36

right-to-life /ˌraɪttəˈlaɪf/ *adjectival phrase* ⊗ 🔳

Especially in US English, concerned to protect the rights of the unborn child and therefore opposed to allowing a woman to choose whether or not to have an abortion; pro-life. (A positive alternative to ANTI-CHOICE.) Also, seeking to protect the rights of the terminally ill, people on life-support machines, etc.

Formed from the noun phrase *the right to life*; the focus of the movement is the right of the unborn child to quality of life and the moral responsibility of those who already have life to safeguard the rights of those who cannot speak for themselves. The model for this formation already existed in *right-to-die* (a similar movement against artificially prolonging the life of those who, because of illness or accident, are unable to have any quality of life).

For the history of the anti-abortion debate in the US, see PRO-. *Right-to-life* fits into this picture as one of three terms for the anti-abortion lobby, and has been commonly used in the US since the mid seventies. A supporter of this position is a **right-to-lifer**. Similar moral issues apply to the debate over the artificial 'life' of those who exist for years on life-support machines, and the movement has also concerned itself with this issue.

The right-to-lifers had to pretty much settle for a mad bomber repping their cause.

Movie Winter 1989, p. 8

RISC /rɪsk/ *acronym* Also written **Risc** or **risc** 🖳

Short for **reduced instruction set computer**, a type of computer designed to perform a limited set of operations, and therefore having relatively simple circuitry and able to work at high speed. Also (short for **reduced instruction set computing**), computing using this kind of computer; the simplified environment in which it operates.

The initial letters of *Reduced Instruction Set Computer* (or *Computing*).

Research into the viability of a *RISC* and its advantages over the traditional approach (*complex instruction set computing* or *CISC*) began in the early eighties, and by 1983 had produced the first commercial products based on this principle. It soon became clear that the greatest advantage was speed, with *RISC* working at twice the speed of *CISC*. The acronym *RISC* is nearly always used attributively (in **RISC architecture**, **RISC chip**, **RISC processor**, **RISC system**, etc.); systems, machines, etc. are often described as **RISC-based**, while the software products with which *RISC* is used are known as **RISCware**.

By incompatible microprocessors, I mean the Risc chips: Sparc, Mips, 88000 and 80860 for starters.

PC Magazine July 1989, p. 130

For the same dollar, CISC will deliver only half the performance of RISC.

New York Times 28 Sept. 1989, section D, p. 2

To set standards for RISC-based workstations, MIPS is challenging Sun Microsystems Inc.

New York Times 10 Dec. 1989, section 3, p. 10

The RISCware Product Directory lists 245 software products.

UnixWorld Apr. 1990, p. 91

ritual abuse 🚺 see CHILD ABUSE

rock /rɒk/ *noun* 💉

In the slang of drug users, a crystallized form of cocaine which is smoked for its stimulating effects; an earlier name (especially on the West coast of the US) for CRACK. Also, a piece of crack in its prepared form, ready for smoking.

Named after its *rock*-like appearance and consistency.

Despite suggestions that *rock* has been in use among drug users for some time as a name for a piece of crystallized cocaine, the word did not begin to appear in the newspapers or become known to the general public until the middle of the eighties. Then a number of West-coast newspapers reported raids on **rock houses** (the same as *crack houses*: see CRACK). By 1986, *crack* had become established as the name for the drug itself, and *rock* seemed to be dying out in this sense, but it remained current as the name for a piece of the drug ready for smoking.

Four people were arrested and a small cache of weapons and ammunition seized at an Inglewood 'rock house', where cocaine in hardened form was being sold, Los Angeles police announced.

Los Angeles Times 11 Jan. 1985, section 1, p. 2

The 'rock' is ... put in a pipe and smoked, with far more potent effects than inhaling the powder.

Daily Telegraph 1 Mar. 1985, p. 15

It's amazing now. You walk around Notting Hill or Stonebridge and you can hardly score ganja any more. All you see is rock and smack ... There are certain geezers who go up to someone who's never touched it, give him a rock, and build him up 'til he gets a habit.

Sunday Correspondent 8 Apr. 1990, p. 4

rocket fuel 🗡 see ANGEL DUST

rockumentary /rɒkjʊ'mɛntərɪ/ *noun* Also written **rock-umentary** ✖ ♪

In informal use, a documentary film dealing with the history of rock music or the lives of rock musicians.

Formed by replacing the first syllable of *documentary* with *rock*, making use of the rhyme to form a punning blend.

The word was coined by Ernie Santosuosso, a Boston critic, in a review of *Beatlemania* in 1977, at a time when the craze for films and television programmes about rock stars of the sixties was a favourite means of bringing rock music to a wider audience. It has remained principally an American word, but by the mid eighties had also been used in British and Australian film criticism. By 1984, *rockumentaries* were so numerous that a completely fictional one (*This is Spinal Tap*, about a British HEAVY METAL group) was made as a parody of the genre.

> Spinal Tap lives: the famed 'henge' sequence from the classic 'rockumentary' was recently re-enacted.
> *Music Making* July 1987, p. 6

> SBS at 7.30 has what it is billing a 'rock-umentary'—an account of Australian singer Jeannie Lewis' last trip to Mexico.
> *Courier-Mail* (Brisbane) 15 Nov. 1988, p. 28

role-playing game /'rəʊlpleɪŋ ˌgeɪm/ *noun phrase* ✖ 🎮

A game in which players take on the roles of imaginary characters who take part in adventures in a (usually fantastical) setting.

Formed by compounding: a *game* which involves the *playing* of a *role*.

The concept of the *role-playing game* (often abbreviated to **RPG**) brings together two much older ideas. The planning of real military campaigns with the aid of boards and counters led earlier this century to an interest in re-enacting famous historical battles, and even completely fictitious ones, in a similar manner—an activity known as *war-gaming*, which became particularly popular in the years following the Second World War. The second idea grew out of psychotherapy, which also enjoyed something of a vogue in the sixties: the technique of **role-playing**, devised by the Viennese psychiatrist J. L. Moreno in the forties, whereby people were encouraged to act out dramatic roles. The technique spread to other fields, and the phrase became generally familiar, so that in the later seventies, when several games appeared which allowed players to immerse themselves more fully in the imaginary setting than had been possible in conventional war-gaming, the name *role-playing game* came readily to mind. Perhaps the best known of these is *Dungeons and Dragons*, which like many such games has a fantasy setting. What makes such games distinctive, however, is not the setting—other games draw on science fiction, ancient Rome, and even gangster novels for their inspiration—but the extent to which the adventure is made as 'realistic' as possible: the setting is painstakingly created in great detail, often by a referee or *Dungeon Master*, and the behaviour of the players' assumed characters is controlled by a welter of rules designed to make the experience believable. Players do not necessarily 'win'—the enjoyment derives from vicariously 'living' another, more exciting life. During the early eighties **gaming** of this sort was consequently condemned as escapism, but it has flourished despite such criticism; indeed, the appearance of the home computer, and of software allowing still more realistic **role-play**, has vastly increased its popularity.

> With role-playing games, the position is different. The rules explain how to generate characters.
> *White Dwarf* Oct.–Nov. 1981, p. 8

> CoC [*Call of Cthulhu*] is a classic RPG . . . casting its shadow over the whole gaming industry.
> *GM* Nov. 1989, p. 18

roof tax /'ru:f ˌtæks/ noun 〰 🏠

In the UK, a derogatory nickname for any property-based replacement for the COMMUNITY CHARGE or POLL TAX.

Formed by compounding; whereas the *poll tax* is a tax on heads (see POLL TAX), the *roof tax* taxes people on the *roof* over their heads.

The nickname *roof tax* first arose as a Conservative retort to Labour politicians' attacks on the community charge and their insistence on calling it a *poll tax*; any Labour government, they said, would remove the community charge only to replace it with an even more unfair *roof tax*, based on the same principles as the old rating system. When the Conservative government announced its review of the community charge in April 1991 and it became clear that the proposed new *council tax* was likely to be based—at least in part—on property ownership, Labour politicians were able to turn the taunt back on the taunters, calling the council tax a *roof tax* (as well as a great many other names).

> The worst outcome would be Labour's roof tax which, by combining a property tax with one on incomes, really could be used to squeeze the rich. *The Times* 8 Mar. 1990, p. 14

roots /ru:ts/ plural noun 🎵

Ethnic origins seen as a basis for cultural consciousness and pride, especially among Blacks; often used attributively as though it were an adjective: expressing this cultural identity, ethnically authentic.

The word *root* has been used in the plural to mean 'one's social, cultural, or ethnic origins or background' since the twenties; the shift in meaning that has led to the word's association with (specifically Black) cultural heritage probably arose from the popularity of Black American author Alex Haley's family chronicle *Roots* (1976), based on research into his own family history and African origins, which won a special Pulitzer prize in 1977.

This more specific sense of *roots* developed during the late seventies, perhaps as a direct result of the success of the Haley book. At about the same time it started to be used attributively, especially in **roots reggae** (a style of music originating in Jamaica which was designed to express Jamaican cultural identity) or **roots music** (sometimes meaning the same as *roots reggae*, but often applied more generally to any music which expresses the cultural identity of a particular ethnic group—ETHNIC music—or has the authentic sound associated with Black cultural origins).

> For the DJ, crossing over is more than simply a move from roots to respectability or even from black to white audiences. *City Limits* 16 Oct. 1986, p. 41

> Biddy's will continue its prior booking policy—an eclectic blend of oldies acts, roots music, world beat and other styles. *Chicago Tribune* 25 Aug. 1989, section 7, p. 8

rootsy /'ru:tsɪ/ adjective 🎵 🎸

(Of music) down-to-earth; in a rudimentary, uncommercialized style which allows traditional or ethnic roots to show through.

Formed by adding the adjectival suffix *-y* to *roots*.

Rootsy shares its early history with ROOTS above, but developed a rather broader meaning during the eighties, moving outside the narrow context of Black or West Indian cultural awareness. Any music (or sometimes another area of culture) can be described as *rootsy* if it has an authentic feel, without the rough edges having been smoothed off by commercialism.

> I'm not here to put any new innovations on you . . . I'm still using things that are already there: the basic American rootsy sound with country and blues and so forth.
> *Los Angeles Times* 21 May 1986, section 6, p. 2

> He went from the depth-charged super-funk of 'Head', straight into the buoyant and rootsy pop of 'When You Were Mine'. *The Times* 26 July 1988, p. 14

Royal Free disease ⊗ see ME

RPG ✖ 🔲 see ROLE-PLAYING GAME

Rubik /ˈruːbɪk/ *noun* 🔳

Part of the name of a number of mathematical puzzles devised by Hungarian teacher E. Rubik; originally **Rubik's cube**, the trade mark of a puzzle consisting of a cube built round a double fulcrum from 26 smaller cubes of which each visible face shows one of six colours, each layer of nine cubes being capable of rotation in its own plane, the task being to restore each face of the larger cube to a single colour after the uniformity has been destroyed by rotating any of the layers.

The surname of the inventor.

Rubik's cube was first marketed under this name in 1980 (it had originally been called the *Magic Cube*), and immediately enjoyed great commercial success, sparking off a craze of similar proportions to the ones later caused by *Transformers* and TURTLEs. Rubik's puzzles (the cube was later followed by **Rubik's triangle** and other puzzles on the same principle) attracted adults as well as children.

> Bűvös Kocka—the Magic Cube, also known as Rubik's Cube—has simultaneously taken the puzzle world, the mathematics world and the computing world by storm.
>
> *Scientific American* Mar. 1981, p. 14

> The life of the modern toy designer is an unending search for the next ... Rubik's Cube, the next teenage Mutant Ninja Turtle.
>
> *Smithsonian* Dec. 1989, p. 73

Rule 43 /ˌruːl ˌfɔːtɪˈθriː/ *noun* ⬛

A prison regulation in the UK whereby an inmate considered to be at risk from the rest of the prison community (for example, because of the nature of the offence that he or she has committed) may be placed in solitary confinement for his or her own protection. Also, a prisoner isolated under this rule (sometimes abbreviated to **43**).

The paragraph number of the rule in the Prison Rules.

The rule has been in force since at least the early seventies; what brought the question of segregation under *Rule 43* to public attention was discussion in the media of the prison riots at Strangeways Prison in Manchester in April 1990, when it became clear that rioting prisoners had quickly broken into the segregated areas where *Rule 43* prisoners were kept in order to attack them.

> Do not suppose that 43s are necessarily the most evil. They may be, they may not be. What is unique to them is their fear.
>
> *Daily Telegraph* 3 Apr. 1990, p. 16

> Most violence was aimed at the vulnerable Rule 43 prisoners ... [Sexual offenders] make up to 70 per cent of the Rule 43 prisoners.
>
> *Independent* 3 Apr. 1990, p. 2

Rust Belt /ˈrʌst ˌbɛlt/ *noun* 〰

Colloquially in the US, the declining industrial heartland of the Midwest and North-East United States, especially the former steel-producing areas such as Pittsburgh.

Humorously formed by compounding: a *belt* or zone where once-profitable industry (in particular the metals industry) is left to *rust* away.

The coinage of the term is often attributed to US Democratic politician Walter Mondale, who opposed Ronald Reagan in the presidential election of 1984. Attacking Mr Reagan's economic policies, Mr Mondale said

> His ... policies are turning our great industrial Midwest and the industrial base of this ... country ... into a rust bowl.

This was picked up in the media and repeated as *Rust Belt*. Although Mr Mondale's presidential campaign was unsuccessful, the plight of the American *Rust Belt* remained a political issue in the US. The term is often used attributively.

> We might look upon the glory of our Rust Belt states, where there are hundreds of vast steel mills that are at least 40 years out of date and also spew smoke that causes acid rain.
>
> *New York Times Book Review* 29 Oct. 1989, p. 48

• •

S

sab /sæb/ *noun* and *verb* ✖ 🔦

Colloquially in the UK,

noun: An opponent of blood sports who disrupts a fox- hunt as a form of protest, a hunt saboteur; also known more fully as a **hunt sab**. Also, any animal rights campaigner who engages in sabotage.

transitive or *intransitive verb*: To disrupt (a hunt) as a hunt saboteur; to go on a **sabbing** expedition.

Formed by abbreviating *saboteur* to its first three letters.

The word arose among hunt saboteurs as a name for themselves and started to appear in print towards the end of the seventies. As the movement against blood sports grew during the eighties, so the terms *sab* and *hunt sab* became increasingly common in the newspapers. It was also sometimes used more generally for animal rights campaigners whose action involved sabotaging scientific experiments etc.

> The battle between the hunters and the 'sabs' is now an integral part of the hunting scene. He is a veteran of countless sabbing missions.   *Sunday Times* 6 Mar. 1983, p. 11

> The sabs made a point of photographing their quarry in the lab before rescuing them, and on publication, these heart-rending photographs of dogs, being experimented on . . . raised a public outcry.
>
> *Illustrated Weekly of India* 13 July 1986, p. 44

> For two seasons I went and 'sabbed' my local hunt.   *Peace News* 19 Sept. 1986, p. 9

safe /seɪf/ *adjective* 💀

In young people's slang: good, sound, having street cred (see CRED¹).

A sense shift which possibly arises from the sensitivity of young people involved in street culture to peer pressure, and in particular to ridicule from peers: a person or thing that is *safe* is one that meets with approval.

Safe became a popular adjective of general approbation towards the end of the eighties, especially in the phrase **well safe**. As a piece of slang used among a small group of people it was naturally limited largely to spoken use, and rarely appeared in printed sources.

> British Knights, Nike Jordans and Nike SEs are 'well safe', but copies like Nicks are the object of pure derision.   *New Statesman* 16 Feb. 1990, p. 12

safe sex /ˌseɪf ˈsɛks/ *noun* Also in the form **safer sex** ✗ 🔦

Sexual activity in which precautions are taken to ensure that the risk of spreading sexually transmitted diseases (especially AIDS) is minimized.

Formed by compounding: *sex* which is *safe* as regards the risk of contracting or spreading Aids. This combination of words was probably already in use in relation to contraception: *sex* which was *safe* from the point of view of unwanted pregnancy. However, it was only in relation to Aids that it became a fixed phrase with a specific meaning.

The concept of *safe sex* (or *safer sex*, as some preferred to call it) arose in the mid eighties, as first American society, and later other societies as well, started to face up to the threat of Aids and think of ways in which it might be controlled. Awareness of the need for *safe sex* and general publicity about it were commonest at first among the gay community, but by the second half of the decade the message was being put across deliberately to all sections of society through health advertising. The main elements of *safe sex* as highlighted in government advertising campaigns were avoidance of promiscuity (by having a single partner) and the use of a CONDOM as a barrier to the exchange of 'body fluids' during intercourse.

> While the city's major bathhouses and clubs . . . are still in business, . . . a few of the owners have been helpful in educating clients about safer sex.   *New York* 17 June 1985, p. 52

> The gay community . . . is now practicing safe sex so conscientiously that the rate of newly infected homosexual men in cities like San Francisco and New York has fallen dramatically.
> *Life* Fall 1989, p. 135

> Part-parody, part safe-sex education, her presentation uses a combination of home movies, slides, vignettes.   *Mediamatic* Summer 1990 (Edge 90: Special Issue), p. 230

sailboard, sailboarder, sailboarding 🔀 see BOARDSAILING

salmonella-free ⊗ see -FREE

sampling /ˈsɑːmplɪŋ/ *noun* 🔊 🖳

In electronic music, the technique or process of taking a piece of digitally encoded sound and re-using it, often in a modified form, as part of a composition or recording.

A specialized use of *sampling*, which would normally be used in the context of quality control or the taking of statistical samples.

Sampling became an important technique in musical composition (especially in popular music) in the mid eighties, as a direct result of the advances in electronics and musical technology which followed from the development of the synthesizer. The music which developed from these techniques (including ACID HOUSE, HOUSE, and TECHNO) has a patchwork quality, since it is formed from many different sequences of modified sound. Associated terms include **sample** (a noun and verb), the adjective **sampled** (used of a sound or a whole sequence of music), and the noun **sampler** (the electronic instrument—actually a musical computer—which is used to sample sounds).

> With new-romanticism, techno-pop, the revival of disco and growth of synthesized sound, from sampling to scratch, the potential for live performance waned.   *Guardian* 11 Aug. 1989, p. 24

> Advanced Midi Amiga Sampler, High Quality Sound Sampler & Midi interface including all necessary Software . . . The sound is stunning, too. All effects are sampled, and very atmospheric.
> *CU Amiga* Apr. 1990, pp. 27 and 43

SAT 🎓 see NATIONAL CURRICULUM

satanic abuse 🎓 see CHILD ABUSE

satellite /ˈsætəlaɪt/ *noun* 🔀 🖳

Short for **satellite broadcasting** or **satellite television**, the transmission of television programmes via an artificial satellite; a special television service using this technique and receivable by subscribers who have paid a fee and own the appropriate **satellite dish** or other antenna.

Formed by abbreviating *satellite television*; the *satellite* is the link in space between the broadcaster and the subscriber, with signals being beamed up to it and the antennas or *dishes* so positioned that they can receive the re-transmitted signal.

Satellite television (at first known officially by the more cumbersome name *direct broadcasting by satellite*) was first tried experimentally in the late sixties, and the use of satellites for broadcasting became commonplace during the following decade. When, in the eighties, communications satellites were launched with the express purpose of providing a television service to compete with network television, the term *satellite television* and its abbreviated forms *satellite TV* and *satellite* came to be applied specifically to these competing services, while *direct broadcasting by satellite* (or simply *direct broadcast*) had to be used for the technique when employed by network stations. *Satellite* was introduced in the UK in the late eighties by two competing stations, *Sky TV* and *BSB*, later merged as *BSkyB*. The unsightliness of the parabolic *dishes* used to receive *satellite* programming led to their being banned by some local authorities and there were moves to use *cable* (see CABLE TELEVISION) to 'pipe' the programmes from a central reception point to individual homes in these areas.

> There are also several monthly magazines with a mix of technical information and features about the films and other programmes on satellite and cable. *Which?* Sept. 1989, p. 444

> While the dollarless majority [in Poland] live in half-finished apartment blocks [and] walk to queue in zloty shops . . . the *wydeos* live in ugly villas, drive to shop at Pewex and display satellite dishes in their garden. *Correspondent Magazine* 29 Oct. 1989, p. 37

SBS Ⓧ see SICK BUILDING

scratch /skrætʃ/ *noun* and *verb* 🎵 📷

noun: A technique, often used in RAP music, in which a record is briefly and repeatedly interrupted during play and manually moved backwards and forwards to produce a rhythmic scratching effect; also, the style of music characterized by this (known more fully as **scratch and rap** or **scratch music**). Also used in other compounds, including:

scratch-mix, a style of popular music in which several records are intercut with each other as they are played, using the **scratch technique** to create a 'collage' of sound; also used as a verb or as an action noun **scratch-mixing**;

scratch video, a technique or game of video-making, in which a number of short, sharp images are cut and mixed into a single film and fitted to a synchronized sound track (usually of rap music); a video made by this method.

transitive or *intransitive verb*: To manipulate (a record) using the scratch technique; to play scratch music or act as a **scratch disc jockey**.

A reference to the *scratching* effect of the original technique.

Scratch music originated in rap and HIP HOP culture in the early eighties; a Scratch 'n' Rap revue was put on in New York in 1982, and the technique was also popularized by disc jockeys who used it in a number of New York clubs. The same principle was applied to video by 1985, giving *scratch video*, itself sometimes abbreviated to *scratch* alone.

> On Tuesday, Mr. Hancock and a band that included the 'scratch' disk jockey Grand Mixer D. Street appeared at the Ritz. *New York Times* 25 Dec. 1983, section 1, p. 47

> Brad Shapiro . . . produces her outrageous records and stage show, backed by a fine funk outfit, flavored with horns and the latest scratch and synth sounds.
> *Washington Post* 27 Apr. 1984, Weekend section, p. 37

> The Rockit Band includes Grandmixer D. ST., whose instrument is a turntable and who makes sounds by 'scratching' records back and forth. *New York Times* 17 June 1984, section 2, p. 28

> Scratch is a playful reaction to the endless offerings and noise of 'the media'. It interrupts the normal passive flow of TV, bends it a bit. *Honey* June 1985, p. 18

> A simple scratch can be built up by recording the chosen music/sound onto the audio channel of the video recorder then switch between channels as the vision is being recorded.
> *Photographer* May 1986, p. 26

Pete Shelley's move from The Buzzcocks to a 12" gay classic 'Homo-Sapiens' and John Lydon's re-arranged public image, appearing with scratch-mix pioneer Africa Bambaattaa, the self-proclaimed Zulu warrior of the hip hop scene, compounded the drift. *New Musical Express* 14 Feb. 1987, p. 27

The 12" dance record is an inevitable liaison with the hi-technology of synthesisers and the rough treatment of rap and scratch. *New Musical Express* 14 Feb. 1987, p. 27

scrunch /skrʌntʃ/ *transitive verb* 🔲

To style (hair) by squeezing or crushing with the hands to give a tousled look. Often in the verbal phrase **scrunch-dry**, to blow-dry (hair) while squeezing or crushing it in this way, in order to set it with a crinkled or tousled effect.

Probably a blend of *squeeze*, *crumple*, *crush*, and *crunch*, originally intended to sum up the action and sound of screwing up a piece of paper in the palm of the hand.

Scrunch first started appearing in hairdressing magazines in about 1983; the technique of **scrunch-drying** followed from about 1985. Both terms spread outside the professional hair-dressing press to general-interest magazines during the second half of the eighties.

Rod just used mousse and a scrunch-drying technique to give it more body and to make it . . . more modern. *Good Housekeeping* May 1986, p. 43

To style, he used mousse and his hands to scrunch her hair into a beautiful halo of curls. *Hairdo Ideas* July 1987, p. 58

Scud /skʌd/ *noun* Sometimes written SCUD 🔥

The NATO code-name (more fully **Scud missile**) for any of a class of long-range surface-to-surface guided missiles originally developed in the Soviet Union, capable of carrying different kinds of warhead, and launchable from a mobile launcher.

Although sometimes written in capitals, *Scud* is not an acronym; the word *scud* was chosen as part of a series of NATO code-names for Soviet surface-to-surface missiles, all of which conventionally began with *s*: other examples included *Savage*, *Sandal*, *Scapegoat*, and *Scrooge*. Similar series of names (beginning with *g*, *k*, and *a* respectively) were chosen for surface-to-air, air-to-surface, and air-to-air missiles.

The *Scud* missile system (first the **Scud A**, and later the **Scud B**) was designed and made in the Soviet Union in the late fifties and early sixties and was soon exported to the Warsaw Pact and other countries friendly to the Soviet Union. *Scuds* were used in the conflict in Afghanistan in the second half of the eighties, and were sometimes mentioned in news reports; what really brought the *Scud* into the news in English-speaking countries, though, was its deployment by Iraq during the Gulf War of January–February 1991. *Scuds* were launched against allied forces in Saudi Arabia and, more controversially, against Israel (a state not otherwise involved in the conflict). Since the *Scud* is capable of carrying conventional, chemical, or biological warheads, *Scud* attacks were seen as a significant threat to the civilian population in Israel and Saudi Arabia; in the event only conventional warheads were used, but there were significant numbers of civilian casualties, especially in Israel. The fact that the missiles were launched from mobile launchers made it difficult for allied air power to locate and destroy the sources of the attacks; their effectiveness was minimized, however, by the success of PATRIOT missiles in intercepting and destroying many of them before they reached their targets. By February 1991 there was already a little evidence to suggest that *Scud* would develop a figurative sense, 'a devastating or unpredictable attack', much as EXOCET had done after the Falklands War.

Now, bad weather in the region and the failure to knock out the Scuds had prolonged the aerial campaign. *Newsweek* 28 Jan. 1991, p. 17

The Sacks/Williams of the film is what Pauline Kael of the *New Yorker*, in one of her critical Scud missile moods, describes as 'another Robin Williams benevolent eunuch role'. *Independent on Sunday* 17 Feb. 1991, p. 21

scuzz /skʌz/ *noun* Also written **scuz** 🔲

In young people's slang (originally in the US): a disgusting person or thing; something or someone considered **scuzzy**.

Probably an abbreviated form of *disgusting* (representing the actual sounds pronounced in the second syllable when the word is drawn out to emphasize the speaker's revulsion); it has been suggested that it might however be a blend of *scum* and *fuzz*.

Scuzz has been in spoken use among US teenagers since the sixties; it seems it first appeared in print in 1968, while the corresponding adjective *scuzzy* was recorded a year later. During the eighties *scuzz* became the basis for a number of compounds, proving that it had become established in the language: the most important of these were **scuzzbag**, **scuzzball**, and **scuzzbucket**, all nouns meaning 'a contemptible or despicable person' and also used as general terms of abuse. All of these variations on the same theme appeared during the mid eighties and started to become known outside the US in the late eighties. The quality of being *scuzzy* is **scuzziness**.

> He calls a minister a 'scuzzbag'. *Time* 11 July 1983, p. 72

> In the larger picture, we're just a little green scuzz on the surface.
> Margaret Atwood *Cat's Eye* (1988; 1989 ed.), p. 230

> Her cheating husband, Ernie, a crotch-grabber who brings new meaning to the word 'scuzzbucket'.
> *Newsday* 17 Sept. 1989, TV Plus section, p. 85

scuzzy 🔲 see SCUZZ and GRODY

SDI 🔲 see STAR WARS

SEAQ 🔲 see BIG BANG

Securitate /sɪk(j)uːrɪˈtɑːteɪ/ *noun* 🔲

The internal security force (until December 1989) of the Socialist Republic of Romania.

A direct borrowing from Romanian *securitate* 'security'; this in turn is a colloquial abbreviation in Romanian of the official name, *Departamentul pentru Securitatea Statului* 'Department for State Security' (the *Securitate* was a Department of the Ministry of the Interior).

Securitate was the colloquial name in Romanian of the feared Communist secret police under the Ceauşescu regime (and before—the *Departamentul pentru Securitatea Statului* was set up in 1948). The word was only rarely used in English during the sixties and seventies; what really brought it into the news and gave it some currency in English was the overthrow of that regime in December 1989. News reports from Romania in late 1989 covered popular demonstrations against the *Securitate* and attempts to ransack its offices and destroy its files. The *Securitate* was officially disbanded in December 1989 and a National Salvation Front decree ratified this on 1 January 1990; in March 1990 a new security service was set up under the direct control of the President, and this was named *Serviciul Român de Informaţii* 'Romanian Information Service'. This organization took over the duties of the Intelligence section of the old *Securitate*, but subject to formal guarantees that there would be no abuses of power such as those seen under the *Securitate* itself.

> The beliefs that they are constantly watched by the regime's political police, the Securitate, more than suffices to convince Rumanians to keep their thoughts to themselves.
> *New York Times* 24 Nov. 1989, section A, p. 17

> The Ceauşescus' execution weakened the resistance of the hated secret police, the Securitate, who had been mounting indiscriminate attacks on army units and civilians in an unsuccessful attempt to crush the revolution. *The Annual Register 1989* (1990), p. 127

sellathon /'sɛləθɒn/ *noun* Also written **sell-a-thon**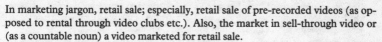

In marketing jargon (especially in the US): a concentrated attempt to sell, as in an extended cut-price sale, a television programme entirely devoted to the advertisement of a sponsor's products, or a marketing convention.

Formed by adding the suffix *-athon* (as in *marathon*) to the verb *sell*.

This is an American coinage of the second half of the seventies which has been applied in a wide variety of contexts, although almost exclusively within the US. Essentially, it seems, any marketing 'marathon' can be a *sellathon*.

> Anyone else embarking on such a sellathon, should run a few VTR screen tests before making their final choice of presenter. *Broadcast* 29 May 1978, p. 20

> [The] marketing program for 1989 was outlined to Nugget Distributors members at the group's January Sellathon in Honolulu. *Institutional Distribution* Mar. 1989, p. 48

sell-by date /'sɛlbaɪ ˌdeɪt/ *noun phrase* ✖

A date marked on food packaging (usually preceded by the words 'sell by') to indicate the latest recommended date of sale, especially for perishable goods. (The British equivalent of the US *pull-by date*.)

Formed by compounding: the *date by* which the retailer should *sell* or discard the goods.

For history, see BEST BEFORE DATE and USE-BY DATE. Like *best before date*, *sell-by date* has occasionally been used in a transferred context or figurative sense.

> Socialism: the package that's passed its sell-by date.
> headline in *Daily Telegraph* 13 Mar. 1987, p. 16

> New legislation is to be introduced to replace sell-by dates with more helpful use-by dates ... More than eight out of ten people in our survey said they never buy food after its sell-by date has passed; only two per cent said they frequently do. *Which?* Apr. 1990, p. 205

sell-through /'sɛlθruː/ *noun* Also written **sell through** or **sellthrough** ✖ ✖

In marketing jargon, retail sale; especially, retail sale of pre-recorded videos (as opposed to rental through video clubs etc.). Also, the market in sell-through video or (as a countable noun) a video marketed for retail sale.

Formed by turning the verbal phrase *sell through* into a compound noun: the principle of *selling* right *through* to the end user, rather than to a rental outlet.

Sell-through was already in use in marketing in the more abstract sense of the level of retail sale (turnover) in the late seventies. The more specific sense in the video market developed as a direct result of the video boom of the first half of the eighties, followed by a slackening of interest in the second half: video manufacturers were forced to put greater effort into marketing their product through retail outlets once interest in video rental started to fall off. From about 1985 onwards, *sell-through* was frequently used attributively in relation to video, in **sell-through market**, **sell-through video**, etc. By about 1987 *sell-through video* was being further abbreviated to *sell-through* alone, resulting eventually in the use of *sell-through* as a countable noun.

> Gregory is convinced that many less obvious outlets could be stocking sell through video profitably. 'Hi-fi shops which sell hardware should have a lot of potential for stocking sell through', he says.
> *Music Week* 20 June 1987, p. 36

> Slackening sales of pre-recorded video cassettes for rental purposes have forced many small video publishing companies to sharpen their focus on 'sell-throughs'. *Sun* (Brisbane) 11 May 1988, p. 39

> Some of the best are currently available on sell-through video ... Doubtless others will appear on sell-through before long. *Empire* Sept. 1989, p. 93

Semtex /'sɛmtɛks/ *noun* 💥

A very malleable, odourless plastic explosive.

The name given to the product by its manufacturer; probably formed from the first part of *Semtín* (the name of the village in East Bohemia, Czechoslovakia, near the *Semtex* factory) and *-ex* (perhaps standing for the initial syllable of *explosive* or *export*).

Semtex was originally a secret Czech military invention and was probably first made during the seventies, although not under this name. Its manufacture was taken over by the East Bohemia Chemical Works in Pardubice-Semtín; it has been known as *Semtex* to intelligence sources outside Czechoslovakia since about 1982. *Semtex* had a number of non-aggressive uses, for example in the construction industry; however, its lack of odour and its malleability made it a favoured explosive for terrorist bombs as well, since it could be concealed easily and was difficult for sniffer dogs to detect. It was this use by terrorists which brought the word *Semtex* into the news in English-speaking countries from about the middle of the eighties.

> Police officials told Agence France-Presse that the explosive might have been Semtex, which they called the 'signature' explosive of Middle Eastern terrorist groups.
> *New York Times* 9 Dec. 1985, section A, p. 7

> The Czechs were replying to a Foreign Office request for help in fighting terrorism and in tracing the growing consignments of Semtex reaching the IRA from Col Gaddafi of Libya.
> *Daily Telegraph* 27 Aug. 1988, p. 1

Senderista /ˌsɛndəˈriːstə/ *noun and adjective* 📕

noun: A member of the revolutionary Peruvian guerrilla organization **Sendero Luminoso** (sometimes abbreviated to **Sendero** or translated *Shining Path*).

adjective: Of or belonging to *Sendero Luminoso* or its members.

A borrowing from Spanish *Senderista*. The Spanish name is formed by adding the suffix *-ista* (equivalent to English *-ist*) to the stem of *Sendero* 'path'; *Sendero Luminoso*, which literally means 'shining path', is taken from the writings of an earlier Peruvian ideologist, José Carlos Mariátegui: 'Marxism-Leninism will open the shining path to revolution.'

Sendero Luminoso, a neo-Marxist Peruvian revolutionary movement, was founded in 1970 as the Communist Party of Peru, but subsequently became a clandestine guerrilla organization which was active throughout the eighties. The activities of the *Senderistas* were reported in the newspapers, especially in the US, from about 1982 onwards.

> Shouting Senderista slogans and songs, the peasants escorted the group to the community meeting hall.
> *New York Times Magazine* 31 July 1983, p. 20

> Deriving their communist ideology from the teaching of Mao Tse-tung, the Senderistas are led by Abimael Guzman (nom de guerre, Col. Gonzalo), a hermit-like former professor of philosophy at the University of Ayacucho.
> *Maclean's* 25 Feb. 1985, p. 44

> Unlike other revolutionary movements . . . Sendero hasn't opened itself to journalists: there have been no clandestine interviews with leaders, no conducted tours of areas under Sendero control.
> *New Yorker* 4 Jan 1988, p. 35

> The treasury is so empty that the government . . . certainly cannot pay all the soldiers needed to protect candidates around the country from the fanatical Sendero Luminoso guerillas.
> *Observer* 1 Apr. 1990, p. 17

sensitive 🌱 see ENVIRONMENTALLY

sequencer /'siːkwənsə(r)/ *noun* 🎵 🖥

A programmable electronic instrument which can store sequences of musical notes, chords, or other signals and reproduce them when required, usually as part of a musical composition.

A specialized sense of *sequencer*, which had been used since the fifties for a number of electronic devices that put information in *sequence*.

Sequencers first became available in the mid seventies, but it was not until the early eighties and the development of MIDI that they started to be widely used. The *sequencer* proved an essential piece of equipment for the electronic music styles of the eighties, with their patchwork or collage-like quality; HOUSE music, in particular, relied heavily on this technology.

> The Synclavier also has a 'sequencer', which is like a word processor for music: you can use it to program the machine to play 'Chopsticks' for you.   *Listener* 24 Oct. 1985, p. 43

> Musicians create their rhythm patterns in the sequencer rather than on the drum machine.
>   *Rhythm* Mar. 1989, p. 30

serious[1] /'sɪərɪəs/ *adjective* 🌦

In business jargon: considerable, worth taking seriously. Used especially in **serious money**, a large sum of money.

A development of sense which relies on a kind of shorthand: it is not the money, the commodity, etc. that is *serious*, but the intention of the person offering it. Thus a *serious offer* of money, for example, became *serious money*.

This is a well-established US business usage (it has been in colloquial use for several decades). It became current in other English-speaking countries in the second half of the eighties and increasingly found its way into print. According to some business executives, the fixed phrase *serious money* can be tied down to a figure containing a specified number of noughts; whether or not in this phrase, *serious* tends to be preceded by the verb *talk*, used transitively.

> Bankability: Serious money. Recent two-book deal with Viking earned him more than £150,000.
>   *Correspondent Magazine* 29 Oct. 1989, p. 66

> She wore these three-inch heels ... I'm talking serious stiletto.
>   Alice Walker *Temple of My Familiar* (1989), p. 244

serious[2] /'sɪərɪəs/ *adverb* 🗿

In young people's slang: very, truly, absolutely. Used especially in **serious bad**, really bad.

Formed by using the adjective *serious* in place of its corresponding adverb *seriously*, in much the same way as *real* had been shifted from adjective to adverb qualifying another adjective several decades previously.

Serious used as a general intensifier, especially to qualify the adjective *bad*, seems to have originated among US Blacks and has been recorded in print since the mid eighties (although it almost certainly goes back further in speech). In the phrase *serious bad* it possibly has the function of alerting the hearer to the fact that *bad* is being used in its traditional or *serious* sense, rather than the opposite slang sense 'good' (for which see BAD).

> With his top lip curled to signify contempt, he goaded an imaginary hapless friend: 'You a *lame* chief, well lame, *serious* lame!'   *New Statesman* 16 Feb. 1990, p. 12

> Those of you who have been popping pills and smoking dope are doing the same thing Len Bias did. Those are serious bad shots you're taking boys, serious poor judgements that you're using with your body and mind.   *New York Times* 20 Aug. 1990, section C, p. 6

shareware 🖳 see -WARE

shark repellent 🌦 see POISON PILL

shell suit /'ʃɛl ˌs(j)uːt/ *noun* Also written **shellsuit** ❄

A double-layered track suit with a showerproof outer nylon shell and a soft cotton lining.

Formed by compounding: a *suit* with an outer *shell*.

The *shell suit* suddenly became a fashion garment for general leisure wear (whether or not this involved any kind of sport) at the end of the eighties, a time when HIP HOP culture had already popularized casual sportswear and turned the running-shoe or *trainer* into a status symbol. The *shell suit* has the advantage of doing away with the need for outdoor clothing, since the outer nylon shell is showerproof and moderately windproof, and the trapping of air between the layers makes for warmth. *Shell suits* are typically brightly coloured, with panels or flashes of different colours across the sleeves, legs, and front.

> Shell suit by Adidas. Strong nylon outer. Hardwearing suit features two side pockets, attractive contrast piping.
> *Burlington Home Shopping Catalogue* Autumn–Winter 1989/90, p. 554

> With the trainers go a garish array of track suits—known as 'shell suits'.
> *Daily Telegraph* 9 June 1990, p. 13

shiatsu Ⓧ see ACUPRESSURE

Shining Path ⌂ see SENDERISTA

shopaholic /ˌʃɒpəˈhɒlɪk/ *noun* ✖ ⫚

Colloquially, a compulsive shopper.

Formed by adding the suffix *-aholic* (as in *workaholic*, ultimately on the model of *alcoholic*) to the verb *shop*.

The phenomenon of the *shopaholic* and the associated social problem of **shopaholism**, or compulsive shopping, came to light as a result of the credit boom of the early eighties and were first so named in the US during the mid eighties. Both terms have remained predominantly American, although the problem they describe is not limited to the US. *Shopaholic* is a considerably better-known word than *shopaholism*.

> [The rumour] that Diana is a 'shopaholic' ... was described as 'absolute rubbish'.
> *Washington Post* 11 Sept. 1984, section C, p. 3

> Shopaholism has been described as being like alcoholism, affecting people from all walks of life ... One finance adviser said some consumers who rang up huge credit card bills, far above their financial limit, knew how to budget but simply did not want to. *Sunday Sun* (Brisbane) 22 Mar. 1987, p. 39

> Studies show that perhaps as many as 24 million Americans, fully 10% of the population, can be classified as 'hard-core shoppers'. These shopaholics shop for shopping's sake.
> *Forbes* 11 Jan. 1988, p. 40

shopping-bag lady ⫚ see BAG PEOPLE

shopping-bag stuffer � ✖ see BAGSTUFFER

shuttle /ˈʃʌt(ə)l/ *noun* 🔲

More fully, **space shuttle**: a rocket-launched space vehicle with wings, enabling it to land like an aircraft and be used repeatedly.

A specialized use of *shuttle* in the sense of 'transport which plies backwards and forwards between two points'; the spacecraft is designed to be able to *shuttle* between the Earth and a space station or other destination in space.

The first mention of a *shuttle* to take people to and from space was a fictional one: in a story in *New Worlds* in 1960, John Wyndham wrote:

> The acceleration in that shuttle would spread you all over the floor.

It was at the end of the sixties that the US space agency NASA first started to plan a real *space shuttle*, a re-usable and relatively inexpensive spacecraft that could be used to ferry people and materials to and from a space station. The idea was that the *shuttle* would be fired vertically, but would shed its fuel tanks in space and would then re-enter the atmosphere and glide to a horizontal landing on a runway like that used by an aircraft. The *shuttle* which resulted from

NASA's programme (officially known as the *Space Transportation System* or *STS*) made its maiden flight in 1981, and looked very much as had been envisaged at the beginning of the project: an aircraft-like winged orbiter, protected by heat-resistant materials so that it did not burn up on re-entering the atmosphere, and riding 'piggyback' on the fuel tank and booster rockets. During the eighties four US *shuttles* (*Columbia*, *Challenger*, *Discovery*, and *Atlantis*) were put into service—principally to launch and repair orbiting satellites and to carry out experiments in the Spacelab—and news reports of *shuttle* flights became commonplace. When, in 1986, *Challenger* exploded shortly after take-off, killing the seven astronauts on board, the US *shuttle* programme was temporarily halted, but it was resumed towards the end of the decade. A number of other countries developed *shuttle* programmes during the eighties.

> The NASA concept for an advanced shuttle . . . could bridge the gap between the present fleet and the horizontally-launched National Aerospace Plane single stage to orbit (SSTO) vehicle planned for the next century.
> *Physics Bulletin* Mar. 1987, p. 91

> Mac . . . argued . . . against NASA's space monopoly and its 40 percent subsidy to users of the space shuttle.
> Robert & Elizabeth Dole *Unlimited Partners* (1988), p. 261

sick building /ˌsɪk 'bɪldɪŋ/ *noun* ⊗

A building in which the environment is a health risk to its occupants, especially because of inadequate ventilation or air conditioning; used especially in **sick building syndrome**, the set of adverse environmental conditions found in a sick building; also, the set of symptoms (such as headaches, dizziness, etc.) experienced by the people who live or work there.

Formed by compounding. There is both a figurative and an elliptical quality to the use of *sick* here: architects and designers try to treat the symptoms caused by poor design, although it is not the *building* that is *sick*, but the people who use it.

Architects first wrote about large, centrally ventilated buildings as *sick buildings* in the early eighties and the set of vague symptoms suffered by people who used such buildings had become known as *sick building syndrome* (sometimes abbreviated to **SBS**) by the mid eighties. Commonly reported symptoms included headaches, dizziness, nausea, chest problems, and general fatigue; most could be attributed to poor air quality or actual air pollution within the building. New buildings in particular tend to make the most efficient use of energy, avoiding unnecessary intake of air from the outside which might increase fuel costs; the result is a building which is airtight to fresh air not forming part of the ventilation system, and in which the same dirty or contaminated air can be circulated over and over again. Such a building is also known as a *tight building*, and an alternative name for *sick building syndrome*—especially when it is attributable entirely to such a limited air supply—is *tight building syndrome* (abbreviated to *TBS*).

> For lack of documentation, employers considered that the collection of symptoms that now go under the label Tight Building Syndrome (TBS)—or Sick Building Syndrome—were psychosomatic. Not too surprising, since TBS's raspy throats, persistent coughs, burning eyes, headaches, dizziness, nausea and midafternoon drowsiness tend to disappear a half-hour after sufferers leave work.
> *Canadian Business* Apr. 1987, p. 58

> This is a book that affects to loathe the modern world. Modern architecture is dismissed in three words ('sick building syndrome') and barely redeemed by another ('Baubiologie'—the architectural sprig of west German green consumerism).
> *Green Magazine* Dec. 1989, p. 18

> Airtight and chemical-laden, office environments may cause 'sick building syndrome', a condition characterized by fatigue, nausea, and respiratory illness.
> *Garbage* Nov. 1990, p. 43

signature ▓ see DESIGNER

single market /ˌsɪŋg(ə)l 'mɑːkɪt/ *noun* ∿

A free trade association allowing for a common currency and largely unrestricted movement of goods, capital, personnel, etc. between countries; specifically, such a free market as the basis for trade between member states of the EC (also known more fully as the **single European market**), planned for full implementation by the end of 1992.

Formed by compounding; a *market* in which, instead of trading co-operatively but individually, the member states would come together to form a *single* unit. The word *single* seems to have been substituted consciously for the *common* of *Common Market* in the sixties, before an actual plan for NINETEEN NINETY-TWO was put forward.

The removal of barriers to trade has been an important aim of the European Community since its creation, but it was not until the Milan summit of June 1985 that a definite target was set for the creation of a single market by 31 December 1992. From about 1989 onwards, there was a concerted government advertising compaign in the UK, urging companies to make themselves aware of the implications of the *single market* and to take advantage of the opportunities it offered for growth and enterprise.

French officials now see the pillars to France's European policy as being: the development of the single European market, with the further opening of frontiers providing an important spur to economic growth, [etc.]. *Financial Times* 24 Mar. 1987, section 1, p. 3

In favour of a total ban are the state monopoly producers—Italy, France, Spain and Portugal. It is in their interests to block tobacco imports and protect their national products, against the spirit of the Single Market. *Marketing* 17 May 1990, p. 1

SITCOM ⟦ see DINK

ska house ♪ ⟦ see HOUSE

skateboarding /'skeɪtbɔːdɪŋ/ *noun* ⟦ ⟦

The use of a **skateboard** (a small board mounted on roller-skate wheels) for sport and recreation.

Formed on the noun *skateboard*, which was formed by compounding after the example of *surfboard*: a *board* which relies on *skates* to provide mobility.

Skateboards first appeared in the early sixties in California, where they originally provided a substitute for surfing when the ocean conditions were unfavourable. In the mid seventies they enjoyed a short-lived worldwide craze, during which numerous **skateparks** were built in which skateboarders could practise tricks and manoeuvres in safety. The pastime never completely died out, and by the second half of the eighties had become fashionable again, perhaps because of its appearance in such films as *Back to the Future* (1985). Some of the special language used by **skateboarders** (such as *gleaming the cube* for 'pushing oneself to the limits') also enjoyed wider popularity as a result of films about *skateboarding*.

Surfing, skateboarding and snowboarding looks have been so mass-marketed that the purists feel betrayed. *Los Angeles Times* 12 Sept. 1990, section S, p. 4

slasher /'slæʃə(r)/ *noun* ⟦ ⟦

Used attributively (in **slasher film**, **slasher movie**, etc.) of horror films or videos which depict vicious or violent behaviour.

So named because the attacker is shown *slashing* the victims with a knife or carrying out similarly violent attacks.

Slasher was first used as the name for a violent horror film in the mid seventies; however, the genre really became established in the mid eighties, with ever more gory horror films being released for rental through video clubs. *Slasher films* came in for a good deal of criticism in the mid eighties, as people started to make a connection between the fashion for them and rising levels of violent crime.

Paramount's low-budget slasher film Friday the 13th Part 3 in 'super 3-D' was roundly thrashed by critics ('Trash', said Newsweek). *Forbes* 27 Sept. 1982, p. 176

Instead of the breakdance and slasher movies aimed at the teen market, you have more thought-provoking films like *Rain Man* and *Dangerous Liaisons*. *Sunday Telegraph* 19 Mar. 1989, p. 11

sleazebag /ˈsliːzbæg/ *noun*

In young people's slang (originally in the US): a sordid or despicable person (especially someone considered morally reprehensible); a 'scuzzbag' (see SCUZZ).

Formed by compounding, from *sleaze* 'squalor, sordidness' (in use since the late sixties) and *-bag* (as in *windbag* etc.).

Sleazebag was the first of a number of compounds based on *sleaze* to be coined in the US during the eighties, appearing at the beginning of the decade as a general term of abuse, but used especially in political contexts to imply that a person had low standards of honesty. It was closely followed in the mid eighties by **sleazeball** (which essentially means the same as *sleazebag*) and **sleaze factor**, the sleazy or sordid aspect of a situation (applied especially, in US politics, to scandals and alleged corruption involving officials of the Reagan administration). *Sleaze factor* was a term coined in 1983 by American journalist Laurence Barrett, as a chapter heading in his book *Gambling with History*; it remained current throughout the Reagan administration, pointing to scandals, resignations, and alleged malpractice which nevertheless largely failed to 'stick' to the President himself (see TEFLON). After the end of the Reagan administration, *sleaze factor* had become a sufficiently familiar expression to survive in other contexts, and was even occasionally used in politics outside the US as well.

We are not giving away any principles, because we do have a few on this side of the House, unlike the sleazebags over there. *National Times* (Australia) 22 Nov. 1985, p. 7

It was stated in court by X's sleaze-ball lawyer. Richard Ford *The Sportswriter* (1986), p. 13

Among the people, places and things making indelible entrances [in the eighties]: ... PCs. Rambo. Sleaze factor. *Life* Fall 1989, p. 13

Slim /slɪm/ *noun* Also written **slim** ⊗

(More fully, **Slim disease**): the name used in Africa for AIDS.

So named because of the severe weight loss associated with the disease.

For history, see AIDS. The disease probably originated in Africa and reached epidemic proportions in some African countries during the eighties, but the problems of these countries were less widely publicized in the West than the corresponding difficulties of the US and decommunized countries like Romania in dealing with Aids.

A new disease has recently been recognised in rural Uganda. Because the major symptoms are weight loss and diarrhoea, it is known locally as slim disease. *Lancet* 19 Oct. 1985, p. 849

Because it is the skilled élite ... who have most money to spend on womanising, it is this group which is suffering the worst ravages of Slim. *Independent on Sunday* 1 Apr. 1990, Sunday Review section, p. 10

Sloane Ranger /ˌsləʊn ˈreɪndʒə(r)/ *noun* and *adjective*

noun: An upper-class and fashionable but conventional young person, especially one who lives in London. (Also abbreviated to **Sloane** or **Sloanie**.)

adjective: Characteristic of this class of person; adopting the style of dress, manner, or lifestyle of a Sloane.

Formed by replacing the *Lone* of *Lone Ranger* (a well-known hero of western stories and films) with *Sloane* (part of the name of *Sloane Square* in London, in or near which many young people of this background live). The formation takes advantage of the shared sound /ləʊn/ to make a blend of the two names.

This allusive name for a social group was coined by Peter York in *Harpers & Queen* magazine in 1975:

> The Sloane Rangers ... are the nicest British Girl.

Although not exclusively limited to young women, the term *Sloane Ranger* was at first mostly associated with the stereotype of the upper-class young woman who had been to one of the best schools, shopped at the smartest shops, and socialized in the 'right' circles (that is, with people whose wealth was inherited rather than earned). By 1982 the nickname had proved successful enough for an *Official Sloane Ranger Handbook* to be published (providing a British counterpart for the American *Preppie Handbook*), and the term started to be applied more widely to the whole class of people (including young men, otherwise known as *Hooray Henries*) who enjoyed the *Sloane* lifestyle. *Sloane Ranger* was abbreviated to *Sloane* in the original Peter York article; *Sloanie* followed in the early eighties. The quality of being like a *Sloane Ranger* is **Sloaneness**. By the end of the eighties the idea of the *Sloane Ranger* already seemed a little dated; however, the type continued to exist, and the name had started a fashion for humorous terms for social types that lived on through the eighties and into the nineties, starting with YUPPIE and still generating new variations.

> Sloane Rangers hesitate to use the term 'breeding' now (of people, not animals) but that's what background means.   Ann Barr & Peter York *The Official Sloane Ranger Handbook* (1982), p. 10
>
> She has to be literally beaten by her mother into marrying Cary Elwes-Guildford—who resembles a low-grade Sloanie with a taste for whores and bad liquor.   *Listener* 5 June 1986, p. 35
>
> Jeremy Taylor, one-time organiser of the Gatecrasher's Ball—a Sloanie teenage rave—was behind the party.   *Independent* 3 July 1989, p. 3

slomo /'sləʊməʊ/ *noun* Also written slo-mo 🗙

Colloquially in the film and video industry (especially in the US): slow motion; a slow-motion replay or the facility for playing back in slow motion (as, for example, on a video recorder).

A clipped compound, formed by combining the first three letters of *slow* with the initial syllable of *motion*.

Slomo was an American coinage which was probably in spoken use in the film industry long before it first appeared in print in the late seventies. It was popularized more widely as a result of the success of video in the early eighties.

> The NFL Films ... had it in slo-mo, and in overheads.   *Washington Post* 16 Sept. 1979, section M, p. 4
>
> Producer to slomo operator: 'Go back to where you were before I told you to go where I told you to go.'   *Broadcast* 7 July 1980, p. 10
>
> Apart from the Hi-Fi facility there's a 14-day, six-event timer, advanced trick frame with five-speed slomo (1/36, 1/24, 1/15, 1/10 or 1/6).   *Which Video?* Jan. 1987, p. 4

smart /smɑːt/ *adjective* 🖳 🌾

Of a machine: able to react to different conditions, computerized, intelligent (see INTELLIGENT¹). Used especially in:

smart bomb (or **missile, weapon**, etc.): a bomb (or other weapon) which is able to track and 'lock on to' its target; a laser-guided weapon;

smart card, a plastic bank card or similar device with an embedded microprocessor, used in conjunction with an electronic card-reader to authorize or provide particular services, especially the automatic transfer of funds between bank accounts;

smart house, a house with a central computer providing integrated control of environmental services such as heating; an intelligent building (see INTELLIGENT²);

smart rock, a code-name for an intelligent weapon planned for the STAR WARS programme.

A figurative use of *smart* in the sense 'clever': compare INTELLIGENT¹.

Smart is a word with a similar history to ACTIVE except that it immediately preceded *active* in the fashionable language of advertising and product names. It was picked up by marketers in the early eighties and by the end of the decade (as the *New York Times* quotation below shows) seemed to be applicable to almost any product with a measure of computerization. The concept of *smart* bombs which could home in on a target with very high levels of accuracy dates from the early seventies, but enjoyed considerable exposure during the Gulf War of 1991.

> The dream of many proponents of precision-guided munitions, very tiny and effective smart weapons, will founder on the need to carry heavy electronic shielding.   *Atlantic* Mar. 1987, p. 28

> The beauty of the algorithm ... is that it can be built into hardware that will fit even on 'smart cards', and enables the identity of end-users to be checked in less than a second.
> *The Times* 23 Feb. 1988, p. 30

> The ultimate manifestation of the 'smart' house ... was the Smart Seat, a microprocessor-controlled bidet attachment for the toilet.   *New York Times* 25 Jan. 1990, section C, p. 6

> With eerie precision, 'smart' bombs dropped down air shafts and burst through bunker doors.
> *Newsweek* 28 Jan. 1991, p. 15

Smart Art ✖ see NEO-GEO

smiley /'smaɪlɪ/ *noun* ✖

(More fully **smiley face** or **smiley badge**): a round cartoon-style representation of a smiling face (usually black on yellow), used as a symbol in youth culture, especially in connection with ACID HOUSE.

Formed by abbreviating *smiley face* to its first word and treating this as a noun.

The black-on-yellow *smiley* first appeared as a late hippie symbol of peace and happiness in the early seventies. Towards the end of the seventies it enjoyed a revival among young people in the US (especially in California), but it was really its association with *acid house*, and in particular the suggestion that it was being used as the symbol of drug users, that brought it into the news in about 1988. As is often the way with young people's fashions, it became unfashionable almost as soon as it had been brought to public notice in this way. The *smiley* symbol has been used in many ways as are connected neither with youth culture nor with drugs: for example, it was the official symbol of the Lord Mayor of London's theme 'Service with a Smile' in 1985–6, and seems to be becoming accepted as a general symbol of approval (shorthand for 'I like this', for example written by the teacher on a child's schoolwork). A *smiley* with black features on white and another in reverse video are part of the standard ASCII character set for microcomputers.

> Brad's eye roved the room, which had recently taken on a second identity as an art gallery and was filled with murals depicting the deconstruction of the smiley face.
> David Leavitt *The Lost Language of Cranes* (1986; 1987 ed.), p. 198

> In the crowd you may also spot the odd man in navy Top Man tracksuit, immaculate new trainers and strange accessories such as bandanas or Smiley badges—these are plain-clothes policemen or tabloid journalists.   *The Face* Dec. 1989, p. 63

> Glasgow's close association with the Mr Smiley logo predates acid house by several years, his happy face harnessed in 1983 to sell the world the PR legend, 'Glasgow's Miles Better'.
> *The Face* June 1990, p. 100

smoothie /'smu:ðɪ/ *noun* ✖

A smooth thick drink consisting of fresh fruit (especially banana), puréed with milk, yoghurt, or ice cream.

So named because of its smooth consistency.

The *smoothie*, a variation on the traditional milkshake, is a drink which is best known in the US, Australia, and New Zealand.

There are some definite winners among the selections: Freshly made onion rings, a yogurt and fruit drink called a 'smoothie', [etc.].   *Washington Post* 2 June 1977, section F, p. 12

In New York now, there are entire bars which cater for trendy non-drinkers. They serve nothing but a selection of mineral waters, soft drinks and non-alcoholic cocktails (called 'smoothies').
Sunday Telegraph Magazine 7 June 1987, p. 30

It's worth noting that the shop underneath makes ripper soymilk smoothies. Buy yourself a strawberry job with frozen yoghurt.   *Sunday Mail* (Brisbane) 1 Jan. 1989, p. 34

snuff /snʌf/ *noun* 🔲 📻

Used attributively of an illegal film or video (in **snuff video**, etc.): depicting scenes of cruelty and killing in which the victim is not an actor, but is actually tortured or killed.

A reference to the horrific *snuffing out* of life which these videos portray.

Privately circulated *snuff videos* have allegedly been known to the police since the seventies. They figured briefly in the news in 1990, when police claimed to have cracked a paedophile ring which had been involved in the production of these films, and linked the crimes with the disappearances of a number of young boys in the UK during the eighties.

New York City police detective Joseph Horman said . . . that the 8-millimetre, eight-reel films called 'snuff' or 'slasher' movies had been in tightly controlled distribution.
Whig-Standard (Kingston, Ontario) 2 Oct. 1975, p. 3

As police in east London continued investigations into the disappearance of young boys, Mr Waddington, Home Secretary, yesterday expressed his 'absolute horror' at the possibility that some of them may have been murdered during the making of pornographic 'snuff' videos.
Daily Telegraph 28 July 1990, p. 3

See also NASTY and SLASHER

soca /'səʊkə/ *noun* Also written **sokah** 🎵 📻

A variety of calypso, originally from Trinidad, which incorporates various elements of soul music, especially its sophisticated instrumental arrangements.

A clipped compound, formed from the first two letters of *soul* and the initial syllable of *calypso*.

Soca (at first called *soul calypso*) originated in Trinidad during the early seventies and by the end of the decade had spread to the world of American and British popular music. The spelling *sokah* relates to the title of an early *soca* record, *Sokah, Soul of Calypso* (1977) by 'Lord Shorty', a founding influence on the genre.

The banned 'Soca Baptist' by Blue Boy . . . brought out the real Carnival spirit from southerners.
Trinidad Guardian 11 Feb. 1980, p. 1

Few people would guess that some soca, reggae, lovers'-rock and, particularly, soul and dance music sometimes outsell 'chart' records.
Sue Steward & Sheryl Garratt *Signed, Sealed & Delivered* (1984), p. 12

The records that fueled it—French Antillean and Trinidadian soca sides . . . from the nearby Guianas.
Village Voice (New York) 30 Jan. 1990, p. 83

soft lens 🔵 see LENS

software package 💾 see PACKAGE

solvent abuse 🔫 see ABUSE

-something 📻 see THIRTYSOMETHING

soul calypso 🎵 📷 see SOCA

sound 🌿 see ENVIRONMENTALLY

sound bite /'saʊnd ˌbaɪt/ *noun* 🔀 📷

A short, pithy extract from a recorded interview, speech, etc. used for maximum punchiness as part of a news or party political broadcast; also, a one-liner deliberately produced to be used in this way.

Formed by compounding. The use of *bite* here both puts across the idea of a snatch of sound-track taken from a longer whole and includes undertones of the high-tech approach to units of information (*bytes*).

The term has been in use among radio and television journalists in the US for some time, and first appeared in print in the early eighties. Perhaps because of developments in television newscasting techniques in the eighties, it has become more and more prevalent, reflecting the view that the public will not follow more than a few seconds of speech from any single inter-view, although several minutes from a reporter will be fine. (In television journalism *sound bites* are often interspersed with a reporter's précis of a speaker's words as a voice-over to a sound-less film of the speaker.) The technique, as well as the term, came to public notice during the US presidential campaign of 1988, when *sound bites* were used to great effect on the campaign trail and in televised debates between the protagonists.

> Remember that any editor watching needs a concise, 30-second sound bite. Anything more than that and you're losing them. *Washington Post* 22 June 1980, section 1, p. 1

> This has been the election of the 'sound-bite', the 20-second film clip on the evening television news which defines most Americans' view of the day's campaigning. The Bush campaign . . . has been consistently out-biting the Dukakis camp. *Independent* 24 Sept. 1988, p. 10

sounding 📷 see DISS

space shuttle, Space Transportation System 🖥 see SHUTTLE

-speak see -BABBLE

specialog(ue) 〰 🔀 see MAGALOG

speed /spiːd/ *noun* 🎵 📷

A variety of HEAVY METAL rock music that is very similar to THRASH; also known more fully as **speed metal**.

> The latest branch on rock's American tree is a phenomenon tagged Speed Metal, the place where HM supposedly mates with hardcore. *New Musical Express* 14 Feb. 1987, p. 7

spin doctor /'spɪn ˌdɒktə(r)/ *noun* 📷

In the jargon of US politics, a senior political spokesperson employed to promote a favourable interpretation of events to journalists; a politician's FLAK.

Formed by compounding. In US politics, *spin* is interpretation, the bias or slant put on infor-mation when it is presented to the public or in a press conference; all information can have a positive or negative *spin*. This in turn is a sporting metaphor, from the *spin* put on the ball, for example by a pitcher in baseball. *Doctor* comes from the various figurative uses of the verb *doc-tor* (ranging from 'patch up, mend' to 'falsify'), perhaps under the influence of *play doctor* 'a writer employed to improve someone else's play'.

The phrase *spin doctor* was first used in print in October 1984 in an editorial in the *New York Times* about the aftermath of the televised debate between US presidential candidates Ronald Reagan and Walter Mondale:

A dozen men in good suits and women in silk dresses will circulate smoothly among the reporters, spouting confident opinions. They won't be just press agents trying to impart a favorable spin to a routine release. They'll be the Spin Doctors, senior advisers to the candidates.

The term started to crop up quite frequently in political journalism in the mid eighties, and became a real buzzword during 1988. It is used both in relation to electoral campaigns and of other events, such as top-level international summits and disarmament negotiations. There is only a subtle distinction between the job of the *flak* and that of the *spin doctor*: the former tries to turn negative publicity, criticism, or failure to advantage, while the latter is trying to impart the right *spin* from the outset, so that there is no DAMAGE LIMITATION exercise to be done. The activity of a *spin doctor* is **spin doctoring**.

We were treated to the insights of Elliott Abrams, ... the administration's most versatile spin doctor on Nicaraguan affairs. *Maclean's* 2 Apr. 1990, p. 11

The resultant emphasis on the British end of things is more than so much 'spin doctoring'. *Delaware Today* July 1990, p. 76

spoiler[1] /'spɔɪlə(r)/ *noun* 🔲

An electronic device incorporated into a piece of recording equipment so as to prevent unauthorized recording (for example from a CD on to DAT), by means of a **spoiler signal** which cannot be heard during normal playing, but which ruins any subsequent recording; also, the signal itself.

A specialized use of *spoiler* 'something which spoils'.

The first *spoilers*, really a form of electronic jamming, were developed experimentally in the late seventies. During the eighties, demand for some kind of **spoiler system** was quite intense in the EC and the US as the introduction of DAT approached; manufacturers of CDs in particular expressed their *DATphobia* (see DAT) by lobbying governments to require DAT tape decks to carry some kind of built-in *spoiler* to prevent widespread pirating of their recordings.

CBS recently tried to introduce a 'spoiler' system called Copycode. This, it was claimed, would prevent any CD/DAT recording. *Which?* July 1988, p. 345

spoiler[2] /'spɔɪlə(r)/ *noun* 🔲

In media jargon, something that is published to spoil the impact of, and divert attention from, a similar item published elsewhere.

Another specialized sense of *spoiler*; a media piece which *spoils* the success of the original.

An aspect of the intense competition of the newspaper world, the *spoiler* depends on good intelligence sources and may be a complete publication such as a newspaper, or simply an individual article designed to minimize the sucess of another publisher's scoop.

Lord Rothermere, who had always claimed the Evening News was more than a temporary spoiler, said yesterday the paper and its staff had fought well. *Financial Times* 31 Oct. 1987, section 1, p. 4

The speech made the front pages of The Daily Mail, The Times and The Daily Telegraph ... The Independent ... treated it as a spoiler for Paddy Ashdown's 'green' speech to his party conference a couple of days later. *Daily Telegraph* 30 Dec. 1989, Weekend section p. v

spud 🔲 🔲 see COUCH POTATO

squeaky clean /ˌskwiːkɪ 'kliːn/ *adjectival phrase* Also written squeaky-clean 🔲

(Of hair) washed and rinsed so clean that it squeaks, completely clean; hence used figuratively (especially in political contexts): above criticism, beyond reproach.

Formed by combining the two adjectives *squeaky* and *clean*; normally an adjective would not qualify another adjective in this way in English, so some speakers might prefer *squeakily clean*.

The phrase seems to have come originally from shampoo or detergent advertising, although it has also been suggested that it was used by army sergeant majors of boots and other surfaces that had to be so highly polished that they squeaked. The first figurative uses date from the mid seventies. To describe a politician or some other public figure as *squeaky clean* is perhaps not altogether a compliment: it can certainly imply disappointment on the part of the person using it that the personality concerned is unlikely to be the subject of any scandal, and sometimes it also implies an image that is hard to believe, or 'too good to be true'.

> Squeaky-clean in body and mind, the Preppy is the class swot and jolly-good-all-rounder all grown up. *Sunday Express Magazine* 17 Sept. 1989, p. 18

> Mr Pearson maintained . . . control over every aspect of his children's rise to fame as squeaky clean pop group Five Star. *Punch* 13 July 1990, p. 33

SRINF 💥 see INF

Stalkergate 📷 see –GATE

standard assessment task ◖◗ see NATIONAL CURRICULUM

starch blocker /ˈstɑːtʃ ˌblɒkə(r)/ *noun* ⊗ 💥

A dietary preparation that supposedly affects a person's metabolism of starch so that it does not contribute to a gain in weight.

Formed by compounding: supposedly a *blocker* of *starch* metabolism.

Starch blockers were first introduced in the US in 1981 and for a short time provoked a good deal of journalistic interest. However, the scientific basis of the claims made for these products was soon debunked.

> Slimmers who use starch blockers . . . are wasting their money. . . . Experts . . . say they do not affect the quantity of starch digested and could have unpleasant effects if they did work. *Daily Telegraph* 14 Apr. 1983, p. 6

> 1982: The FDA cracked down on starch blockers, a diet fad that purportedly prevented the body from digesting starch calories. *Life* Fall 1989, p. 64

start-up /ˈstɑːtʌp/ *noun* Also written startup 〰

A business enterprise that is in the process of starting up. Also known more fully as a **start-up company**.

A more concrete use of the noun *start-up*, which previously meant 'the action or process of starting up'.

This usage arose in US business writing in the second half of the seventies. With the encouragement of small businesses in the UK which marked the ENTERPRISE CULTURE of the early eighties, it also became a feature of British business language. *Start-up* is often used attributively (in **start-up loan**, **start-up scheme**, etc.), but in these cases it is used in its older sense of 'the process of starting up (a business)'.

> The company is a relatively rare thing for Europe: a successful high-technology start-up. *Economist* 24 Mar. 1990, p. 129

> Nixdorf supported the development of a loosely coupled, fault-tolerant multiprocessor technology at a New Jersey-based startup named Auragen Systems. *UnixWorld* Apr. 1990, p. 39

Star Wars /ˈstɑː ˌwɔːz/ *noun* Also written star wars 💥

A colloquial nickname for the programme known officially as the *Strategic Defense Initiative* (abbreviation *SDI*), a military defence strategy proposed by US President Reagan in 1983, in which enemy weapons would be destroyed in space by lasers, antiballistic missiles, etc., launched or directed from orbiting military satellites.

A nickname based on the title of a popular science-fiction film released in 1977 and involving similar weapons; this film was, according to *Halliwell's Film Guide*, 'a phenomenon and one of the top grossers of all time', and it was therefore prominent in the public consciousness at the time when President Reagan made his proposals.

The nickname *Star Wars* was applied to President Reagan's proposals for a high-technology space-based defence system almost as soon as he had made them in a nationwide television address in March 1983. At first it was used somewhat scathingly, pointing to the fact that the technology required for such a system had not yet been developed and expressing the view that it might prove as fictional as the film. Funding for the project was eventually voted through Congress by the middle of the decade, but there was enduring criticism of the whole idea, especially since it appeared to contravene existing antiballistic missile treaties and seemed more likely to contribute to the arms race than to end it (as President Reagan had supposed).

> The first question is one of commitment: whether Ronald Reagan understands what it takes to nudge a doubting, cash-short nation into serious consideration of his star wars defense concept.
>
> *Time* 4 Apr. 1983, p. 19

> The only reason Star Wars happened is that the staff erred and allowed Edward Teller and a small group of conservatives from the Heritage Foundation who were behind it to get to Reagan.
>
> *Life* Fall 1989, p. 56

Stasi /ˈʃtɑːzɪ/ *noun* Sometimes written STASI 🔒

The internal security force (until 1989–90) of the German Democratic Republic.

Formed from two of the syllables of the full name, *STAatsSIcherheitsdienst* 'State Security Service'.

Stasi was the colloquial name in German of the feared East German secret police for a number of decades before it became popularly known in English. It was used in spy novels etc. written in English during the sixties and seventies, but ironically it was its demise in 1989–90 that really brought it into the headlines and gave it a wider currency. News reports of the breakdown of the Communist system in the GDR included coverage of popular demonstrations against the *Stasi* and demands for its abolition; its offices were reduced and many of its employees dismissed in December 1989 (more than 100,000 agents had been sacked by February 1990) and by March 1990 the Spy section was being cut down drastically as well.

> The mood has become tense in the past week with mounting warning strikes and calls for the Stasi to be rooted out for good. *The Times* 16 Jan. 1990, p. 1

> He had received information that CDU leader de Maizière had himself been a Stasi informer.
>
> *Maclean's* 2 Apr. 1990, p. 31

statesperson /ˈsteɪtspɜːs(ə)n/ *noun* 🔒

A statesman or stateswoman. (Invented as a generic term to avoid sexism.)

Formed by substituting the non-sexist *-person* for *-man* or *-woman*.

The term was invented by the media in the second half of the seventies, and at first was in practice more or less limited to references to stateswomen: Indira Gandhi and, a little later, Margaret Thatcher were the people most often referred to as *statespersons*. By the end of the eighties, though, it was starting to be used of statesmen as well.

> Contributors to the diary's current competition (see below) may like to know that somebody wants Our Greatest Statesperson to have some free history lessons. Namely, Mike Harris, a Labour member of Barnet council, embracing Mrs Thatcher's seat (Finchley, that is, I rush to point out).
>
> *Guardian* 10 Aug. 1989, p. 19

> Genscher has become Europe's senior statesperson. *New Yorker* 23 Oct. 1989, p. 104

Stealth /stɛlθ/ *noun* 🎖

A branch of military technology in the US concerned with making aircraft and weapons hard for the enemy to detect by radar or other sensing systems; usually used attributively, in **Stealth aircraft**, **Stealth bomber**, **Stealth technology**, etc.

A specialized use of an old sense of *stealth* 'furtive or underhand action, an act accomplished by eluding observation or discovery' (a sense which survives in modern English mainly in the phrase *by stealth*).

The development of *Stealth* technology (known more formally as *low observable technology*) first gained official backing in the US in the second half of the seventies. Its most famous example, the *Stealth bomber* or *B2 bomber*, was developed amid great secrecy during the eighties and was first seen in operation by the general public during the Gulf War of January–February 1991. Detection is avoided by the use of a shape with proportions and angles that are not easily visible on radar, materials which evade infrared sensing, etc.

> Key technologies that have been identified are the following: Stealth technology. Engines and fuels. Avionics.　　　　　　　　　　　　　　　　　　　　　　　　　*Aviation Week* 29 Jan. 1979, p. 121

> Microprose produced an F-19 simulation on the PC at a time when the B2 stealth bomber hadn't even been glimpsed.　　　　　　　　　　　　　　　　　　　　　　*CU Amiga* Apr. 1990, p. 12

steaming /'sti:mɪŋ/ *noun* 🎐

In British teenagers' slang: the activity of passing rapidly in a gang through a public place, robbing bystanders by force of numbers.

Probably related to the Cockney slang phrase *steam in* 'to start or join a fight'; it has been claimed that the term came from US street slang, but there is little evidence to support this.

The phenomenon of *steaming* first started to be reported in the newspapers in the UK in 1987–8, when there was a spate of incidents of this kind on trains and buses, and also at large public gatherings such as street carnivals. The verb **steam** (which is used intransitively or transitively) has been back-formed from the noun; a person who takes part in *steaming* is a **steamer**.

> Video tapes of the two-day carnival are being studied in an attempt to trace 'steamers', who ran *en masse* through the crowds, stealing at random.　　　　　　　　　*The Times* 9 Sept. 1987, p. 7

> Frightening for its victims, steaming is also proving to be a difficult crime to prevent, and very expensive, in both manpower and financial terms, to stamp out.
> 　　　　　　　　　　　　　　　　　　　　　*Sunday Times* 21 Feb. 1988, section A, p. 18

Stinger /'stɪŋə(r)/ *noun* 🐝

The name (more fully **Stinger missile**) of a lightweight, shoulder-launched anti-aircraft missile developed in the US and incorporating an infrared homing device.

Presumably a figurative use of *stinger* in the sense 'something that stings or smarts'.

The *Stinger* missile system was developed by General Dynamics and other contractors in the US in the second half of the seventies. Being light in weight and shoulder-launched, it proved an ideal form of anti-aircraft missile for guerrilla warfare. The use of *Stingers* by rebels against Soviet and Afghan government aircraft in Afghanistan brought them into the news in the second half of the eighties.

> The Pentagon told Congress Wednesday it intends to sell Saudi Arabia 400 ground-to-air Stinger missile systems along with 1,200 missiles.　　　　*Christian Science Monitor* 2 Mar. 1984, p. 2

> The transfer of the Stingers to the counter-revolutionary bands, which use these missiles to down civilian aircraft, is simply immoral and totally unjustifiable.
> 　　　　　　　　Mikhail Gorbachev *Perestroika* (English translation, 1987), p. 177

store option card 〰 see CARD¹

storming /'stɔːmɪŋ/ *adjective* 🌀

In British slang: outstanding in vigour, speed, or skill; 'cracking'.

Formed on the verb *storm*, probably as a transferred use of the military sense 'to make a vigorous assault on; to take by storm'.

This sense of *storming* was a feature mainly of sport reports and tabloid journalism from the seventies onwards; in the same sources, a **stormer** was anything that could be described in the superlative: something very large, very successful, or very good. When, during the Gulf War of early 1991, the tabloid papers in the UK described the US Commander General Norman Schwarzkopf as *Stormin' Norman*, they were taking advantage of both the rhyme and the pun with the military sense of *storm* from which this adjective derives.

The outstanding performer in the open was Stuart Evans who had a storming game.

Rugby News Mar. 1987, p. 2

There are conflicting views on whether Gen Schwarzkopf . . . deserves the nickname 'Stormin' Norman', which he detests.

Independent 18 Feb. 1991, p. 3

Strategic Defense Initiative 💥 see STAR WARS

street cred 🎞 see CRED¹

string 💻 see SUPERSTRING

STS 💻 see SHUTTLE

sugar-free ⊗ ✖ see -FREE

suit /s(j)u:t/ *noun* 〰️ 📑

In business jargon, a manager or boss; someone who wears a suit to work (rather than overalls, a uniform, etc.). Also in political contexts (especially in the phrase **men in (dark** or **grey) suits**), a faceless bureaucrat; an elder statesman or senior civil servant who acts as a political adviser.

In both cases, a reference to the fact that the characteristic dress of these people singles them out for what they are (although, of course, many other people wear a *suit*!).

Suit was a slang term for a member of the management or officialdom which in the mid eighties took on a new lease of life in a number of phrases to do with *men in suits*. The idea of the *men in grey suits* who ultimately had the power to bring about the downfall of a Prime Minister was made much of by journalists in connection with the leadership contest within the Conservative Party and the eventual resignation of Margaret Thatcher in December 1990.

Major's spectacular ordinariness—the Treasury is now led by 'a man in a suit' whose most distinguishing feature is his spectacles.

Observer 29 Oct. 1989, p. 28

Blaming the 'suits' is a national pastime. If a traffic cop has a faulty search warrant or a flat tyre, he curses the 'suits' at headquarters.

The Times 14 Mar. 1990, p. 16

I claim paternity of 'the men in suits' from an *Observer* column of the mid-1980s. Not, you may notice, the men in dark suits, still less those in grey ones, which gives quite the wrong idea.

Alan Watkins in *Spectator* 1 Dec. 1990, p. 7

Margaret Thatcher was brought down by a brief, tacit alliance of 'men in grey suits' and Thatcher loyalists.

Sunday Telegraph 25 Nov. 1990, p. 23

suitor /'s(j)u:tə(r)/ *noun* 〰️

In financial jargon, a prospective buyer of a business corporation; a person or institution making a take-over bid.

A figurative use of *suitor* in the sense of 'a person who seeks a woman's hand in marriage'. Such metaphors are common in the financial world: compare DAISY CHAIN¹, DAWN RAID, POISON PILL, and WHITE KNIGHT.

Originally an American colloquial usage of the seventies, *suitor* had spread into British use by 1980 and during the eighties became a standard way of referring to a prospective buyer, no longer thought of as colloquial in financial circles. Its use in the newspapers and the media generally brought it to a wider and more popular audience.

> Lifting the veil of secrecy was ordinarily enough to kill a developing buyout in its cradle: once disclosed, corporate raiders or other unwanted suitors were free to make a run at the company before management had a chance to prepare its own bid.
>
> Bryan Burrough & John Helyar *Barbarians at the Gate* (1990), p. 8

superparticle 💻 see SUPERSTRING

superstring /ˈsuːpəstrɪŋ/ *noun* 💻

In physics, the form taken by sub-atomic particles according to **superstring theory**, a theory devised to account for the interactions of particles by viewing them as one-dimensional objects resembling tiny pieces of string.

Formed by adding the prefix *super-* in the sense 'supersymmetric' to *string* (see below).

Quantum theory and general relativity are two major developments which have taken place in physics this century: the former enables us to see particles and waves as different aspects of the same entity, while the latter paved the way for such concepts as black holes and the curvature of space–time. However, theoretical physicists found considerable difficulty in reconciling the two theories to produce a unified theory of *quantum gravity* (so called because the explanation of gravity is a central aspect of the theory). Most current models of the nature of the elementary particles which make up the universe supplement the familiar four dimensions of space and time with up to seven other (not directly observable) dimensions: one way of simplifying the resulting complexity is to view different particles as in some sense derived from the same *superparticle*—a proposal known as *supersymmetry*. Some of the other inconsistencies of unified models can be eliminated by replacing points in space–time conceptually by 'loops' or short 'lengths' of 'string', likewise observable only in more than four dimensions (some theories postulate as many as 26). In 1982 a way of combining these two approaches was developed which became known as *superstring theory*. Its acceptability as a possible *TOE* ('theory of everything') remains debatable, but its possibilities in this direction have fascinated physicists for most of the past decade.

> Superstrings are entities in ten dimensions (nine space-like, one time-like) which are expected to behave like ordinary particles when the ten dimensions are collapsed to four.
>
> *Nature* 3 Jan. 1985, p. 9

> Michael Green . . . won the honour for his work on superstring theory. He is one of those who believe that everything in the cosmos . . . is made of these incredibly tiny objects.
>
> *Daily Telegraph* 20 Mar. 1989, p. 23

supersymmetry 💻 see SUPERSTRING

supertitle 🎬 🎵 see SURTITLE

surf /sɜːf/ *intransitive* or *transitive verb* 🏄

To ride on the roof or outside of a train, as a dare or for 'kicks'; to ride (a train) as though it were a surfboard.

A figurative use of *surf*: the youngsters concerned use the trains for sport, to get excitement and thrills, just as richer youngsters in coastal areas use the waves.

The practice of **surfing** (sometimes known more fully as **train surfing**) seems to have begun among poor youngsters in Rio de Janeiro and by the late eighties had spread to some US cities as well. In the late eighties it also started to become a problem in the UK, with a number of incidents in which young people were killed engaging in this extremely dangerous 'sport'.

> What has become known as 'train surfing' is killing 150 teenagers a year in Rio, and injuring 400 more.
>
> *Chicago Tribune* 5 May 1988, p. 28

A verdict of misadventure was recorded yesterday on an 18-year-old student who fell to his death while 'surfing' on a 70mph Tube train.   *Daily Telegraph* 1 Dec. 1988, p. 5

surrogacy /'sʌrəgəsɪ/ *noun* Ⓧ ▌▌

The practice (also known as **surrogate motherhood** or **surrogate mothering**) in which a woman carries and bears a child for another, either from her own egg, fertilized outside the womb by the other woman's partner and then re-implanted, or from a fertilized egg from the other woman.

A specialized use of *surrogacy*, which formerly meant 'the office of deputy' (a *surrogate* being a person who stands in for another).

The practice of *surrogacy*, which first took place in the US in the late seventies, was the subject of heated moral and legal debate both in the US and in the UK during the eighties. The central question concerned the ethics of an arrangement in which a woman agreed to carry and bear a child for others in return for a fee, on condition that she would hand over the baby to the couple 'employing' her after the birth. In a famous case in the US (known as the case of Baby M), the **surrogate mother** was reluctant to relinquish the baby after bonding with her at birth, and a court battle for custody of the child ensued. In the UK a committee chaired by Dame Mary Warnock considered the ethics of *surrogacy* and recommended in its report (published in July 1984) that it be made illegal. The continuing debate in the US has led to a distinction between **host surrogacy** (in which the fertilized egg is the product of both the 'employing' parents, and the *surrogate mother* is providing no more than an incubator for the embryo during gestation) and *surrogacy* in which the *surrogate mother* is biologically involved by supplying the egg for fertilization.

Is surrogate mothering class exploitation? Even the gift of life can come wrapped in ethical quandaries.   *Life* Fall 1989, p. 104

A surrogate mother ... can be impregnated with his sperm artificially and she can even be impregnated by the sperm *and* the ovum of the infertile couple (a process known as 'host' surrogacy). Providing the surrogate mother does not have intercourse with her partner before the embryo 'takes', the infertile couple will be presented with a baby which is genetically all their own.
                                          She Aug. 1990, p. 6

surtitle /'sɜːtaɪt(ə)l/ *noun and verb* ✖ ♪

noun: A caption which is projected on to a screen above the stage during the performance of an opera, giving a translation of the libretto or some other explanation of the action.

transitive verb: To provide (a stage production) with surtitles.

Formed by adding the prefix *sur-* in the sense 'above' to *title*; consciously altering *subtitle* (as used in films etc.) to put across the idea that these captions appear above rather than below the action.

Opera *surtitles* were first so called by the Canadian Opera Company in 1983, when they were used to provide an English translation of Hugo von Hofmannsthal's German libretto to Richard Strauss's *Elektra*; the company went on to register the name *Surtitles* as a trade mark in Canada in July 1983. Within three years they had spread to opera productions all over the English-speaking world, although some producers chose to call them *subtitles* despite the fact that they appear above the stage. By the end of the eighties the term *surtitle* had become established and had been applied to stage productions of foreign plays as well as opera. Among opera buffs the provision of these captions caused some controversy, both because some people found them intrusive and because it was claimed that the word was badly formed and should actually have been *supertitle* (the name in fact used by US opera companies). The verbal noun used to describe the practice is **surtitling**; the adjective to describe productions in which it is used is **surtitled**.

The Australian Opera will use surtitles at all performances in languages other than English in 1985.
Courier-Mail (Brisbane) 12 Dec. 1984, p. 24

Glyndebourne . . . faced an angry response when it surtitled a touring production in 1984.
The Times 23 June 1986, p. 3

survivalism /səˈvaɪvəlɪz(ə)m/ *noun*

The practising of outdoor survival skills as a sport or hobby.

Formed by adding the suffix *-ism* to *survival* in the sense 'the ability to survive under harsh or war-like conditions'.

Survivalism as a word for the pastime of perfecting **survival techniques** or **survival skills** dates only from the second half of the eighties, although **survival** (in the sense of acquiring and using these skills) had started to become a popular pastime during the seventies. At first this developed through such channels as territorial army training and other military reserves, 'outward bound' courses, etc., but by the early eighties people were beginning to pursue it as a hobby in its own right; such a person became known as a **survivalist** from about 1982 onwards. The growth of *survivalism* as a hobby was already causing some public concern because of the proliferation of dangerous weapons with which it was associated when, in August 1987, a keen *survivalist* called Michael Ryan ran amok in the town of Hungerford in Berkshire (southern England), shooting and wounding people apparently at random, and eventually shooting himself. Fourteen people were killed outright and two died later as a result of their wounds. The circumstances of this incident were, of course, unique, and do not reflect upon *survivalism* as a whole; however, the public perception of *survivalists* was no doubt affected by it, and indeed many only became aware of the hobby at all because of this tragedy.

Soldier of Fortune is a . . . militaristic publication packed with vitriol and ordnance . . . It has . . . touched a nerve with many Vietnam veterans as well as with survivalists who want to arm themselves to the teeth. *New York Times* 15 Oct. 1982, section A, p. 12

Apart from the growth of martial arts clubs, much of this self-arming is taking place under the auspices of . . . the newish and very fast-growing fad called Survivalism. *Spectator* 27 Sept. 1986, p. 9

sustainable /səˈsteɪnəb(ə)l/ *adjective*

In environmental jargon: (of an activity, use of a resource, etc.) able to be sustained over an indefinite period without damage to the environment; (of a resource) that can be used at a given level without permanent depletion, renewable.

A specialized use of *sustainable* in the sense 'able to be maintained at a certain rate or level', itself a sense which only entered the language in the sixties.

The adjective *sustainable* has been used in relation to wildlife conservation since the seventies; especially in the phrase **sustainable development**, it became one of the environmental buzzwords of the eighties as the GREEN movement succeeded in focusing public attention on the long-term effects of energy use and industrial processes in Western societies. The corresponding adverb **sustainably** and the noun **sustainability** also became popular in environmental contexts: governments were urged to use energy sources *sustainably* and to consider the *sustainability* of processes, for example.

It was host . . . to an environmental meeting in Bergen at which ministers from ECE's member countries discussed practical steps to promote 'sustainable development'. *EuroBusiness* June 1990, p. 64

The conference . . . was the first . . . ever to discuss the potential, as well as the problems, of conserving rainforests by sustainably exploiting non-timber resources. *Earth Matters* Summer 1990, p. 3

Suzuki see BASUCO

sweep /swiːp/ noun 🎬

In the US, a survey of the popularity of radio and television shows, especially for the Nielsen index of popularity ratings; often in the plural, as **the sweeps**: the designated times during the year when these surveys are carried out.

A specialized use of *sweep* in its established figurative sense of 'a comprehensive search or pass over something'.

The Nielsen rating system for radio and television programmes in the US dates from the early fifties, but the practice of carrying out a *sweep* during particular weeks of the year (called a **sweep week**) does not seem to have started until the seventies. In the second half of the seventies and the eighties, there was considerable public interest in the *sweeps*, especially since certain channels appeared to be putting on the best and most popular shows at this time (a practice which is actually against the rules, but difficult to prove).

> Channel 7 . . . dominated the local Nielsen news ratings during the May 'sweeps'.
> *New York Times* 2 June 1982, section C, p. 26

> Demographic ratings for children 2 to 11 will not be available until after the November sweeps.
> *Advertising Age* 10 Nov. 1986, p. 32

swipe 🖲 see CARD[1]

switch /swɪtʃ/ noun 〰️

A computerized link between financial institutions and points of sale, enabling goods to be paid for by debit card using EFTPOS; in the UK, a computerized EFTPOS system set up in 1988 and used by a number of banks.

Switch in computing already meant 'a program instruction that selects one or other of a number of possible paths according to the way that it is set'; in the context of EFTPOS, the choice of the name *switch* was probably also influenced by *packet-switching*, a standard mode of data transmission in which a message is broken down into parts or *packets*.

The first point-of-sale computer systems to incorporate *switches* as the link between retail outlets and financial services was set up in the US in the second half of the seventies, when the State of Iowa established a statewide *switch* network. The debit card system actually known as *Switch* in the UK was launched by the Midland Bank, NatWest, and the Royal Bank of Scotland in 1988. Using this system, shoppers need only a plastic debit card (see CARD[1]) called a **Switch card** to pay for goods; the *switch* ensures that the appropriate sums are transferred electronically from the purchaser's account to the retailer's. For this reason, the *switch* was thought of in the early eighties as the herald of a cashless society in which a debit card would be all anyone would need to carry; although the *switch* systems are reasonably successful, in the early nineties this result still appears a long way off.

> Though similar systems have been tried on a much smaller scale by Hy Vee and Dahl's, both in Iowa, Publix is the first supermarket company to own not only the in-store terminals but also the crucial switch that channels the messages from varied sites to the appropriate banks.
> *Supermarket News* 2 July 1984, p. 1

> Barclays and Lloyds are pushing their debit cards hard. So are National Westminster, Midland and Royal Bank of Scotland, which have jointly developed the Switch debit card system. Their standard cheque guarantee cards double as Switch cards; there are now 10 million Switch cards in circulation.
> *Independent* 27 Jan. 1990, p. 8

T

tablet /'tæblɪt/ *noun* 🖥️

In computing, a flat rectangular plate or pad over which a stylus or MOUSE is moved to input graphics or alter the position of the CURSOR on a VDU screen.

A specialized use of *tablet* in its original meaning of 'a small, flat, and comparatively thin piece of hard material fashioned for a particular purpose'.

The *tablet*, which essentially digitizes information about the position of the stylus or mouse, was developed by the Rand corporation in the US in the mid sixties. At first it was used mainly for inputting graphic images, using a stylus which could be moved around on the *tablet* like a pen on a pad of paper, the resulting 'lines' being instantly translated into images on the VDU screen. With the boom in personal computing and the increasing popularity of WIMPS (see WIMP²) in the eighties, the *tablet* reached a wider market of users and became a commonplace piece of computing equipment. The *tablet* is often known more fully as a **data tablet, electronic tablet**, etc.; one designed for use with the fingers instead of a stylus or mouse is a **touch-tablet** (or *touchpad*).

> A graphics tablet allowing sophisticated computer graphics facilities to be added at low cost to a wide variety of microcomputers has just been announced. *Computing Equipment* Sept. 1985, p. 16

> To get the most out of drawing options, I strongly recommend the use of a mouse, joystick or touch-tablet. *Personal Computer World* Nov. 1986, p. 191

tack /tæk/ *noun* ✖️

In slang: gaudy or shoddy material, rubbish, 'tat'; also, cheap-and-nastiness, kitsch.

Formed by abbreviating the adjective *tacky* 'cheap and nasty, vulgar' (itself a piece of US slang which dates from the nineteenth century).

A media word of the second half of the eighties, especially beloved of arts critics, who also like to use the punning form **hi(gh)-tack** (see HIGH-TECH).

> The king of cinematic trash and tack turns his attentions to the written word. *Arena* Autumn/Winter 1988, p. 198

> Clubbers would turn up wearing exceptionally 'high tack' smiley-faced T-shirts. *Q* Oct. 1988, p. 66

> There's no point in being snooty about hi-tack shows of this sort. We may as well admit that they have an elemental pull on our psyche and submit gracefully. *Time Out* 4 Apr. 1990, p. 54

> Leonard Cohen presents the tale of 'Elvis's Rolls Royce' in a lugubrious deadpan that effortlessly conveys all the sleaze, tack and warped majesty of the subject. *Independent* 13 July 1990, p. 15

tactical /'tæktɪk(ə)l/ *adjective* 🗳️

Of voting: involving a switch of electoral allegiance for strategic purposes (especially so as to prevent a particular party or candidate from succeeding). Also of a voter: operating on this principle.

A specialized use of *tactical*; a person voting on this basis is using a *tactic* designed to ensure that the candidate he or she favours least is not elected.

Voting designed to keep one's least favoured candidate out was first described as *tactical* in the mid seventies. The practice—and therefore also the name—became widespread in British general elections and (especially, perhaps) by-elections during the eighties. An elector living in a constituency where his or her favoured party has no hope of success is most likely to vote **tactically**, so as to confound the opposition.

> There was glee in Government quarters at Labour's predicament. Mr Rifkind, Scottish Secretary,

said Labour had lost one of its safest seats and said Tory tactical voting had contributed to the swing to the SNP. *Daily Telegraph* 12 Nov. 1988, p. 1

Taffia /'tæfɪə/ *noun* Also written **Tafia**

Humorously in the UK, a supposed nepotistic network of prominent Welsh people; a Welsh 'Mafia'.

Formed by telescoping *Taffy* (a nickname for a Welshman) and *Mafia* to make a blend.

A humorous coinage which has been attributed to the Welsh satirical paper *Rebecca* during the seventies. By the early eighties, the word had begun to appear in the national newspapers as well.

I heard murmurings from the London Welsh network (otherwise known as the 'Tafia') on the subject of Sir Geoffrey's repudiation of true Welshness. Tim Heald *Networks* (1983), p. 160

A benevolent, nepotistic gang at the top, who make sure that good jobs are kept in the 'family'. Who … could imagine that the Welsh 'Taffia' would ever have let a juicy growth industry like cultural management get into English hands? *Observer* 28 Aug. 1988, p. 11

tag¹ /tæg/ *noun* and *verb*

noun: An electronic marker which makes it possible to track the whereabouts of the person or thing to which it is attached.

transitive verb: To mark (a person or thing) with an electronic tag so as to control or monitor movement.

A specialized sense of *tag* which represents a metaphorical extension of the meaning 'a label attached to something'.

Electronic *tags* have been used to control shoplifting since the end of the seventies; usually they take the form of a heavy plastic label which must be detached from the goods by a shop assistant using a special machine before the goods can be removed from the shop without setting off an alarm. Similar tags for people had been tried in mental institutions in the US during the sixties. In the late eighties this idea was extended to prisoners and people on parole. In this **tagging** system a small electronic beacon was attached by a band to the person's wrist or ankle; the signals from the beacon could be monitored by a central computer so that the whereabouts of any person wearing the *tag* (also known as an **offender's tag**) would always be known.

A determined-enough shoplifter can remove any electronic tag—but not readily. Tags have been found gnawed in half and left bloodied on fitting-room floors. *Fortune* 25 Feb. 1980, p. 115

The tag, designed for the petty criminal, can be fitted to the leg, neck or wrist. It is controlled by a central computer, which rings the offender at home at random intervals. *The Times* 9 Feb. 1988, p. 5

The latest statistics point to a majority of people working with offenders as being in favour of tagging as a potential reducer of the prison population and hence of crime.

Daily Telegraph 20 Dec. 1989, p. 14

tag² /tæg/ *noun* and *verb*

In HIP HOP culture,

noun: A graffito, usually consisting of a decorated nickname, word, or initial, made by a graffiti artist as a personal 'signature'.

transitive verb: To decorate (a place or object) with graffiti; to leave (one's graffiti signature) in a public place.

Another figurative use of *tag* in the sense of 'label'.

Graffiti *tags* first started to appear in the streets of New York during the first half of the seventies, but the practice of **tagging** did not spread far outside large American cities until the mid eighties. Then it was the popularization of hip-hop culture as a whole that involved youngsters outside the US in constructing these highly decorated nicknames, often on very visible public buildings. The person who paints a *tag* is known as a **tagger**; graffiti artists often work in teams or CREWS and a particular *tag* can belong to a **tag team** or **tag crew** rather than to an individual **tag artist**. A more elaborate graffito is known as a *piece* (short for *masterpiece*).

> The proliferation of 'writing' ... along with its spectacular development from scrawled felt-tip 'tags' on city walls to spray-can 'pieces' ... has been a visible part of New York's daily life.
>
> *New Yorker* 26 Mar. 1984, p. 98

> Vandals have imported graffiti materials from America to ape New York 'tag teams'—gangs who vie to leave their personal trademarks in daring or eye-catching places.
>
> *Daily Telegraph* 3 May 1990, p. 4

talkline ⟨⟨ see -LINE

tamper /'tæmpə(r)/ *intransitive verb* ✖ ⟨⟨

To interfere with the packaging of consumer goods, especially so as to engage in consumer terrorism. Used especially to form compound adjectives:

tamper-evident, of the packaging of foodstuffs, medicines, etc.: having a visible seal or other device which makes obvious any opening of the packet between manufacture and sale;

tamper-resistant, so constructed as to make tampering with the product difficult or impossible.

The search for *tamper-resistant* packaging, especially to prevent young children from harming themselves by mistaking adult medicines for sweets, had already been going on for some time before the first major case of consumer terrorism in the US in 1982. In this incident, cyanide was added to the contents of Tylenol pain-killing capsules and several people were killed after taking them. Later in the eighties, consumer terrorists tampered with baby foods and other foodstuffs in the US and the UK. This new area of crime led to the concept of *tamper-evident* packaging, incorporating some feature (such as shrink-wrapping or a seal which changed colour on contact with the air) to make it obvious if the package had been opened since leaving the factory.

> He said the firm had been checking products item by item since the first Tylenol poisonings in the fall of 1982, but that it 'quickened' its pace to put tamper-evident packaging on its products in the wake of the second Tylenol poisoning incident earlier this year.
>
> *Chicago Tribune* 2 May 1986, p. 2

Tankie /'tæŋkɪ/ *noun* Also written **Tanky** ▣

In British slang, a hard-line Communist who unquestioningly supported Soviet policies.

Said to be so named because of the *Tankies'* reluctance to condemn Soviet military intervention (*tanks*) in Afghanistan (or, long before that, in Czechoslovakia).

The split of British Communism into a Eurocommunist (see EURO¹) and a Sovietist or *Tankie* branch dates from the second half of the seventies, although the dismissive nickname *Tankie* did not start to appear in print until the mid eighties. The hard-line *Tankies* were associated particularly with the *Morning Star* newspaper by users of the nickname.

> The New Communist Party of Britain, the Battersea Sovietist splinter off the old bloc, has issued this guidance to the world's press. 'Please do not describe the NCP as "Stalinists" or "Tankies" ... If you insist on using this misleading shorthand, please make it clear you are talking about "Stalinists and Tankies" who support glasnost and perestroika.'
>
> *Guardian* 28 Apr. 1988, p. 23

tar ⟋ see BLACK TAR

taxflation ⩘ see KIDFLATION

TBS ⊗ see SICK BUILDING

techno /'tɛknəʊ/ *adjective* and *noun* 🎵 📷

adjective: Of popular music, making heavy use of technology (such as synthesized and sampled sounds, electronic effects, etc.).

noun: A style of popular music with a synthesized, technological sound and a dance beat.

Formed by abbreviating *technological*; compare ELECTRO.

Techno is one of the sounds of the second half of the eighties, taking the electronic revolution in modern music to its limits. The word is also used in combination with other popular-music terms, notably in **techno-funk**, **techno-fusion**, **techno-pop**, and **techno-rock**, as well as in derived words such as **technofied**.

> 'Musical Melody' comes across like a technofied version of a rare groove.
> *Music Technology* Apr. 1990, p. 76

> The endemic mistrust of dance music that makes it a rock and roll island also means that the new noises of the Eighties—hip hop, house, techno et al—have been, at best ignored, at worst patronised.
> *The Face* June 1990, p. 48

> Marillion with Hogarth are now a band, not four musicians playing backing to a rampant ego, and the only 'old' track that survives the transition to embryonic techno rock band is the excellent 'Freaks'.
> *Sounds* 28 July 1990, p. 34

technobabble 💻 see -BABBLE

technopunk 🧩 see CYBERPUNK

technostress /'tɛknəʊstrɛs/ *noun* Also written **techno-stress** ⊗

Stress arising from working in a technological environment (especially with computer technology); a psychiatric illness whose main cause is difficulty in adapting to new technology.

Formed from *techno-* (the combining form of *technological*) and *stress*.

Technostress was first identified in the US in the mid eighties, as people's working environments were changed out of all recognition by the technological revolution. In 1984 US psychologist Craig Bord devoted a whole book to the subject, subtitled *The Human Cost of the Computer Revolution*. A person suffering from *technostress* is described as **technostressed** or even **technostressed out**; both terms can refer either to problems of adaptation, or simply to the special stresses of spending the day at a computer which might fail. In California, psychologists recommend *electrobashing* (literally taking one's frustrations out on a computer) to release these tensions.

> An assortment of 'technostressed-out' humans delighted in hurling malfunctioning televisions, telephone answering machines ... and video cameras off a balcony to oblivion.
> *The Times* 18 May 1990, p. 1

> Throughout modern society, humans are enslaved by the machines that seem to empower them. Symptoms include paranoia, fatigue, low self-esteem, flagging libido, anxiety, headaches, and over-stimulation. Collectively, they are 'technostress'.
> *The Australian* 29 May 1990, p. 47

Teenage Mutant Hero (or Ninja) Turtle 🧩 📷 see TURTLE

Teflon /'tɛflɒn/ *noun* 📷

Used attributively of a politician or political administration, in **Teflon politician**,

Teflon presidency, Teflon president, etc.: able to shrug off scandal or misjudgement and deflect criticism on to others, so that nothing 'sticks'.

A metaphorical use of the trade mark *Teflon*, a non-stick polymer coating used on saucepans and other cooking utensils.

This sense was invented by US Congresswoman Pat Schroeder in August 1983, when she said in Congress:

> After carefully watching Ronald Reagan he is attempting a breakthrough in political technology—he has been perfecting the Teflon coated Presidency. He sees to it that nothing sticks to him.

The imagery proved very successful in political life, and was later applied to a number of other politicians—at local and national level—who somehow managed to ensure that someone else was blamed for any scandals or misjudgements involving their administration.

> The Mayor is celebrated for ... distancing himself as far as possible from whatever may have gone wrong ... The executive director of the largest local public-employees' union has called him 'the Teflon mayor'.   *New Yorker* 28 Jan. 1985, p. 74

> Presidential assistant Richard Darman told me that the so-called Teflon phenomenon—the fact that blame never seemed to stick to President Reagan, even after such disasters as the Beirut suicide bombing ... —was directly related to journalists' tendency to emphasize personality over substance.   Mark Hertsgaard *On Bended Knee* (1988), p. 67

tele- /tɛlɪ/ *combining form* 🖳

Widely used as the first element of compounds relating to telecommunications, particularly in words for concepts which have been transformed by the use of telecommunications and information technology.

Originally from Greek *tele* 'afar, far off': the first two syllables of *telephone, television*, etc.

Every innovation in telecommunications during the twentieth century seems to have set off its own explosion of words formed on *tele-*, which of course has a far longer history in the more general sense of 'at a distance'. It is the continuous improvement in **telematics**, the long-distance transmission of computerized information, which lies behind many of the new *tele*-words formed during the eighties. This proliferation began in the mid seventies, when such services as CEEFAX and ORACLE began to be referred to collectively as **teletext**. The later extension of this idea to text transmission via the telephone network, combined with a facility enabling the domestic user to transmit as well as receive text, created the conditions for a variety of services: **teleordering** (the ordering of books direct from publishers by booksellers) was followed by **teleshopping** (shopping conducted from home using a computer and a telephone), **telebanking, telebroking,** and even **telebetting**. The telecommunications revolution also had its effect on working practices: the **teleconference** (or **telemeeting**), an idea dating from the fifties, became more practical, and some office workers began to **telecommute,** or work from home while communicating with the office and elsewhere via data links (a process also known as **teleworking**). From Scandinavia in the second half of the decade came the concept of the **telecottage**: a room in a rural area filled with equipment for teleworking, available for shared use by local residents; working from one of these is known as **telecottaging**. Alongside all of this new technology, the old technologies continued to give rise to *tele*- compounds: **telemarketing,** the marketing of goods or services through unsolicited telephone calls (carried out by **telemarketers**), became an established selling technique, while television journalism produced many humorous nonce-words such as **telepundit** and fund-raising extravaganzas such as the **Telethon** (an old concept, but one which was given a new lease of life in the eighties).

> France provided the impetus by seeing the smart card as a way of modernising the country's telephone and banking systems with card-based payphones and telebanking and teleshopping facilities which rely on home computers connected to a telephone.   *New Scientist* 11 Feb. 1989, p. 64

> The appeal of telecommuting lies in its ability to extend office functionality beyond the confines of the office.   *UnixWorld* Sept. 1989, p. 102

Nynex intends to make the country a high-tech show-place, with fiber-optics and other digital technologies, video teleconferencing and high-speed facsimile services.
New York Times 10 Dec. 1989, section 3, p. 9

In Scandinavia around 200 rural 'Telecottages' have been set up for business use in the last five years. *Daily Telegraph* 11 Apr. 1990, p. 32

ITV Telethon '90: ... A mass tap dance ..., plus a celebrity tug o' war, ditto It's A Knockout, a giggle of comedians ... and a flying visit from the RAF. *Guardian* 28 May 1990, p. 30

Alan Denbigh, Acre's teleworking adviser, predicts that the telecottage movement will soon begin to grow fast. *Daily Telegraph* 5 Jan. 1991, Weekend section, p. iii

telespud ❌ ▐▌ see COUCH POTATO

televangelist /tɛlɪ'vændʒəlɪst/ *noun* ❌ ▐▌

An evangelical preacher who uses television or other mass media to promote his or her doctrines.

Formed by telescoping *television* and *evangelist* to make a blend. The unblended forms **television evangelist** and **TV evangelist**, and the compound **tele-evangelist**, also occur, but are less common.

Television, especially on channels devoted to religious broadcasting, was first used by some evangelical Christian denominations as an effective means of preaching the Christian gospel as long ago as the fifties, when the first *pray-TV* channel was set up in the US. Evangelists with a gift for mass communication, such as Billy Graham, became world-famous, but **televangelism** as such remained a predominantly American phenomenon for some years after the words *televangelist* and *televangelism* started to be used in the mid seventies. With the renewed fashion for fundamentalist doctrine during the early eighties, however (see FUNDIE), *televangelists* such as Pat Robertson, Jim Bakker, Jimmy Swaggart, and Oral Roberts (who even founded a university named after himself) achieved considerable fame and political influence. In the later eighties, a succession of scandals involving the financial and sexual affairs of certain *televangelists* brought them into the news in a more negative way.

A study ... performed by the A. C. Nielsen Co. found that 34 million people watched one of the top 10 tele-evangelists during the month studied. *Washington Post* 5 Feb. 1986, section C, p. 11

Televangelist Jimmy Swaggart smugly cast stones at adulterous PTL (Praise the Lord) head Jim Bakker—until his own voyeuristic trysts with a New Orleans hooker came to light in 1988.
Life Fall 1989, p. 142

teraflop 🖥 see MEGAFLOP

Tessa /'tɛsə/ *acronym* Also written TESSA 〰

Short for **tax exempt special savings account,** a special type of savings account for those who are exempt from income tax in the UK, from which tax on the interest earned is not deducted at source.

The initial letters of *Tax Exempt Special Savings Account*.

The *Tessa* was announced as a 'wholly new tax incentive' by the then Chancellor John Major in the April 1990 budget; the accounts themselves were not to be operational until January 1991. Until that time, all savers making use of banks and building societies in the UK (including, for example, children and pensioners) were paying tax at source on the interest on their savings, whether or not they were in fact liable to income tax. The *Tessa* scheme allows the holder of one of these special accounts to earn tax-free interest on savings up to a total of £9,000 accumulated over five years. Almost immediately after the Chancellor's announcement, the accounts became known by the pronounceable acronym *Tessa*, which was often treated in advertising as though it were a girl's name. No doubt the full name had been chosen with this in mind (the *s* of *special*, for example, was essential to avoid the pronunciation /'ti:zə/, so the less-than-essential word *special* was included).

> You may already be able to find TESSA-style accounts on the market, even though TESSAs won't officially start until January 1991.   *Which?* May 1990, p. 249

> The first Tessa—or Tax Exempt Special Savings Account—to be launched since the Chancellor introduced them in the Budget, it offers 13 per cent tax-free for 5.5 years.   *Guardian* 9 June 1990, p. 12

test-tube baby Ⓧ 🖳 see IVF

Thatcher /'θætʃə(r)/ *noun* 🏛

The name of Margaret Thatcher, British Prime Minister 1979–90, used in **Thatcher's Britain** to summarize the effects of her policies, and as the basis for derivatives such as **Thatcherism, Thatcherite**, etc.

The surname of one of Britain's longest-serving Prime Ministers.

Thatcherism and *Thatcherite* (a noun or adjective) both date from the second half of the seventies, when Mrs Thatcher was rising through the ranks of Tory MPs and her policies were becoming influential. At the end of the seventies another adjective, **Thatcheresque**, was coined: this essentially means 'akin to Mrs Thatcher or her policies, Thatcher-like', but has been used particularly in relation to public spending cuts and unwavering resolve in carrying out a policy. Within two years of the start of Mrs Thatcher's administration, journalists started to use the term *Thatcher's Britain* as a shorthand for British society as it was supposedly affected by Mrs Thatcher's policies; although some uses were positive, the emphasis tended to be on the economic effects or on the social divisions which Conservative policies of the past decade were seen to have produced. Providing a positive counterbalance to all this, an admirer or devotee of Mrs Thatcher is called a **Thatcherphile**.

> When one of them said 'make sure you tell them what Thatcher's Britain has done to young people,' I agreed with that young person wholeheartedly.   *Guardian Weekly* 5 June 1988, p. 2

> Pauling manifested a quite unnerving certitude and Thatcheresque disregard for even the remotest possibility that he might be wrong.   *New Scientist* 9 Dec. 1989, p. 55

> Christopher Hogwood and ... Barry Tuckwell are some of the *weltklasse* artists converging on a corner of England that looks, unfortunately, forever Thatcher.   *20/20* July 1990, p. 99

> Because of the Thatchers ... a Chinese couple has already been to look round, and an American Thatcherphile has booked a visit.   *Daily Telegraph* 18 Dec. 1990, p. 15

theme park /'θi:m ˌpɑːk/ *noun* ▓

An amusement park organized on a particular theme or based on a unifying idea, with each attraction linked in some way to the theme.

Formed by compounding: a *park* organized round a *theme*.

The first *theme parks* were modelled on the American Disneyland in the sixties. During the late sixties and seventies, several such parks were set up outside the US, but it was not until the late seventies and early eighties that the leisure industry took up the idea in a big way in the UK and started to apply it in other contexts. In the mid eighties, for example, the principle was applied to catering outlets in the UK, resulting in the **theme pub** and **theme restaurant**, in which each aspect of design and atmosphere was related to a particular unifying theme. *Theme parks* came in for some criticism from environmentalists in the late eighties, since they take up large tracts of countryside and are felt by some to be an eyesore.

> Grand Metropolitan's Host Group ... is to spend well over £100m over the next three years on converting its outlets to a wide range of theme pubs.   *The Times* 4 Nov. 1983, p. 17

> Local conservationists are even more horrified by a new proposal—including a Disney-style theme park—covering 1,000 acres.   *Holiday Which?* Sept. 1989, p. 176

The long-awaited plan is the product of months of work by Disney 'imagineers', who conjured up Port Disney, a complex of waterfront dining, a marina, a theme park that explores the 'mysteries of the sea', and steel cages under water where tourists can swim with sharks.

San Jose Mercury 1 Aug. 1990, section B, p. 8

theory of everything 🖳 see SUPERSTRING

Third Ager 🍴 see WOOPIE

thirtysomething /'θɜːtɪˌsʌmθɪŋ/ *noun* and *adjective* 🔀

noun: An indeterminate age between thirty and forty; a person of this age, especially a BOOMER who reached this age during the eighties.

adjective: Of or belonging to such a person or the group as a whole; characteristic of baby boomers and their lifestyle in the eighties.

The form *-something* could always be added to a number such as *twenty, thirty, forty*, etc. to indicate uncertainty as to the precise age of a person (or indeed the precise number of something else), so the word *thirtysomething* had existed for some time, used when the context demanded; what brought it into public focus and led to its being used widely to refer to the boomer generation was a popular US television series called *Thirtysomething* (also shown outside the US), which from 1987 recounted the ups and downs and family lives of a group of boomers who had reached their thirties in the eighties.

The success of the television series *Thirtysomething* can in part be attributed to the fact that a large proportion of its viewing public was able to identify directly with the characters; it also came at a time when the trend analysts and marketers in the US had been focusing their efforts on meeting the demands of this very group. The word very quickly came to be used as a noun and adjective not directly alluding to the programme, but to the whole socio-economic grouping; within months this also gave rise to an explosion of other uses of *-something* to refer to other groups belonging to a different generation (**twentysomething, fortysomething**, etc.: see the examples below). The fashion for such formations continued into the early nineties.

At least 83 of the 121 films that leading distributors are opening in the New York area promise to be intellectually respectable enough for bright fortysomethings. *Newsday* 11 Sept. 1988, section 2, p. 3

This comic strip collection chronicles the demands of a 'thirtysomething' career woman.

Publishers Weekly 11 Aug. 1989, p. 373

Are you ready for seventysomething rock? John Lee Hooker (b. Clarkesdale, Mississippi in 1917) is the most thoroughly unreconstructed Delta bluesman still practising. *Q* Dec. 1989, p. 127

Rosen was a lawyer from the 'Thirtysomething' crowd: ... the kind of early 1970s rabble-rouser embarrassed to tell his Swarthmore class reunion he now made millions sniffing out tax loopholes for corporate takeovers. Bryan Burrough & John Helyar *Barbarians at the Gate* (1990), p. 406

thrash /θræʃ/ *noun* 🎵 🍴

A style of rock music (also known more fully as **thrash metal**) which includes elements of HEAVY METAL combined with the violence and spirit of punk rock.

A development of *thrash* in the sense of 'a short, energetic (and usually fast and loud) passage of popular music or jazz', which developed in spoken use among jazz musicians and was itself first recorded in print in the sixties.

Thrash, which often features images of horror and violence expressed in the harsh style of heavy metal, developed out of the more shocking aspects of punk rock in the early eighties. The emphasis on morbid themes led to the alternative name *death metal*, while its relentlessly fast rhythms gave rise to a third name for essentially the same style of music, *speed metal*. *Thrash* is often used attributively, in **thrash band**, etc. This style of rock enjoyed a vogue in the closing years of the eighties, but by 1990 was already beginning to wane in popularity.

Avoiding solo virtuosity and theatrical excesses, the new bands deliver a buzzsaw thrash that is as hard, fast and loud as possible ... The success of the likes of Metallica and Anthrax suggests that thrash metal is about to find itself in a conundrum, coping with commercial success born from a noise designed to outrage. *Guardian* 20 Mar. 1987, p. 19

Totally happening Melbourne based glam thrash all-girl rock n roll phenomenon searching for wild drummer ... Come on girls! Grab this chance. *Time Off* (Brisbane) 19 Feb. 1988, p. 15

tight building syndrome ⊗ see SICK BUILDING

timeframe /ˈtaɪmfreɪm/ *noun* Also written time frame

In US English: a period of time, an approximate time (originally a limited period during which something could be achieved).

Formed by compounding; in the original meaning, there was a sense of constraints forming a *frame* round the *time* during which something could be done.

The term *timeframe* was originally used in the sixties, with very specific reference to a period of time on which definite starting and finishing constraints had been set, for example the schedule within which certain work was to be achieved. By the eighties, though, it had become a fashionable synonym for 'period' in general and started to spread outside US English. Thus a shipbuilder interviewed in a television documentary who said 'We built this ship in the 1976–7 timeframe' meant not that the ship had to be built to that schedule but that it was built in about 1976 or 1977. The result is that the word has changed its meaning from a very specific to an approximate period.

Hubbard told us the MSO's plans will not impact his intention to launch his service in the 1991–93 timeframe. *Satellite News* 12 Feb. 1990, p. 3

Timeline see -LINE

Tinkie ⚏ see DINK

TOE ▣ see SUPERSTRING

tonepad /ˈtəʊnpæd/ *noun* Also written tone pad ▣

An electronic device similar in size and shape to the remote control handset of a television set and used for the transfer of data to a central computer, often over a telephone line.

Formed by compounding: a box the size of a *pad* of paper, used to transmit data by means of electronic *tones*.

Unlike its competitors, the TSB Speedlink requires only a tonepad ... and an ordinary telephone. After punching out a code number similar to those used in automated teller machines and the account number given on the customer's cheque book, he or she simply tells the voice-activated computer which services are required. *Daily Telegraph* 15 Apr. 1987, p. 27

Until payphones are converted, they allow anyone with a readily available gadget called a 'tone pad' to make free calls of unlimited duration anywhere in the world. *New Scientist* 9 June 1990, p. 27

tossing ⚏ see OUT

total body scanner ⊗ see BODY-SCANNER

totally ⚇ see AWESOME, TUBULAR, and VALSPEAK

touchpad, touch-tablet ▣ see TABLET

toyboy /ˈtɔɪbɔɪ/ noun 🎎

In British media slang, an attractive young man who is 'kept' as a lover by an older person.

Formed by compounding, taking advantage of the rhyming syllables: a *boy* who is the plaything or *toy* of an older partner.

The concept of the *toyboy*—socially the male equivalent of the BIMBO—arose in the early eighties and soon became established as a regular feature of the language of the tabloids. Normally the *toyboy* is the younger lover of a mature woman, but the word has also been applied to gay relationships; often it is used attributively, with the implication that the person being described is young and attractive. The term has even begun to generate variations: for example, the rock star Madonna was punningly described as the **boy toy** because of the motto on her belt-buckle and the overtly sexy image that she cultivated, and this was later applied in a transferred sense to other female stars in the same mould.

> At 48 she is like a teenage girl again—raving it up with four different lovers including a toyboy of 27!
> *News of the World* 15 Nov. 1987, p. 32

> Rock's richest pop-tart [Madonna], the Boy Toy who made lingerie-and-crucifixes fashionable.
> *Life* Fall 1989, p. 84

> Olivia . . . has been wearing out her toy boy hubby! At 31, Macho Matt Lattanzi is 11 years younger than his famous wife.
> *People* 11 Mar. 1990, p. 3

train surfing 🏄 see SURF

triple A /ˌtrɪp(ə)l ˈeɪ/ noun 🎖️

In military jargon: anti-aircraft artillery.

A form representing the way in which many people would say *AAA*, itself the initial letters of *Anti-Aircraft Artillery*.

In the form *AA* or *AAA*, the abbreviation has been in use since the First World War among the military. What brought it into public focus in particular was its use by journalists reporting the Iraqi response to allied air attacks on Baghdad and other Iraqi cities during the Gulf War of 1991. It seems it was only in newspaper reporting of the wars of the previous decade that the form *triple A* started to be written down rather than being a way of speaking *AAA*.

> There was an awful lot of triple-A (antiaircraft artillery) in the area and that was a surprise.
> *Christian Science Monitor* 8 Dec. 1983, p. 52

> Viewers heard debriefing pilots say *triple* A, or A.A.A. . . . in reference to cannons and machine guns but not surface-to-air missiles.
> *New York Times Magazine* 3 Feb. 1991, p. 8

triple witching hour /ˌtrɪp(ə)l ˈwɪtʃɪŋ ˌaʊə(r)/ noun 〰️

Colloquially, the unpredictable final hour of trading on the US stock exchange before three different kinds of options simultaneously expire.

The *witching hour* is traditionally midnight, a time when the witches are supposed to come out and anything can happen; the *triple witching hour* is so called because the market can easily be thrown into turmoil (especially by computer-driven changes) when options are all expiring at once, and anything could happen to the Dow–Jones index.

The term has been in use among traders on Wall Street since at least the sixties, but was not much heard outside their jargon until the ARBs stared to exploit the gaps between the price of stock index futures and the actual level of the market in the mid eighties. This and the increasing use of PROGRAM TRADING brought the term into the daily papers, especially when one of the quarterly *triple witching hours* was approaching; they occur on the third Friday of the final month of each quarter and involve stock options, stock index options, and stock index futures.

Several days before last Friday's 'triple witching hour', many professional stock traders again braced for a wild final 60 minutes in the life of three key market forces... and a wild 60 minutes it was.

New York Times 24 June 1985, p. 5

Wall Street also responded to concerted action by the major U.S. financial markets to close down programme trading . . . which became notorious because of the so-called triple witching hour volatility. *Jordan Times* 21 Oct. 1987, p. 1

triple zero option 🖳 see ZERO

trivia /'trɪvɪə/ *plural noun* ✖

Miscellaneous (often unusual or peripheral) facts about something; a quiz game in which the object is to answer questions eliciting such facts.

Originally the name of one such quiz game; it refers to the peripheral or trivial nature of many of the facts included in the game.

The craze for *trivia* quiz games began in the late sixties, but really took off only with the invention in 1982 of **Trivial Pursuit** (a trade mark), a board game devised in Canada by two journalists, Chris Haney and Scott Abbott. This game combined the quiz element with the traditional board game format, with each player acquiring credits by answering general knowledge questions in six subject areas, represented by different squares on the board. The game was enormously successful throughout the world and was followed by many imitations using the word *trivia* somewhere in their name. As a result, many people associate the word *trivia* not with 'matters of little importance' (its original meaning) but with quizzes and the arcane facts that it is always useful to know when competing in these games.

Here's a question even a three-year-old could answer: What was the best-selling new board game of the 1980s? Trivial Pursuit. *Life* Fall 1989, p. 64

Doing a column on presidential trivia is like volunteering to be the victim in a dunking booth at the country fair. *Baltimore Sun* 7 Mar. 1990, section A, p. 15

Sounds readers may prefer to wait for the paperback to appear, by which time most mistakes will have been ironed out. But anyone buying it *will* find it invaluable for answering tricky trivia questions. *Sounds* 28 July 1990, p. 20

Trojan /'trəʊdʒ(ə)n/ *noun* Also written **trojan** 🖳

A computer program which (like a VIRUS or WORM) is designed to sabotage a computer system, but which usually breaks the security of the system by appearing to be part of a legitimate program, only starting to erase or retrieve data once it has been carried successfully into the system. Also known more fully as a **Trojan horse**.

A reference to the *Trojan horse* in the Greek epic tradition: a hollow wooden horse in Virgil's *Aeneid* in which Greek soldiers concealed themselves to enter and capture the town of Troy. Since the nineteenth century, the term *Trojan horse* had been applied figuratively to any person or device concealed as a trick to undermine something from within. The computing sense was the first to abbreviate this further to *Trojan* (and it is perhaps surprising that this happened even in the computing sense, since *Trojan* is the trade mark of a well-known brand of contraceptive sheath in the US).

Under the name *Trojan horse*, the *Trojan* was first developed in the seventies by hackers (see HACK) wanting to gain access to other people's systems or carry out computer frauds involving the transfer of funds by computer. By the second half of the eighties, *Trojans* were considered an important hazard and special systems had been set up to detect and block them. The *Trojan* may be no more than a few lines of code inserted into another (apparently useful) program; it cannot replicate itself, but once the program is running it can start carrying out its under-cover activities, copying or destroying data as required. In many ways, a *Trojan* is similar to a LOGIC BOMB except that it does not usually require a specific set of conditions to obtain before it can be activated.

Among the dozens of trojans in circulation, some begin their destruction within minutes.

The Times 26 May 1987, p. 26

A perfect place to plant a Trojan horse. By changing a couple [of] lines of code in our *telnet* program, he could make a password grabber. Whenever my scientists connected to a distant system, his insidious program would stash their passwords into a secret file.

Clifford Stoll *The Cuckoo's Egg* (1989), p. 154

tubular /tjuːbjʊlə(r)/ *adjective*

In young people's slang, originally in the US: excellent, wonderful, very good or exciting, AWESOME. Often in the phrase **totally tubular**, superlative.

Originally from Californian surfers' slang, in which a *tubular* wave was one which was well curved (and so shaped like a *tube*); a hollow, well-curved wave was the best and most exciting kind to ride on, so *tubular* soon came to mean no more than 'very good'.

Tubular originated in the slang of Californian surfers in the seventies; in its more general sense it was one of the words taken up by VALSPEAK in the early eighties and spread to a whole generation of American youngsters. Although already considered a little passé by teenagers, in the second half of the decade it acquired a new currency among younger children (partly as a result of its use by the TURTLEs and other screen idols). This later vogue extended to British English, at least among children.

It would be nice to be able to say that last night's opening round of *The Story of English* (BBC-2) was 'tubular', 'the max' or just 'totally'. It was not up to that standard. But it was quite exciting.

Daily Telegraph 23 Sept. 1986, p. 14

Hey Ron, you and Nancy were totally tubular, dude. I'm talking radical to the bone, buddy. Nobody can beat your admin, you know what I'm saying? Oh man, you were awesome, the best.

USA Today 11 Jan. 1989, section A, p. 7

Donatello [one of the Turtles] is totally tubular when he's jamming on his hand-held keyboards.

Daily Star 23 Oct. 1990, p. 19

Turtle /ˈtɜːt(ə)l/ *noun* Also written **turtle**

In full, (in the US) **Teenage Mutant Ninja Turtle** or (in the UK) **Teenage Mutant Hero Turtle**: any of a group of four fantasy characters for children, in the form of terrapins who have supposedly been mutated through being covered in radioactive slime in a New York sewer. In the plural, **Teenage Mutant Ninja Turtles**, the trade mark of a series of children's stories, programmes, games, and toys based on the exploits of these characters.

An abbreviated form of the full name, *Teenage Mutant* etc.; in US English, *turtle* is the standard word for all the animals of the order Chelonia, which in British English are known variously as terrapins, tortoises, and turtles.

The pizza-loving *Turtles* were the invention of American comic-book artists Kevin Eastman and Peter Laird in 1988 and early in their history as comic-book figures were apparently used by a New York pizza house as a way of providing amusement for children while they were waiting to be served with their pizzas. The idea proved so successful that soon a whole range of *Turtle* licensed products appeared on the market, including computer games, toys, stationery, and a television series. The craze for *Turtle* licensed products was particularly intense in the US in 1989 and in the UK in 1990; so intense, in fact, that it became known as **turtlemania**. The *Turtles*, also known in the merchandising hype as the *awesome foursome* or the *heroes in a half shell*, helped to popularize a version of Californian youngsters' slang heavily influenced by VAL-SPEAK and surfers' talk; this language, including the cry of COWABUNGA and adjectives such as AWESOME, RAD, TUBULAR, etc., has been called **turtlespeak**. In the US the *Turtles* were known in full as *Teenage Mutant Ninja Turtles*, but when they were introduced to the UK market the name was changed to *Teenage Mutant Hero Turtles* in some cases (presumably because the word NINJA was felt to be too unfamiliar to British ears). The name is often abbreviated to **Ninja Turtle** rather than simply *Turtle* (even in the UK).

> Actors wearing mutated-turtle outfits and hired to sign autographs at a toy store outdrew President Reagan, who made an appearance in town on the same day. *New Yorker* 11 Dec. 1989, p. 142

> Their new line of cereals includes Teenage Mutant Ninja Turtles, Nintendo Cereal System . . . and Batman, as well as Breakfast with Barbie. *People* 19 Feb. 1990, p. 9

> Turtlemania! headline in *The Sun* (Brisbane) 5 Apr. 1990, p. 24

> Hollywood declined to fund a full-length Ninja Turtles feature, thus missing the chance to cash in on this extraordinary craze. *20/20* July 1990, p. 21

> Now the rock world is reeling from the most awesome teenage heart-throbs of the lot—the Turtles. *Daily Star* 23 Oct. 1990, p. 19

tweak[1] /twiːk/ *noun* 🖳

A minor modification to a computer system or some other mechanism; hence, an inessential but desirable enhancement, an optional extra.

A figurative sense development based on the idea of giving a mechanical device a *tweak* or fine-tuning twitch into shape; the corresponding verb has been in use in a number of technical contexts since the mid sixties.

Originally a feature of US English, this sense became associated particularly with the world of computing and with the design and manufacture of large consumer items such as cars and motorcycles in the second half of the eighties.

> Some tweaks were necessary. He had to adjust the screen code to accommodate the different sizes of the DEC and personal computer displays. *Computerworld* 18 Dec. 1989, p. 35

> The game is very neat and the ability to edit the levels is an additional tweak. *Your Amiga* Mar. 1990, p. 25

tweak[2] /twiːk/ *intransitive verb* ✂

In the slang of drug users, especially in the US: to suffer from nervous twitching, mental disturbance, etc. as a result of addiction to a drug.

Formed by using what would normally be a transitive verb intransitively; a reference to the involuntary twitching associated with withdrawal from drugs, as though the person were being *tweaked*. An earlier sense in drugs slang had been 'to inject heroin', and heroin users are sometimes known as *tweakers*.

Although no doubt in spoken use among drug users for some years, this sense of *tweak* only began to appear in print in the late eighties as a result of media interest in the growing drugs problem in the US.

> Redneck, tweaking as the coke wears off, erupts when he hears that. He begins smashing his right hand into a wall. *Newsweek* 25 Apr. 1988, p. 64

> Then there are wounds inflicted with knives, baseball bats and other weapons when drug users are 'tweaking', the street jargon for the volatile behavior that accompanies crack. *New York Times* 6 Aug. 1989, section 1, p. 1

U

UDMH see ALAR

unban /ʌn'bæn/ *transitive verb*

To remove a ban from (an organization, activity, etc.); to legitimize.

Formed by adding the prefix *un-* (indicating reversal) to the verb *ban*; the fact that the verb *ban* itself has negative meaning makes the addition of *un-* to it rather unexpected and means that *unban* has a droll effect for some people.

The word *unban* has existed since at least the late sixties, but most people were probably unaware of it until discussion of the possible lifting of the South African government's ban on the African National Congress became a feature of the news in the second half of the eighties. This *unbanning* actually took place in February 1990, providing a concentration of uses in journalism at that time and helping to establish the noun *unbanning* and the adjective **unbanned**. All three forms have since been applied in other contexts.

> He announced that he was unbanning the long-outlawed African National Congress and would soon free its aging leader.
> *People* 19 Feb. 1990, p. 57

> The unbanning of foreign investment in Finnish markka bonds has taken place but has not encouraged a flood of interest.
> *European Investor* May 1990, p. 63

> Now that Dr Boesak has forsaken his power base in the church, now that Nelson Mandela and his colleagues are free and the unbanned African National Congress is talking with the government, will there be a role centre-stage for him?
> *Independent on Sunday* 29 July 1990, p. 21

unbundle /ʌn'bʌnd(ə)l/ *transitive* or *intransitive verb*

In financial jargon, to divide (a company or group, its assets, products, etc.) into a core company and a number of smaller businesses, usually so as to sell off the smaller companies to finance a take-over. Occasionally used intransitively: to carry out this kind of activity.

A specialized figurative sense of a verb which was already in use in the business world in the sense 'to charge separately for (items previously treated as a group)'.

The activity of **unbundling** was first practised under this name in the US in the seventies, but many financiers see it as no more than a more up-to-date term for asset-stripping (see ASSET). In the UK, the whole process is specially associated with Sir James Goldsmith and his dealings with the BAT Industries conglomerate at the end of the eighties: in fact, he became so famous as an **unbundler** that he acquired the nickname 'the great unbundler' for his attempts to deal with CORPOCRACY in large conglomerates. A conglomerate to which this process has been applied may be described as **unbundled**.

> In practical terms, companies are learning to 'unbundle', to move away from the classic idea of the traditional package of equity, technology, and management.
> *American Banker* 28 July 1982, p. 20

> Conglomerates, who needs 'em? That sums up the prevailing attitude following the bid for BAT Industries by Sir James Goldsmith and friends. The immediate response is that Sir James certainly doesn't need them. If there were no conglomerates to 'unbundle' he would no doubt argue in favour of the concept and buy companies to create a conglomerate.
> *Guardian* 8 Aug. 1989, p. 11

> Since the demerger forced on it by the Great Unbundler and Co, its simplified business has not been properly understood.
> *Independent on Sunday* 29 July 1990, Business on Sunday section, p. 2

undink ⟨ see DINK

unfriendly[1] /ʌnˈfrɛndlɪ/ *noun*

A hostile person or thing; in military jargon, an enemy.

Formed by treating the adjective *unfriendly* as a noun; in the military usage there could be some influence from the adjective FRIENDLY meaning 'fighting on one's own side'.

Unfriendly was first used as a noun in the seventies. Apart from the military usage recorded here, it has been used to refer to any hostile person or thing (for example, a HOSTILE take-over bid or an attacking rogue program such as a VIRUS).

> The old model [missiles] you can buy ... Makes a big difference if the friendlies or the unfriendlies get 'em, and what kind of encoding hardware, computer directors, and so on go with 'em.
> S. F. X. Dean *Such Pretty Things* (1982), p. 146

> We violated the sovereign nation's borders with our troops; shot and killed 'unfriendlies' as well as that nation's civilians.　*Charlotte Observer* 2 Jan. 1990, section A, p. 5

unfriendly[2] /ʌnˈfrɛndlɪ/ *adjective*

Unhelpful or harmful; used especially as a combining form in compound adjectives in which the preceding noun names the person or thing hindered or harmed, including:

environment-unfriendly, harmful to the environment (see ENVIRONMENT[1]); not ECOLOGICAL;

ozone-unfriendly, contributing to ozone depletion; not ozone-friendly (see OZONE);

user-unfriendly, unhelpful to the user; not USER-FRIENDLY; also as a noun **user-unfriendliness**.

Formed by adding the prefix *un-* to *friendly*: see -FRIENDLY.

The idea of this kind of **unfriendliness** arose from the success of the term USER-FRIENDLY in the world of computing: see the history given under that heading and at -FRIENDLY. Searching for a word to serve as the opposite of *friendly* in this sense, some people chose *hostile* (see under USER-FRIENDLY) and others preferred *unfriendly*. In general, *unfriendly* was the more successful and productive choice (especially as a combining form) in writing on environmental issues since about the middle of the eighties, while *-hostile* enjoyed almost equal success in computing. *Unfriendly* presented some of the same grammatical problems as *-friendly*, especially when printed without a preceding hyphen: as a free-standing adjective it could not be combined with another adjective to form a compound, so the parallel form **environmentally unfriendly** developed alongside *environment-unfriendly*.

> One of the most popular general-purpose benchmarks is the Sieve of Eratosthenes, probably the most user-unfriendly title in the business.　*Byte* Feb. 1984, p. 160

> A useful document for anyone campaigning on the ozone issue or wishing to avoid ozone-unfriendly packaging.　*Green Line* Oct. 1988, p. 5

> Chemical reactions take place ... transforming ... 'friendly' non-destructive chlorine and bromine into an 'unfriendly' radical form that destroys ozone.　*Boston Globe* 23 Jan. 1989, p. 30

> Denmark, which also has strict environmental regulations, heavily taxes environment-unfriendly products.　*Chemical Week* 6 Sept. 1989, p. 30

ungreen /ʌnˈɡriːn/ *adjective*

Of a person: not concerned about the environment (see ENVIRONMENT[1]); of a product or activity: harmful to the environment, not ecologically AWARE.

Formed by adding the prefix *un-* to *green*, an adjective which would not normally have an opposite.

An inevitable development of the GREEN revolution, *ungreen* first started appearing in print in the second half of the eighties and quickly became established. In political life *grey* has also been tried as the opposite of *green*, but it is less transparent in meaning and so perhaps unlikely to be taken up in popular use.

> It [BAT industries] is one of the three biggest tobacco companies in the world ... The trouble is that its core business is in the ungreen area of cigarettes.   *Guardian Weekly* 30 July 1989, p. 23

> It is the worst example of an ungreen commercial development in Britain; a concept of the seventies with a fundamental purpose of maximising private investment at the expense of the environment.
> *Green Magazine* Dec. 1989, p. 12

uniquely abled 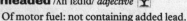 see ABLED

unleaded /ʌnˈlɛdɪd/ *adjective*

Of motor fuel: not containing added lead.

Formed by adding the prefix *un-* to *leaded*.

Unleaded motor fuel has been available since the sixties, but did not really come into the news in the UK until the late eighties, when motorists were actively encouraged to have their vehicles converted to use it. This encouragement, which included price incentives, arose from the high profile of the GREEN movement and widespread concern about the effects of pollution on the atmosphere: unleaded fuel produces less harmful exhaust emissions and reduces engine deposits. This kind of fuel is also called *lead-free* (see *-FREE*); both adjectives can be used on their own, as though they were nouns meaning 'unleaded fuel'.

> Reader offers ... included free weekend breaks, the prize of a house in France and the post-Budget free offer to convert readers' cars to unleaded petrol.   *Today* 12 Mar. 1990, p. 2

> Running a car will cost you more this year—but if you're 'environment-friendlier' the change won't hit as hard. Duty on petrol went up by about 10 per cent—an extra 11p per gallon for leaded petrol, 9p for unleaded.   *Which?* May 1990, p. 249

> The chain claimed its petrol is now Britain's cheapest at 198.7p a gallon for four star unleaded.
> *Sun* 20 Oct. 1990, p. 2

unsafe /ʌnˈseɪf/ *adjective*

Of a conviction or verdict at law: open to appeal, liable to be challenged or overturned. Especially in the phrase **unsafe and unsatisfactory**.

Formed by adding the prefix *un-* to *safe* in its legal sense, which is in turn related to the more general sense 'sure in procedure, not liable to fail'.

This term has been in use in the law for many decades, but acquired popular currency in the late eighties, especially as a result of the controversy over the allegedly *unsafe* convictions of a number of people for terrorist crimes in the UK in the seventies. In the case of the 'Guildford Four', four people convicted of IRA bombings at Guildford and sent to jail in 1975, the discovery that the convictions were in fact *unsafe* eventually led to their release in October 1989. This case helped to suggest a distinction between *unsafe* and *unsatisfactory*: in the opinion of the Appeal Court judges, the convictions were *unsafe* because they were founded on a prosecution case which was later shown to have been unreliable (evidence vital to the defence had been suppressed and false confessions obtained). The convictions therefore had to be quashed regardless of whether they were *unsatisfactory* (in other words, without regard to the original question of the guilt or innocence of the people concerned). In his judgment, Lord Lane said:

> Any evidence which casts real doubt on the reliability or veracity of the officers responsible for the various interrogations has to mean that the whole foundation of the prosecution case disappears, and the convictions will be unsafe.

However, this distinction was once again questioned in the courts in early 1991 in connection with the appeal and eventual release of the 'Birmingham Six' (another group of people jailed for terrorist bombings in the seventies), and the legal conclusion seemed to be that no court had ever separated the two entirely and that the distinction between them might anyway be impossible to draw.

> The manner in which the inquest was conducted by the coroner ... made the jury's verdict ... unsafe and unsatisfactory. *Financial Times* 30 Mar. 1983, p. 14

> While agreeing that the verdict was unsafe and unsatisfactory, he said that the judgment made no finding about whether the new evidence justified the conclusions of deliberate fabrication.
> *Guardian* 18 July 1989, p. 24

unsymmetrical dimethylhydrazine 🗝 see ALAR

unwaged /ʌn'weɪdʒd/ *noun* 🏴

Of a person: unemployed, not currently earning a wage. Often as a collective noun, in the form **the unwaged**: unemployed people and non-earners considered as a group.

Formed by adding the prefix *un-* to *waged*; the adjective *unwaged* had existed since the sixteenth century in the sense 'not recompensed with wages' (of work), but was not applied to people until the early eighties.

This is a term of the eighties which has often been interpreted as a euphemism for 'out of work', but which is actually designed to recognize the contribution and financial difficulties of other groups (such as full-time mothers) whose work goes unpaid in our society.

> The cost will be £2 per line for waged persons or £1 per line for those who are unwaged.
> *Library Association Record* (Vacancies Supplement) 30 Nov. 1982, p. cxlviii

> Dream analyst Sophia Young's workshop is at the Koestler Foundation, 484 King's Road, World's End, Chelsea on June 23, from 2pm to 6pm. It is free to the unwaged, and £3 for others.
> *Guardian* 19 June 1990, p. 21

use-by date /'juːzbaɪ ˌdeɪt/ *noun phrase* 🎌

A date marked on a food package or other perishable goods (usually preceded by the words 'use by') to show the latest time by which the contents should be used to avoid risk of deterioration.

Formed by compounding: the *date by* which the contents should be *used*.

Use-by dates have been in use on food packages in the US since at least the beginning of the eighties, and started to replace BEST BEFORE DATES in the UK in the middle of the decade. The *use-by date* is considered less ambiguous than a best before date in that it sounds more imperative (implying that the food will not only be less enjoyable after the date, but could actually constitute a health risk). For this reason, stricter legislation on the use of *use-by dates* was proposed in the UK in 1990 as part of a range of measures designed to allay public fears about food safety in the late eighties.

> The food is delivered the day it is made and marked with a 'use-by' date four days from preparation, although unsold items are pulled two days after being delivered to the kiosk.
> *Washington Post* 17 Feb. 1985, section K, p. 5

> New legislation is to be introduced to replace sell-by dates with more helpful use-by dates.
> *Which?* Apr. 1990, p. 205

user-friendly /ˌjuːzə'frɛndlɪ/ *adjective* Also written **user friendly** 💻

Easy for the user to operate; designed with the needs of the non-technical user in mind. Also, displaying a customer-conscious image; emphasizing public relations.

Formed by adding the combining form -FRIENDLY to *user*; such systems are meant to display a *friendly* attitude to the *user* rather than perplexing him or her with complicated instructions and cryptic error messages.

User-friendly was a coinage of the late seventies which started purely as a computing term to describe systems which incorporated a user interface geared to the needs of the non-specialist. As such, it became one of the computing buzzwords of the early eighties, ever-present in computer advertising and reviews. Within five years it had proved so successful in summing up the whole concept of accessibility to the ordinary person that it was already being applied in a variety of other contexts outside computing. This transferred sense itself developed further in the mid and late eighties, with the *-friendly* part being interpreted more literally again (especially in advertising), so that in some contexts it now means no more than the literal sum of its parts, 'friendly to the user/customer'. The same is largely true of the corresponding noun **user-friendliness**. The model of *user-friendly* has given rise to a multitude of other formations ending in *-friendly*: these are described under the heading -FRIENDLY. The success of *user-friendly* created the motivation for an adjective which would describe the opposite characteristics, those of inaccessibility and inscrutability for users: in the early eighties both *user-unfriendly* (see UNFRIENDLY²) and **user-hostile** developed in this sense and also soon became popular outside computing.

Every computer manufacturer now claims its products are 'user friendly'.

Which Micro? Dec. 1984, p. 3

'They should never be placed near flammable materials, and damaged bulbs should be cooled at least five minutes before they can be changed safely.' With such user-hostile tendencies, it's not surprising that fixtures recently became available with heavier bases and glass shields to protect both the consumers and the bulbs. *Chicago Tribune* 20 Sept. 1987, section 15, p. 3

Claimants were not getting paid. On top of everything else, the sytem was user-hostile. It took a long time to input information, and it was even harder to retrieve. *Best's Review* Jan. 1989, p. 90

It's so user-friendly that you can adjust it to suit any player. *CU Amiga* Apr. 1990, p. 11

A trip to the user-friendly Brandywine Zoo is also a good idea for an outing.

Delaware Today July 1990, p. 47

• •

V

vaccine /'væksiːn/ *noun*

A program which protects a computer system against being attacked by malicious software such as a VIRUS or WORM.

A figurative sense of *vaccine*; an extension of the *virus* metaphor, moving on one step further than INFECT.

This is a usage of the late eighties, used at first in the names of individual *antivirus* programs, but soon extended to the group as a whole. The metaphor is also extended to derived forms such as **vaccinate** and **vaccination**.

The vaccine program scans data and program files and triggers an alarm if operating instructions or data have been modified ... Other vaccines screen the commands that programs send to the computer's operating system ... Researchers have taken several approaches to block virus entry or 'vaccinate' computers so that users are notified when a virus is at work.

New York Times 30 May 1989, section C, pp. 1 and 9

Valdez Principles /væl‚diːz 'prɪnsɪp(ə)lz/ *noun phrase* 🌱

A set of guidelines, drawn up in the US in 1989, which is designed to regulate and monitor the conduct of corporations in relation to the environment.

Formed by compounding: part of the name of the oil tanker *Exxon Valdez*, which ran aground off *Valdez* in Alaska and spilled millions of gallons of oil into Prince William Sound in March 1989, combined with *principles* (because these were environmental *principles* which were already being considered and were finally agreed as a direct result of the disaster).

The *Valdez Principles* started as an environmental charter drawn up by CERES (the Coalition for Environmentally Responsible Economies), an organization representing American environmentalists and investment groups. It existed in draft form early in 1989, before the *Exxon Valdez* disaster had occurred, and acquired the name *Valdez Principles* among CERES staff as soon as it became clear that this was to be one of the US's worst environmental disasters and one from which corporations promised to learn lessons about environmental responsibility. This colloquial name was made official when the Principles were publicly announced in September 1989. The *Valdez Principles* themselves deal with broader issues than the problems raised by the oil spillage: they cover protection of the biosphere from pollutants, SUSTAINABLE use of renewable resources, the reduction and safe disposal of waste, energy conservation, the health and safety of employees, the marketing of environmentally sound products and services, compensation for victims of pollution, freedom of information about hazards, and provision for audit procedures.

> Information about whether a company has signed a pledge to follow the Valdez Principles will be disseminated to shareholders.
> *Newsday* 7 Sept. 1989, p. 77

> Ecologist Barry Commoner sees the beginning of a revolution in the idea of 'corporate responsibility' and the 'Valdez Principles', ... introduced by a coalition of environmental organizations and investment groups.
> *Boston Globe* 22 Apr. 1990, p. 28

Valspeak /'vælspiːk/ *noun* 🔲

A variety of US slang which originated among teenage girls from the San Fernando valley in California and was later taken up more widely by youngsters in the US.

A contraction of *Valleyspeak*, itself formed from the *Valley* of *San Fernando Valley* and *-speak* 'language', modelled on George Orwell's *Newspeak* and *Oldspeak* in the novel *1984*.

Valspeak, the language of the **Valley girl**, originated at the end of the seventies and was popularized under this name—or as **Valleyspeak**, **Valley talk**, or **Valley Girl talk**—from about 1982 onwards, especially by Frank Zappa's daughter Moon Unit. It is characterized by frequent repetition of certain 'filler' words (especially *like* and *totally*), emphasis on a small group of adjectives of approval or disapproval (see AWESOME, RAD, TUBULAR, and GRODY), abbreviation of words to a single syllable (see, for example, MAX), set phrases such as *grody to the max* and *gag me with a spoon*, and a dizzy, giggly, schoolgirl style of delivery.

> On the record, in pure, uncut Valspeak, Moon laments in bubbly staccato that, 'Like my mother like makes me do the dishes. It's like so *gross*.'
> *People* 13 Sept. 1982, p. 90

vapourware 🔲 see -WARE

VCR /viːsiːˈɑː(r)/ *abbreviation* 🔲 🔲

A video cassette recorder. The abbreviation is also used as a verb: to record (a television programme) on video.

The initial letters of *Video Cassette Recorder*.

Sales of *VCRs* reached the one million mark in the US in 1981, heralding the beginning of a video boom. The abbreviation *VCR* became widely used in the US at the beginning of this boom, but is less well known than *video* in the UK. Even though it is an abbreviation in which all the initials have to be pronounced separately, it acquired derivatives such as the verb defined above, the adjective **VCR'd** (provided with a *VCR*, recorded on *VCR*), and the noun **VCR-ing**.

> It's tempting to conclude that docs are automatically big draws in a four-TV channel (although heavily VCR'd) nation [the UK].
> *Los Angeles Times* 13 Nov. 1986, section 6, p. 10

The VCR-ing of America: videocassettes have fast-forwarded into our lives.

headline in *Los Angeles Times* 28 Dec. 1986, calendar section, p. 2

Nothing they do in the Winter Olympics reminds me of the torture I went through in phys ed class. So I'll be watching or VCRing every minute. *People* 15 Feb. 1988, p. 9

vegeburger /ˈvɛdʒɪbɜːgə(r)/ *noun* Also written **veggie burger** ✖

A flat savoury cake (similar in form to a hamburger but containing vegetables or soya protein rather than meat), sometimes served in a bread bun.

Formed by replacing the first syllable of *hamburger* with the first two syllables of *vegetable*. As in the case of *beefburger*, the formation is based on the false assumption that the *ham-* of *hamburger* names a kind of meat, whereas in fact it is a shortening of *Hamburger steak* and comes from the place-name *Hamburg*. The form *veggie burger* probably represents 'vegetarian burger', since in US English *veggie* is a well-known colloquial abbreviation of *vegetarian*.

The *vegeburger* was 'invented' in the early seventies and by 1980 had been registered as a trade mark in a number of different spellings. At first, this kind of burger tended to be available only in health-food outlets, but the success of the animal rights and green movements meant that a meat-free diet became more generally acceptable during the eighties, and the *vegeburger* more widely available.

Free festivals are market-places for everything hippies most like to sell, from hashish to vege-burgers. *Listener* 12 June 1986, p. 16

Fantastic Foods ... offers everything from instant soups sans meat to veggie burger mix, vegetarian chili and tofu stroganoff. *Chicago Tribune* 9 Aug. 1990, section 7, p. 4

venture /ˈvɛntʃə(r)/ *noun* 〰

In business jargon, enterprise that involves a substantial degree of risk or speculation, particularly the financing of small new businesses. Used especially in compounds:

venture arbitrage, risk arbitrage; the activity of an ARB;

venture buyout, a buyout financed by risk capital;

venture capital, risk capital; money that is put up for speculative investment;

venture capitalism, the system or practice of investment based on risk capital, especially in new and innovative high-capital projects; the activity of a **venture capitalist**.

A business or enterprise that has a substantial risk of loss as well as gain has been known as a *venture* since the sixteenth century; the compounds defined here extend that concrete sense into something more abstract: the whole practice of founding business on risk and speculation.

The idea of *venture capital* is not at all new—the term has been used since the forties—but the whole area of *venture capitalism* grew and developed in a new way in the US during the sixties and seventies and the UK during the early eighties, giving rise to new uses for *venture* in compounds. The main reasons for the change were the growth of risk arbitrage (for history, see under ARB) and the official encouragement of small businesses (see ENTERPRISE CULTURE) which took place at this time. For the first time, *venture capitalism* became a profession in its own right, with individuals and institutions which specialized in it alone; this happened first in the US and was mirrored in the UK and Australia a decade or so later. Organizations providing *venture capital* were seen as the foundation on which business growth could be built, since it was these organizations that funded the small firms trying to market the results of the technological revolution.

A shoeshine boy had been working the crowd near their table ... 'This is venture capitalism, Warren. Be supportive.' William Garner *Rats' Alley* (1984), p. 146

'Venture capitalism is basically placing equity-oriented capital in businesses that have prospects for high and rapid capital expansion,' explained the businesswoman.

Chicago Tribune 28 Oct. 1985, p. 20

Following the MBO has come, for example, the venture buyout and the buy-in.
Daily Telegraph 30 Oct. 1989, Management Buyouts Supplement, p. vi

The wider issues that are generally ignored in the brutal world of town planners and venture capitalists.
Vogue Sept. 1990, p. 376

video nasty ⚂ ❪ see NASTY

vidspud ⚂ ❪ see COUCH POTATO

viewdata /ˈvjuːdeɪtə/ *noun* Also written **Viewdata** 🖥

A system allowing for a normal television set to be linked to a computer database and for information to be passed in both directions between the two, making use of a telephone line as the communication link.

Formed by compounding: the system allows the user to *view* alphabetic characters and other computer *data* which could not normally be displayed on a television screen.

The first experiments with *viewdata* were carried out in the mid seventies. Towards the end of the decade, the British Post Office tried unsuccessfully to register the name as a trade mark for its telephone service providing this facility; this explains to some degree why it is often written with a capital initial (since people suppose it to be a trade mark). After choosing instead the name *Prestel*, the Post Office promoted the word *viewdata* as a general term for this kind of data display (competing with *teletext*, for which see TELE-).

Telematics regards its entry as timely because of the rise in such dissemination systems as viewdata and teletext.
Computerworld 23 May 1983, p. id-14

Last week British Telecom took over Micronet, the six year old micro-orientated user group on its Prestel viewdata service.
Guardian 27 July 1989, p. 25

virus /ˈvaɪərəs/ *noun* 🖥

A computer program or section of programming code which is designed to sabotage a computer system by causing itself to be copied into other parts of the system, often destroying data in the process.

A figurative use of *virus* based on the ability of the computer *virus* to replicate itself within the computer system, just as a biological *virus* multiplies within an organism.

Like the WORM, the computer *virus* was originally a concept of science fiction: it was used in David Gerrold's book *When Harlie was One* (1972), and also in John Brunner's *The Shockwave Rider* in 1975 (see the inset quotation under WORM). The first real *virus* was the subject of a computer science experiment in November 1983, presented by American computer scientist F. Cohen to a seminar on computer security. When Cohen had introduced the concept to the seminar, the name *virus* was apparently suggested by Len Adleman, and the results of the experiment were demonstrated a week later:

The initial infection was implanted in 'vd', a program that displays Unix structures graphically, and introduced to users via the system bulletin board ... The virus was implanted at the beginning of the program so that it was performed before any other processing ... In each of five attacks, all system rights were granted to the attacker in under an hour.

By the second half of the eighties the *virus* had become a serious hazard to individual and corporate computer users; because the code copies itself into the computer's memory and then causes havoc, it became advisable to avoid using floppy discs which might conceivably contain a *virus*—freeware and discs supplied by clubs, for example. Considerable financial loss was suffered as a result of the epidemic, not to mention research time and valuable data: in one famous incident, London's Royal National Institute for the Blind temporarily lost six months' worth of research after being attacked by a virus contained in files on a floppy disc. A number of software companies began to offer **virus detection** programs and 'good' viruses which could guard against infection (this kind of virus was sometimes known as a **vigilante virus**).

It's easy to build malicious viruses which duplicate themselves and then erase data files. Just a[s] easy to create a virus that lies dormant for months and then erupts some day in the future.

Clifford Stoll The Cuckoo's Egg (1989), p. 2[

The debate over vigilante viruses is part of a broader discussion now taking place among some com[puter researchers and programmers over what is being termed 'forbidden knowledge'.

New York Times 7 Oct. 1989, p. 3[

Comprehensive virus detection and removal features to protect your software investment. Work[with all presently known viruses. *CU Amiga* Apr. 1990, p. 7[

See also LOGIC BOMB and TROJAN

visualization /ˌvɪzjʊəlaɪˈzeɪʃ(ə)n/ *noun* ▓

The technique of forming a mental picture or vision of something (particularly of hoped-for event or outcome to a situation) as a psychological aid to confidence an[d] achievement.

Formed by adding the noun suffix *-ation* to the verb *visualize* 'make visible, form an image of'

As a psychological term, *visualization* has been in use for most of the twentieth century, but ha[s enjoyed a particular fashion in the fields of sports psychology and NEW AGE philosophy in th[eighties.

A crystal that, combined with visualization, can be used like a pair of scissors or a knife, is the lase[r] wand. *Soozi Holbeche The Power of Gems & Crystals* (1989), p. 9[

Most competitors down the years have thought roughly about what they intended to do . . . Nov[visualisation of what is going to happen from the moment of arrival at the arena, through the warm[up process and then through every throw or jump is part of the detailed preparation by Backley an[May. Backley describes it as self-hypnosis. *Guardian* 5 Aug. 1989, p. 1[

Vodafone /ˈvəʊdəˌfəʊn/ *noun* Also written **Vodaphone** ▓

The trade mark of a CELLULAR telephone system, one of two originally operating i[n the UK. Also, the equipment itself; a cellular telephone handset.

Formed by combining the first two letters of *voice*, the first two letters of *data*, and a respelle[d version of *phone*.

The *Vodafone* system was introduced by Racal in the mid eighties.

Optional extras include an eardrum-shattering quadrophonic in-car stereo, car phone and con[stantly bleeping radiopager. It's not unusual for the biggest poseurs to be blabbing into their Voda[phones with one hand and snapping away [taking photographs] with the other.

Guardian 26 July 1989, p. 2[

vogueing /ˈvəʊgɪŋ/ *noun* Also written **voguing** ▓

A type of dance or mime performed to popular music (usually HOUSE) and designe[d to imitate the characteristic postures of a fashion model on a catwalk; a form of club entertainment based on this.

Named after the fashion magazine *Vogue*: the idea is to pose and posture as if having one's pic[ture taken for *Vogue* magazine.

Vogueing originated in the Black and Puerto Rican gay community of New York, and started t[o be enjoyed as a more widespread form of club entertainment in 1988, spreading outside the US to Europe and the UK. It involves very little actual movement—the feet remain more or less o[n the same spot while different poses of the body, arms, legs, and face are taken up every fev[beats—and is often competitive, with 'judges' assessing the effect.

Willie Leake . . . directed the Voguing segment of 'An Evening Devoted to House Music and Voguing at El Museo del Barrio . . . 'Voguing,' the program notes explained, 'is an underground club form o[entertainment which appropriates and subverts the images, fashion and music prevalent in main[stream culture.' *New Yorker* 16 Jan. 1989, p. 2[

voice over /ˈvɔɪs ˌəʊvə(r)/ *transitive verb*

To provide (a television programme, commercial, etc.) with a commentary spoken by an unseen narrator (often a famous actor or other person whose voice is well known); to dub over (a soundtrack) with another, more famous voice.

A phrasal verb formed from the noun *voice-over*, which has been used in the entertainment world since the forties for film or television narration which is not accompanied by a picture of the speaker.

The television *voice-over*, especially by a famous actor, is a well-known feature of advertising in the eighties. Although perhaps used as a technical term in the entertainment industry for almost as long as the noun, the verb *voice over* only started to enter popular writing at the beginning of the eighties. The corresponding adjective may be **voiced-over** or **voice-overed**.

> Every single report or interview that she did for that programme was subsequently 'voiced-over' by a man.
> *Listener* 21 Aug. 1980, p. 229

> The jet-setting Lady Penelope in *Thunderbirds* (voiced over by ex-wife/business partner Sylvia Anderson).
> *The Times* 6 Oct. 1983, p. 12

> The first three parts of my report are . . . taped, edited, voice-overed, commentary written, everything.
> George V. Higgins *Penance for Jerry Kennedy* (1985), p. 230

• •

W

wack /wæk/ *adjective*

In young people's slang (especially in the US): bad, unhip, harmful.

Possibly derived from *wacky* or *wacko* 'crazy, mad' (the former in slang use since the turn of the century, the latter a variant of the late seventies and eighties and apparently a favourite with New York mayor Ed Koch). The connection with drugs can be seen in *wacky tabacky*, a slang name for the drug of the sixties, marijuana. The implication is both that drugs affect the mind, and (in the case of the present use) that it is *mad* to get involved with them.

Wack seems to have arisen in the street slang of US cities in the second half of the eighties, especially in connection with the spread of CRACK. It has been used in writing especially in the anti-drug slogan **crack is wack** (or **crack be wack, jack**) notably in a number of mural paintings in New York and other cities.

> Another inscription . . . warned, 'Crack is wack. You use crack today, tomorrow you be bumming. That's word experience talk.'
> *Atlantic* Sept. 1989, p. 75

> Blacks and Jews have a lot more in common than most American ethnic groups . . . Cultured Americans . . . know a bad that's good from a bad that's bad. So who's perfect already? Fly maybe, dope maybe, def maybe, and down by law, but perfect—oy gevalt! What wack, farmished, loc-ed-out dreck.
> *Interview* Mar. 1990, p. 148

Waldsterben /ˈvældʃtɛəbən/, in German /ˈvaltʃtɛrbən/ *noun* Also written **waldsterben**

A type of environmental disaster in which trees and other vegetation in a forest become diseased and die, usually as a result of pollution.

A direct borrowing from German *Waldsterben*, literally 'forest death'.

The process of *Waldsterben* was first noticed in fir trees in Germany in the seventies; by the early eighties, the effect had spread to other species of tree as well, and there was considerable alarm in Central and Northern Europe at the prospect of whole tracts of forest perhaps disappearing as a result of pollution. The German term has been used in English since about 1983, and is applied to the death of forests from environmental causes whether or not the forests are in Germany.

> A survey conducted in mid-summer by the Allensbach Institute revealed that 99 per cent of those asked had heard of *Waldsterben*—the death of Germany's forests.
>
> *Financial Times* 19 Nov. 1983, p. 1

> Although the industrial areas are the worst affected, pollution damage has spread throughout Poland and beyond. Half the trees are showing signs of waldsterben, or 'forest dieback'.
>
> *EuroBusiness* June 1990, p. 1

Walkman /'wɔːkmən/ *noun* 🔲

The trade mark of a type of personal stereo system consisting of a small battery operated cassette player with headphones (often also incorporating a radio).

So named because it can be used while *walking* or cycling along the street, in public transport etc., ostensibly without causing a disturbance to other people (although the noise which does escape, a tinny hiss, is considered a nuisance by many).

The *Walkman* was first made available under this name in the West by the Japanese company Sony in 1979 and proved to be one of the marketing success stories of the eighties. By the middle of the decade, personal stereos were in widespread use on the streets (even, dangerously, by cyclists), in buses and trains, and in other public places such as libraries. So popular were they that the word *Walkman* started to go the way of *Hoover* and other household names which are really trade marks: many people, in speech at least, use it as a generic term, although *personal stereo* should properly be used when it is not Sony's product that is being discussed. Some people have tried to get round this problem by describing a personal stereo or other miniaturized device as **walkmanlike**; there have been other derivatives, too (usually one-offs), such as **walkmanized**, an adjective to describe someone who is using a *Walkman*—and doctors have even identified **alopecia walkmania**, loss of hair from wearing *Walkman* headphones all the time! The plural form causes some confusion, with almost equal numbers of instances of *Walkmans* and *Walkmen*. In the mid eighties Sony called a similar portable system which plays CDs instead of cassettes by the trade mark **Discman**; in 1990 this was followed by the **Data Discman**, a type of electronic book.

> Professional men who once commuted in acceptable style, comfort and company, in the first class carriages of friendly steam trains, now have to make do with grubby corners in semi-graffitied Tube compartments, sandwiched, as like as not, between Walkmanised typists and heavily tattooed skin heads.
>
> *Punch* 15 July 1987, p. 43

> In any civilised society, Crazyhead would . . . come hissing from the Walkmans of every librarian on the tube.
>
> *New Musical Express* 25 Feb. 1989, p. 1

> Wherever you go nowadays, you find people with Walkmen, listening to a drizzle of pop music. Has anyone yet investigated the effects of this on the brain, and on capacity for concentration on words'
>
> *Weekend Guardian* 8 July 1989, p. 1

> Sony Corp. came out with its famous Walkman cassette player. In 1984, it unveiled the Discman . . . Now comes Sony's Data Discman, a device for reading books recorded on 3-inch optical disks that are capable of storing 10,000 pages each.
>
> *Business Week* 4 June 1990, p. 110

WAN /wæn/ *acronym* 🔲

Short for **wide area network**, a computer network (see NETWORK²) in which computers over a wide area are enabled to communicate and share resources.

The initial letters of *Wide Area Network*.

The *wide area network* was developed in the early eighties to perform a similar function to the *local area network* (or LAN) but over longer communication links. *WAN* seems to have been used almost immediately as a pronounceable acronym, probably under the influence of the pre-existence of LAN.

A 'WAN'—wide area network—facility so that your organisation can talk to the computers of other organisations.   *Your Business* Mar. 1986, p. 47

One only has to have lived through a few disasters to know that an effective network management system can quite literally be worth as much as the network itself. This is why the transition to a corporatewide, LAN/WAN network can leave many LAN administrators feeling like they're living their worst nightmare.   *InfoWorld* 14 Jan. 1991, Enterprise Computing Supplement, p. 6

wannabe /'wɒnəbɪ/ *noun* and *adjective* Also written wannabee

In young people's slang (originally in the US):

noun: An avid fan or follower who hero-worships and tries to emulate the person he or she admires, modelling personal appearance, dress, etc. on this person. Also, more generally, anyone who wants to be someone else.

adjective: Aspiring, would-be; like a wannabe; inspired by envy.

A respelling of *want to be* (as in the sixties song *I Wan'na Be Like You* by Richard M. and Robert B. Sherman), treated as a single word which can operate as a noun (someone whose appearance etc. seems to say 'I wanna be like you') or an adjective.

The noun was first used in the mid eighties to refer to White youths in the US who dressed and behaved like members of Black gangs, but were actually relatively harmless. It was probably most widely popularized, though, by its application to the female fans of the rock star Madonna, many of whom adopted a style of dress and make-up which almost turned them into Madonna look-alikes. There are also the sporting *wannabes*, the people who own all the kit that goes with the sport and manage to look the part, but have not yet the ability to fulfil the role. The adjective *wannabe* developed during the second half of the eighties.

Scores of Samantha Fox and Linda Lusardi wannabees raided British lingerie shops for skimpy lace and satin undies recently.   *Australasian Post* 23 Apr. 1988, p. 16

Madonna's appeal to adoring wannabes rests less on her ... personal life than her music, a blend of tweaking lyrics ... and a beat that dares you not to dance.   *Life* Fall 1989, p. 84

Today, whose in-house motto is 'Green and Greed' (it loves environment stories as well as 'wannabe' lifestyle ones) thought up a cheeky wheeze for last week's world conference in Bergen.   *Observer* 20 May 1990, p. 49

-ware /wɛə(r)/ *combining form*

Part of the word *software*, widely used as a combining form in computing, in words whose first element describes some characteristic of the software under discussion. Used especially in:

courseware, software specifically designed for educational use;

fontware, typesetting software or other software designed to enable the use of unusual printing fonts and alphabets;

freeware, software distributed free to users, without support from its developer;

groupware, a related set of software; software belonging to a group of related packages or designed for use by a work-group;

middleware, programs which function between an operating system and applications software;

shareware, software developed specifically for the purpose of sharing it in the computing community (in practice usually the same thing as *freeware*, although there is some attempt to register users and provide them with basic support such as a manual and contact with other users, and a fee may be charged for continued use);

vapourware, software that as yet only exists in the plans of its developers.

Formed by splitting the word *software* into its constituent parts (the adjective *soft* and the noun *ware* 'merchandise, goods') and then reapplying *-ware* in new but similar combinations.

These variations on the theme of *hardware* and *software* started to develop in the early seventies with the concept of *middleware*. In practice, most have been names for particular types of software, although at first it appeared that *-ware* would be used for 'hard' components and other items necessary for the functioning of a computer system as well. In the slang of computer scientists, **liveware** and **wetware** survive as humorous names for the human element—the people needed to keep the system running—and the human brain which makes software development possible. (**Liveware** has also been proposed as the name for a benign type of computer virus, which usefully updates itself each time a disc is loaded.) There was an explosion of new *-ware* formations in the second half of the eighties (including many of those listed above), partly as a result of the personal computing boom which followed the development of the IBM PC. By the end of the decade the inventors of these terms almost seemed to be competing with each other to create more ingenious and graphic names.

> The key to good design . . . was to start thinking about 'liveware' (human beings) along with the hardware and software.
>
> *Independent* 1 May 1987, p. 19

> It's useful to think of groupware as a class of products—similar to a toolbox containing tools for diverse tasks.
>
> *Byte* Dec. 1988, p. 275

> A third principle is that the ministry does not license vapourware. There has to be at least a preproduction prototype of the software and associated documentation, which can be used and tested before any money changes hands.
>
> *Guardian* 13 July 1989, p. 29

> Company president David Miller referred to 'dBASE/SQL' as 'the ultimate vapourware, since it's unannounced, undesigned, undeveloped, unknown, has no marketing plan, . . . nor any release date or pricing.'
>
> *Australian Personal Computer* Oct. 1989, p. 26

> Since groupware began to appear about 18 months ago, most of the programs . . . try to deliver some new, whizzy benefit to users, such as organizing communications among work-group members.
>
> *PC World* Oct. 1989, p. 49

> FormBase includes Bitstream fontware and supports Postscript, Hewlett-Packard Graphic Language printers. The program can print reports, forms with or without data.
>
> *Daily Telegraph* 5 Mar. 1990, p. 27

See also RISC

warehouse /ˈwɛəhaʊs/ *noun* 🎵 👹

Mostly in **warehouse party**: a large, illicitly organized party (usually held in a warehouse or some other spacious building) at which the main entertainment is dancing to popular (especially HOUSE) music; similar to an ACID HOUSE party.

Formed by compounding: the parties involve such large numbers of people that a building the size of a *warehouse* is needed to accommodate them. A connection is sometimes made with the *Warehouse* club in Chicago (see HOUSE), but parties were already being held in warehouses before the fashion for house music started.

Large parties were held in warehouses in the UK from the early eighties onwards; as the craze for house music spread from Chicago across the US and the Atlantic to the UK in the mid eighties, they became associated with this youth cult in particular. Because of the large concentrations of people at the parties and police suspicions that they were used for drug-pushing, the arrangers tended to keep the details secret until the last moment: see ACID HOUSE. Although usually in the combination *warehouse party*, *warehouse* is sometimes used on its own to refer to the culture of house music, parties, and dancing as a whole.

> Ten people . . . were arrested during a drugs raid on a derelict school building in Cowley, Oxford, yesterday after leaflets advertising an 'Acid Warehouse Party' were seized.
>
> *Daily Telegraph* 10 Oct. 1988, p. 3

> There are also secretive murmurings of a possible jazz warehouse party.   *The Face* Jan. 1989, p. 38

The only way the warehouse scene can survive is to get small again like it was a few years back and offer something special. *Q* Nov. 1989, p. 16

washing machine ♪ 🔲 see ACID HOUSE

waxed jacket /ˌwækst ˈdʒækɪt/ *noun* 🔲

An outdoor jacket similar in style to an anorak and made of waterproof waxed cotton.

Formed by compounding: a *jacket* of *waxed* material.

Waxed jacket is the generic term for this garment (the best known brand being the BARBOUR jacket). Once the chosen outdoor wear—along with green wellies—mainly of aristocratic country-dwellers, the *waxed jacket* became a fashion item in the eighties, in keeping with the emphasis on casual wear generally.

They had been there a week, and had gone for long tramps along the Downs in all weathers, well-protected with high boots, waxed jackets and portable parkas.

Antonia Byatt *Possession* (1990), p. 487

well safe 🔲 see SAFE

well woman /ˌwɛl ˈwʊmən/ *noun* 🔲

A woman who undergoes screening tests to ensure that she is healthy; used especially as an attributive phrase in **well woman clinic**, a clinic for women which concentrates on preventing disease by carrying out such screening.

Well has meant 'sound in health' since the sixteenth century; it is the construction in which it is used here, rather than the meaning, that is new.

Although the idea of a *well-baby clinic* had been thought of and put into practice as long ago as the twenties, the same principle was not applied to women's health until the late sixties. Throughout its short history *well woman* has caused some confusion when applied as an attributive phrase in the plural, with many writers opting for *well women* in these cases. Soon after *well woman* tests and clinics had been set up there was a move towards greater emphasis on preventive medicine generally, giving rise to the **well man** and **well person** clinics in the eighties as well.

Saturday's session included a motion urging establishment of 'well women clinics' to help specifically with women's medical problems, underrated in a medical profession still dominated by MCPs.

New Statesman 27 Sept. 1985, p. 7

The college also wants to see special funds made available to enable practices to offer preventive and educational services such as well-woman and well-man clinics, together with stop-smoking groups. *Daily Telegraph* 10 Feb. 1987, p. 2

Our nurses do all the immunisations, run the Well person clinics, and do most of the family planning work. *Which?* Oct. 1989, p. 483

Three weeks ago she had made an appointment to take her breast lump to a doctor. But she had done it in a very peculiar way: she had booked in to a private Well Woman Clinic under an assumed name. Sara Maitland *Three Times Table* (1990), p. 155

Westlandgate 🏠 see –GATE

wetware 🔲 see –WARE

wheat-free 🔲 🔲 see –FREE

wheel clamp /ˈwiːl ˌklæmp/ *noun* and *verb* Also written **wheel-clamp** or **wheelclamp** ⚿

noun: A clamp designed to be locked to one of the wheels of an illegally parked vehicle, thus immobilizing it until the appropriate fine has been paid and the clamp is removed.

transitive verb: To immobilize (a vehicle) by attaching one of these clamps; to CLAMP. Also, by extension, to subject (a person) to the experience of having his or her car clamped.

Formed by compounding.

The *wheel clamp* was first used in the city of Denver, Colorado, allegedly as long ago as 1949. At that time, though, it was not known as a *wheel clamp*: from the late sixties, the device was nicknamed the *Denver boot* or *Denver shoe*, and it was not until the eighties, when the idea was widely taken up in the UK, that *wheel clamp* started to be used as a neutral name for these objects. The metal clamp prevents one of the wheels of the car from turning, and sometimes also positions a sharp spike above the front of the car to deter attempts to drive out of it. Although very unpopular, **wheel clamping** is very effective and therefore seems likely to remain a part of everyday life in car-based societies.

Right now the world is in a dreadful state what with terrorists, famine and wheel clamping.
Comic Relief Christmas Book (1986), p. 103

His powers of forbearance had been severely stretched the night before when he found himself wheel-clamped outside a restaurant. 'I said something unpleasant to this man and afterwards I felt absolutely awful.' *Sunday Express Magazine* 1 Feb. 1987, p. 18

Wheel clamps have recently been introduced in Rome in a move against illegal parking.
Holiday Which? Mar. 1990, p. 73

wheelie bin /ˈwiːlɪ ˌbɪn/ *noun* ⚿

A large refuse bin on wheels; a Eurobin (see EURO-).

Formed by compounding.

The *wheelie bin* first appeared in the UK in about 1986, but both the object and the name seem to have been used in Australia for some years before that. The bins are designed to cut refuse collection costs (an important consideration in view of the privatization of local government services in the eighties); since they are on wheels, members of the public can move them to the front of their properties on the appropriate day for refuse collection in their area, thus saving dustmen thousands of trips to the side or back of properties and removing the unsightliness of black plastic sacks left out for collection. However, a *wheelie bin* is usually quite large—up to five feet tall—and this has meant that the whole idea has come in for criticism on two counts: that the elderly and infirm cannot manage them, and that they encourage people to throw away material which could otherwise be recycled.

To all the freedom fighters who chucked their enthusiastic weight into my battle against the wheelie-bins; ... my warmest thanks. *The Times* 29 Dec. 1989, p. 16

whistle ⌨ see BELLS AND WHISTLES

white knight /ˌwaɪt ˈnaɪt/ *noun* ⚔

In financial jargon, a company that comes to the rescue of one facing a HOSTILE take-over bid.

A figurative use that is perhaps a mixed metaphor: on the one hand it relies on the fairy-tale image of the knight on the white charger who appears at the last moment to rescue the damsel in distress, on the other on the imagery of *black* (bad) and *white* (good). The *white knight* is also a character in Lewis Carroll's *Alice Through the Looking Glass* who is full of enthusiasm but has little common sense:

> He was dressed in tin armour, which seemed to fit him very badly, and he had a queer-shaped little deal box fastened across his shoulders upside down, and with the lid hanging open . . . 'I see you're admiring my little box,' the Knight said in a friendly tone. 'It's my own invention—to keep clothes and sandwiches in. You see I carry it upside down so that the rain can't get in.'

By the end of the nineteenth century the term *white knight* was already in figurative use in English to refer to a person who, like Carroll's character, is enthusiastic but ineffectual (but this seems unconnected with the present development). It had acquired the secondary meaning 'a hero or champion' in more general contexts in the early 1970s before being taken up in this specialized financial sense.

The first uses of *white knight* in the context of corporate take-overs date from the very beginning of the eighties; once established, the term was applied specifically to a corporate counter-bidder who comes into play to force a bid battle with the company trying to take over. As the decade progressed, so did the imagery: by 1987 the term **white squire** had been coined, for an individual who buys a large shareholding in a company facing a take-over so as to make it less attractive to the bidder. (The *squire* is a little less powerful than the *knight*, and enters the fray at the first rumour of a take-over, whereas the *knight* charges in at the last moment to save the day.)

> Much speculation surrounds the future of the near-40 p.c. equity stake held by the 'white squires' who helped Standard see off Lloyds Bank's £1.3 billion bid two years ago.
>
> *Daily Telegraph* 15 Aug. 1988, p. 22

> Adia . . . launched a hostile bid for Hestair . . . When Hestair found a white knight, BET, Adia refused to enter a bidding war.
>
> *Business* Apr. 1990, p. 81

whole-body scanner ⊗ 🖵 see BODY-SCANNER

wicked /'wɪkɪd/ *adjective* 🕮

In young people's slang: excellent, great, wonderful.

A reversal of meaning: compare BAD. In this case, there might first have been a catch-phrase or advertising slogan *so good it's wicked* which was later abbreviated to *wicked* alone; however, it is not unusual for an adjective to be used as an 'in' word in the opposite sense to its usual one among a limited group of people, and then pass into more general slang.

In US slang, *wicked* has been used in the sense 'formidable' since the end of the nineteenth century (compare *mean* in British English). A famous example occurs in F. Scott Fitzgerald's *This Side of Paradise* (1920), when Sloane calls for music and announces

> Phoebe and I are going to shake a wicked calf.

It was only in the early eighties, though, that *wicked* was taken up by young people (including, and perhaps especially, young children) as a fashionable term of approval, often preceded by the adverb *well*. This usage, unlike the earlier slang use, spread outside US English to enjoy a vogue among British and Australian youngsters as well. A children's weekend television programme in the UK took up the theme in its title, *It's Wicked!*

> I've been to loads of Acid House parties. We have a wicked time but never, not never, do we take any drugs.
>
> *Time Out* 18 Oct. 1989, p. 9

> This boy looked in wonder at the polyurethane and leather marvel and offered it the coolest of street compliments. 'Well wicked,' he breathed.
>
> *Daily Telegraph* 9 June 1990, p. 13

wide area network 🖵 see WAN

widening /'waɪd(ə)nɪŋ/ *noun* 🗎

In relation to the EC: the policy of extending membership of the Community to more countries (possibly including the countries of Eastern Europe).

A specialized use of the figurative sense of *widening*, adopted by analogy with *deepening* (see below).

A word which has been used especially in connection with the debate over European integration in the second half of the eighties, and is often presented as the opposite approach from the Delors plan for EMU¹ (otherwise known as *deepening*). A person who favours *widening* in the Community is known as a **widener**.

> Some of the wideners have gone to the other extreme, arguing that the Community must now abandon much of its cohesion ... There is no need for widening to conflict with deepening. Indeed, every widening has brought more deepening.
> *Independent* 13 Dec. 1990, p. 22

wilding /'waɪldɪŋ/ *noun* 🔌

In US teenagers' slang: the activity of going on a wild rampage in a group through the streets, often involving mugging or otherwise attacking innocent bystanders.

Apparently a reference to a rap version of the pop song *Wild Thing*, which the original gang had been chanting. This might be an example of a new word created entirely by misunderstanding; it is not clear whether the teenagers concerned were already using the word in their own street slang to mean 'going on a spree', or whether they only started doing so after newspaper reports of the original case expressed interest in the word that journalists thought the accused had been saying (when in fact he had only been muttering 'wild thing').

The activity of *wilding* (which, whatever its name, had occurred in US cities before) came to public notice as a result of a series of reports of gang violence culminating in the assault, rape, and attempted murder of a young woman in New York's Central Park in April 1989. The gang consisted of more than thirty youngsters, mostly of school age, who went on a two-hour rampage during which they attacked joggers, shoppers, and other passers-by. The case was widely reported and may have provoked a number of similar incidents which occurred soon afterwards.

> There has been little response by the city government to the wide-spread concern over wilding in general ... The police should begin to gather intelligence on wilding attacks, identify the schools and subways where they are most likely to occur and beef up their presence there.
> *New York Times* 13 Jan. 1990, p. 27

See also STEAMING

wimmin /'wɪmɪn/ *plural noun* Also written **womyn** 🏠 🔌

In writing by or about feminists: women.

A respelling of *women* which is meant to reflect its pronunciation and is expressly intended to remove from it the 'word' *men*. The spelling *womyn* is an attempt to preserve the historical continuity of the word to some extent, in answer to criticism of the purely phonetic *wimmin*.

The first examples of *wimmin* used in print date from the late seventies. According to a feminist dictionary, in August 1979 a feminist magazine 'for, about, and by young wimmin' explained the motivation for the new spelling:

> We have spelt it this way because we are not wo*men* neither are we fe*male* ... You may find it trivial—it's just another part of the deep, very deep rooted sexist attitudes.

By the mid eighties, the spelling had come to be particularly associated, in the UK at least, with militant feminism and with the **peace wimmin** or **Greenham wimmin**, feminist peace campaigners who from 1981 picketed the US airbase at Greenham Common in Berkshire to protest about the deployment of nuclear weapons at this and other bases. The spelling *womyn*, which developed in the second half of the eighties, offers the possibility of a singular form (much rarer than the plural).

> Wimmin rewrite Manglish herstory.
> headline in *Sunday Telegraph* 3 Nov. 1985, p. 13

> According to Jane's Defence Weekly, the authoritative British defence journal, women members of the Spetsnaz forces have been mingling with the Greenham 'wimmin' ... The Greenham 'wimmin' laugh at this suggestion.
> *Daily Telegraph* 23 Jan. 1986, p. 18

> Why are these (ignorant) gay men (and sadly sometimes wimmin) stereotyping gayness? ... Next time you see a feminine looking womyn ... don't show hostility toward her.
> *Pink Paper* 17 Nov. 1990, p. 19

wimp¹ /wɪmp/ *noun* 🏛️

In slang, a feeble, cowardly, or ineffectual person; especially, a public servant who has a grey or weak public persona.

Probably ultimately related to *whimper*. In the twenties *wimp* was Cambridge University undergraduates' slang for 'a young woman'; when first applied to young men in US slang, it certainly had implications of effeminacy.

A word with a many-stranded history. The present sense seems to have had some currency among college students in the US from about the mid sixties; to them, a *wimp* was a weedy or effeminate man. During the second half of the sixties this sense became more widespread, passing into British English as well. By the late seventies a slightly different sense had cropped up in US teenagers' slang: to describe someone as a *wimp* was to imply that this person was old-fashioned, especially in dress and appearance. The two meanings came together in US slang in connection with the vice-presidential and presidential campaigns of George Bush at the end of the eighties: when a number of journalists seemed to be trying to gain him a reputation as a *wimp*, there was some discussion of the implications of the label, from which it emerged that it was as much his background and appearance (typical of the 'Preppie') as his grey image that had prompted it. So frequently was this taunt used that it even came to be referred to as the **W-word** (by analogy with *F-word*) in some sources; Mr Bush sought to counter it in his READ MY LIPS speech and policy. *Wimp* has a number of derivatives, mostly connected with the connotations of cowardice and spinelessness: for example, the adjective **wimpish** and the nouns **wimpery** and **wimpishness**. In the US during the late seventies and eighties, a phrasal verb with *out* also developed: to **wimp out** is to 'chicken out' or fail to face up to a situation; the corresponding noun is **wimp-out**.

'We thought the Brits might wimp out. After Libya we hoped that the United States would not have to go out in front again,' said a senior American intelligence official.
Sunday Telegraph 26 Oct. 1986, p. 40

Vice President George Bush is a preppy, despite many mouse-brained journalists' continued attempts to hang the wimp label on him.
Maledicta 1986–7, p. 23

Bush and Jesse Jackson . . . are battling serious image problems that forced Bush to declare he is not a 'wimp'.
Kuwait Times 18 Oct. 1987, p. 5

That word 'wimp', when used by an American about Mr Bush, is partly a euphemism for upper class.
Sunday Telegraph 12 June 1988, p. 22

WIMP² /wɪmp/ *acronym* Also written **Wimp, wimp,** or **WIMPS** 💻

In computing jargon, a user interface incorporating a set of software features and hardware devices (such as windows (see WINDOW¹), ICONS, mice (see MOUSE), and pull-down menus) that are designed to make the computer system simpler or less baffling for its user.

Formed on the initial letters of *Windows, Icons, Mice*; the fourth initial is variously explained as standing for *Program, Pointer,* or *Pull-down*.

WIMPs were developed by Rank Xerox during the seventies and became commercially available in the first half of the eighties. The package of features—in which different tasks are allocated to different portions of the screen (windows), with small symbolic pictures (icons) and lists of options (menus) representing the different operations which may be selected by clicking on them with the mouse—has come to be associated particularly with Apple computers but was a general feature of the popular computing boom of the mid eighties. By the end of the decade, the idea of *WIMP* was already thought a little outdated by computer scientists, who had moved on to the excitements of *GUI* (graphical user interface), an even more advanced interface which would be needed for the development of multimedia.

An intriguing WIMPS (Windows, Icons, Mouse and Pointer-based System) implementation that does a creditable job of imitating the workings of the Apple Macintosh.

Which Computer? July 1985, p. 35

The Apple Lisa is generally credited for being the first machine to make use of wimps. In fact the idea first originated in the Palo Alto, California laboratories of Rank Xerox, but it was the Lisa which turned it into a marketable product. *The Australian* 13 May 1986, p. 45

With Presentation Manager the Wimp . . . will find its way onto the desks of millions of office workers. *Computer Weekly* 28 Apr. 1988, p. 26

Using the term GUI is stretching things more than a little, although the no longer fashionable WIMP tag just about applies. *Personal Computer World* July 1990, p. 128

window¹ /ˈwɪndəʊ/ *noun* and *verb*

noun: In computing, an area of the VDU screen which can be sectioned off for a particular purpose so that different functions can be carried out and viewed simultaneously in different parts of the screen.

transitive verb: To place (data) in a window; to divide (the screen) into windows.

One of a long line of figurative applications of the word *window* for things which in some way resemble a window in appearance or function; in this case, the effect of so dividing the screen is to give the user the possibility of looking (as if through a *window*) into a number of different areas of memory at once.

The earliest uses of *window* in computing relate to the facility for 'homing in' on a part of a drawing or other graphics so as to display only a portion of it on the screen; this was developed during the sixties. The idea of sectioning the screen for simultaneous display of different sets of data was worked on by Rank Xerox in the seventies (see WIMP² above); the first references to call such an area of the screen a *window* date from the mid seventies. For a short time in the seventies and early eighties, the term *viewport* (adopted from science fiction) was also used for a *window* in which a clipped portion of a drawing, or a formatted set of data, was viewed; by the second half of the eighties, though, *window* seemed to have taken over at least in popular usage. The adjective **windowed** and action noun **windowing** are also used.

Thanks to my windowed terminal, I am simultaneously editing the source code in a second window.

Datamation 1 Dec. 1984, p. 17

The screen can be windowed, and the cursor moved between two windows.

Practical Computing Dec. 1985, p. 83

Thursday's . . . module opens with Mel Slater . . . talking on dynamic window management, multiple window nesting and the implications for hardware. *Invision* Oct. 1988, p. 26

window² /ˈwɪndəʊ/ *noun*

A period of time, usually of limited duration; used especially in international relations and politics to refer to a limited period during which something may be achieved (a **window of opportunity**) or during which forces, weapons, etc. are vulnerable to enemy attack (a **window of vulnerability**). Also, by extension, a gap in one's timetable; a spare moment which can be earmarked for a particular activity.

Another figurative use of *window*, this time based on the idea that a *window* represents an opening in an otherwise solid wall. This sense grew out of a figurative use of *window* in space exploration: since the sixties, the short period of time during which a rocket or satellite can be launched if it is to reach the required orbit has been known as a *launch window*.

The phrases *window of opportunity* and *window of vulnerability* date from the beginning of the eighties, when both were used by US negotiators in relation to the arms race between the US and the Soviet Union; both acquired a wider currency as catch-phrases during the eighties. This perhaps explains why, during the second half of the eighties, the word *window* became a fashionable piece of executives' jargon for a space in one's diary or FILOFAX; but it is possible that this is just a piece of visual imagery (referring to the small white space surrounded by the many appointments written in on the page).

> After the list come the cold calls, which White makes during the crucial half-hour 'window' from 11.45am to 12.15, when some of the initial frenzy has burned off the London markets.
>
> *Sunday Express Magazine* 26 Oct. 1986, p. 17

> Instead of fixing the meeting, you are allowed to issue the delicious Coastal phrase, 'I'll leave you a window.' This hole in your schedule can then be cancelled a few days before the event, and you go through the motions all over again. *Sunday Telegraph Magazine* 19 July 1987, p. 39

> Unexpected changes in price or volatility might provide sudden and short-lived windows of opportunity to reduce costs or generate profits. *Energy in the News* Third Quarter 1988, p. 10

windowed¹ 🖳 see WINDOW¹

windowed² /'wɪndəʊd/ *adjective* 〰

Of the security thread in a banknote: woven into the paper so that it is visible only in short stretches.

A figurative use of *windowed*, alluding to the fact that the thread is partially embedded and partially visible.

Windowed threads were introduced in Bank of England notes in the mid eighties.

> It is ... the only means of incorporating security threads in the 'windowed' form which has become a feature of Bank of England £20 and £10 notes in recent years. *New Scientist* 3 Dec. 1988, p. 84

windsurfing /'wɪndsɜːfɪŋ/ *noun* Also written wind surfing 🎌

The sport of sailing on a board similar to a surfboard, but using wind in a small sail rather than waves for its power.

Formed by compounding: *surfing* in which it is the *wind* in the sail, rather than the waves, that supplies the power.

The special board used in *windsurfing* (known by the trade mark **Windsurfer**) came on to the US market in 1969 and caused a craze on the West coast of the US in the seventies. By the beginning of the eighties the sport was well-known outside the US; it first featured as a demonstration sport in the Olympic games of 1984. By that time, though, it had been decided that it should be known officially as BOARDSAILING. Despite this fact, *windsurfing* remains the name by which most people know the sport and the one which crops up most frequently in printed sources. The agent noun **windsurfer** and verb **windsurf** also remain frequent.

> It combines lifestyle and adventure with wind surfing to make it more than just a sports magazine. He takes his cameras and windsurfers to exotic locations. *Auckland Metro* Feb. 1986, p. 18

> It is *the* event in the Windsurfing calendar with a spectacular display of the latest in watersports equipment ... and fashion from jetskis and paraskis for the active enthusiast to dayglo surf shorts for those who just want to don the look. *Woman's Journal* Mar. 1990, p. xiv

witching hour 〰 see TRIPLE WITCHING HOUR

wok /wɒk/ *noun* 🎌

A bowl-shaped pan used in Chinese cookery, especially for stir-fry dishes.

A direct borrowing from Cantonese.

The *wok* (and the Chinese cooking for which it is used) enjoyed a vogue in the Western world in the late seventies and early eighties and by the end of the eighties the *wok* had come to be regarded as a standard piece of kitchen equipment.

> Fry the peanuts in the oil in a large saucepan or wok for 4–5 minutes, until lightly browned. *Green Cuisine* Feb./Mar. 1987, p. 24

'Where would you put it?' Vic inquires, looking round at the kitchen surfaces already cluttered with numerous electrical appliances—toaster, kettle, coffee-maker, food-processor, electric wok, chip-fryer, waffle-maker . . . 'I thought we could put the electric wok away. We never use it. A microwave would be more useful.'

David Lodge *Nice Work* (1988), p. 10

wolf pack /'wʊlf ˌpæk/ *noun* 🆔

In the US, a gang of marauding young men who engage in mugging or WILDING.

A new figurative application for a compound which literally means 'a group of wolves who work together when hunting etc.'; during the Second World War the term was applied figuratively to an attacking group of German submarines.

Wolf pack has been in use in this figurative sense in the US for fifteen years or more; it was also the term used by New York police to describe the marauding gang of youngsters from Harlem who were involved in the case of wilding in April 1989 (see WILDING). This incident caused considerable debate in the US as a result of which the term *wolf pack* became quite widely known there and was popularized outside the US as well.

In terms of group attacks, the No. 1 crime that we've seen among juveniles . . . is robbery 2—that is, aided robberies, the wolf-pack robberies . . . I guess it became a little easier to knock the old lady over and just grab the bag rather than to reach into the pocket and hope you came out with something. So things have gotten a lot rougher in the city with respect to wolf packs.

New York Times 25 Apr. 1989, section B, p. 1

The *New York Post* observed that calling the gang a 'wolf pack' was libellous to wolves.

Economist 29 Apr. 1989, p. 31

womanist /'wʊmənɪst/ *noun* 🆔

In the US: a Black feminist or feminist of colour. Also, a woman who prefers the company and culture of women, but who is committed to the wholeness of the entire people.

Formed by adding the suffix *-ist* (as in *feminist*) to *woman*, on the model of a Black English word *womanish* meaning 'wilful, grown up (or trying to be too soon)', as in an expression which Black mothers might use to their daughters: 'You acting womanish.' *Womanist* had been independently formed several hundred years ago in the sense 'a womanizer', but this usage did not catch on.

The word *womanist* was coined by the American Black woman writer Alice Walker as a deliberate attempt to challenge the racist implications of the feminist movement, which found it necessary to speak of a separate category of 'Black feminism' and which thereby excluded Black women from mainstream feminism. Some of the followers of **womanism** see in it a more general challenge to the content of radical White feminism as well, offering a less aggressive and more positive view of womanhood as contributing to the community as a whole. As Alice Walker has written in *In Search of Our Mothers' Gardens* (1983):

Women who love other women, yes, but women who also have concern, in a culture that oppresses all black people (and this would go back very far), for their fathers, brothers, and sons, no matter how they feel about them as males. My own term for such women would be 'womanist' . . . It would have to be a word that affirmed connectedness to the entire community and the world.

Womanist is to feminist as purple to lavender.

Alice Walker *In Search of Our Mothers' Gardens* (1983), p. xii

I've been female so long that I'd be stupid not to be on my own side but if I have to be an 'ist' at all I'd rather be a womanist. The feminists lost me because they can't laugh at themselves.

Maya Angelou in *Daily Telegraph* 26 Oct. 1985, p. 11

I suppose I forgot I was talking to a womanist.   Alice Walker *Temple of My Familiar* (1989), p. 320

woopie /'wʊpɪ/ *noun* Also written **WOOP** or **woopy**

A well-off older person; a member of a socio-economic group composed of retired people who are still sufficiently affluent to have an active lifestyle and to be significant consumers.

Formed on the initial letters of *Well-Off Older Person* and the diminutive suffix *-ie*, after the model of YUPPIE.

One of many humorous terms for social groupings that followed in the wake of *yuppie* in the second half of the eighties. The fact that the acronym is still nearly always explained when the word is used suggests that it has not really gained a place in the language. However, in view of the increasing numerical importance of retired people in Western societies (and consequently their significance as consumers) it might yet prove an important word. Other attempts to categorize (or acronymize) more or less the same social group have included *GLAM* (Greying Leisured Affluent Middle-aged), *Zuppie* (Zestful Upscale People in their Prime) and *Third Ager* (in the sense in which *Third Age* is used in *University of the Third Age* etc.: the years of retirement).

> Mrs Edwina Currie . . . claimed that many pensioners were well off . . . 'We're in the age of the "woopy"—the well-off old person—and it is about time we all recognised that fact, planned for our own future and helped them to enjoy theirs,' she said. *Daily Telegraph* 23 Apr. 1988, p. 1

> Woopies will stimulate demand into the 1990s says Connell.
> headline in *Property Weekly* (Oxford) 22 June 1989, p. 1

> Dick Tracy gets everybody, from the fast-growing pensioner market who remember the old comic strip, to the WOOPS (Well-Off Older Persons) and Baby-boomers who want to see Warren Beatty in a hit movie again. *Guardian* 24 May 1990, p. 30

world music /'wɜːld ˌmjuːzɪk/ *noun* Also written **World Music**

In the jargon of the popular-music industry, any music that incorporates elements of local or ethnic tradition (especially from the developing world) and is promoted on the UK or US pop market.

Formed by compounding: *music* from the wider *world*.

The phrase *world music* has been in use since the late seventies in a general sense; it became a label for a category of popular music in the late eighties, as a number of ETHNIC sounds were incorporated into Western rock. As promoters raced to 'discover' groups from around the world and bring their music to a wider audience, *world music* became symptomatic of the increasingly blurred dividing line between folk music and commercial pop. *World music* (or simply **world**) is also sometimes used attributively or as an adjective to categorize an artist, group, etc. as belonging to *world music*.

> There are those who dismiss the growing interest in World Music as a passing fashion.
> *Tower Records' Top* Feb. 1988, p. 28

> 'Songhai', four stars, strong world music interest, file under jazz. *The Face* Jan. 1989, p. 52

worm /wɜːm/ *noun*

A computer program which (like a VIRUS) is designed to sabotage a computer or network of computers and can replicate itself without first being incorporated into another program (compare TROJAN).

So called because it operates like a parasitic *worm* in an animal host; it can *worm* its way into a network without first having to be copied into another program, breeds extra segments, and cannot easily be killed off.

The concept was invented by John Brunner in the science fiction novel *The Shockwave Rider* in 1975; his *worm* is the computing equivalent of a parasitic tapeworm, generating new segments for itself in all the machines of a network and therefore unstoppable. In the novel he uses the word *worm* interchangeably with *tapeworm*:

> Am I right in thinking Hearing Aid is defended by a tapeworm?... If I'd had to tackle the job... I'd have written the worm as an explosive scrambler, probably about half a million bits long, with a backup virus facility and a last-ditch infinitely replicating tail. It should just about have been possible to hang that sort of tail on a worm by 2005.

Although this type of program was beyond the capability of programmers at the time, a group of research scientists at the Rank Xerox laboratories in Palo Alto, California, attempted to develop a set of benign *worm* programs in the early eighties as a means of distributing computing operations across a number of different machines in a network, with the program finding spare computing capacity for itself and copying the necessary segment on to any machine that it was going to use. What really brought the *worm* into the news, though, was the *worm* which temporarily disabled more than three thousand computers at universities, businesses, and research establishments on the Internet network in the US in November 1988. Robert T. Morris, a research student at Cornell University, was later convicted of releasing the *worm* into the system.

> One year after an Ivy League graduate student unleashed a computer 'worm' that brought a national scientific and defense computer network to its knees for a day, experts say the threat of computer worms and viruses is greater than ever. *Boston Globe* 30 Oct. 1989, p. 29

> About 180 companies in the U.S. market offer services and software to stymie worms and viruses, which can alter or destroy data in a corporation's information systems.
> *American Banker* 1 Aug. 1990, p. 10

wrinklie /'rɪŋklɪ/ *noun* Also written wrinkly 🔠

In young people's slang: a middle-aged or old person (younger than a CRUMBLIE).

Formed by treating the adjective *wrinkly* as a noun; the metaphor homes in on *wrinkles* as one of the visible signs of advancing age.

A word of much the same vintage and history as *crumblie*, now well known to the older generation to which it refers.

> Mayotte, who is leading the way as the wrinklies strike back, has an uncomplicated theory as to why the teenagers are performing so well. 'There has been a lot of talk about big rackets and stuff. I think the truth is that training is better and there's a lot of money to be made, so there's a lot of people interested in tennis these days.' *Guardian* 4 July 1989, p. 14

WYSIWYG /'wɪzɪwɪg/ *acronym* Also written wysiwyg or (erroneously) wysiwig 🔠

Short for **what you see is what you get**, a slogan applied to computer systems in which what appears on the screen exactly mirrors the eventual output.

The initial letters of *What You See Is What You Get*.

A feature of advanced high-resolution VDU displays, *WYSIWYG* first appeared on the mass computing scene in the early eighties and became increasingly important as the DESK-TOP publishing boom gained momentum in the middle of the decade.

> True Wysiwig would show bold, extended and italic characters ... on the screen and the only way that will happen is with a very high resolution display (which in turn will normally require a graphics card). *Daily Telegraph* 8 Oct. 1990, p. 27

X

XTC see ECSTASY

Y

yah /jɑː/ *noun* Also written **ya**

A SLOANE RANGER or YUPPIE; someone who says 'yah' instead of 'yes'.

Formed by converting their characteristic pronunciation of *yah* ('yes') into a noun. This mannerism had apparently been noted as long ago as 1887 in a student newspaper.

Despite the fact that *yah* has evidently been a well-known affected pronunciation of *yes* for some time, the word was not used to characterize a social type until the early eighties. By the early nineties most people probably associated loud and repetitive use of *yah* more with the brash executive or yuppie type than with the upper classes.

> Pursuing my researches into the social make-up of the university [of St Andrews] with daughter and friends, I am reminded that the rich set are known as the Ya's, derived from their loud affirmations.
> *Sunday Telegraph* 17 July 1983, p. 9

yappie /'jæpɪ/ *noun*

Either a young affluent parent *or* a young aspiring professional.

A variation on the theme of YUPPIE, using the initial letters of *Young Affluent Parent* or *Young Aspiring Professional* for the 'root'.

Like GUPPIE, this is really a stunt word, jumping on the bandwagon of *yuppie* but in a rather ad hoc fashion. The word *yappie* has been used by journalists in a variety of contexts and meanings—including 'a talkative yuppie', 'a yuppie dog-owner', 'young Asian-American professional', and 'young athletic participant'—but it is the two meanings given in the definition above that at present hold the majority. The word seems unlikely to survive in the language unless it becomes established in one of these two meanings.

> The yappies are the creation of the Henley Centre, the research organisation which plots changes in social and spending trends. They are the young professional people who were possibly yuppies in the 1980s... When children come on the scene yappies spend most of their time in the more prosaic roles of 'parent' and 'provider'.
> *Financial Times* 19 Apr. 1990, section 1, p. 9

Yardie /'jɑːdɪ/ *noun* and *adjective*

In British slang:

noun: A member of any of a number of Jamaican or West Indian gangs (see POSSE) which engage in organized crime throughout the world, especially in connection with illicit drug-trafficking. In the plural, **Yardies**: these gangs as a whole or the criminal subculture that they represent.

adjective: Of or belonging to the Yardies.

The name is derived from the Jamaican English word *yard* (or *yaad*) which originally meant 'a house or home' and came to be used by Jamaicans living outside Jamaica for the home country. The suffix *-ie* is common in nicknames for people from a particular place: compare *Aussie* or *Ozzie* for an Australian.

Although probably active in the UK for some time, the *Yardies* only began to feature in the news towards the end of the eighties, when they were associated with the spread of drug-related crime in the UK in much the same way as the drug *posses* were in the US.

> The Yard was responding to claims that a Caribbean gang—ironically called The Yardies—has moved into London's Brixton area and is now setting up its own network of pushers to sell the so-called champagne-drug.
> *Today* 9 July 1986, p. 9

> The Yardies is a loose association of violent criminals, most of whom originated in Kingston, Jamaica and whose principal interest is the trafficking and sale of cocaine. In Britain they are perceived as a new phenomenon. In America, however, their counterparts, the 'posses', are said to have been responsible for up to 800 drug-related murders since 1984.
> *Daily Telegraph* 13 Oct. 1988, p. 13

> Many of the Shower who escaped the raid have fled abroad, some of them perhaps heading for Britain to join their 'yardie' colleagues. But more young Jamaican recruits will soon leave the tranquillity of the Caribbean for the mean streets of Washington DC.
> *Sunday Telegraph* 27 Nov. 1988, p. 10

yo /jəʊ/ *interjection* 🔲

Among young people (especially in the US): an exclamation used in greeting or to express excitement etc., and associated particularly with RAP and HIP HOP culture; hey!

Yo has been used as an exclamation to attract attention (especially when warning of some danger) since the fifteenth century, and is familiar to many in the sailor's *yo-ho-ho*; the present use is a re-adoption of the old word in a new context by a limited group of people, who use it as a cult expression.

Yo started in Black street slang in the US, probably during the late seventies, and was popularized through the spread of rap and hip hop to White youth culture during the eighties. By the end of the eighties it had become a fashionable greeting among youngsters in the UK as well as the US; a fashion which was reinforced, perhaps, by its use in the popular television series *The Simpsons* and in a number of films featuring Sylvester Stallone.

> During the holiday, wherever he roamed in his Watts neighborhood, congratulations rained down. 'Yo, Hagan! Nice job, man!'
> *Sports Illustrated* 25 Dec. 1989, p. 45

> Yo, man, quit lookin' at 'em! You got detec written all over you.
> *Village Voice* (New York) 30 Jan. 1990, p. 35

> The Guardian Angels ... applauded him with a meaty sound. Great fists, many gloved, bashed into each other. 'Yo,' they shouted, rather than anything English.
> *Independent* 16 May 1990, p. 6

yuppie /ˈjʌpɪ/ *noun* and *adjective* Also written **Yuppie** or **yuppy** 🔲

noun: A young urban (or upwardly mobile) professional; a humorous name for a member of a socio-economic group made up of professional people working in cities.

adjective: Of or characteristic of a yuppie or yuppies in general; of a kind that would appeal to a yuppie.

Formed from the initial letters of *Young Urban Professional* (or *Young Upwardly mobile Professional*) and the suffix *-ie*.

Yuppie was probably the most important buzzword of the mid eighties, an extraordinarily successful coinage which somehow succeeded in summing up a whole social group, its lifestyle and aspirations, in a single word. In an article on the writer John Irving in 1982, the American critic Joseph Epstein described them as

> People who are undecided about growing up: they are college-educated, getting on and even getting up in the world, but with a bit of the hippie-dippie counterculture clinging to them still—yuppies, they have been called, the YUP standing for young urban professionals.

At first (in 1982–4) *yuppie* competed with the form **yumpie** (which included the *m* of upwardly-*m*obile), but this form was perhaps too close to the verb *yomp*, with its military route-march associations, to succeed. A measure of the popularity of *yuppie* was the speed with which it generated derivatives: the nouns **yuppiedom, yuppieism,** and **yuppi(e)ness** all appeared within two years of the coinage of *yuppie,* closely followed by the adjective **yuppyish.** By the middle of the decade there was also an awareness of the way in which *yuppie* culture pervaded and changed its surroundings, a process known as **yuppification** (with an associated verb, **yuppify,** and adjective **yuppified**). Perhaps more telling even than the derivatives were all the variations on the theme of *yuppie* that journalists turned out in the second half of the decade, including *yuffie* (young urban failure), *yummie* (young upwardly-mobile mommy), and those listed under BUPPIE, GUPPIE, WOOPIE, and YAPPIE. The second half of the eighties saw the rise in popularity of NEW AGE culture and of a more environmentally aware lifestyle which made the *yuppie* approach seem already a little outdated, but it was by then so familiar that it could safely be abbreviated to **yup** without fear of misunderstanding. Even the abbreviated form acquired derivatives: the language of *yups* was **Yuppese** or **Yupspeak,** a young female *yup* was a **yuppette** (compare HACKETTE), their preferred type of car was a **yupmobile,** and so on.

> Yuppies have come in for some revisionist thinking lately. The yup backlash is such that many people will no longer speak the 'Y word' and others are spurning pesto for pot pies.
> *Adweek* 17 June 1985

> Who are the yuppies? Gee acknowledges that young urban professionals 'who once thought nothing of jumping in the old Bimmer [BMW] and heading down to the local gourmet grocer for some Brie' are keeping a lower profile, fearing they may be called 'too yup'.
> *Los Angeles Times* 5 May 1986, section 4, p. 2

> Their 'bashers' (shacks) will be forcibly removed by police to make way for developers who want to 'yuppify' the Charing Cross area.
> *Observer* 16 Aug. 1987, p. 3

> What Dickens is describing, I suddenly realised, is yuppification. The trendies were moving in.
> *Independent* 17 Sept. 1987, p. 18

> 'The yupskies are coming!' said Mr Baker ... in Leningrad yesterday after being impressed by the new breed of young upwardly-mobile Soviet entrepreneurs.
> *Daily Telegraph* 8 Oct. 1988, p. 32

> There is a risk of forced selling breaking out in the yuppier sections of London's housing market.
> *Arena* Autumn/Winter 1988, p. 99

> Married yuppette Kathy is knee deep into her affair with ... Tom.
> *Independent* 16 May 1989, p. 29

> How will the eighties be labelled? The Yuppie decade? The Thatcher miracle/disaster? The years when pop and rock got a conscience? The dawning of the breakdown of communism?
> *Guardian* 22 Nov. 1989, p. 43

> You didn't think yuppies liked poetry. Don't be vulgar and simplistic, dear Val.
> Antonia Byatt *Possession* (1990), p. 417

> These sound like thoroughly well-organised chaps who would take to the executive life like yuppies to bottles of Perrier water.
> *Punch* 20 Apr. 1990, p. 9

yuppie flu / ˌjʌpɪ ˈfluː/ *noun* ⊗

A colloquial nickname for myalgic encephalomyelitis (see ME).

So named because it attacks high achievers (*yuppie* types), and mimics or follows an attack of flu.

A popular nickname which reflects the scepticism of doctors and public alike about this illness until quite recently: see the entry for ME.

> Graham ... told Mr Patrick Cuff, the coroner, that his mother had suffered for several years from ME—myalgic encephalomyelitis, known as Yuppie Flu.
> *Daily Telegraph* 8 Feb. 1990, p. 3

> For many years, it has been called 'yuppie flu', because most of the estimated 1 to 5 million who suffer from the disorder are affluent professional women from 25 to 45.
> *Chicago Tribune* (North Sports Final edition) 19 Nov. 1990, p. 6

Z

zap /zæp/ *intransitive or transitive verb* 🔲

In media slang, to move quickly through the commercial break on a recorded video-tape, either by using the fast-forward facility or by switching through live channels. Also, to avoid the commercials in live television by using the remote control device to switch through other channels until they are over.

Zap began as an onomatopoeic word in comic strips for the sound of a ray gun, bullet, laser, etc.; as a verb it has meant either 'to kill' or 'to move quickly and vigorously' since the sixties. The sense defined here is essentially a specialized application of the second of these two branches of meaning, but when applied to live television it is influenced by the first branch—the remote control device is used like a ray gun, and the effectiveness of the advertisements is destroyed if people *zap* through other channels while they are on.

This sense of *zap* arose in the mid eighties, when many television sets became available with remote control (in other words, they became **zappable**) and there were the first signs of a boom in domestic video. The action noun **zapping** arose at about the same time; at first, a **zapper** was a person who did this, but by the end of the decade it had also become a standard name for the remote control device itself.

For the ITV companies there is the additional problem of 'zapping' to contend with—the habitual use of the fast-forward button to bypass the commercial breaks in recorded material.

Listener 9 Feb. 1984, p. 14

The television remote controller or 'thingy' which Christopher Croft (letter, 18 January) is at a loss to name, is the enabling device for the practice of 'zapping', whereby Channel 4 News and Wogan can be viewed simultaneously. In our household the thingy is called 'Frank', after the eponymous rock star, Frank Zappa.

Independent 19 Jan. 1989, p. 27

The decade was also marked by gizmos that accelerated our daily lives: food was nukable; TVs, zap-pable; mail, faxable.

Life Fall 1989, p. 13

The remote control is small and handy . . . It's almost identical to Tatung's Astra-box zapper.

What Satellite July 1990, p. 120

zero /'zɪərəʊ/ *adjective* 🔲

In the names of disarmament proposals:

zero option, a proposal made in the early eighties for the US to cancel plans to deploy longer-range theatre nuclear weapons in Europe if Soviet longer-range weapons were also withdrawn;

zero zero option (or **double zero option** or simply **double zero**), a proposal made by the Soviet Union for the withdrawal from Europe of all NATO and Soviet shorter- and longer-range nuclear weapons (made a reality in 1987 under the terms of the INF treaty);

triple zero option (or simply **triple zero**), a proposal to include short-range tactical weapons as well.

All based on the idea of *zero* as representing 'nothing', although, strictly speaking, none of the proposals would do away with all weapons.

The original *zero option* dates from the beginning of the eighties, when some European countries felt very uneasy about the build-up of theatre nuclear weapons on both sides of the Iron Curtain; the term was revived in relation to the control of these longer-range INF weapons in the mid eighties. *Double zero* was a Soviet proposal of 1986–7, made at a time when the cold war was visibly thawing under Mr Gorbachev's administration in the Soviet Union; it was essentially put into practice (for Europe at least) by the INF treaty. There remained some pressure to move on to the **global double zero**, which would extend the provisions to weapons held outside Europe. *Triple zero* involved even shorter-range weapons, which some European countries saw as a worrying threat.

> If Pershing II and Cruise are ... to be negotiated away under the zero-zero option, and if Polaris is truly obsolescent ... then the Labour Party 'unilateral' policy seems to differ very little in substance from that of the Alliance. *New Scientist* 16 Apr. 1987, p. 49

> If we said yes to zero option, we said yes, yes to double zero option, and who knows, there may be a triple zero option involved in tactical neutral weapons. *MacNeil/Lehrer NewsHour* 22 Apr. 1987

> The further offer was formalised in Moscow last March, when Mr Gorbachev proposed to Mr George Schultz that all SRINF category weapons be removed from Europe. Because the LRINF proposal had been called the 'zero option', the joint scheme has come to be called the 'double zero'. 'Double zero' is, nonetheless, an inexact term, because 'single zero' would leave the superpowers with 100 missiles each, as long as they were held in Asiatic Russia and the continental United States respectively. *Daily Telegraph* 21 May 1987, p. 16

> Eduard Shevardnadze emphasised that in the Soviet Union the fact is appreciated that Spain was among the first West European States which supported the double zero for Europe and then also the global double zero. *BBC Summary of World Broadcasts* 22 Jan. 1988, p. SU/A7

Zidovudine /zɪˈdɒvjʊdiːn/ *noun* Also written **zidovudine** ⊗

The approved name of the anti-viral drug AZT, used in the management of AIDS.

The first part, *zido-*, and the ending, *-dine*, are taken from the chemical name *azidodeoxythymidine*, but it is not clear why the syllable *-vu-* was added.

The name *Zidovudine* has been in use since 1987, but the drug remains popularly known as *AZT* (see the comments at AZT). *Zidovudine* itself is sometimes abbreviated to **ZDV**.

> Acyclovir is already in use, in combination with Zidovudine (formerly AZT), for Aids patients. *Guardian* 7 July 1989, p. 3

> Every week I watch AIDS patients deteriorate and waste away despite Zidovudine (ZDV) therapy. *Nature* 14 June 1990, p. 574

ZIFT /zɪft/ *acronym* Also written **Zift** ⊗ 🖳

Short for **zygote intra-fallopian transfer**, a technique for helping infertile couples to conceive, in which a zygote (a fertilized egg which has been allowed to begin developing into an embryo) is re-implanted into one of the woman's Fallopian tubes after fertilization with her partner's sperm outside the body.

The initial letters of *Zygote Intra-Fallopian Transfer*. In scientific terms, a *zygote* is a cell formed by the union of two gametes (see GIFT).

The technique was developed during the second half of the eighties as a further refinement of GIFT, offering greater certainty of establishing a pregnancy. However, unlike GIFT, it takes fertilization outside the body once again, and is therefore open to the same ethical or religious objections as IVF.

> A new variation, zygote intrafallopian transfer (ZIFT), may further improve GIFT's odds. The egg is fertilized in a petri dish, and the embryo is placed in the fallopian tube about 18 hours later. ZIFT has been tried on fewer than 50 couples, so it is too soon to measure its success. *US News & World Report* 3 Apr. 1989, p. 75

> On this occasion, I was being treated with a variation of Gift, called Zift (Zygote intrafallopian transfer), in which the eggs and sperm are mixed outside the body and then replaced in the tube. *Independent* 15 Jan. 1991, p. 17

zouave /zwɑːv/ *adjective and noun*

adjective: Of trousers for women: cut wide at the top, with folds of material at the hips, and tapered into a narrow ankle.

noun: (In the plural **zouaves**) women's trousers of this design.

Named after the Algerian *Zouave* regiment of the French army, who wore a uniform with trousers of this shape (known as *peg-top* trousers) in the middle of the nineteenth century.

This is an example of an old word which has been revived in modern fashion and applied in a slightly different context. In the late nineteenth century there was a fashion for garments of various kinds (particularly women's short jackets and men's peg-top trousers) which copied the uniform of the Zouave regiment and were known as *Zouave jacket*, *Zouave trousers*, etc. When wide-topped, draped trousers became a fashion item for *women* in the 1980s, the word was reapplied to them, and this time round also came to be used as a noun in its own right.

First came the ankle-length Zouaves, looking a bit like baggies gone berserk, worn under two layers of fitted, belted coats with full skirts, Russian peasant hats with tassels and ankle-high boots. Then came the shorter Zouaves, like knee-length bloomers. *Washington Post* 22 Apr. 1981, section B, p. 3

Zouave pants with elasticated waist and two pockets.
Grattan Direct Catalogue Spring–Summer 1989, p. 218

zouk /zuːk/ *noun*

An exuberant style of popular music originating in Guadeloupe in the French Antilles and combining ETHNIC and Western elements.

Reputedly a borrowing from Guadeloupean creole *zouk*, a verb meaning 'to party', possibly influenced by US slang *juke* or *jook* 'to have a good time'.

Zouk was developed by Guadeloupean musicians in Paris at the end of the seventies as a deliberate attempt to construct a distinctive Antillean style of popular music which could hold its own against Western pop. It was also designed to compete with disco music, especially in Paris, where its main proponents (a group named *Kassav*) have been popularizing it during the eighties. It was only towards the end of the decade that *zouk* started to get exposure in the UK and the US. *Zouk* is often used attributively, especially in **zouk music**, and occasionally forms the basis for derivatives such as **zoukish**.

His latest, 'Kilimandjaro' (AR1000) nosedives into held-back zoukish rhythms that never let go, wimpy vocals and over the top arrangements. *Blues & Soul* 3 Feb. 1987, p. 27

Tonight, the first ever zouk on British soil kicks off this year's Camden Festival International Arts programme . . . Zouk, especially Kassav, is the pulse of Paris streets and the soundtrack for her nightclubs. *Guardian* 24 Mar. 1987, p. 11

Zuppie see WOOPIE

zygote intra-fallopian transfer see ZIFT